Cell Signaling

Cell Signaling

Edited by Peter Madison

STATES
ACADEMIC PRESS
www.statesacademicpress.com

States Academic Press,
109 South 5th Street,
Brooklyn, NY 11249, USA

Visit us on the World Wide Web at:
www.statesacademicpress.com

ISBN: 978-1-63989-099-6

Cataloging-in-Publication Data

Cell signaling / edited by Peter Madison.
p. cm.
Includes bibliographical references and index.
ISBN 978-1-63989-099-6
1. Cellular signal transduction. 2. Cell interaction. 3. Cellular control mechanisms. I. Madison, Peter.
QP517.C45 C45 2022
571.74--dc23

Table of Contents

Preface

The ability of the cells to receive, process and transmit signals with its environment as well as with itself is termed as cell signaling. Extracellular signals are the signals which originate from outside the cells. Various physical agents can be responsible for extracellular signals such as voltage, mechanical pressure, light, temperature, etc. The transformation of a signal into a chemical form marks the beginning of signal transduction. This can either directly activate an ion channel or initiate a second messenger system cascade, which conveys the signal through a cell. This book aims to shed light on some of the unexplored aspects of host-pathogen interactions and the recent researches in this field. It provides significant information to help develop a good understanding of this discipline. For all readers who are interested in this field, the case studies included in this book will serve as an excellent guide to develop a comprehensive understanding.

After months of intensive research and writing, this book is the end result of all who devoted their time and efforts in the initiation and progress of this book. It will surely be a source of reference in enhancing the required knowledge of the new developments in the area. During the course of developing this book, certain measures such as accuracy, authenticity and research focused analytical studies were given preference in order to produce a comprehensive book in the area of study.

This book would not have been possible without the efforts of the authors and the publisher. I extend my sincere thanks to them. Secondly, I express my gratitude to my family and well-wishers. And most importantly, I thank my students for constantly expressing their willingness and curiosity in enhancing their knowledge in the field, which encourages me to take up further research projects for the advancement of the area.

Editor

1

Purinergic Receptors: Key Mediators of HIV-1 Infection and Inflammation

Talia H. Swartz[1], George R. Dubyak[2] and Benjamin K. Chen[1]*

[1] Division of Infectious Diseases, Icahn School of Medicine at Mount Sinai, New York, NY, USA, [2] Department of Physiology and Biophysics, Case Western Reserve University, Cleveland, OH, USA

Edited by:
Diana Bahia, Universidade Federal de Minas Gerais, Brazil

Reviewed by:
Dirk Dittmer, University of North Carolina at Chapel Hill, USA
Hridayesh Prakash, University of Hyderabad, India

***Correspondence:**
Talia H. Swartz
talia.swartz@mssm.edu

Human immunodeficiency virus type 1 (HIV-1) causes a chronic infection that afflicts more than 30 million individuals worldwide. While the infection can be suppressed with potent antiretroviral therapies, individuals infected with HIV-1 have elevated levels of inflammation as indicated by increased T cell activation, soluble biomarkers, and associated morbidity and mortality. A single mechanism linking HIV-1 pathogenesis to this inflammation has yet to be identified. Purinergic receptors are known to mediate inflammation and have been shown to be required for HIV-1 infection at the level of HIV-1 membrane fusion. Here, we review the literature on the role of purinergic receptors in HIV-1 infection and associated inflammation and describe a role for these receptors as potential therapeutic targets.

Keywords: P2X, P2X7, HIV, inflammation mediators, inflammasome, inflammatory cytokines

INTRODUCTION

Human immunodeficiency virus type 1 (HIV-1) disease afflicts more than 30 million individuals worldwide. The infection remains incurable despite the advent of antiretroviral therapies. Individuals who are infected with HIV-1 can live long lives without infectious complications; however, they experience non-infectious comorbidities known as non-AIDS-associated comorbidities. These are thought to be due to a process of chronic inflammation that occurs despite virologic suppression (1). This phenomenon may account for a wide variety of comorbidities including cognitive decline, cardiovascular disease, and thrombotic disease (2–8). A unifying mechanism has not been identified; however, an emerging literature implicates the role of purinergic receptors, proinflammatory signaling mediators, as important regulators of HIV-1 productive infection. Because these receptors are required for HIV-1 entry, it is hypothesized that they may additionally play a key role in inflammation and underlie comorbidities that shorten the life expectancy of HIV-infected individuals. An understanding of how these receptors may be involved in HIV-1 infection and inflammation would enable the production of novel therapeutics that both antagonize HIV-1 entry and inflammation associated with HIV-1 infection.

HIV-1 AND INFLAMMATION

Patients with HIV-1 infection have experienced a tremendous leap in life expectancy due to the advent of effective antiretroviral therapy (ART). The result has been that individuals are living longer and now experiencing comorbidities similar to disease processes found in the general population. In fact, a study in 2008 demonstrated that only 10% of deaths in HIV-infected individuals were related to AIDS-defining illnesses while other causes included non-AIDS-defining malignancies, cardiovascular disease, liver disease, and others (9). There are certain conditions that appear to develop in HIV-infected individuals at an earlier age than the general population. This phenomenon has been referred to as "accelerated aging" and is thought to relate to chronic inflammation and immunosenescence. There are multiple possible explanations that may include ART toxicity, lifestyle (i.e., tobacco, alcohol, and IV drug abuse), as well as HIV-1 infection itself (10, 11).

How might HIV-infected individuals develop comorbidities associated with chronic inflammation? There are multiple possible explanations. HIV-1 infection causes a chronic viral infection that results in selective $CD4^+$ T cell depletion which has a major impact on lymphocytes in the gastrointestinal tract (12, 13). Chronic HIV-1 infection leads to reduced integrity of the mucosal epithelium causing bacterial translocation. This process is proposed to play a key role in chronic inflammation in HIV-1 disease (14–16). High bacterial lipopolysaccharide (LPS) levels in HIV-infected individuals are associated with elevated inflammatory biomarkers (17). Abnormally high levels of T cell activation can persist despite years of virologic suppression (18), and these individuals have lower levels of $CD4^+$ reconstitution (19, 20). Elevated soluble inflammatory biomarkers are detected in these individuals including markers of type I interferon (1), monocyte activation (21), and inflammation and coagulation (22). Specifically, levels of IL-6, hsCRP, and D-dimer persist at elevated levels in HIV-infected patients. There are multiple proposed mechanisms that may elevated immune activation even when virus is suppressed (1). One is that low levels of viral replication may continue, but with viral loads below the limit of detection. These levels may stimulate systemic inflammation; however, no studies support a role for intensification of therapy to reduce inflammation (23, 24). The importance of this chronic inflammation lies in its associations with comorbidities that account for the major mortality in individuals infected with HIV-1. These include cardiovascular disease, neurological decline, end organ dysfunction, and thrombotic events (2–8). Even with highly active antiretroviral therapy, which is effective at achieving virologic suppression, individuals who are chronically infected with HIV-1 have elevated inflammatory biomarkers and display innate immune activation and immune dysfunction that does not normalize with therapy (25). No unifying mechanism thus far has connected HIV-1 infection to the regulation of proinflammatory signaling.

OVERVIEW OF PURINERGIC RECEPTORS

Purinergic receptors are ubiquitously expressed in mammalian cells. In 1970s, extracellular nucleotide became recognized as an important mediator of cellular signaling (26). A large literature describes the role of purinergic receptors as detectors of extracellular adenosine and adenosine triphosphate (ATP) that activate intracellular signaling events (27). These receptors can be characterized into two classes, the P1 adenosine receptors and P2 ATP/ADP receptors. P2 receptors are further divided into two categories: the P2X and P2Y subtypes. P2X receptors are ATP-gated plasma membrane channels that can be formed by a trimeric assembly of seven different subunits (P2X1–P2X7) which assemble as homotrimers or heterotrimers (28, 29).

P2X receptors are key regulators of a number of important physiological processes including neuronal synaptic and modulation, cell death and proliferation, cell and organ motility, and infection and inflammation (30–34). P2X receptors are ATP-gated non-selective cation channels. An agonist, ATP or other nucleotides, binds to the extracellular portion, inducing conformational changes that triggers channel opening and cation flux (28, 35). Some of these receptors can dilate to a larger pore, thus increasing permeability to large organic molecules (36–38). This occurs with prolonged exposure to agonist; other subtypes, notably P2X1, undergo fast desensitization with prolonged ATP exposure, resulting in closure of the channel (39, 40). Functional P2X receptors assemble as either homotrimers or heterotrimers, each subunit of which contains two transmembrane domains, a large extracellular loop containing 10 conserved cysteine residues and glycosylation sites, and intracellular N and C termini containing consensus phosphorylation sites (29, 31, 41–43). P2X7 specifically has a large pore that assembles as a homotrimer, by contrast to some other P2X receptors. Roles for the P2X7 subtype are well-characterized in the innate immune response and include proinflammatory cytokine activation, antigen presentation, and lymphocyte proliferation and differentiation (44–46). P2X7 receptor activation requires submillimolar ATP concentrations which are only transiently released extracellular compartments in response to acute cell death or injury (28). Sustained activation of P2X7 can result in large pore opening which enables passage of molecules up to 900 Da that to eventually induce cell death (31, 47).

P2Y receptors function widely across diverse physiological systems and have roles in clotting, hormone secretion, vasodilatation, neuromodulation, cell migration, cell proliferation and cell death, wound healing, and immune response (26, 27, 34, 37, 48–50). The P2Y subtypes are of G protein-coupled receptors. They consist of seven transmembrane domains with an extracellular N-terminus and an intracellular C-terminus (48, 51, 52). Activation of the receptor results in G protein dissociation into α and $\beta\gamma$ subunits which activates downstream effector molecules. A large sequence diversity encodes for diverse pharmacological profiles among these receptors (53). There are eight P2Y monomer subtypes. P2Y1, P2Y2, P2Y4, and P2Y6 couple to G_q to activate phospholipase C and P2Y12, P2Y13, and P2Y14 couple to G_i to inhibit adenylyl cyclase and activate GIRK-family K^+ channels. P2Y11 can couple to both G_q and G_s and trigger increases in intracellular Ca^{2+} and in cAMP levels.

PURINERGIC SIGNALING IN INFLAMMATION

Purinergic receptors can be found in a wide variety of leukocyte sub-types, notably lymphocytes, monocyte/macrophages, and dendritic cells (29, 44, 54–56). They are critical mediators of the innate immune response in a variety of different disease states including rheumatoid arthritis, transplant rejection, and inflammatory bowel disease (57–60). Nucleotides are known mediators of innate immune cell function including cell migration (61, 62). Extracellular nucleotides, such as ATP, are released by metabolic stress, ischemia, hypoxia, and inflammation that leads to cell death and the further release of intracellular contents into surrounding tissue (26, 33). Release of ATP of through channels can signal through purinergic receptors to modify cellular orientation, cytoskeletal rearrangement, chemotaxis, and cell migration (63, 64). Studies have supported a role for signaling of these receptors in immune function of macrophages, neutrophils, B lymphocytes, and T lymphocytes. Thymocytes can undergo programed cell death in response to purinergic activation and nucleotides have been implicated in fate-determination during T cell development (65). Extracellular ATP bind these to purinergic receptor which can activate T cells through extracellular calcium influx, p38 MAPK activation, and IL-2 secretion (66–69). ATP can also activate γδ T cells through the P2X4 receptor, while P2X7 activation can promote differentiation of T into proinflammatory TH_{17} effector cells (45, 70). The P2X1, P2X4, and P2X7 subtypes are most highly expressed on leukocytes, and literature implicates the P2X7 subtype specifically in inflammatory signaling (71–74).

P2X7 receptors are the most highly expressed P2X receptor subtype in innate immune cells (44, 71, 75). They activate proinflammatory cytokine production (44, 76) and can trigger activation of the inflammasome. The inflammasome is a central scaffold protein complex that serves to coordinate interaction with caspase molecules which cleave precursor protein substrates into immunomodulatory products. Activation of P2X7 results in massive K^+ efflux, and this change in ionic strength signals to the processing of procaspase-1 (77). Mature caspase-1 cleaves prointerleukin-1β (pro-IL-1β) into interleukin-1β (IL-1β) which is released into the cytoplasm (23, 30). Signaling takes place as part of a two component signal which requires an initial signal, such as a toll-like receptor activation via bacterial or viral ligands. Activation of P2X7 can serve as a second signal, inducing assembly of the inflammasome complex which activates caspase-1, with consequent cleavage of pro-IL-1β to mature secretory IL-1β (78). Inflammasome activation is also known to mediate pyroptosis, a mode of inflammatory programed cell death in myeloid and lymphoid leukocytes (79–81).

PURINERGIC RECEPTORS IN HIV-1 INFECTION

Because purinergic receptor signaling can clearly mediate inflammatory responses, these receptors are likely to be activated in response to infections. As HIV-1 is a viral infection marked by chronic inflammation, this signaling pathway might serve as an important intersection between viral infection and chronic inflammation. Purinergic signaling is involved in several infectious processes (82, 83), including bacterial and mycobacterial [*Mycobacterium tuberculosis* (84–87) and *Chlamydia* infections (88)], protozoal infections including *Leishmania* (89, 90) and *Toxoplasma* (91, 92), and viral infections including respiratory viral infections (93, 94), hepatitis B and hepatitis delta virus (95, 96), hepatitis C virus (97, 98), *Cytomegalovirus* (99), and HIV-1 (100–105).

Adenosine receptors have been implicated in HIV pathogenesis as Nikolova et al. reported an association between CD39 expression and AIDS progression (106). CD39 is an ectoenzyme that breaks down ATP to AMP which in turn, is hydrolyzed by CD73 to generate adenosine that signals through purinergic A1/2-type receptors. T_{reg} inhibition was shown to be mitigated by CD39 downregulation with associated elevated levels of A2A receptor on T cells of infection patients. The authors also noted that T_{reg} CD39 expansion was associated with elevated immune activation and that a CD39 gene polymorphism was associated with reduced CD39 expression and a delay in the onset of AIDS.

A role for extracellular ATP signaling has been proposed in HIV-1 infection. Sorrell et al. observed that treatment with a non-selective P2X antagonist reduced neurotoxic effects of opiates with generated in the context of HIV Tat activity which suggested that P2X receptors might modulate neurotoxicity. Those authors proposed that P2X inhibitors may serve to reduce neuroinflammation and neurodegeration in neuro-AIDS in the context of opiate abuse (107). Tovar and colleagues found that ATP released from HIV-infected macrophages can reduce dendritic spine density through purinergic-dependent glutamate receptor down-modulation. They proposed that neuronal injury in HIV-infected patients may relate to purinergic signaling and ATP release from macrophages that can impact on glutamate regulation (108).

Recent studies have raised the possibility that purinergic receptors as host proteins may be directly related to HIV-1 pathogenesis. Seror et al. demonstrated that infection of human lymphocytes with HIV-1 can induce ATP release and that this event is required for infection (104). Pharmacologic inhibition of purinergic receptors reduced HIV-mediated cell death and HIV infection. Non-selective purinergic receptors antagonists inhibited CCR5 and CXCR4-tropic HIV-1 productive infection in lymphocytes and CCR5-tropic virus in dendritic cells and macrophages. This study found that the selective depletion of P2Y2 with small interfering RNA diminished the HIV-induced inflammatory response and also resulted in mildly elevated levels of P2Y2 in HIV-infected patient tissue compared with uninfected control tissue. Immunofluorescence analyses indicated that P2Y2 and the ATP-release channel pannexin-1 appeared to polarize to the virologic synapse; the latter is the interface between an infected donor cell and an uninfected target cell where cell-to-cell transfer and infection takes place (109, 110).

Hazleton et al. demonstrated a key role for purinergic receptors in HIV-1 replication in macrophages (102). Macrophages are critical to HIV-1 pathogenesis as they may represent key reservoirs and can mediate immune responses through production of proinflammatory cytokines. The authors demonstrated that selective pharmacologic inhibition of P2X1, P2X7, and P2Y1 resulted in dose-dependent inhibition of HIV-1 infection. Using a beta-lactamase fusion assay, they observed a requirement for P2X1 in HIV-1 fusion in macrophages and that activation of P2X1 results in calcium flux that enables HIV-1 entry (111). More recently, Giroud et al. described a role for P2X1 (112) that involved block age of binding of HIV-1 to the chemokine receptors CCR5 and CXCR4. The group corroborated findings that inhibition of P2X1 with an inhibitor did not interfere with attachment but did inhibit fusion downstream of CD4 binding prior to coreceptor engagement.

Swartz et al. demonstrated that non-selective P2X receptor inhibitors inhibit HIV-1 infection of CD4+ lymphocytes by cell-to-cell and cell-free mechanisms (105). Using a systematic pharmacologic screening approach, it was found that only antagonists of a P2X subclass of purinergic receptors mediated inhibition of HIV-1 viral membrane fusion and productive infection of T cells. Because P2X inhibitors are a major focus of current pharmaceutical development for chronic inflammation, pain, and depression (59, 113, 114), this drug class has variants that may be assessed for both HIV inhibitory and inflammation inhibitory activities.

Orellana and colleagues observed that the function of the pannexin-1 ATP-release hemichannel was transiently increased during early infection with both R5 and X4 tropic HIV-1 and that HIV-1 envelope binding to CD4 and coreceptors (both CXCR4 and CCR5) activates pannexin-1 channel opening as a feed-forward signal which can enable HIV-1 internalization in CD4+ T cells (103). This study highlights the pannexin-1 hemichannel and associated factors, i.e., purinergic receptors as host factors that play important roles in early stages of HIV-1 entry (115). The role of purinergic inhibitors in HIV-1 disease is currently being investigated (116).

Most recently, Graziano et al. demonstrated that extracellular ATP induced rapid release of HIV-1 particles from human monocyte-derived macrophages that was P2X7 dependent (117). They hypothesized that virion egress may be additionally regulated by P2X7 function.

Definitive data are still lacking regarding which P2X receptor(s) are specifically required by HIV-1 and how purinergic signaling facilitates HIV-1 entry. Additionally, it is unknown whether HIV-1 infection activates other P2X7 signaling pathways, notably those involved in the NLRP3 inflammasome, which mediates IL-1β release. Elevated IL-1β is observed in HIV-infected patients (118–121), although these studies do have not directly link HIV-1 infection to inflammasome activation. Intriguing studies in CD4+ T cells found that pathogen sensor IFI-16 recognition of HIV-1 DNA can activate the inflammasome that induces proinflammatory lymphocyte programed cell death known as pyroptosis (122–125). This may represent a mechanism for CD4+ T cell depletion in HIV-1 disease and AIDS (126, 127). We present a model for the role of HIV-1 and purinergic signaling in **Figure 1**. This posits that HIV-1 entry results in the activation of P2X receptors and facilities fusion. This event may also trigger

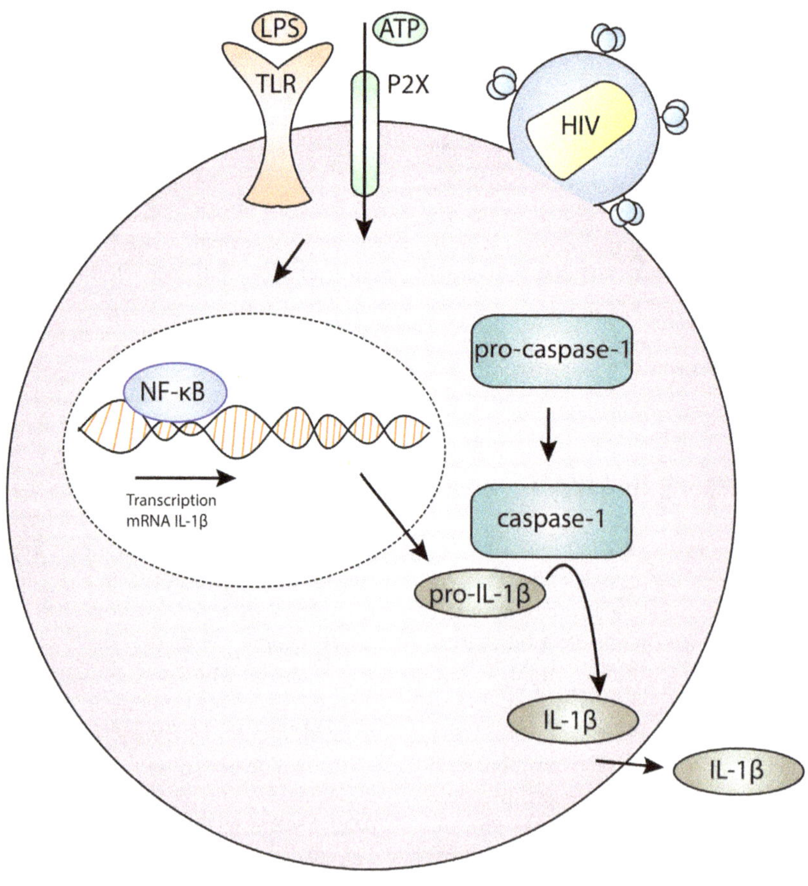

FIGURE 1 | Model for HIV infection and purinergic receptor signaling in a lymphocyte or macrophage/monocyte. HIV-1 attaches to a cell, and this is associated with P2X activation which results in cation and potentially large molecule flux. Concurrent toll-like receptor (TLR) activation by ligands, such as bacterial lipopolysaccharide (LPS), results in gene regulation through NF-kB. These two signals – TLR and P2X – are required for inflammasome activation which results in cleavage of procaspase-1 to caspase-1 which activates IL-1β which is then secreted.

TABLE 1 | P2X7 inhibitors in clinical trials.

Drug	Company	Phase	Endpoint	Reference
EVT 401	Evotec	I	Safety inhibition of ATP-stimulated IL-1β release	(133)
AZ9056	AstraZeneca	IIa	Safety, ACR20[a]	(58)
CE-224,535	Pfizer	IIa	Safety, ACR20	(59, 134)
GSK1482160	GlaxoSmithKline	I	Safety	(113, 135)

[a]American College of Rheumatology 20% response criteria (136).

inflammasome activation which results in maturation and release of IL-1β; this in turn drives inflammation and inflammatory cell death, thus depleting neighboring CD4+ T cells and contributing to systemic inflammation.

Novel antiretroviral therapies that target both HIV-1 productive infection as well as inflammation would be helpful in treating HIV-associated comorbidities. Because targeting purinergic receptors appears to be equivalently effective at blocking cell-free and cell-to-cell infection, these are attractive targets; inhibition of cell-to-cell infection with some ART can exhibit diminished efficacy (128–130). Finally, a recent study suggests that nucleoside reverse transcriptase inhibitors can inhibit inflammasome activation and reduce levels of IL-1β production (131). This suggests an important connection between HIV-1 pathogenesis and underlying inflammation through inflammasome activation.

Well-studied purinergic compounds in advanced stages of therapeutic development are the P2X7 antagonists. Various inhibitors, such as KN-62, PPADS, oxidized ATP, brilliant Blue G, AZ9056, A 740003, and A 438079, have been tested in inflammatory and neurological diseases (132). Several highly selective P2X7 receptor antagonists have been tested in clinical trials for safety for inflammatory pain conditions, specifically rheumatoid arthritis (**Table 1**). All drugs tested have demonstrated safety but have yet to show efficacy at reducing inflammatory pain.

While these drugs have not demonstrated efficacy in reducing neuropathic pain, there is potential for their applications in the modulation of inflammation related to infection. Of note, suramin is a well-described antiprotozoal agent that also has reverse transcription inhibitor activity *in vitro* against HIV-1 (137). In the 1980s, suramin was proposed as an ART and was given to 98 patients with AIDS (138). The study was ineffective at demonstrating survival advantage in the treated patients, largely because the patients had a high burden of disease and because the compound is toxic. Current drug development aims for compounds with a lower molecular weight that are moderately lipophilic (132). We propose that testing these agents may yield novel classes of anti-infective drugs that can function both to reduce viral replication and associated inflammation. An important goal in this area is to clarify how purinergic receptor antagonists block HIV-1 entry and to determine the role of purinergic signaling pathways in HIV-1 pathogenesis. The development of drugs that target these pathways may aid in treatment and prevention of HIV-1 disease and associated comorbidities.

CONCLUSION

Human immunodeficiency virus type 1 disease remains incurable, and as the affected population ages, patients will experience sequelae of chronic inflammation. As ART is still ineffective at eliminating this inflammation, novel therapies and a clearer understanding of the mechanisms that induce inflammation are necessary for improving long-term health of HIV-infected patients. An intriguing convergence of purinergic signaling with HIV-1 infectious pathways and inflammatory pathways indicates that these pathways may be central to disease pathogenesis. Understanding of the mechanisms that underlie such inflammation may enable targeted therapies that are more effective at enhancing the survival of HIV-infected patients by reducing chronic HIV-induced inflammation.

FUNDING

This work was funded by the National Center for Advancing Translational Science KL2TR00006 and the National Institute of General Medical Sciences GM36387.

REFERENCES

1. French MA, King MS, Tschampa JM, da Silva BA, Landay AL. Serum immune activation markers are persistently increased in patients with HIV infection after 6 years of antiretroviral therapy despite suppression of viral replication and reconstitution of CD4+ T cells. *J Infect Dis* (2009) **200**:1212–5. doi:10.1086/605890
2. Tenorio AR, Zheng Y, Bosch RJ, Krishnan S, Rodriguez B, Hunt PW, et al. Soluble markers of inflammation and coagulation but not T-cell activation predict non-AIDS-defining morbid events during suppressive antiretroviral treatment. *J Infect Dis* (2014) **210**:1248–59. doi:10.1093/infdis/jiu254
3. Hunt PW, Sinclair E, Rodriguez B, Shive C, Clagett B, Funderburg N, et al. Gut epithelial barrier dysfunction and innate immune activation predict mortality in treated HIV infection. *J Infect Dis* (2014) **210**:1228–38. doi:10.1093/infdis/jiu238
4. Cain LE, Logan R, Robins JM, Sterne JA, Sabin C, Bansi L, et al. When to initiate combined antiretroviral therapy to reduce mortality and AIDS-defining illness in HIV-infected persons in developed countries: an observational study. *Ann Intern Med* (2011) **154**:509–15. doi:10.7326/0003-4819-154-8-201104190-00001
5. Guaraldi G, Orlando G, Zona S, Menozzi M, Carli F, Garlassi E, et al. Premature age-related comorbidities among HIV-infected persons compared with the general population. *Clin Infect Dis* (2011) **53**:1120–6. doi:10.1093/cid/cir627
6. Ghiringhelli F, Apetoh L, Tesniere A, Aymeric L, Ma Y, Ortiz C, et al. Activation of the NLRP3 inflammasome in dendritic cells induces IL-1beta-dependent adaptive immunity against tumors. *Nat Med* (2009) **15**:1170–8. doi:10.1038/nm.2028
7. Bassett IV, Freedberg KA, Walensky RP. Two drugs or three? Balancing efficacy, toxicity, and resistance in postexposure prophylaxis for occupational exposure to HIV. *Clin Infect Dis* (2004) **39**:395–401. doi:10.1086/422459
8. Sigel K, Dubrow R, Silverberg M, Crothers K, Braithwaite S, Justice A. Cancer screening in patients infected with HIV. *Curr HIV/AIDS Rep* (2011) **8**:142–52. doi:10.1007/s11904-011-0085-5
9. Lifson AR; INSIGHT Cause of Death Writing Group, Belloso WH, Carey C, Davey RT, Duprez D, et al. Determination of the underlying cause of death

in three multicenter international HIV clinical trials. *HIV Clin Trials* (2008) **9**:177–85. doi:10.1310/hct0903-177
10. Aberg JA. Aging, inflammation, and HIV infection. *Top Antivir Med* (2012) **20**:101–5.
11. Deeks SG. HIV infection, inflammation, immunosenescence, and aging. *Annu Rev Med* (2011) **62**:141–55. doi:10.1146/annurev-med-042909-093756
12. Brenchley JM, Schacker TW, Ruff LE, Price DA, Taylor JH, Beilman GJ, et al. CD4+ T cell depletion during all stages of HIV disease occurs predominantly in the gastrointestinal tract. *J Exp Med* (2004) **200**:749–59. doi:10.1084/jem.20040874
13. Mehandru S, Poles MA, Tenner-Racz K, Horowitz A, Hurley A, Hogan C, et al. Primary HIV-1 infection is associated with preferential depletion of CD4+ T lymphocytes from effector sites in the gastrointestinal tract. *J Exp Med* (2004) **200**:761–70. doi:10.1084/jem.20041196
14. Hazenberg MD, Otto SA, van Benthem BH, Roos MT, Coutinho RA, Lange JM, et al. Persistent immune activation in HIV-1 infection is associated with progression to AIDS. *AIDS* (2003) **17**:1881–8. doi:10.1097/00002030-200309050-00006
15. Deeks SG, Kitchen CM, Liu L, Guo H, Gascon R, Narvaez AB, et al. Immune activation set point during early HIV infection predicts subsequent CD4+ T-cell changes independent of viral load. *Blood* (2004) **104**:942–7. doi:10.1182/blood-2003-09-3333
16. Brenchley JM, Price DA, Douek DC. HIV disease: fallout from a mucosal catastrophe? *Nat Immunol* (2006) **7**:235–9. doi:10.1038/ni1316
17. Vassallo M, Mercie P, Cottalorda J, Ticchioni M, Dellamonica P. The role of lipopolysaccharide as a marker of immune activation in HIV-1 infected patients: a systematic literature review. *Virol J* (2012) **9**:174. doi:10.1186/1743-422X-9-174
18. Valdez H, Connick E, Smith KY, Lederman MM, Bosch RJ, Kim RS, et al. Limited immune restoration after 3 years' suppression of HIV-1 replication in patients with moderately advanced disease. *AIDS* (2002) **16**:1859–66. doi:10.1097/00002030-200209270-00002
19. Hunt PW, Martin JN, Sinclair E, Bredt B, Hagos E, Lampiris H, et al. T cell activation is associated with lower CD4+ T cell gains in human immunodeficiency virus-infected patients with sustained viral suppression during antiretroviral therapy. *J Infect Dis* (2003) **187**:1534–43. doi:10.1086/374786
20. Massanella M, Negredo E, Perez-Alvarez N, Puig J, Ruiz-Hernandez R, Bofill M, et al. CD4 T-cell hyperactivation and susceptibility to cell death determine poor CD4 T-cell recovery during suppressive HAART. *AIDS* (2010) **24**:959–68. doi:10.1097/QAD.0b013e328337b957
21. Burdo TH, Lentz MR, Autissier P, Krishnan A, Halpern E, Letendre S, et al. Soluble CD163 made by monocyte/macrophages is a novel marker of HIV activity in early and chronic infection prior to and after anti-retroviral therapy. *J Infect Dis* (2011) **204**:154–63. doi:10.1093/infdis/jir214
22. Neuhaus J, Jacobs DR Jr, Baker JV, Calmy A, Duprez D, La Rosa A, et al. Markers of inflammation, coagulation, and renal function are elevated in adults with HIV infection. *J Infect Dis* (2010) **201**:1788–95. doi:10.1086/652749
23. Dinoso JB, Kim SY, Wiegand AM, Palmer SE, Gange SJ, Cranmer L, et al. Treatment intensification does not reduce residual HIV-1 viremia in patients on highly active antiretroviral therapy. *Proc Natl Acad Sci U S A* (2009) **106**:9403–8. doi:10.1073/pnas.0903107106
24. Hatano H, Hayes TL, Dahl V, Sinclair E, Lee TH, Hoh R, et al. A randomized, controlled trial of raltegravir intensification in antiretroviral-treated, HIV-infected patients with a suboptimal CD4+ T cell response. *J Infect Dis* (2011) **203**:960–8. doi:10.1093/infdis/jiq138
25. Lederman MM, Calabrese L, Funderburg NT, Clagett B, Medvik K, Bonilla H, et al. Immunologic failure despite suppressive antiretroviral therapy is related to activation and turnover of memory CD4 cells. *J Infect Dis* (2011) **204**:1217–26. doi:10.1093/infdis/jir507
26. Burnstock G. Purinergic signalling: its unpopular beginning, its acceptance and its exciting future. *Bioessays* (2012) **34**:218–25. doi:10.1002/bies.201100130
27. Burnstock G. Purinergic signalling: pathophysiology and therapeutic potential. *Keio J Med* (2013) **62**:63–73. doi:10.2302/kjm.2013-0003-RE
28. North RA. Molecular physiology of P2X receptors. *Physiol Rev* (2002) **82**:1013–67. doi:10.1152/physrev.00015.2002
29. Burnstock G, Knight GE. Cellular distribution and functions of P2 receptor subtypes in different systems. *Int Rev Cytol* (2004) **240**:31–304. doi:10.1016/S0074-7696(04)40002-3
30. Dubyak GR. Signal transduction by P2-purinergic receptors for extracellular ATP. *Am J Respir Cell Mol Biol* (1991) **4**:295–300. doi:10.1165/ajrcmb/4.4.295
31. Surprenant A, North RA. Signaling at purinergic P2X receptors. *Annu Rev Physiol* (2009) **71**:333–59. doi:10.1146/annurev.physiol.70.113006.100630
32. Khakh BS, North RA. P2X receptors as cell-surface ATP sensors in health and disease. *Nature* (2006) **442**:527–32. doi:10.1038/nature04886
33. Burnstock G, Kennedy C. P2X receptors in health and disease. *Adv Pharmacol* (2011) **61**:333–72. doi:10.1016/B978-0-12-385526-8.00011-4
34. Burnstock G. Purinergic signalling and disorders of the central nervous system. *Nat Rev Drug Discov* (2008) **7**:575–90. doi:10.1038/nrd2605
35. Khakh BS. Molecular physiology of P2X receptors and ATP signalling at synapses. *Nat Rev Neurosci* (2001) **2**:165–74. doi:10.1038/35058521
36. Khakh BS, Bao XR, Labarca C, Lester HA. Neuronal P2X transmitter-gated cation channels change their ion selectivity in seconds. *Nat Neurosci* (1999) **2**:322–30. doi:10.1038/7233
37. Khakh BS, Lester HA. Dynamic selectivity filters in ion channels. *Neuron* (1999) **23**:653–8. doi:10.1016/S0896-6273(01)80025-8
38. Virginio C, MacKenzie A, Rassendren FA, North RA, Surprenant A. Pore dilation of neuronal P2X receptor channels. *Nat Neurosci* (1999) **2**:315–21. doi:10.1038/7225
39. Jarvis MF, Khakh BS. ATP-gated P2X cation-channels. *Neuropharmacology* (2009) **56**:208–15. doi:10.1016/j.neuropharm.2008.06.067
40. Rettinger J, Schmalzing G. Activation and desensitization of the recombinant P2X1 receptor at nanomolar ATP concentrations. *J Gen Physiol* (2003) **121**:451–61. doi:10.1085/jgp.200208730
41. Coddou C, Yan Z, Obsil T, Huidobro-Toro JP, Stojilkovic SS. Activation and regulation of purinergic P2X receptor channels. *Pharmacol Rev* (2011) **63**:641–83. doi:10.1124/pr.110.003129
42. Erb L, Liao Z, Seye CI, Weisman GA. P2 receptors: intracellular signaling. *Pflugers Arch* (2006) **452**:552–62. doi:10.1007/s00424-006-0069-2
43. Nicke A, Baumert HG, Rettinger J, Eichele A, Lambrecht G, Mutschler E, et al. P2X1 and P2X3 receptors form stable trimers: a novel structural motif of ligand-gated ion channels. *EMBO J* (1998) **17**:3016–28. doi:10.1093/emboj/17.11.3016
44. Ferrari D, Pizzirani C, Adinolfi E, Lemoli RM, Curti A, Idzko M, et al. The P2X7 receptor: a key player in IL-1 processing and release. *J Immunol* (2006) **176**:3877–83. doi:10.4049/jimmunol.176.7.3877
45. Schenk U, Frascoli M, Proietti M, Geffers R, Traggiai E, Buer J, et al. ATP inhibits the generation and function of regulatory T cells through the activation of purinergic P2X receptors. *Sci Signal* (2011) **4**:ra12. doi:10.1126/scisignal.2001270
46. Junger WG. Immune cell regulation by autocrine purinergic signalling. *Nat Rev Immunol* (2011) **11**:201–12. doi:10.1038/nri2938
47. Browne LE, Compan V, Bragg L, North RA. P2X7 receptor channels allow direct permeation of nanometer-sized dyes. *J Neurosci* (2013) **33**:3557–66. doi:10.1523/JNEUROSCI.2235-12.2013
48. von Kugelgen I, Harden TK. Molecular pharmacology, physiology, and structure of the P2Y receptors. *Adv Pharmacol* (2011) **61**:373–415. doi:10.1016/B978-0-12-385526-8.00012-6
49. Wang ZX, Nakayama T, Sato N, Izumi Y, Kasamaki Y, Ohta M, et al. Association of the purinergic receptor P2Y, G-protein coupled, 2 (P2RY2) gene with myocardial infarction in Japanese men. *Circ J* (2009) **73**:2322–9. doi:10.1253/circj.CJ-08-1198
50. Burnstock G. Purinergic signaling and vascular cell proliferation and death. *Arterioscler Thromb Vasc Biol* (2002) **22**:364–73. doi:10.1161/hq0302.105360
51. Boarder MR, Weisman GA, Turner JT, Wilkinson GF. G protein-coupled P2 purinoceptors: from molecular biology to functional responses. *Trends Pharmacol Sci* (1995) **16**:133–9. doi:10.1016/S0165-6147(00)89001-X
52. Barnard EA. The transmitter-gated channels: a range of receptor types and structures. *Trends Pharmacol Sci* (1996) **17**:305–9. doi:10.1016/0165-6147(96)10041-9
53. von Kugelgen I. Pharmacological profiles of cloned mammalian P2Y-receptor subtypes. *Pharmacol Ther* (2006) **110**:415–32. doi:10.1016/j.pharmthera.2005.08.014

54. Collo G, Neidhart S, Kawashima E, Kosco-Vilbois M, North RA, Buell G. Tissue distribution of the P2X7 receptor. *Neuropharmacology* (1997) **36**:1277–83. doi:10.1016/S0028-3908(97)00140-8
55. Di Virgilio F. Liaisons dangereuses: P2X(7) and the inflammasome. *Trends Pharmacol Sci* (2007) **28**:465–72. doi:10.1016/j.tips.2007.07.002
56. Linden J. Regulation of leukocyte function by adenosine receptors. *Adv Pharmacol* (2011) **61**:95–114. doi:10.1016/B978-0-12-385526-8.00004-7
57. Eltzschig HK, Sitkovsky MV, Robson SC. Purinergic signaling during inflammation. *N Engl J Med* (2012) **367**:2322–33. doi:10.1056/NEJMra1205750
58. Keystone EC, Wang MM, Layton M, Hollis S, McInnes IB, Team DCS. Clinical evaluation of the efficacy of the P2X7 purinergic receptor antagonist AZD9056 on the signs and symptoms of rheumatoid arthritis in patients with active disease despite treatment with methotrexate or sulphasalazine. *Ann Rheum Dis* (2012) **71**:1630–5. doi:10.1136/annrheumdis-2011-143578
59. Stock TC, Bloom BJ, Wei N, Ishaq S, Park W, Wang X, et al. Efficacy and safety of CE-224,535, an antagonist of P2X7 receptor, in treatment of patients with rheumatoid arthritis inadequately controlled by methotrexate. *J Rheumatol* (2012) **39**:720–7. doi:10.3899/jrheum.110874
60. Vergani A, Tezza S, D'Addio F, Fotino C, Liu K, Niewczas M, et al. Long-term heart transplant survival by targeting the ionotropic purinergic receptor P2X7. *Circulation* (2013) **127**:463–75. doi:10.1161/CIRCULATIONAHA.112.123653
61. Idzko M, Dichmann S, Ferrari D, Di Virgilio F, la Sala A, Girolomoni G, et al. Nucleotides induce chemotaxis and actin polymerization in immature but not mature human dendritic cells via activation of pertussis toxin-sensitive P2y receptors. *Blood* (2002) **100**:925–32. doi:10.1182/blood.V100.3.925
62. Mei L, Du W, Gao W, Mei QB. Purinergic signaling: a novel mechanism in immune surveillance. *Acta Pharmacol Sin* (2010) **31**:1149–53. doi:10.1038/aps.2010.128
63. Kronlage M, Song J, Sorokin L, Isfort K, Schwerdtle T, Leipziger J, et al. Autocrine purinergic receptor signaling is essential for macrophage chemotaxis. *Sci Signal* (2010) **3**:ra55. doi:10.1126/scisignal.2000588
64. McDonald B, Pittman K, Menezes GB, Hirota SA, Slaba I, Waterhouse CC, et al. Intravascular danger signals guide neutrophils to sites of sterile inflammation. *Science* (2010) **330**:362–6. doi:10.1126/science.1195491
65. Frascoli M, Marcandalli J, Schenk U, Grassi F. Purinergic P2X7 receptor drives T cell lineage choice and shapes peripheral gammadelta cells. *J Immunol* (2012) **189**:174–80. doi:10.4049/jimmunol.1101582
66. Baricordi OR, Ferrari D, Melchiorri L, Chiozzi P, Hanau S, Chiari E, et al. An ATP-activated channel is involved in mitogenic stimulation of human T lymphocytes. *Blood* (1996) **87**:682–90.
67. Yip L, Woehrle T, Corriden R, Hirsh M, Chen Y, Inoue Y, et al. Autocrine regulation of T-cell activation by ATP release and P2X7 receptors. *FASEB J* (2009) **23**:1685–93. doi:10.1096/fj.08-126458
68. Loomis WH, Namiki S, Ostrom RS, Insel PA, Junger WG. Hypertonic stress increases T cell interleukin-2 expression through a mechanism that involves ATP release, P2 receptor, and p38 MAPK activation. *J Biol Chem* (2003) **278**:4590–6. doi:10.1074/jbc.M207868200
69. Filippini A, Taffs RE, Sitkovsky MV. Extracellular ATP in T-lymphocyte activation: possible role in effector functions. *Proc Natl Acad Sci U S A* (1990) **87**:8267–71. doi:10.1073/pnas.87.21.8267
70. Manohar M, Hirsh MI, Chen Y, Woehrle T, Karande AA, Junger WG. ATP release and autocrine signaling through P2X4 receptors regulate gammadelta T cell activation. *J Leukoc Biol* (2012) **92**:787–94. doi:10.1189/jlb.0312121
71. Dubyak GR. P2X7 receptor regulation of non-classical secretion from immune effector cells. *Cell Microbiol* (2012) **14**:1697–706. doi:10.1111/cmi.12001
72. Franchi L, Kanneganti TD, Dubyak GR, Nunez G. Differential requirement of P2X7 receptor and intracellular K+ for caspase-1 activation induced by intracellular and extracellular bacteria. *J Biol Chem* (2007) **282**:18810–8. doi:10.1074/jbc.M610762200
73. Gudipaty L, Munetz J, Verhoef PA, Dubyak GR. Essential role for Ca2+ in regulation of IL-1beta secretion by P2X7 nucleotide receptor in monocytes, macrophages, and HEK-293 cells. *Am J Physiol Cell Physiol* (2003) **285**:C286–99. doi:10.1152/ajpcell.00070.2003
74. Qu Y, Ramachandra L, Mohr S, Franchi L, Harding CV, Nunez G, et al. P2X7 receptor-stimulated secretion of MHC class II-containing exosomes requires the ASC/NLRP3 inflammasome but is independent of caspase-1. *J Immunol* (2009) **182**:5052–62. doi:10.4049/jimmunol.0802968
75. Idzko M, Ferrari D, Eltzschig HK. Nucleotide signalling during inflammation. *Nature* (2014) **509**:310–7. doi:10.1038/nature13085
76. Volonte C, Apolloni S, Skaper SD, Burnstock G. P2X7 receptors: channels, pores and more. *CNS Neurol Disord Drug Targets* (2012) **11**:705–21. doi:10.2174/187152712803581137
77. Cheneval D, Ramage P, Kastelic T, Szelestenyi T, Niggli H, Hemmig R, et al. Increased mature interleukin-1beta (IL-1beta) secretion from THP-1 cells induced by nigericin is a result of activation of p45 IL-1beta-converting enzyme processing. *J Biol Chem* (1998) **273**:17846–51. doi:10.1074/jbc.273.28.17846
78. Netea MG, Simon A, van de Veerdonk F, Kullberg BJ, Van der Meer JW, Joosten LA. IL-1beta processing in host defense: beyond the inflammasomes. *PLoS Pathog* (2010) **6**:e1000661. doi:10.1371/journal.ppat.1000661
79. Bergsbaken T, Fink SL, Cookson BT. Pyroptosis: host cell death and inflammation. *Nat Rev Microbiol* (2009) **7**:99–109. doi:10.1038/nrmicro2070
80. Jacobs SR, Damania B. NLRs, inflammasomes, and viral infection. *J Leukoc Biol* (2012) **92**:469–77. doi:10.1189/jlb.0312132
81. Lamkanfi M, Dixit VM. Inflammasomes and their roles in health and disease. *Ann Rev Cell Dev Biol* (2012) **28**:137–61. doi:10.1146/annurev-cellbio-101011-155745
82. Miller CM, Boulter NR, Fuller SJ, Zakrzewski AM, Lees MP, Saunders BM, et al. The role of the P2X(7) receptor in infectious diseases. *PLoS Pathog* (2011) **7**:e1002212. doi:10.1371/journal.ppat.1002212
83. Morandini AC, Savio LE, Coutinho-Silva R. The role of P2X7 receptor in infectious inflammatory diseases and the influence of ectonucleotidases. *Biomed J* (2014) **37**:169–77. doi:10.4103/2319-4170.127803
84. Franco-Martinez S, Nino-Moreno P, Bernal-Silva S, Baranda L, Rocha-Meza M, Portales-Cervantes L, et al. Expression and function of the purinergic receptor P2X7 in patients with pulmonary tuberculosis. *Clin Exp Immunol* (2006) **146**:253–61. doi:10.1111/j.1365-2249.2006.03213.x
85. Wu G, Zhao M, Gu X, Yao Y, Liu H, Song Y. The effect of P2X7 receptor 1513 polymorphism on susceptibility to tuberculosis: a meta-analysis. *Infect Genet Evol* (2014) **24**:82–91. doi:10.1016/j.meegid.2014.03.006
86. Xiao J, Sun L, Yan H, Jiao W, Miao Q, Feng W, et al. Metaanalysis of P2X7 gene polymorphisms and tuberculosis susceptibility. *FEMS Immunol Med Microbiol* (2010) **60**:165–70. doi:10.1111/j.1574-695X.2010.00735.x
87. Yi L, Cheng D, Shi H, Huo X, Zhang K, Zhen G. A meta-analysis of P2X7 gene-762T/C polymorphism and pulmonary tuberculosis susceptibility. *PLoS One* (2014) **9**:e96359. doi:10.1371/journal.pone.0096359
88. Darville T, Welter-Stahl L, Cruz C, Sater AA, Andrews CW Jr, Ojcius DM. Effect of the purinergic receptor P2X7 on *Chlamydia* infection in cervical epithelial cells and vaginally infected mice. *J Immunol* (2007) **179**:3707–14. doi:10.4049/jimmunol.179.6.3707
89. Marques-da-Silva C, Chaves MM, Chaves SP, Figliuolo VR, Meyer-Fernandes JR, Corte-Real S, et al. Infection with *Leishmania amazonensis* upregulates purinergic receptor expression and induces host-cell susceptibility to UTP-mediated apoptosis. *Cell Microbiol* (2011) **13**:1410–28. doi:10.1111/j.1462-5822.2011.01630.x
90. Marques-da-Silva C, Chaves MM, Rodrigues JC, Corte-Real S, Coutinho-Silva R, Persechini PM. Differential modulation of ATP-induced P2X7-associated permeabilities to cations and anions of macrophages by infection with *Leishmania amazonensis*. *PLoS One* (2011) **6**:e25356. doi:10.1371/journal.pone.0025356
91. Lees MP, Fuller SJ, McLeod R, Boulter NR, Miller CM, Zakrzewski AM, et al. P2X7 receptor-mediated killing of an intracellular parasite, *Toxoplasma gondii*, by human and murine macrophages. *J Immunol* (2010) **184**:7040–6. doi:10.4049/jimmunol.1000012
92. Miller CM, Zakrzewski AM, Ikin RJ, Boulter NR, Katrib M, Lees MP, et al. Dysregulation of the inflammatory response to the parasite, *Toxoplasma gondii*, in P2X7 receptor-deficient mice. *Int J Parasitol* (2011) **41**:301–8. doi:10.1016/j.ijpara.2010.10.001
93. Paoletti A, Raza SQ, Voisin L, Law F, Pipoli da Fonseca J, Caillet M, et al. Multifaceted roles of purinergic receptors in viral infection. *Microbes Infect* (2012) **14**:1278–83. doi:10.1016/j.micinf.2012.05.010

94. Lee BH, Hwang DM, Palaniyar N, Grinstein S, Philpott DJ, Hu J. Activation of P2X(7) receptor by ATP plays an important role in regulating inflammatory responses during acute viral infection. *PLoS One* (2012) **7**:e35812. doi:10.1371/journal.pone.0035812
95. Lamas Longarela O, Schmidt TT, Schoneweis K, Romeo R, Wedemeyer H, Urban S, et al. Proteoglycans act as cellular hepatitis delta virus attachment receptors. *PLoS One* (2013) **8**:e58340. doi:10.1371/journal.pone.0058340
96. Taylor JM, Han Z. Purinergic receptor functionality is necessary for infection of human hepatocytes by hepatitis delta virus and hepatitis B virus. *PLoS One* (2010) **5**:e15784. doi:10.1371/journal.pone.0015784
97. Manzoor S, Idrees M, Ashraf J, Mehmood A, Butt S, Fatima K, et al. Identification of ionotrophic purinergic receptors in Huh-7 cells and their response towards structural proteins of HCV genotype 3a. *Virol J* (2011) **8**:431. doi:10.1186/1743-422X-8-431
98. Ashraf W, Manzoor S, Ashraf J, Ahmed QL, Khalid M, Tariq M, et al. Transcript analysis of P2X receptors in PBMCs of chronic HCV patients: an insight into antiviral treatment response and HCV-induced pathogenesis. *Viral Immunol* (2013) **26**:343–50. doi:10.1089/vim.2013.0044
99. Zandberg M, van Son WJ, Harmsen MC, Bakker WW. Infection of human endothelium in vitro by *Cytomegalovirus* causes enhanced expression of purinergic receptors: a potential virus escape mechanism? *Transplantation* (2007) **84**:1343–7. doi:10.1097/01.tp.0000287598.25493.a5
100. Pacheco PA, Faria RX, Ferreira LG, Paixao IC. Putative roles of purinergic signaling in human immunodeficiency virus-1 infection. *Biol Direct* (2014) **9**:21. doi:10.1186/1745-6150-9-21
101. Barat C, Gilbert C, Imbeault M, Tremblay MJ. Extracellular ATP reduces HIV-1 transfer from immature dendritic cells to CD4+ T lymphocytes. *Retrovirology* (2008) **5**:30. doi:10.1186/1742-4690-5-30
102. Hazleton JE, Berman JW, Eugenin EA. Purinergic receptors are required for HIV-1 infection of primary human macrophages. *J Immunol* (2012) **188**:4488–95. doi:10.4049/jimmunol.1102482
103. Orellana JA, Velasquez S, Williams DW, Saez JC, Berman JW, Eugenin EA. Pannexin1 hemichannels are critical for HIV infection of human primary CD4+ T lymphocytes. *J Leukoc Biol* (2013) **94**:399–407. doi:10.1189/jlb.0512249
104. Seror C, Melki MT, Subra F, Raza SQ, Bras M, Saidi H, et al. Extracellular ATP acts on P2Y2 purinergic receptors to facilitate HIV-1 infection. *J Exp Med* (2011) **208**:1823–34. doi:10.1084/jem.20101805
105. Swartz TH, Esposito AM, Durham ND, Hartmann BM, Chen BK. P2X-selective purinergic antagonists are strong inhibitors of HIV-1 fusion during both cell-to-cell and cell-free infection. *J Virol* (2014) **88**:11504–15. doi:10.1128/JVI.01158-14
106. Nikolova M, Carriere M, Jenabian MA, Limou S, Younas M, Kok A, et al. CD39/adenosine pathway is involved in AIDS progression. *PLoS Pathog* (2011) **7**:e1002110. doi:10.1371/journal.ppat.1002110
107. Sorrell ME, Hauser KF. Ligand-gated purinergic receptors regulate HIV-1 tat and morphine related neurotoxicity in primary mouse striatal neuron-glia co-cultures. *J Neuroimmune Pharmacol* (2014) **9**:233–44. doi:10.1007/s11481-013-9507-z
108. Tovar YRLB, Kolson DL, Bandaru VV, Drewes JL, Graham DR, Haughey NJ. Adenosine triphosphate released from HIV-infected macrophages regulates glutamatergic tone and dendritic spine density on neurons. *J Neuroimmune Pharmacol* (2013) **8**:998–1009. doi:10.1007/s11481-013-9471-7
109. Jolly C, Kashefi K, Hollinshead M, Sattentau QJ. HIV-1 cell to cell transfer across an Env-induced, actin-dependent synapse. *J Exp Med* (2004) **199**:283–93. doi:10.1084/jem.20030648
110. Hubner W, Chen P, Del Portillo A, Liu Y, Gordon RE, Chen BK. Sequence of human immunodeficiency virus type 1 (HIV-1) Gag localization and oligomerization monitored with live confocal imaging of a replication-competent, fluorescently tagged HIV-1. *J Virol* (2007) **81**:12596–607. doi:10.1128/JVI.01088-07
111. Marin M, Du Y, Giroud C, Kim JH, Qui M, Fu H, et al. High-throughput HIV-cell fusion assay for discovery of virus entry inhibitors. *Assay Drug Dev Technol* (2015) **13**:155–66. doi:10.1089/adt.2015.639
112. Giroud C, Marin M, Hammonds J, Spearman P, Melikyan GB. P2X1 receptor antagonists inhibit HIV-1 fusion by blocking virus-coreceptor interactions. *J Virol* (2015) **89**:9368–82. doi:10.1128/JVI.01178-15
113. Ali Z, Laurijssens B, Ostenfeld T, McHugh S, Stylianou A, Scott-Stevens P, et al. Pharmacokinetic and pharmacodynamic profiling of a P2X7 receptor allosteric modulator GSK1482160 in healthy human subjects. *Br J Clin Pharmacol* (2013) **75**:197–207. doi:10.1111/j.1365-2125.2012.04320.x
114. Arulkumaran N, Unwin RJ, Tam FW. A potential therapeutic role for P2X7 receptor (P2X7R) antagonists in the treatment of inflammatory diseases. *Expert Opin Investig Drugs* (2011) **20**:897–915. doi:10.1517/13543784.2011.578068
115. Paoletti A, Raza SQ, Voisin L, Law F, Caillet M, Martins I, et al. Editorial: pannexin-1 – the hidden gatekeeper for HIV-1. *J Leukoc Biol* (2013) **94**:390–2. doi:10.1189/jlb.0313148
116. Velasquez S, Eugenin EA. Role of Pannexin-1 hemichannels and purinergic receptors in the pathogenesis of human diseases. *Front Physiol* (2014) **5**:96. doi:10.3389/fphys.2014.00096
117. Graziano F, Desdouits M, Garzetti L, Podini P, Alfano M, Rubartelli A, et al. Extracellular ATP induces the rapid release of HIV-1 from virus containing compartments of human macrophages. *Proc Natl Acad Sci U S A* (2015) **112**:E3265–73. doi:10.1073/pnas.1500656112
118. Allers K, Fehr M, Conrad K, Epple HJ, Schurmann D, Geelhaar-Karsch A, et al. Macrophages accumulate in the gut mucosa of untreated HIV-infected patients. *J Infect Dis* (2014) **209**:739–48. doi:10.1093/infdis/jit547
119. Shive CL, Mudd JC, Funderburg NT, Sieg SF, Kyi B, Bazdar DA, et al. Inflammatory cytokines drive CD4+ T-cell cycling and impaired responsiveness to interleukin 7: implications for immune failure in HIV disease. *J Infect Dis* (2014) **210**:619–29. doi:10.1093/infdis/jiu125
120. Liu CM, Osborne BJ, Hungate BA, Shahabi K, Huibner S, Lester R, et al. The semen microbiome and its relationship with local immunology and viral load in HIV infection. *PLoS Pathog* (2014) **10**:e1004262. doi:10.1371/journal.ppat.1004262
121. Guo H, Gao J, Taxman DJ, Ting JP, Su L. HIV-1 infection induces interleukin-1beta production via TLR8 protein-dependent and NLRP3 inflammasome mechanisms in human monocytes. *J Biol Chem* (2014) **289**:21716–26. doi:10.1074/jbc.M114.566620
122. Galloway NL, Doitsh G, Monroe KM, Yang Z, Munoz-Arias I, Levy DN, et al. Cell-to-cell transmission of HIV-1 is required to trigger pyroptotic death of lymphoid-tissue-derived CD4 T cells. *Cell Rep* (2015) **12**:1555–63. doi:10.1016/j.celrep.2015.08.011
123. Doitsh G, Cavrois M, Lassen KG, Zepeda O, Yang Z, Santiago ML, et al. Abortive HIV infection mediates CD4 T cell depletion and inflammation in human lymphoid tissue. *Cell* (2010) **143**:789–801. doi:10.1016/j.cell.2010.11.001
124. Doitsh G, Galloway NL, Geng X, Yang Z, Monroe KM, Zepeda O, et al. Cell death by pyroptosis drives CD4 T-cell depletion in HIV-1 infection. *Nature* (2014) **505**:509–14. doi:10.1038/nature12940
125. Monroe KM, Yang Z, Johnson JR, Geng X, Doitsh G, Krogan NJ, et al. IFI16 DNA sensor is required for death of lymphoid CD4 T cells abortively infected with HIV. *Science* (2014) **343**:428–32. doi:10.1126/science.1243640
126. Steele AK, Lee EJ, Manuzak JA, Dillon SM, Beckham JD, McCarter MD, et al. Microbial exposure alters HIV-1-induced mucosal CD4+ T cell death pathways ex vivo. *Retrovirology* (2014) **11**:14. doi:10.1186/1742-4690-11-14
127. Jorgensen I, Miao EA. Pyroptotic cell death defends against intracellular pathogens. *Immunol Rev* (2015) **265**:130–42. doi:10.1111/imr.12287
128. Alvarez RA, Barria MI, Chen BK. Unique features of HIV-1 spread through T cell virological synapses. *PLoS Pathog* (2014) **10**:e1004513. doi:10.1371/journal.ppat.1004513
129. Agosto LM, Uchil PD, Mothes W. HIV cell-to-cell transmission: effects on pathogenesis and antiretroviral therapy. *Trends Microbiol* (2015) **23**:289–95. doi:10.1016/j.tim.2015.02.003
130. Agosto LM, Zhong P, Munro J, Mothes W. Highly active antiretroviral therapies are effective against HIV-1 cell-to-cell transmission. *PLoS Pathog* (2014) **10**:e1003982. doi:10.1371/journal.ppat.1003982
131. Fowler BJ, Gelfand BD, Kim Y, Kerur N, Tarallo V, Hirano Y, et al. Nucleoside reverse transcriptase inhibitors possess intrinsic anti-inflammatory activity. *Science* (2014) **346**:1000–3. doi:10.1126/science.1261754
132. Carroll WA, Donnelly-Roberts D, Jarvis MF. Selective P2X(7) receptor antagonists for chronic inflammation and pain. *Purinergic Signal* (2009) **5**:63–73. doi:10.1007/s11302-008-9110-6

133. Evotec. (2015). Available from: https://www.evotec.com/archive/en/Press-releases/2008/Evotec-Announces-Phase-I-Initiation-with-P2X7-Antagonist/1527/1
134. Duplantier AJ, Dombroski MA, Subramanyam C, Beaulieu AM, Chang SP, Gabel CA, et al. Optimization of the physicochemical and pharmacokinetic attributes in a 6-azauracil series of P2X7 receptor antagonists leading to the discovery of the clinical candidate CE-224,535. *Bioorg Med Chem Lett* (2011) **21**:3708–11. doi:10.1016/j.bmcl.2011.04.077
135. Gao M, Wang M, Green MA, Hutchins GD, Zheng QH. Synthesis of [(11)C]GSK1482160 as a new PET agent for targeting P2X7 receptor. *Bioorg Med Chem Lett* (2015) **25**:1965–70. doi:10.1016/j.bmcl.2015.03.021
136. Felson DT, LaValley MP. The ACR20 and defining a threshold for response in rheumatic diseases: too much of a good thing. *Arthritis Res Ther* (2014) **16**:101. doi:10.1186/ar4428
137. De Clercq E. Suramin: a potent inhibitor of the reverse transcriptase of RNA tumor viruses. *Cancer Lett* (1979) **8**:9–22. doi:10.1016/0304-3835(79)90017-X
138. De Clercq E. Suramin in the treatment of AIDS: mechanism of action. *Antiviral Res* (1987) **7**:1–10. doi:10.1016/0166-3542(87)90034-9

2

Cytosolic Innate Immune Sensing and Signaling upon Infection

Lilliana Radoshevich[1,2,3] and Olivier Dussurget[1,2,3,4]**

[1] Unité des Interactions Bactéries-Cellules, Institut Pasteur, Paris, France, [2] Institut National de la Santé et de la Recherche Médicale, U604, Paris, France, [3] Institut National de la Recherche Agronomique, USC2020, Paris, France, [4] Cellule Pasteur, Université Paris Diderot, Sorbonne Paris Cité, Paris, France

Edited by:
Heinrich Korner, Menzies Research Institute Tasmania, Australia

Reviewed by:
Mirko Trilling, University Duisburg-Essen; Essen University Hospital, Germany
Rance E. Berg, University of North Texas Health Science Center, USA

***Correspondence:**
Lilliana Radoshevich lilliana.radoshevich@pasteur.fr;
Olivier Dussurget olivier.dussurget@pasteur.fr

Cytosolic sensing of pathogens is essential to a productive immune response. Recent reports have emphasized the importance of signaling platforms emanating from organelles and cytosolic sensors, particularly during the response to intracellular pathogens. Here, we highlight recent discoveries identifying the key mediators of nucleic acid and cyclic nucleotide sensing and discuss their importance in host defense. This review will also cover strategies evolved by pathogens to manipulate these pathways.

Keywords: immunity, pathogen, STING, MAVS, DNA, RNA, interferon

INTRODUCTION

Surveillance of the cytosol by host pathogen recognition receptors is essential to activation of the innate immune response and pathogen clearance. It has long been appreciated that cell surface receptors can identify pathogen-associated molecular patterns to initiate the innate immune response. More recently, a concept is emerging in the context of intracellular sensing of pathogens. In particular, if molecules specific to pathogens or shared by pathogens and host are located in an inappropriate cellular compartment (i.e., DNA in the cytosol), this triggers innate immune signaling. Subsequent signaling downstream of pathogen recognition is also compartmentalized to organelle-based signaling platforms on the surface of the endoplasmic reticulum (ER), the mitochondria and peroxisomes. In addition, in the ongoing arms race between pathogen and host, the various nodes of pathogen surveillance and signaling are frequently targeted by a variety of bacterial and viral effectors. In this review, we have chosen to focus on cytosolic sensors of DNA, RNA, and cyclic dinucleotides and their subversion by bacterial and viral pathogens.

DNA CYTOSOLIC SURVEILLANCE PATHWAYS

DNA Sensing through STING

In 2006 cytosolic DNA from the bacterial pathogen *Listeria monocytogenes* was first reported to lead to a type I interferon (IFN) response (Stetson and Medzhitov, 2006). More recently the field of cytosolic DNA and cyclic dinucleotide (CDN) sensing has made significant conceptual and mechanistic advances. Initially, STING (Stimulator of Interferon Genes) was cloned and characterized as an ER-localized transmembrane protein that was an essential signaling adaptor to IFN production (Ishikawa and Barber, 2008). Subsequent work identified the minimal DNA motifs required for IFN production and the effect of STING on *in vivo* infection and T cell mediated immunity (Ishikawa et al., 2009). A parallel breakthrough occurred when several groups determined that cytosolic CDNs – small nucleotides that are unique to bacteria – from a variety of pathogens provoke an IFN response (Karaolis et al., 2007a,b; Mcwhirter et al., 2009; Woodward et al., 2010). STING's central role in this process was highlighted through a forward genetic screen which identified a mutant of STING that was unresponsive to CDNs and through a second study which proved a direct biochemical interaction between STING and cyclic-diguanylate (c-di-GMP; Burdette et al., 2011; Sauer et al., 2011). Most intriguingly, mutants of STING that could bind CDNs but did not activate downstream signaling were still capable of responding to cytosolic DNA. These data as well as the observation that STING expression alone in 293T cells could not complement the deficiency of this cell line in sensing cytosolic DNA suggested a missing DNA sensor (Burdette et al., 2011). One report posited that STING could directly bind single stranded or double stranded DNA, however, the same region of STING was important for CDN binding (Abe et al., 2013), data that is difficult to reconcile with earlier studies that highlighted the capacity of mutant STING (that could not signal following CDN binding) to induce IFN downstream of cytosolic DNA.

Over the past two years several groups have identified a cyclic dinucleotide eukaryotic second messenger, 2′3′-cyclic guanosine monophosphate–adenosine monophosphate (cGAMP), which led to the discovery of the cyclic GMP-AMP synthase (cGAS) as the missing cytosolic DNA sensor linking STING activity to IFN production (**Figure 1**). Using a cell free system, cGAMP was generated and identified following exposure of the cytosol to DNA or infection with a DNA virus (Wu et al., 2013b). Interestingly, this second messenger has a unique structure which differentiates it from bacterial CDNs and potentially renders it less susceptible to phosphodiesterases (Ablasser et al., 2013; Diner et al., 2013; Gao et al., 2013a; Zhang et al., 2013). This was a major conceptual leap since structurally cGAS could directly interact with DNA and then through a unique enzymatic function, it could generate a second messenger capable of activating STING (Sun et al., 2013). In addition, cGAS deletion phenocopies STING deletion with regards to IFN production following activation, a property which all of the other previously described DNA sensors lack. The initial structures of STING bound to c-di-GMP were rather perplexing, as binding did not lead to a conformational change (Cai et al., 2014). Binding of the cGAS product, cGAMP, on the other hand leads to a conformational change of STING (Gao et al., 2013b; Zhang et al., 2013).

Other cytoplasmic DNA sensors involved in innate immune responses are the DNA-dependent protein kinase (DNA-PK) complex and the MRN complex. The two complexes are both sensors and effectors of the DNA damage response (DDR). DNA-PK is a heterotrimeric complex composed of the heterodimer Ku70/Ku80 and the catalytic subunit DNA-PKcs. The MRN complex consists of meiotic recombination 11 (Mre11), which has been linked to STING activation, radiation-sensitive 50 (Rad50) and Nijmegen breakage syndrome 1 (Nbs1) proteins. DNA sensing by DNA-PK and MRN is critical for innate immune responses and DDR, inhibiting viral DNA replication.

A few other proteins, including DNA-dependent activator of IFN-regulatory factors (DAI) and interferon-inducible protein 16 (IFI16), have been previously suggested to be sensors of DNA. However, consensus on how and whether they act upstream of STING has not been reached. In addition, none of these proteins has a comparably profound effect on IFN signaling as cGAS when deleted (Paludan and Bowie, 2013). Therefore, understanding how the DNA-sensing pathways intersect and complement each other in various cell types or tissues is an important future direction.

Cellular Responses

Following STING activation on the surface of the ER by CDNs, STING interacts with the TNFR-associated NF-κB kinase (TANK)-binding kinase 1 (TBK1) and relocalizes to cytoplasmic puncta. These puncta are also microtubule-associated protein 1 light chain 3 (LC3), autophagy-related gene 9 (ATG9) and p62 positive (Saitoh et al., 2009). TBK1 can subsequently phosphorylate interferon regulatory factor 3 (IRF3) leading to IFN production. Chen et al. (2011) have recently highlighted a key role played by signal transducer and activator of transcription 6 (STAT6) in this process. Interestingly, STAT6 becomes phosphorylated and active following viral infection in a Janus kinase (JAK)-independent manner that requires STING and leads to a different subset of STAT6 target genes than canonical STAT6 activation (Chen et al., 2011). Several recent papers have highlighted early IFN-independent induction of interferon-stimulated genes (ISGs) following viral or bacterial infection. The cytosolic 3′ repair exonuclease 1 (TREX1) degrades excess cytosolic DNA and mutations in this gene are associated with the autoimmune disease Aicardi–Goutières syndrome (Stetson et al., 2008). Hasan et al. (2013) took advantage of TREX1-deficient cells to discover a cell-intrinsic IFN-independent cytosolic surveillance pathway in response to viral infection. While IFN production was the same for RNA viruses in wild-type or TREX1-deficient cells, viral load was much lower due to a baseline increase in ISGs. Interestingly, this increase was dependent on STING/TBK1/IRF3 and IRF7 and also led to increased lysososomal biogenesis (Hasan et al., 2013). Our own group independently found an IFN-independent induction of ISG15 upon infection of non-phagocytic cells with *L. monocytogenes*. This IFN-independent induction required STING/TBK1/IRF3 and IRF7 and interestingly could be recapitulated by *Listeria* genomic DNA but not bacterial cyclic di-AMP (Radoshevich et al., 2015). An exciting future direction will be to determine varying phenotypic outcomes based on the subset of ISGs that are induced during a variety of infections. Another critical theme that has emerged from several recent studies is the link between STING/TBK1/IRF3 signaling and autophagy. In particular, one surprising finding is that the membrane-limited pathogen *Mycobacterium tuberculosis* activates the cytosolic surveillance pathway to a comparable degree as the intracellular cytosolic pathogen *L. monocytogenes* (Manzanillo et al., 2012). These data imply that *M. tuberculosis* DNA is being sensed by cGAS, which in turn activates the STING/TBK1/IRF3 signaling axis and anti-bacterial autophagy to help with pathogen clearance. In fact, three papers have recently demonstrated the crucial role of cGAS during *M. tuberculosis* infection (Collins et al., 2015; Wassermann et al., 2015; Watson et al., 2015). Future questions of interest to the field will be to determine whether the bacterium is secreting its own DNA or lysed bacteria are the source of the DNA signal. In addition it will be informative to compare and contrast signaling outcomes from eukaryotic versus bacterial

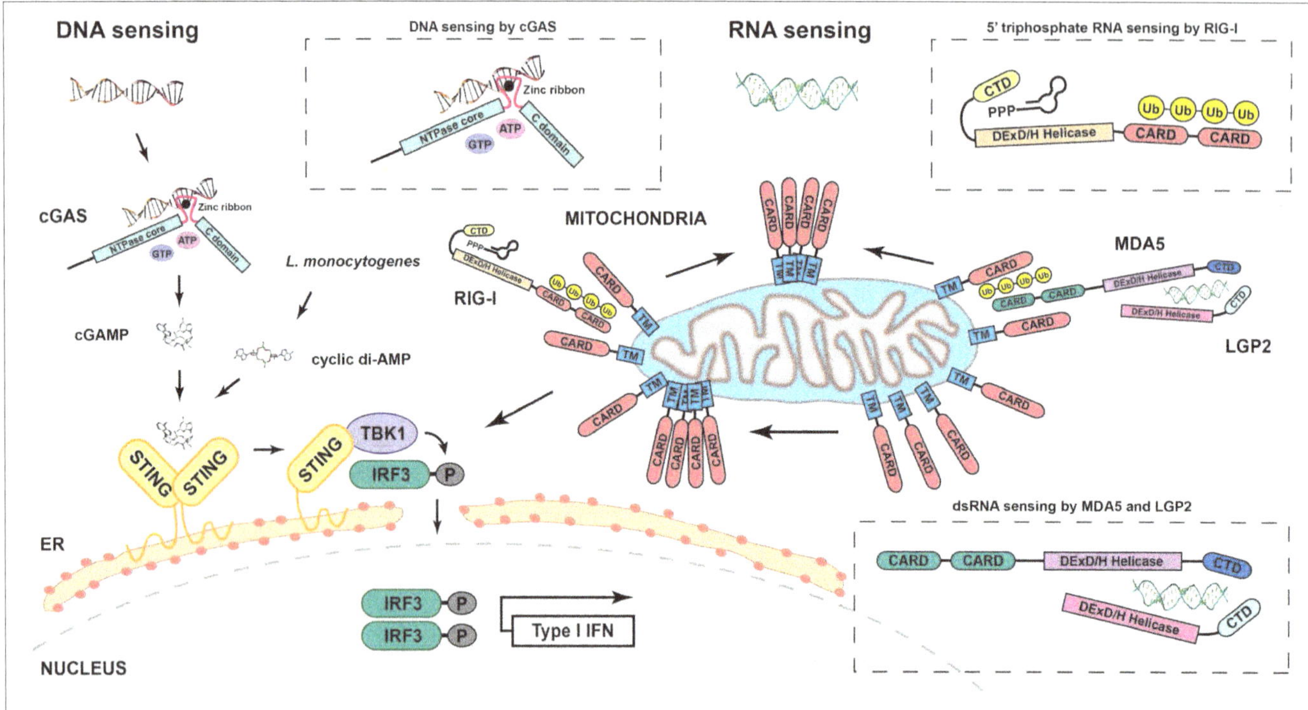

FIGURE 1 | Schematic representation of major cytosolic DNA and RNA sensing pathways. Cytosolic DNA is sensed by cGAS leading to the production of cGAMP and subsequent STING/TBK1/IRF3 signaling, and type I IFN production; bacterial cyclic dinucleotides (like *L. monocytogenes* cyclic di-AMP) also activate STING. 5′ triphosphate RNA and double stranded RNA are sensed by RIG-I and either MDA-5 alone or MDA-5/LGP2. This leads to polymerization of mitochondrial MAVS and also activates TBK1 and IRF3.

CDNs in terms of which ISGs are induced and whether this has a functional outcome on the adaptive immune response. In the case of *Listeria* for example cyclic di-AMP secretion negatively affects T-cell mediated immunity but it would be interesting to dissect the role of eukaryotic cGAMP in this context (Archer et al., 2014).

RNA CYTOSOLIC SURVEILLANCE PATHWAYS

RNA Sensing through MAVS

Since viral infection generates pathogen-associated molecular patterns such as double stranded RNA, the field of RNA sensing made earlier advances than the field of DNA sensing following the search for cytosolic sensors of these pathogen-specific molecules. Several landmark papers elucidated a role for TBK1 and IκB kinase-*i* (IKK-*i*) as kinases critical for phosphorylation and activation of IRF3 and IRF7 both *in vitro* and *in vivo* (Fitzgerald et al., 2003; Sharma et al., 2003; Hemmi et al., 2004; Mcwhirter et al., 2004; Perry et al., 2004). Subsequently, two groups demonstrated the ability of known proteins, retinoic acid inducible gene-I (RIG-I) and melanoma differentiation-associated gene-5 (MDA-5), to bind dsRNA (**Figure 1**). RIG-I was identified as a protein that could lead to IRF activity in the presence of dsRNA in a large scale cDNA screen (Yoneyama et al., 2004). Further experiments showed that this activity followed direct dsRNA binding by the cytosolic protein RIG-I. Interestingly, a mutant of RIG-I solely expressing its caspase-recruitment domain CARD was sufficient to spontaneously activate the IFN-producing activity. At the time this suggested, and the authors predicted, that this domain interacts with an as yet unidentified downstream molecule, a hypothesis proven shortly thereafter. Parallel work demonstrated that MDA-5 was a second direct sensor of viral RNA. As for RIG-I, ectopic expression of MDA-5 was sufficient to induce IFN signaling (Andrejeva et al., 2004). RIG-I and MDA-5 play complementary but separate roles in the response to viral infection. More specifically, genetic ablation of MDA-5 was shown to block the response to poly(I:C) whereas loss of RIG-I abrogated sensing of *in vitro* transcribed dsRNA (Kato et al., 2006). Several groups went on to identify the ligand for RIG-I as uncapped 5′ triphosphate RNA and to biochemically characterize the interaction (Hornung et al., 2006; Pichlmair et al., 2006; Schlee et al., 2009). Interestingly, the 5′ triphosphate RNA must form internal base pairs with a blunt end to be stimulatory (Schmidt et al., 2009). Finally, the structure of RIG-I in complex with dsRNA was solved in 2011 (Jiang et al., 2011; Kowalinski et al., 2011). In the case of the ligand for MDA-5, 2′-*O*-methylation normally occurs in the cap of eukaryotic RNA molecules and is mimicked by viruses. Deletion of enzymes required for this modification leads to increased IFN production dependent on MDA-5 sensing, suggesting that RNA that lacks this modification is the MDA-5 ligand (Züst et al., 2011). The structure of this interaction was finally solved in 2013 (Wu et al., 2013a), shedding light on the physical basis for distinct ligand sensing mediated by homologous regions of RIG-I and MDA-5. LGP2 (Laboratory of genetics and physiology 2) is the third member of the RIG-I like receptor (RLRs) family however, much less is known about its mechanism of action. It too can bind dsRNA, but seems to repress RIG-I or compete with it, while it can activate MDA-5 (Li et al., 2009; Pippig et al., 2009). A recent hypothesis is that LGP2 synergizes with MDA-5, since LGP2 has superior dsRNA binding capacity but reduced signaling efficacy whereas MDA-5 is a relatively weaker binder of dsRNA but signals much more efficiently than LGP2 (Bruns et al., 2014). Thus, a pertinent future direction would be to dissect the interplay between the three molecules, potentially using a genetic deletion model.

Following the aforementioned prediction of a downstream signaling effector of RIG-I, a flurry of activity led to the cloning and identification of the MAVS (mitochondrial-antiviral signaling) protein by four separate groups. Two of the groups took a bioinformatics approach mining the genome for CARD domains that resembled that of MDA-5 and RIG-I (Meylan et al., 2005; Seth et al., 2005). The other two groups identified MAVS using unbiased high-throughput screens for IFN production (Kawai et al., 2005; Xu et al., 2005). MAVS has a CARD domain, a leucine rich repeat and a transmembrane domain, which tethers the protein to the mitochondrial membrane (Seth et al., 2005). MAVS can physically interact with RIG-I and is epistatic to TBK1 and IKK-*i*. Most importantly, MAVS plays a functional role in the immune response to viruses, i.e., its depletion leads to increased viral load and its overexpression reduces viral load. More recently, the molecular details of the interactions between RIG-I, MAVS, and viral RNA were elucidated using a novel cell-free system that incorporates isolated mitochondria (Zeng et al., 2010). Other work had previously demonstrated the role of K63-linked ubiquitin modification in RIG-I signaling (Zeng et al., 2009). More specifically, critical signaling molecules in innate immune pathways, such as RIG-I, become modified by polyubiquitin chains linked via the K63 residue of ubiquitin (Gack et al., 2007). Rather than leading to degradation of the target protein, this linkage enhances the signaling capacity of the target. However, using the cell free system, the role of free polyubiquitin K63 chains emerged as a requirement in RIG-I/MAVS signaling (Zeng et al., 2010). Another seminal paper from the Chen lab put forth the concept that MAVS activation accompanies a conformational change in and polymerization of the protein which travels across the mitochondria in waves, reminiscent of a prion-type conversion (Hou et al., 2011). While the cell-free system has illuminated a number of novel concepts in RIG-I signaling, it will be important to validate these findings within cells and potentially *in vivo* as well. The clustering and conversion of MAVS also begs the question whether or not it can be restored to its pre-viral state or must it be removed and how does this affect mitochondrial fitness, morphology and clearance?

Cellular Responses

Similar to the activation of STING-mediated signaling, clustering of MAVS leads to interferon production following activation of IRF3 and IRF7 and NF-κB activation, helping cells to establish an anti-viral state. While the vast majority of MAVS is anchored in the mitochondrial outer membrane, a fascinating paper identified that MAVS can also be peroxisomal (Dixit et al., 2010). Interestingly, this affects the downstream signaling response. Peroxisomal MAVS leads to an early interferon-independent induction of ISGs that requires IRF1, whereas mitochondrial MAVS contributes to the canonical type I IFN response. Subsequent work determined that in response to a number of viruses and at least one intracellular bacterium, peroxisomal MAVS signaling led to the induction of type III interferon (IFN-λ) and that this signaling was downstream of RIG-I sensing (Odendall et al., 2014). Interestingly, 5′ triphosphate RNA from bacterial pathogens such as *L. monocytogenes* can be sensed by RIG-I leading to IFN production in non-phagocytic cells (Abdullah et al., 2012; Hagmann et al., 2013). It is possible that the bacteria actively secrete these RNA molecules as a secretion deficient-mutant of the pathogen elicits a dampened immune response and 5′ triphosphate RNA is detectable at early time points of infection in the cytosol. These studies raise the question of whether early detection of pathogenic RNA affects T cell-mediated immunity as bacterially produced cyclic di-AMP does.

MICROBIAL STRATEGIES FOR SUBVERSION OF CYTOSOLIC SURVEILLANCE PATHWAYS

Microbial Strategies for Subversion of DNA Cytosolic Surveillance Pathways

Since STING is such a critical protein for innate immune sensing, it and its downstream signaling components are strategic targets of a number of pathogens. *Shigella flexneri* type III effector invasion plasmid antigen J (IpaJ) is a cysteine protease that cleaves the *N*-myristoylated glycine of lipidated ADP ribosylation factor (ARF)-family GTPases, disrupts the Golgi apparatus, thereby inhibiting host protein secretion (Burnaevskiy et al., 2014). Recently, it has been shown that IpaJ antagonizes STING-mediated IFNβ activation by blocking STING translocation from the ER to ER-Golgi intermediate compartments in mouse embryonic fibroblasts (Dobbs et al., 2015). This inhibition of immune detection is important for *Shigella* pathogenesis (Burnaevskiy et al., 2014). In contrast, other intracellular pathogens activate STING to inhibit the T cell-mediated immune response. For example, *L. monocytogenes* secretes c-di-AMP to activate STING, leading to production of IFNβ and inhibition of cell-mediated immunity (Woodward et al., 2010; Archer et al., 2014). Another trigger of IFNβ expression is *L. monocytogenes* DNA, which operates through IFI16, cGAS and STING in human macrophages (Hansen et al., 2014). Similarly, *M. tuberculosis* DNA, possibly secreted through the type VII secretion system ESX-1, associates with cGAS to stimulate a type I interferon response, which favors bacterial pathogenesis and disease progression (Wassermann et al., 2015).

Likewise viral DNA is recognized by STING and STING-dependent cytosolic sensors and triggers antiviral responses. Like bacterial pathogens, viruses have evolved sophisticated mechanisms to manipulate host DNA sensing. In order to evade the type I interferon response, the dengue virus (DENV) expresses the NS2B3 protease complex, which specifically cleaves STING in human dendritic cells (Aguirre et al., 2012). Interestingly, DENV NS2B3 inhibits type I interferon production in human but not in mouse cells, since the proteolytic complex does not degrade murine STING. Hence, STING restricts DENV replication in mouse cells. Several viral proteins can antagonize STING-dependent DNA sensing. Kaposi's sarcoma-associated herpesvirus ORF52 protein directly inhibits cGAS enzymatic activity (Wu et al., 2015). Oncoproteins E7 from human papillomavirus and E1A from adenovirus bind to STING and antagonize the cGAS-STING pathway, preventing the antiviral response (Lau et al., 2015). Human cytomegalovirus tegument protein pUL83 binds IFI16 pyrin domain, blocking its oligomerization upon DNA sensing, preventing the expression of antiviral cytokines (Li et al., 2013). Herpes simplex virus-1 (HSV-1) ICP0 E3 ubiquitin ligase targets nuclear IFI16 for degradation, inhibiting nuclear innate immune sensing, cytoplasmic STING activation and antiviral responses (Orzalli et al., 2012). Murine cytomegalovirus protein M45 inhibits signaling mediated through DAI, which is required for DNA sensing and type I interferon production in some cells (Rebsamen et al., 2009).

Cytosolic foreign DNA not only activates STING-mediated innate immune responses, but it is also able to stimulate inflammasome activation, leading to IL-1β and IL-18 maturation and secretion. As such, some pathogens antagonize inflammasome signaling pathways. *M. tuberculosis*, but not non-virulent *Mycobacterium smegmatis*, inhibits AIM2 (absent in melanoma 2) inflammasome activation (Shah et al., 2013). While the *M. tuberculosis* AIM2 inhibitor is as yet unknown, it seems to be secreted by the ESX-1 secretion system.

The importance of inhibiting the inflammasome to promote viral infection has been demonstrated by revealing the critical role of the Poxvirus M13L immunomodulatory protein in infection of monocytes and lymphocytes and in disease (Johnston et al., 2005). Poxvirus M13L is a member of the pyrin domain-containing superfamily of proteins. It interacts with host pyrin domain protein ASC-1 (apoptosis-associated speck-like protein containing CARD-1), inhibiting caspase-1 activation in monocytes, thereby disrupting the intracellular pathways leading to IL-1β processing and secretion.

Other important DNA sensors such as DNA-PK and MRN are also targets of viral antagonists. Peters et al. (2013) have shown that vaccinia virus inhibits the innate immune response of infected fibroblasts by expressing the C16 protein. C16 directly binds to the heterodimer Ku70/Ku80, impairing DNA-PK binding to DNA and downstream signaling cascades. Adenovirus type 5 (Ad5) antagonizes the MRN-dependent DDR by two mechanisms. First, it triggers MRN proteasomal degradation upon ubiquitination by an E3 ligase complex composed of Ad5 E1B-55K and E4-ORF6 proteins associated with host proteins (Stracker et al., 2002; Weitzman and Ornelles, 2005). Second, it inhibits MRN activity by expressing E4-ORF3. Ad5 E4-ORF3 recruits nuclear proteins, such as promyelocytic leukemia (PML) and PML nuclear body-associated proteins, to sequester MRN proteins in the nucleus of infected cells (Stracker et al., 2002; Weitzman and Ornelles, 2005). These distinct mechanisms induce efficient inactivation of cellular DDR and antiviral defense.

Microbial Strategies for Subversion of RNA Cytosolic Surveillance Pathways

Because cytoplasmic RNA sensors RIG-I, MDA-5 and LGP2 confer resistance to intracellular pathogens, many viruses have evolved strategies to control the RLRs. Some viruses are able to suppress RIG-I RNA sensing step. Indeed, both Marburg virus and Ebola virus VP35 proteins and severe acute respiratory syndrome coronavirus (SARS-CoV) nucleocapsid protein N antagonize RIG-I signaling by binding and masking dsRNA (Cárdenas et al., 2006; Lu et al., 2011; Bale et al., 2012, 2013). Of note, Ebola virus VP35 protein also binds dsRNA and inhibits the antiviral activity of the dsRNA-dependent protein kinase R (Feng et al., 2007). The vaccinia virus E3 protein binds to poly(A-U) RNA generated from viral dsDNA by RNA polymerase III, preventing RIG-I activation (Valentine and Smith, 2010). DENV is able to elude RIG-I recognition, possibly by formation of convoluted membranes tightly associated with the NS4A viral protein and convoluted-membrane vesicles containing dsRNA (Welsch et al., 2009). Some negative-stranded RNA viruses, such as Hantaan virus, Crimean–Congo hemorrhagic fever virus and Borna disease virus, cleave the 5′-triphosphate group of their genome, which therefore becomes invisible to RIG-I (Habjan et al., 2008). Arenaviruses 3′-5′ exoribonucleases, such as Lassa fever virus nucleocapsid protein NP, process viral dsRNA to prevent its recognition by RLRs (Jiang et al., 2013). Arenaviruses also produce decoys, i.e., short 5′-ppp dsRNA containing an overhanging GTP nucleotide, which trap RIG-I in inactive dsRNA complexes (Marq et al., 2011).

Some viral proteins bind to RLRs and inhibit their downstream signaling pathways, dampening the production of type I interferon. For example, influenza virus non-structural protein 1 (NS1) binds to RIG-I containing complexes and blocks RIG-I activation (Pichlmair et al., 2006). HSV-1 RNA binding tegument protein US11 binds to RIG-I and MDA-5 to antagonize the IFNβ pathway and promote pathogenesis (Xing et al., 2012). Z proteins from New World arenaviruses Guanarito virus, Junin virus, Machupo virus and Sabia virus, but not Old World arenavirus lymphocytic choriomeningitis virus or Lassa virus, bind to RIG-I and prevent the IFNβ response (Fan et al., 2010). The V proteins of paramyxoviruses bind MDA-5 and block its activity (Andrejeva et al., 2004; Motz et al., 2013). The V proteins of Nipah virus and Hendra virus bind LGP2 and induce the formation of stable LGP2/RIG-I complexes unable to recognize dsRNA (Parisien et al., 2009; Childs et al., 2012).

Other viral proteins abrogate RIG-I ubiquitination which has been shown to be critical for its antiviral activity (Gack et al., 2007). Influenza A virus NS1 specifically blocks multimerization of the TRIM25 ligase, thereby inhibiting RIG-I ubiquitination (Gack et al., 2009). Hepatitis C virus NS3-4A proteases target the Riplet ubiquitin ligase which is required for TRIM25-mediated activation of RIG-I (Oshiumi et al., 2013). Kaposi's sarcoma associated herpes virus ORF64 tegument protein is a deubiquitinase, which inhibits RIG-I activation by suppressing its ubiquitination and reduces interferon signaling (Inn et al., 2011). A similar mechanism of immune escape has been evolved by arterivirus and nairovirus which express ovarian tumor domain-containing deubiquitinases targeting RIG-I (Frias-Staheli et al., 2007; van Kasteren et al., 2012), and by hepatitis E virus which expresses the ORF1 protein bearing a papain-like cystein protease domain responsible for RIG-I deubiquitination (Nan et al., 2014).

Another mechanism of RLR subversion is the inhibition of RNA-sensor dephosphorylation, which is necessary for their activation. Measles virus can suppress RIG-I and MDA-5 activation by binding to DC-SIGN (dendritic cell-specific intercellular adhesion molecule-3-grabbing non-integrin), which triggers activation of the kinase Raf-1 (Rapidly accelerated fibrosarcoma-1), association of protein phosphatase 1 (PP1) inhibitor I-1 with GADD34-PP1 phosphatases and ultimately inhibition of RIG-I and MDA-5 dephosphorylation (Mesman et al., 2014).

Not surprisingly, MAVS is also targeted by several viral proteins to antagonize type I interferon responses. SARS-CoV ORF3b, ORF6, NP and M proteins and influenza A virus NS1, PB1, PB2, PA, and PB1-F2 proteins inhibit production of type I interferon, in most cases through direct interaction with MAVS (Kopecky-Bromberg et al., 2007; Freundt et al., 2009; Siu et al., 2009; Iwai et al., 2010; Varga et al., 2012; Zinzula and Tramontano, 2013). Other viruses abrogate MAVS signaling indirectly. For example, vaccinia virus K7 protein and hepatitis C virus core protein interact with DEAD box protein 3 (DDX3) to interfere with MAVS signaling (Schröder et al., 2008; Oshiumi et al., 2010).

CONCLUSION AND PERSPECTIVES

The field of cytosolic innate immunity has progressed tremendously in the past decade. A multiplicity of DNA and RNA cytosolic sensors and signaling pathways has been discovered, revealing an unexpected complexity of the response to nucleic acids of pathogens. Key questions remain to be answered. What is the role of each of these pathways in different cell types and tissues? What is their respective importance in clearance of infection in different animal species, including humans? On the pathogen side, a plethora of immune escape mechanisms developed by viruses have been unraveled over the years. Fewer bacterial determinants inhibiting cytosolic immune sensing have been clearly elucidated so far and future work will surely lead to the identification of many more. Microbial proteins targeting cytosolic sensing have been repeatedly shown to be virulence factors important for pathogenesis. Further characterization of these factors could lead to the development new preventive or therapeutic antimicrobial strategies.

AUTHOR CONTRIBUTIONS

All authors listed, have made substantial, direct and intellectual contribution to the work, and approved it for publication.

FUNDING

The authors' work was supported in part by Institut Pasteur, Institut National de la Santé et de la Recherche Médicale, Institut National de la Recherche Agronomique, Université Paris Diderot, Agence Nationale de la Recherche (Grant Proantilis, Laboratoire d'Excellence ANR-10-LABX-62-IBEID), European Research Council, Howard Hughes Medical Institute, Fondation Le Roch and Louis Jeantet Foundation. LR is a long term fellow of the Human Frontier Science Program and European Molecular Biology Organization.

ACKNOWLEDGMENTS

The authors thank Pascale Cossart and Mikael Koutero for critical reading of the manuscript and apologize to colleagues whose work could not be cited owing to space limitation.

REFERENCES

Abdullah, Z., Schlee, M., Roth, S., Mraheil, M. A., Barchet, W., Boettcher, J., et al. (2012). RIG-I detects infection with live *Listeria* by sensing secreted bacterial nucleic acids. *EMBO J.* 31, 4153–4164. doi: 10.1038/emboj.2012.274

Abe, T., Harashima, A., Xia, T., Konno, H., Konno, K., Morales, A., et al. (2013). STING recognition of cytoplasmic DNA instigates cellular defense. *Mol. Cell* 50, 5–15. doi: 10.1016/j.molcel.2013.01.039

Ablasser, A., Goldeck, M., Cavlar, T., Deimling, T., Witte, G., Röhl, I., et al. (2013). cGAS produces a 2′-5′-linked cyclic dinucleotide second messenger that activates STING. *Nature* 498, 380–384. doi: 10.1038/nature12306

Aguirre, S., Maestre, A. M., Pagni, S., Patel, J. R., Savage, T., Gutman, D., et al. (2012). DENV inhibits type I IFN production in infected cells by cleaving human STING. *PLoS Pathog.* 8:e1002934. doi: 10.1371/journal.ppat.1002934

Andrejeva, J., Childs, K. S., Young, D. F., Carlos, T. S., Stock, N., Goodbourn, S., et al. (2004). The V proteins of paramyxoviruses bind the IFN-inducible RNA helicase, mda-5, and inhibit its activation of the IFN-beta promoter. *Proc. Natl. Acad. Sci. U.S.A.* 101, 17264–17269. doi: 10.1073/pnas.0407639101

Archer, K. A., Durack, J., and Portnoy, D. A. (2014). STING-dependent type I IFN production inhibits cell-mediated immunity to *Listeria monocytogenes*. *PLoS Pathog.* 10:e1003861. doi: 10.1371/journal.ppat.1003861

Bale, S., Julien, J.-P., Bornholdt, Z. A., Kimberlin, C. R., Halfmann, P., Zandonatti, M. A., et al. (2012). Marburg virus VP35 can both fully coat the backbone and cap the ends of dsRNA for interferon antagonism. *PLoS Pathog.* 8:e1002916. doi: 10.1371/journal.ppat.1002916

Bale, S., Julien, J.-P., Bornholdt, Z. A., Krois, A. S., Wilson, I. A., and Saphire, E. O. (2013). Ebolavirus VP35 coats the backbone of double-stranded RNA for interferon antagonism. *J. Virol.* 87, 10385–10388. doi: 10.1128/JVI.01452-13

Bruns, A. M., Leser, G. P., Lamb, R. A., and Horvath, C. M. (2014). The innate immune sensor LGP2 activates antiviral signaling by regulating MDA5-RNA interaction and filament assembly. *Mol. Cell* 55, 771–781. doi: 10.1016/j.molcel.2014.07.003

Burdette, D. L., Monroe, K. M., Sotelo-Troha, K., Iwig, J. S., Eckert, B., Hyodo, M., et al. (2011). STING is a direct innate immune sensor of cyclic di-GMP. *Nature* 478, 515–518. doi: 10.1038/nature10429

Burnaevskiy, N., Fox, T. G., Plymire, D. A., Ertelt, J. M., Weigele, B. A., Selyunin, A. S., et al. (2014). Proteolytic elimination of N-myristoyl modifications by the *Shigella* virulence factor IpaJ. *Nature* 496, 106–109. doi: 10.1038/nature 12004

Cai, X., Chiu, Y.-H., and Chen, Z. J. (2014). The cGAS-cGAMP-STING pathway of cytosolic DNA sensing and signaling. *Mol. Cell* 54, 289–296. doi: 10.1016/j.molcel.2014.03.040

Cárdenas, W. B., Loo, Y.-M., Gale, M., Hartman, A. L., Kimberlin, C. R., Martínez-Sobrido, L., et al. (2006). Ebola virus VP35 protein binds double-stranded RNA and inhibits alpha/beta interferon production induced by RIG-I signaling. *J. Virol.* 80, 5168–5178. doi: 10.1128/JVI.02199-05

Chen, H., Sun, H., You, F., Sun, W., Zhou, X., Chen, L., et al. (2011). Activation of STAT6 by STING is critical for antiviral innate immunity. *Cell* 147, 436–446. doi: 10.1016/j.cell.2011.09.022

Childs, K., Randall, R., and Goodbourn, S. (2012). Paramyxovirus V proteins interact with the RNA helicase LGP2 to inhibit RIG-I-dependent interferon induction. *J. Virol.* 86, 3411–3421. doi: 10.1128/JVI.06405-11

Collins, A. C., Cai, H., Li, T., Franco, L. H., Li, X.-D., Nair, V. R., et al. (2015). Cyclic GMP-AMP synthase is an innate immune DNA sensor for *Mycobacterium tuberculosis*. *Cell Host Microbe* 17, 820–828. doi: 10.1016/j.chom.2015.05.005

Diner, E. J., Burdette, D. L., Wilson, S. C., Monroe, K. M., Kellenberger, C. A., Hyodo, M., et al. (2013). The innate immune DNA sensor cGAS produces a noncanonical cyclic dinucleotide that activates human STING. *Cell Rep.* 3, 1355–1361. doi: 10.1016/j.celrep.2013.05.009

Dixit, E., Boulant, S., Zhang, Y., Lee, A. S. Y., Odendall, C., Shum, B., et al. (2010). Peroxisomes are signaling platforms for antiviral innate immunity. *Cell* 141, 668–681. doi: 10.1016/j.cell.2010.04.018

Dobbs, N., Burnaevskiy, N., Chen, D., Gonugunta, V. K., Alto, N. M., and Yan, N. (2015). STING Activation by translocation from the ER is associated with infection and autoinflammatory disease. *Cell Host Microbe* 18, 157–168. doi: 10.1016/j.chom.2015.07.001

Fan, L., Briese, T., and Lipkin, W. I. (2010). Z proteins of New World arenaviruses bind RIG-I and interfere with type I interferon induction. *J. Virol.* 84, 1785–1791. doi: 10.1128/JVI.01362-09

Feng, Z., Cerveny, M., Yan, Z., and He, B. (2007). The VP35 protein of Ebola virus inhibits the antiviral effect mediated by double-stranded RNA-dependent protein kinase PKR. *J. Virol.* 81, 182–192. doi: 10.1128/JVI.01006-06

Fitzgerald, K. A., Mcwhirter, S. M., Faia, K. L., Rowe, D. C., Latz, E., Golenbock, D. T., et al. (2003). IKKepsilon and TBK1 are essential components of the IRF3 signaling pathway. *Nat. Immunol.* 4, 491–496. doi: 10.1038/ni921

Freundt, E. C., Yu, L., Park, E., Lenardo, M. J., and Xu, X.-N. (2009). Molecular determinants for subcellular localization of the severe acute respiratory syndrome coronavirus open reading frame 3b protein. *J. Virol.* 83, 6631–6640. doi: 10.1128/JVI.00367-09

Frias-Staheli, N., Giannakopoulos, N. V., Kikkert, M., Taylor, S. L., Bridgen, A., Paragas, J., et al. (2007). Ovarian tumor domain-containing viral proteases evade ubiquitin- and ISG15-dependent innate immune responses. *Cell Host Microbe* 2, 404–416. doi: 10.1016/j.chom.2007.09.014

Gack, M. U., Albrecht, R. A., Urano, T., Inn, K.-S., Huang, I.-C., Carnero, E., et al. (2009). Influenza A virus NS1 targets the ubiquitin ligase TRIM25 to evade recognition by the host viral RNA sensor RIG-I. *Cell Host Microbe* 5, 439–449. doi: 10.1016/j.chom.2009.04.006

Gack, M. U., Shin, Y. C., Joo, C.-H., Urano, T., Liang, C., Sun, L., et al. (2007). TRIM25 RING-finger E3 ubiquitin ligase is essential for RIG-I-mediated antiviral activity. *Nature* 446, 916–920. doi: 10.1038/nature05732

Gao, P., Ascano, M., Wu, Y., Barchet, W., Gaffney, B. L., Zillinger, T., et al. (2013a). Cyclic [G(2',5')pA(3',5')p] is the metazoan second messenger produced by DNA-activated cyclic GMP-AMP synthase. *Cell* 153, 1094–1107. doi: 10.1016/j.cell.2013.04.046

Gao, P., Ascano, M., Zillinger, T., Wang, W., Dai, P., Serganov, A. A., et al. (2013b). Structure-function analysis of STING activation by c[G(2',5')pA(3',5')p] and targeting by antiviral DMXAA. *Cell* 154, 748–762. doi: 10.1016/j.cell.2013.07.023

Habjan, M., Andersson, I., Klingström, J., Schümann, M., Martin, A., Zimmermann, P., et al. (2008). Processing of genome 5' termini as a strategy of negative-strand RNA viruses to avoid RIG-I-dependent interferon induction. *PLoS ONE* 3:e2032. doi: 10.1371/journal.pone.0002032

Hagmann, C. A., Herzner, A. M., Abdullah, Z., Zillinger, T., Jakobs, C., Schuberth, C., et al. (2013). RIG-I detects triphosphorylated RNA of *Listeria monocytogenes* during infection in non-immune cells. *PLoS ONE* 8:e62872. doi: 10.1371/journal.pone.0062872

Hansen, K., Prabakaran, T., Laustsen, A., Jorgensen, S. E., Rahbaek, S. H., Jensen, S. B., et al. (2014). *Listeria monocytogenes* induces IFN expression through an IFI16-, cGAS- and STING-dependent pathway. *EMBO J.* 33, 1654–1666. doi: 10.15252/embj.201488029

Hasan, M., Koch, J., Rakheja, D., Pattnaik, A. K., Brugarolas, J., Dozmorov, I., et al. (2013). Trex1 regulates lysosomal biogenesis and interferon-independent activation of antiviral genes. *Nat. Immunol.* 14, 61–71. doi: 10.1038/ni.2475

Hemmi, H., Takeuchi, O., Sato, S., Yamamoto, M., Kaisho, T., Sanjo, H., et al. (2004). The roles of two I-kappa-B kinase-related kinases in lipopolysaccharide and double stranded RNA signaling and viral infection. *J. Exp. Med.* 199, 1641–1650. doi: 10.1084/jem.20040520

Hornung, V., Ellegast, J., Kim, S., Brzózka, K., Jung, A., Kato, H., et al. (2006). 5'-triphosphate RNA is the ligand for RIG-I. *Science* 314, 994–997. doi: 10.1126/science.1132505

Hou, F., Sun, L., Zheng, H., Skaug, B., Jiang, Q.-X., and Chen, Z. J. (2011). MAVS forms functional prion-like aggregates to activate and propagate antiviral innate immune response. *Cell* 146, 448–461. doi: 10.1016/j.cell.2011.06.041

Inn, K.-S., Lee, S.-H., Rathbun, J. Y., Wong, L.-Y., Toth, Z., Machida, K., et al. (2011). Inhibition of RIG-I-mediated signaling by Kaposi's sarcoma-associated herpesvirus-encoded deubiquitinase ORF64. *J. Virol.* 85, 10899–10904. doi: 10.1128/JVI.00690-11

Ishikawa, H., and Barber, G. N. (2008). STING is an endoplasmic reticulum adaptor that facilitates innate immune signalling. *Nature* 455, 674–678. doi: 10.1038/nature07317

Ishikawa, H., Ma, Z., and Barber, G. N. (2009). STING regulates intracellular DNA-mediated, type I interferon-dependent innate immunity. *Nature* 461, 788–792. doi: 10.1038/nature08476

Iwai, A., Shiozaki, T., Kawai, T., Akira, S., Kawaoka, Y., Takada, A., et al. (2010). Influenza A virus polymerase inhibits type I interferon induction by binding to interferon beta promoter stimulator 1. *J. Biol. Chem.* 285, 32064–32074. doi: 10.1074/jbc.M110.112458

Jiang, F., Ramanathan, A., Miller, M. T., Tang, G.-Q., Gale, M., Patel, S. S., et al. (2011). Structural basis of RNA recognition and activation by innate immune receptor RIG-I. *Nature* 479, 423–427. doi: 10.1038/nature10537

Jiang, X., Huang, Q., Wang, W., Dong, H., Ly, H., Liang, Y., et al. (2013). Structures of arenaviral nucleoproteins with triphosphate dsRNA reveal a unique mechanism of immune suppression. *J. Biol. Chem.* 288, 16949–16959. doi: 10.1074/jbc.M112.420521

Johnston, J. B., Barrett, J. W., Nazarian, S. H., Goodwin, M., Ricuttio, D., Wang, G., et al. (2005). A Poxvirus-encoded pyrin domain protein interacts with ASC-1 to inhibit host inflammatory and apoptotic responses to infection. *Immunity* 23, 587–598. doi: 10.1016/j.immuni.2005.10.003

Karaolis, D. K. R., Means, T. K., Yang, D., Takahashi, M., Yoshimura, T., Muraille, E., et al. (2007a). Bacterial c-di-GMP is an immunostimulatory molecule. *J. Immunol.* 178, 2171–2181. doi: 10.4049/jimmunol.178.4.2171

Karaolis, D. K. R., Newstead, M. W., Zeng, X., Hyodo, M., Hayakawa, Y., Bhan, U., et al. (2007b). Cyclic di-GMP stimulates protective innate immunity in bacterial pneumonia. *Infect. Immun.* 75, 4942–4950. doi: 10.1128/IAI.01762-06

Kato, H., Takeuchi, O., Sato, S., Yoneyama, M., Yamamoto, M., Matsui, K., et al. (2006). Differential roles of MDA5 and RIG-I helicases in the recognition of RNA viruses. *Nature* 441, 101–105. doi: 10.1038/nature04734

Kawai, T., Takahashi, K., Sato, S., Coban, C., Kumar, H., Kato, H., et al. (2005). IPS-1, an adaptor triggering RIG-I- and Mda5-mediated type I interferon induction. *Nat. Immunol.* 6, 981–988. doi: 10.1038/ni1243

Kopecky-Bromberg, S. A., Martínez-Sobrido, L., Frieman, M., Baric, R. A., and Palese, P. (2007). Severe acute respiratory syndrome coronavirus open reading frame (ORF) 3b, ORF 6, and nucleocapsid proteins function as interferon antagonists. *J. Virol.* 81, 548–557.

Kowalinski, E., Lunardi, T., McCarthy, A. A., Louber, J., Brunel, J., Grigorov, B., et al. (2011). Structural basis for the activation of innate immune pattern-recognition receptor RIG-I by viral RNA. *Cell* 147, 423–435. doi: 10.1016/j.cell.2011.09.039

Lau, L., Gray, E. E., Brunette, R. L., and Stetson, D. B. (2015). DNA tumor virus oncogenes antagonize the cGAS-STING DNA-sensing pathway. *Science* 350, 568–571. doi: 10.1126/science.aab3291

Li, T., Chen, J., and Cristea, I. M. (2013). Human cytomegalovirus tegument protein pUL83 inhibits IFI16-mediated DNA sensing for immune evasion. *Cell Host Microbe* 14, 591–599. doi: 10.1016/j.chom.2013.10.007

Li, X., Ranjith-Kumar, C. T., Brooks, M. T., Dharmaiah, S., Herr, A. B., Kao, C., et al. (2009). The RIG-I-like receptor LGP2 recognizes the termini of double-stranded RNA. *J. Biol. Chem.* 284, 13881–13891. doi: 10.1074/jbc.M900818200

Lu, X., Pan, J., Tao, J., and Guo, D. (2011). SARS-CoV nucleocapsid protein antagonizes IFN-β response by targeting initial step of IFN-β induction pathway, and its C-terminal region is critical for the antagonism. *Virus Genes* 42, 37–45. doi: 10.1007/s11262-010-0544-x

Manzanillo, P. S., Shiloh, M. U., Portnoy, D. A., and Cox, J. S. (2012). *Mycobacterium tuberculosis* activates the DNA-dependent cytosolic surveillance pathway within macrophages. *Cell Host Microbe* 11, 469–480. doi: 10.1016/j.chom.2012.03.007

Marq, J.-B., Hausmann, S., Veillard, N., Kolakofsky, D., and Garcin, D. (2011). Short double-stranded RNAs with an overhanging 5' ppp-nucleotide, as found in *Arenavirus* genomes, act as RIG-I decoys. *J. Biol. Chem.* 286, 6108–6116. doi: 10.1074/jbc.M110.186262

Mcwhirter, S. M., Barbalat, R., Monroe, K. M., Fontana, M. F., Hyodo, M., Joncker, N. T., et al. (2009). A host type I interferon response is induced by cytosolic sensing of the bacterial second messenger cyclic-di-GMP. *J. Exp. Med.* 206, 1899–1911. doi: 10.1084/jem.20082874

Mcwhirter, S. M., Fitzgerald, K. A., Rosains, J., Rowe, D. C., Golenbock, D. T., and Maniatis, T. (2004). IFN-regulatory factor 3-dependent gene expression is defective in Tbk1-deficient mouse embryonic fibroblasts. *Proc. Natl. Acad. Sci. U.S.A.* 101, 233–238. doi: 10.1073/pnas.2237236100

Mesman, A. W., Zijlstra-Willems, E. M., Kaptein, T. M., de Swart, R. L., Davis, M. E., Ludlow, M., et al. (2014). Measles virus suppresses RIG-I-like receptor activation in dendritic cells via DC-SIGN-mediated inhibition of PP1 phosphatases. *Cell Host Microbe* 16, 31–42. doi: 10.1016/j.chom.2014.06.008

Meylan, E., Curran, J., Hofmann, K., Moradpour, D., Binder, M., Bartenschlager, R., et al. (2005). Cardif is an adaptor protein in the RIG-I antiviral pathway and is targeted by hepatitis C virus. *Nature* 437, 1167–1172. doi: 10.1038/nature04193

Motz, C., Schuhmann, K. M., Kirchhofer, A., Moldt, M., Witte, G., Conzelmann, K. K., et al. (2013). Paramyxovirus V proteins disrupt the fold of the RNA sensor MDA5 to inhibit antiviral signaling. *Science* 339, 690–693. doi: 10.1126/science.1230949

Nan, Y., Yu, Y., Ma, Z., Khattar, S. K., Fredericksen, B., and Zhang, Y.-J. (2014). Hepatitis E virus inhibits type I interferon induction by ORF1 products. *J. Virol.* 88, 11924–11932. doi: 10.1128/JVI.01935-14

Odendall, C., Dixit, E., Stavru, F., Bierne, H., Franz, K. M., Durbin, A. F., et al. (2014). Diverse intracellular pathogens activate type III interferon expression from peroxisomes. *Nat. Immunol.* 15, 717–726. doi: 10.1038/ni.2915

Orzalli, M. H., DeLuca, N. A., and Knipe, D. M. (2012). Nuclear IFI16 induction of IRF-3 signaling during herpesviral infection and degradation of IFI16 by the viral ICP0 protein. *Proc. Natl. Acad. Sci. U.S.A.* 109, E3008–E3017. doi: 10.1073/pnas.1211302109

Oshiumi, H., Ikeda, M., Matsumoto, M., Watanabe, A., Takeuchi, O., Akira, S., et al. (2010). Hepatitis C virus core protein abrogates the DDX3 function that enhances IPS-1-mediated IFN-beta induction. *PLoS ONE* 5:e14258. doi: 10.1371/journal.pone.0014258

Oshiumi, H., Miyashita, M., Matsumoto, M., and Seya, T. (2013). A distinct role of Riplet-mediated K63-Linked polyubiquitination of the RIG-I repressor domain in human antiviral innate immune responses. *PLoS Pathog.* 9:e1003533. doi: 10.1371/journal.ppat.1003533

Paludan, S. R., and Bowie, A. G. (2013). Immune sensing of DNA. *Immunity* 38, 870–880. doi: 10.1016/j.immuni.2013.05.004

Parisien, J.-P., Bamming, D., Komuro, A., Ramachandran, A., Rodriguez, J. J., Barber, G., et al. (2009). A shared interface mediates paramyxovirus interference with antiviral RNA helicases MDA5 and LGP2. *J. Virol* 83, 7252–7260. doi: 10.1128/JVI.00153-09

Perry, A. K., Chow, E. K., Goodnough, J. B., Yeh, W.-C., and Cheng, G. (2004). Differential requirement for TANK-binding kinase-1 in type I interferon responses to toll-like receptor activation and viral infection. *J. Exp. Med.* 199, 1651–1658. doi: 10.1084/jem.20040528

Peters, N. E., Ferguson, B. J., Mazzon, M., Fahy, A. S., Krysztofinska, E., Arribas-Bosacoma, R., et al. (2013). A mechanism for the inhibition of DNA-PK-mediated DNA sensing by a virus. *PLoS Pathog.* 9:e1003649. doi: 10.1371/journal.ppat.1003649

Pichlmair, A., Schulz, O., Tan, C. P., Näslund, T. I., Liljeström, P., Weber, F., et al. (2006). RIG-I-mediated antiviral responses to single-stranded RNA bearing 5'-phosphates. *Science* 314, 997–1001. doi: 10.1126/science.1132998

Pippig, D. A., Hellmuth, J. C., Cui, S., Kirchhofer, A., Lammens, K., Lammens, A., et al. (2009). The regulatory domain of the RIG-I family ATPase LGP2 senses double-stranded RNA. *Nucleic Acids Res.* 37, 2014–2025. doi: 10.1093/nar/gkp059

Radoshevich, L., Impens, F., Ribet, D., Quereda, J. J., Nam Tham, T., Nahori, M.-A., et al. (2015). ISG15 counteracts *Listeria monocytogenes* infection. *Elife* 4, e06848. doi: 10.7554/eLife.06848

Rebsamen, M., Heinz, L. X., Meylan, E., Michallet, M.-C. E. C., Schroder, K., Hofmann, K., et al. (2009). DAI/ZBP1 recruits RIP1 and RIP3 through RIP homotypic interaction motifs to activate NF-kappaB. *EMBO Rep.* 10, 916–922. doi: 10.1038/embor.2009.109

Saitoh, T., Fujita, N., Hayashi, T., Takahara, K., Satoh, T., Lee, H., et al. (2009). Atg9a controls dsDNA-driven dynamic translocation of STING and the innate immune response. *Proc. Natl. Acad. Sci. U.S.A.* 106, 20842–20846. doi: 10.1073/pnas.0911267106

Sauer, J.-D., Sotelo-Troha, K., von Moltke, J., Monroe, K. M., Rae, C. S., Brubaker, S. W., et al. (2011). The N-ethyl-N-nitrosourea-induced goldenticket mouse mutant reveals an essential function of Sting in the in vivo interferon response to *Listeria monocytogenes* and cyclic dinucleotides. *Infect. Immun.* 79, 688–694. doi: 10.1128/IAI.00999-10

Schlee, M., Roth, A., Hornung, V., Hagmann, C. A., Wimmenauer, V., Barchet, W., et al. (2009). Recognition of 5′ triphosphate by RIG-I helicase requires short blunt double-stranded RNA as contained in panhandle of negative-strand virus. *Immunity* 31, 25–34. doi: 10.1016/j.immuni.2009.05.008

Schmidt, A., Schwerd, T., Hamm, W., Hellmuth, J. C., Cui, S., Wenzel, M., et al. (2009). 5′-triphosphate RNA requires base-paired structures to activate antiviral signaling via RIG-I. *Proc. Natl. Acad. Sci. U.S.A.* 106, 12067–12072. doi: 10.1073/pnas.0900971106

Schröder, M., Baran, M., and Bowie, A. G. (2008). Viral targeting of DEAD box protein 3 reveals its role in TBK1/IKKepsilon-mediated IRF activation. *EMBO J.* 27, 2147–2157. doi: 10.1038/emboj.2008.143

Seth, R., Sun, L., Ea, C., and Chen, Z. (2005). Identification and characterization of MAVS, a mitochondrial antiviral signaling protein that activates NF-κB and IRF3. *Cell* 122, 669–682. doi: 10.1016/j.cell.2005.08.012

Shah, S., Bohsali, A., Ahlbrand, S. E., Srinivasan, L., Rathinam, V. A. K., Vogel, S. N., et al. (2013). Cutting Edge: *Mycobacterium tuberculosis* but not nonvirulent mycobacteria inhibits IFN-beta and AIM2 inflammasome-dependent IL-1beta production via its ESX-1 secretion system. *J. Immunol.* 191, 3514–3518. doi: 10.4049/jimmunol.1301331

Sharma, S., Tenoever, B. R., Grandvaux, N., Zhou, G.-P., Lin, R., and Hiscott, J. (2003). Triggering the interferon antiviral response through an IKK-related pathway. *Science* 300, 1148–1151. doi: 10.1126/science.1081315

Siu, K.-L., Kok, K.-H., Ng, M.-H. J., Poon, V. K. M., Yuen, K.-Y., Zheng, B.-J., et al. (2009). Severe acute respiratory syndrome coronavirus M protein inhibits type I interferon production by impeding the formation of TRAF3.TANK.TBK1/IKKepsilon complex. *J. Biol. Chem.* 284, 16202–16209. doi: 10.1074/jbc.M109.008227

Stetson, D., and Medzhitov, R. (2006). Recognition of cytosolic DNA activates an IRF3-dependent innate immune response. *Immunity* 24, 93–103. doi: 10.1016/j.immuni.2005.12.003

Stetson, D. B., Ko, J. S., Heidmann, T., and Medzhitov, R. (2008). Trex1 prevents cell-intrinsic initiation of autoimmunity. *Cell* 134, 587–598. doi: 10.1016/j.cell.2008.06.032

Stracker, T. H., Carson, C. T., and Weitzman, M. D. (2002). Adenovirus oncoproteins inactivate the Mre11-Rad50-NBS1 DNA repair complex. *Nature* 418, 348–352. doi: 10.1038/nature00863

Sun, L., Wu, J., Du, F., Chen, X., and Chen, Z. J. (2013). Cyclic GMP-AMP synthase is a cytosolic DNA sensor that activates the type I interferon pathway. *Science* 339, 786–791. doi: 10.1126/science.1232458

Valentine, R., and Smith, G. L. (2010). Inhibition of the RNA polymerase III-mediated dsDNA-sensing pathway of innate immunity by vaccinia virus protein E3. *J. Gen. Virol.* 91, 2221–2229. doi: 10.1099/vir.0.021998-0 van Kasteren, P. B., Beugeling, C., Ninaber, D. K., Frias-Staheli, N., van Boheemen, S., García-Sastre, A., et al. (2012). Arterivirus and nairovirus ovarian tumor domain-containing deubiquitinases target activated RIG-I to control innate immune signaling. *J. Virol.* 86, 773–785. doi: 10.1128/JVI.06277-11

Varga, Z. T., Grant, A., Manicassamy, B., and Palese, P. (2012). Influenza virus protein PB1-F2 inhibits the induction of type I interferon by binding to MAVS and decreasing mitochondrial membrane potential. *J. Virol.* 86, 8359–8366. doi: 10.1128/JVI.01122-12

Wassermann, R., Gulen, M. F., Sala, C., Perin, S. G., Lou, Y., Rybniker, J., et al. (2015). *Mycobacterium tuberculosis* differentially activates cGAS- and inflammasome-dependent intracellular immune responses through ESX-1. *Cell Host Microbe* 17, 799–810. doi: 10.1016/j.chom.2015.05.003

Watson, R. O., Bell, S. L., MacDuff, D. A., Kimmey, J. M., Diner, E. J., Olivas, J., et al. (2015). The cytosolic sensor cGAS detects *Mycobacterium tuberculosis* DNA to induce type I interferons and activate autophagy. *Cell Host Microbe* 17, 811–819. doi: 10.1016/j.chom.2015.05.004

Weitzman, M. D., and Ornelles, D. A. (2005). Inactivating intracellular antiviral responses during adenovirus infection. *Oncogene* 24, 7686–7696. doi: 10.1038/sj.onc.1209063

Welsch, S., Miller, S., Romero-Brey, I., Merz, A., Bleck, C. K. E., Walther, P., et al. (2009). Composition and three-dimensional architecture of the dengue virus replication and assembly sites. *Cell Host Microbe* 5, 365–375. doi: 10.1016/j.chom.2009.03.007

Woodward, J. J., Iavarone, A. T., and Portnoy, D. A. (2010). c-di-AMP secreted by intracellular *Listeria monocytogenes* activates a host type I interferon response. *Science* 328, 1703–1705. doi: 10.1126/science.1189801

Wu, B., Peisley, A., Richards, C., Yao, H., Zeng, X., Lin, C., et al. (2013a). Structural basis for dsRNA recognition, filament formation, and antiviral signal activation by MDA5. *Cell* 152, 276–289. doi: 10.1016/j.cell.2012.11.048

Wu, J., Sun, L., Chen, X., Du, F., Shi, H., Chen, C., et al. (2013b). Cyclic GMP-AMP is an endogenous second messenger in innate immune signaling by cytosolic DNA. *Science* 339, 826–830. doi: 10.1126/science.1229963

Wu, J.-J., Li, W., Shao, Y., Avey, D., Fu, B., Gillen, J., et al. (2015). Inhibition of cGAS DNA sensing by a Herpesvirus virion protein. *Cell Host Microbe* 18, 333–344. doi: 10.1016/j.chom.2015.07.015

Xing, J., Wang, S., Lin, R., Mossman, K. L., and Zheng, C. (2012). Herpes simplex virus 1 tegument protein US11 downmodulates the RLR signaling pathway via direct interaction with RIG-I and MDA-5. *J. Virol.* 86, 3528–3540. doi: 10.1128/JVI.06713-11

Xu, L.-G., Wang, Y.-Y., Han, K.-J., Li, L.-Y., Zhai, Z., and Shu, H.-B. (2005). VISA is an adapter protein required for virus-triggered IFN-beta signaling. *Mol. Cell* 19, 727–740. doi: 10.1016/j.molcel.2005.08.014

Yoneyama, M., Kikuchi, M., Natsukawa, T., Shinobu, N., Imaizumi, T., Miyagishi, M., et al. (2004). The RNA helicase RIG-I has an essential function in double-stranded RNA-induced innate antiviral responses. *Nat. Immunol.* 5, 730–737. doi: 10.1038/ni1087

Zeng, W., Sun, L., Jiang, X., Chen, X., Hou, F., Adhikari, A., et al. (2010). Reconstitution of the RIG-I pathway reveals a signaling role of unanchored polyubiquitin chains in innate immunity. *Cell* 141, 315–330. doi: 10.1016/j.cell.2010.03.029

Zeng, W., Xu, M., Liu, S., Sun, L., and Chen, Z. J. (2009). Key role of Ubc5 and lysine-63 polyubiquitination in viral activation of IRF3. *Mol. Cell* 36, 315–325. doi: 10.1016/j.molcel.2009.09.037

Zhang, X., Shi, H., Wu, J., Zhang, X., Sun, L., Chen, C., et al. (2013). Cyclic GMP-AMP containing mixed phosphodiester linkages is an endogenous high-affinity ligand for STING. *Mol. Cell* 51, 226–235. doi: 10.1016/j.molcel.2013.05.022

Zinzula, L., and Tramontano, E. (2013). Strategies of highly pathogenic RNA viruses to block dsRNA detection by RIG-I-like receptors: hide, mask, hit. *Antiviral Res.* 100, 615–635. doi: 10.1016/j.antiviral.2013.10.002

Züst, R., Cervantes-Barragan, L., Habjan, M., Maier, R., Neuman, B. W., Ziebuhr, J., et al. (2011). Ribose 2'-O-methylation provides a molecular signature for the distinction of self and non-self mRNA dependent on the RNA sensor Mda5. *Nat. immunol.* 12, 137–143. doi: 10.1038/ni.1979

3

Roles of Autophagy and Autophagy-Related Proteins in Antifungal Immunity

*Masashi Kanayama[1] and Mari L. Shinohara[1,2]**

[1] Department of Immunology, Duke University School of Medicine, Durham, NC, USA, [2] Department of Molecular Genetics and Microbiology, Duke University School of Medicine, Durham, NC, USA

Edited by:
Diana Bahia,
Universidade Federal de Minas Gerais, Brazil

Reviewed by:
Eleonora Garcia Vescovi,
Instituto de Biología Molecular y Celular de Rosario, Argentina
Georgios Chamilos,
University of Crete, Greece

***Correspondence:**
Mari L. Shinohara
mari.shinohara@duke.edu

Autophagy was initially characterized as a process to digest cellular components, including damaged cell organelles or unused proteins. However, later studies showed that autophagy plays an important role to protect hosts from microbial infections. Accumulating evidences showed the contribution of autophagy itself and autophagy-related proteins (ATGs) in the clearance of bacteria, virus, and parasites. A number of studies also revealed the molecular mechanisms by which autophagy is initiated and developed. Furthermore, it is now understood that some ATGs are shared between two distinct processes; autophagy and LC3-associated phagocytosis (LAP). Thus, our understanding on autophagy has been greatly enhanced in the last decade. By contrast, roles of autophagy and ATGs in fungal infections are still elusive relative to those in bacterial and viral infections. Based on limited numbers of reports, ATG-mediated host responses appear to significantly vary depending on invading fungal species. In this review, we discuss how autophagy and ATGs are involved in antifungal immune responses based on recent discoveries.

Keywords: autophagy, LC3-associated phagocytosis, fungal infection, phagocytosis, macrophages, *Candida*, *Cryptococcus*, *Aspergillus*

HOST IMMUNITY AGAINST FUNGI

Pathogenic fungi, such as *Cryptococcus, Candida, Aspergillus*, and *Pneumocystis*, are considered to be four major genera of human fungal pathogens (1). A major host risk factor for development of these fungal infections (mycosis) is immunological incompetence, as they are more commonly invasive in patients with immunodeficient disorders and those who receive immunosuppressive treatments (2, 3).

The innate immune system plays a critical role in host protection against fungi. Specific defects in innate immunity, such as neutropenia or functional deficiency of NADPH oxidase, allow hosts to develop invasive aspergillosis, candidiasis, and other mycosis (3, 4). For hosts to initiate antifungal immune responses, fungi have to be detected by pattern recognition receptors (PRRs). Neutrophils and inflammatory macrophages are the first inflammatory cells recruited to the site of infection and the major killers of fungi during an early stage of infection. Dendritic cells (DCs) serve an important role as professional antigen-presenting cells (APCs) to connect innate and adaptive immunity by presenting fungal antigens to prime naïve T cells. As seen in all the microbial infections, coordinating innate immunity is the first step to protect hosts from fungal pathogens.

Defects in the adaptive immune systems are also well-known risk factors in fungal infections. Once primed by DCs, T cells activate and produce inflammatory cytokines, which further recruit innate immune cells to infected sites and facilitate phagocytosis (5, 6). B cells are also involved in fungal clearance (7, 8) by producing antibodies to opsonize fungal spores, i.e., antibodies binds spores and facilitate phagocytosis through stimulating Fc Receptor on phagocytes (9, 10).

In this review, we discuss on autophagy and autophagy-related processes in host cells during fungal infections as a part of the immune responses described above.

AUTOPHAGY

Autophagy is a highly conserved cellular process in eukaryotes to maintain cellular homeostasis by supporting cell survival and regulating inflammation. Autophagy degrades unnecessary or dysfunctional intracellular components, such as abnormal proteins, old organelles, and pathogens, and has been widely studied in various immune cells, including T cell, B cell, macrophages, DCs, and neutrophils. Autophagy eliminates mitochondoria (mitophagy), lipid droplets (lipophagy), ribosomes (ribophagy), protein aggregates (aggrephagy), and intracellular microbes (xenophagy) (11, 12). Multiple roles for autophagy in host defense responses against microbial infections and inflammation have been reported. During a process of autophagy, a spherical double-membrane structure, termed autophagosome, is formed within a cell. A number of autophagy-related proteins (ATGs), together with other proteins, are involved in the process to form autophagosomes; starting from the formation of the autophagy initiation complex to elongating autophagosome membranes. After elongation, the membrane closes and autophagosome formation is completed. [Detailed molecular information on these steps can be found in excellent review articles (13, 14)].

It has been reported that ATGs are associated with human autoimmune disorders, cancer, and various infectious diseases (15). Single-nucleotide polymorphisms (SNPs) in *ATG16L* and *Immunity-related GTPase family M member* (*IRGM*) genes are known to increase the risk of Crohn's disease (16, 17). A SNP of *IRGM* is also associated with susceptibility against *Mycobacterium tuberculosis* infection (18, 19), and SNPs in *ATG5* are associated with risk of systemic lupus erythematosus (20). *UV radiation resistance-associated gene* (*UVRAG*) encoding a promoter of the autophagy pathway, is monoallelically mutated at a high frequency in human colon cancer (21). Recent study demonstrated that *Atg5* in neutrophils protects mice from *M. tuberculosis* infection in autophagy-independent manner (22). Thus, autophagy and autophagy-related genes are suggested to be involved in pathogenesis of wide variety of human diseases.

Recent mechanistic studies have shown that autophagy plays an immunomodulatory role in both innate and adaptive immune responses by selectively targeting signal molecules. For example, autophagy inhibits inflammasome activation in macrophages by degrading inflammasome assemblies as well as reactive oxygen species (ROS)-producing mitochondria, which trigger activation of the NLRP3 inflammasome (23, 24). Autophagy is also known to be required for neutrophil extracellular trap formation (NETosis) (25–27) and immunological training induced by BCG or β-glucan in monocytes (28). In T cells, autophagy suppresses T cell receptor-mediated signaling by degrading BCL10, a downstream molecule of the T cell receptor (29). Autophagy also enhances memory B cell responses (30, 31). Collectively, these findings implicate autophagy in preventing excessive inflammation and protecting the host from collateral damage.

LAP AND AUTOPHAGY

LAP shares some common mechanisms and functions to autophagy; and it has been often difficult to separate LAP and autophagy. For example, LC3 staining cannot differentiate LAP from autophagy, because LC3-associated membranes are formed during both processes. LAP formation requires autophagic proteins, such as ATG5, ATG7, and LC3; therefore, mice or cells lacking one of these proteins cannot undergo both autophagy and LAP (32). Initiation of autophagy and LAP also require ROS and phosphatidylinositol 3-phosphate synthesis (33, 34); although ROS-independent LAP formation was reported in epithelial cells (35), suggesting cell type-specific signaling requirements to induce autophagy or LAP. It is possible that autophagy in non-hematopoietic cells is involved in induction of host antifungal responses. Similarly to autophagy's contribution to microbial clearance by digesting intracellular pathogens, LAP also plays a role in pathogen clearance (36, 37). Despite the similarities, LAP is intrinsically distinct from autophagy in forming the LAPosome with a single membrane structure (32, 38), the requirement of Rubicon and NOX2 (33), and not requiring the autophagy pre-initiation complex comprised of ULK1/2, FIP200, and ATG13 (33, 39, 40).

A very recent article demonstrated that LAP is differentiated from autophagy using Rubicon-deficient mice (autophagy is intact in the mice) (33). Here, we would like to note that previously published studies using *Atg5*- or *Atg7*-deficient mice or cells might reflect impacts of impaired LAP as well as autophagy, though most of them have been published as "autophagy" studies. Therefore, it would be prudent to consider the possible involvement of LAP in previous studies, depending on an experimental condition. In this review, we use "LC3-associated cargoes" for autophagosomes and LAPosomes if responses in referred articles have not identified as autophagy or LAP.

ROLES OF AUTOPHAGY-RELATED PROTEINS IN FUNGAL INFECTIONS

The role of autophagy in antifungal immunity was strongly suggested both in mammals and plants (41). Here, we focus on autophagy in mammalian cells. Autophagy can be induced directly by signaling from fungal-sensing PRRs, and also indirectly by pro-inflammatory cytokines, including TNF-α, IL-6, IL-1, and IFN-γ (42) during fungal infections. In contrast, spores of *Aspergillus fumigatus* induce LAP based on the study using Rubicon-deficient mice (33). LAP is also induced by zymosan, a cell-wall component from *Saccharomyces cerevisiae* (33). Therefore, it is possible that other fungi induce LAP. In the following subsections and **Table 1**, we focus on three major fungal pathogens: *Candida albicans, Cryptococcus neoformans*, and *A. fumigatus* to discuss impacts of ATGs on host antifungal responses and pathogenesis of fungal infections.

Candida albicans

Several studies have described roles of "autophagy" (or possibly LAP) during Candida infection. Candida spores, both live and heat-killed, are potent inducers of LC3 puncta formation and conversion of LC3-I to LC3-II (43–45). However, it appears that live Candida spores are not good at recruiting LC3 around internalized spores (43, 45, 46). These results suggest that live Candida can induce autophagy (and/or LAP), but direct clearance of *Candida* spores within LC3-associated cargoes is not likely to occur. A study by Vyas and colleagues showed that heat-killed *C. albicans* (HKCA) recruited clear LC3-associated cargoes around

TABLE 1 | Impacts of autophagy and LAP on antifungal immunity.

	Approaches	Findings	Reference
A. fumigatus	• *Rubicon*-, *Cybb*-, *Atg4b*-, or *Ulk1*-deficient mice • *Atg3*$^{fl/fl}$*LysM*-Cre mice, *Atg5*$^{fl/fl}$*LysM*-Cre mice, *Atg7*$^{fl/fl}$*LysM*-Cre mice, *Atg12*$^{fl/fl}$*LysM*-Cre mice, *Atg14*$^{fl/fl}$*LysM*-Cre mice, *Becn1*$^{fl/fl}$*LysM*-Cre mice, *Rb1cc1*$^{fl/fl}$*LysM*-Cre mice • shRNA for Rubicon • 3-MA • Rapamycin	• *A. fumigatus* phagocytosed by macrophages induces LAP formation • LAP, but not canonical autophagy, is required for clearance of *A. fumigatus in vitro* and *in vivo*. • Lack of LAP enhances fungi-induced pulmonary inflammation and granulomas • SNP in *Atg16L* does not affect LAP formation	(33)
	• 3-MA • Chloroquine	• Autophagy suppresses inflammasome-mediated inflammation during *A. fumigatus* infection • IL-1R blockade restores autophagy and suppresses fungal growth in CGD mice.	(58)
C. neoformans	• RNAi screening • siRNA for *Atg2a*, *Atg5*, *Atg9*, *Atg12*, and *Map1lc3a* (Coding LC3) • 3-MA	• Knockdown of *Atg5*, *Atg9a*, and *Atg12* decreases phagocytosis of *C. neoformans* by macrophages • Knockdown of *Atg2a*, *Atg5*, *Atg9a*, *Atg12*, and LC3 decreases fungal replication and escape of *C. neoformans* • Autophagy inhibition by 3-MA reduces phagocytosis, fungal replication, and escape of *C. neoformans*	(54)
	• shRNA for *Atg5* • *Atg5*$^{fl/fl}$*LysM*-Cre mice	• Phagocytosed *C. neoformans* are surrounded by LAP in macrophages • *Atg5*-knock down by shRNA does not affect phagocytosis • *Atg5*-knock down by shRNA or *Atg5*-knock-out decreases fungicidal activity in macrophages	(50)
C. albicans	• shRNA for *Atg5* • *Atg5*$^{fl/fl}$*LysM*-Cre mice	• Phagocytosed *C. albicans* are surrounded by LC3 in macrophages • *Atg5*-knockdown decreases phagocytosis and fungicidal activity of macrophages *in vitro* • Lack of ATG5 in myeloid cells enhances susceptibility of mice against systemic *C. albicans* infection	(50)
	• *Atg7*$^{fl/fl}$*LysM*-Cre mice • 3-MA • Cohort study of candidemia patients	• Lack of ATG7 in myeloid cells does not impact on susceptibility of mice against systemic *C. albicans* infection • Autophagy inhibition by 3-MA does not affect phagocytosis and fungicidal activity against *C. albicans* in human monocytes • SNP in autophagy-related genes does not associate with incidence of candidemia	(44)
	• *Atg7*$^{fl/fl}$*LysM*-Cre mice • 3-MA • Rapamycin	• *C. albicans* induces autophagy in macrophages enhances susceptibility against systemic *C. albicans* infection • Phagocytosed *C. albicans* are not surrounded by LC3 • Autophagy does not affect phagocytosis and fungicidal activity of macrophages and neutrophils • Autophagic sequestration of A20 enhances NFκB-mediated chemokine production in tissue-resident macrophages and increases neutrophil recruitment to infected site • Lack of autophagy in myeloid cells	(43)
	• *In vivo* imaging using zebra fish	• Very few LC3^{+} phagosome contain *C. albicans in vivo*	(46)
	• LC3-deficient BMMs	• Dectin-1-induced signaling triggers LAP formation in macrophages • HK *C. albicans* induces LAP in macrophages • Live *C. albicans* induces modest level of LAP in macrophages • LC3-deficiency decreases fungicidal activity of macrophages against *C. albicans*	(45)

spores by *in vitro* observation using the RAW264.7 mouse macrophage cell line. By contrast, when live *C. albicans* was internalized, LC3-associated cargoes around spores were not very clear (45). The result is consistent with an *in vivo* study using zebra fish, in which live *C. albicans* rarely recruited significant levels of LC3 (46). We also could not detect LC3 signal surrounding live *C. albicans* in both primary macrophages and in a macrophage cell line (43). This finding was unexpected, because we speculated that Candida spores were engulfed in LC3-positve cargoes and directly killed by xenophagy or LAP. The expectation came from previous reports showing successful recruitment of LC3 to zymosan particles (33) and around β-glucan-coated polystyrene beads (45, 47). However, it is possible that live Candida spores do not expose enough β-glucan on their cell surface, while HKCA spores do (45, 48, 49). Collectively, these studies suggested that direct killing of Candida spores within LC3-associated cargoes is not very likely.

Despite the unexpectedly poor engulfment of live Candida in LC3-associated cargoes, studies have shown that autophagy (or LAP) protects hosts from Candida infections. Lack of ATG5 or ATG7 in myeloid cells decreased resistance against systemic *C. albicans* infection (43, 50). The protective role of autophagy (or LAP) in *Candida* infection is partly attributed to enhanced fungicidal activity in host myeloid cells, such as expression of ROS and efficiency of phagocytosis (50). Here, we should mention that host protection by autophagy or LAP during *Candida* infection might not be always apparent. Smeekens et al. reported no difference in survival between wild-type and *Atg7* conditional knock-out myeloid cells *(Atg7* CKO), as well as no difference in phagocytosis and killing of *Candida* (44). Although we found that *Atg7* CKO mice are more sensitive to systemic *Candida* infection, we also found no difference in phagocytosis and killing of *Candida* with or without ATG7 (43). Reasons for the discrepancy may be experimental conditions and differential strain usage of *C. albicans*. For example, Smeekens et al. used a different strain of *C. albicans* (MYA-3573) from others (SC-5314, 18804) (43, 45, 50). We used strain 18804 and found that ATG7 in myeloid cells plays a protective role in hosts without enhancing phagocytosis and killing of Candida (43). Same host responses cannot be expected when Candida strains are different. For example, published articles suggested distinct dectin-1 detection towards two different *Candida* strains, SC-5314 and 18804 (51, 52). Taken together, multiple studies suggested the involvement of ATG5 and ATG7 in enhancing resistance to Candida infection.

As shown in human studies, despite moderate influences on pro-inflammatory cytokine production, autophagy genes *ATG16L1* and *IRGM* have a minor impact on the susceptibility to both mucosal and systemic *Candida* infections (53). Other genes were investigated, such as *ATG10, ATG16L2, ATG2A, ATG2B, ATG5,* and *ATG9B,* but a clear correlation between SNPs of the genes and susceptibility to candidemia was not found (44). It was reported that an ATG16L human SNP mutant protein (T316A) expressed in mice decreases starvation-induced autophagy to 50% with no influence on zymosan-induced LAP (33). Nevertheless, impacts of these SNPs on autophagy and LAP in humans are still elusive. Therefore, further studies are needed to understand the consequence of autophagy on host immunity against Candida in humans.

Cryptococcus neoformans

Cryptococcus neoformans is an opportunistic fungus. Cryptococcal yeasts are encapsulated in polysaccharides and, thus, can evade immune detection by hosts. Interestingly, host autophagy supports intracellular survival and dissemination of *C. neoformans* (54). The report demonstrated that ATGs (ATG5, ATG9a, and ATG12) are engaged, but not required, in phagocytosing *C. neoformans* by RAW264.7 macrophages, and the proteins are recruited to the vicinity of vacuoles containing *C. neoformans* (54). At a later time point (15 h after infection), ATG2a, ATG5, ATG9a, ATG12, and LC3 enhance intracellular replication and escape of *C. neoformans* from vacuoles in macrophages (54). Indeed, pharmacological inhibition of autophagy by 3-MA reduced levels of Cryptococcus infection (54). Another article reported LC3 recruitment to internalized cryptococcal spores; and mouse survival from Cryptococcus infection was not altered by *Atg5* CKO in myeloid cells (50). Nevertheless, the CKO mice exhibited reduced lung fungal burdens and protein expression of MIP-1α (CCL3), IP-10 (CXCL10), as well as Type-2 cytokines IL-4 and IL-13 (50). Expression of IFN-γ and IL-17 did not appear to be altered (50). Therefore, further studies are awaited to better understand the impact and roles of autophagy (and LAP) in Cryptococcus infection.

Aspergillus fumigatus

LC3-II recruitment in *A. fumigatus* phagosomes was reported (55). A recent study using Rubicon-deficient mice clarified that LC3 recruitment and clearance of *A. fumigatus* are mediated by LAP, but not by autophagy (33). LAP protects mice from pulmonary aspergillosis by suppressing expression of pro-inflammatory cytokine genes and granuloma formation in the lung (33). LAP also stabilizes the NOX2 NADPH oxidase complex to produce ROS. Macrophages deficient for NOX2 failed to translocate LC3 to *Aspergillus*-containing phagosomes, as well as macrophages deficient for Beclin1, Rubicon, and ATG7 failed to do so (33). It was reported that patients with chronic granulomatous disease (CGD), caused by genetic defects in the NADPH oxidase, do not recruit LC3 around the *Aspergillus*-containing phagosomes (55). CGD patients indeed show an increased susceptibility to aspergillosis (56, 57). In addition, corticosteroid blocks recruitment of ATG (55), and its treatment is considered to be a risk factor for invasive aspergillosis (57). Another report demonstrated that IL-1R blockade protects hosts from invasive aspergillosis by increasing LC3 recruitment to *Aspergillus*-containing phagosomes and inhibiting fungal growth (58). In summary, LC3-associated cargoes appear to play a critical role in clearance of *Aspergillus* by host cells.

IMMUNOMODULATION BY AUTOPHAGY DURING FUNGAL INFECTION

Regulation of Inflammasome-Mediated Immune Responses

Recent studies showed that autophagy controls immune responses during fungal infections. Autophagosomes are known to sequester assembled NLRP3 inflammasome complexes (23, 24), which are crucial for host protection against *C. albicans* and *A. fumigatus* infections (59–61). Another study showed that pharmacological inhibition of autophagy by 3-MA or chloroquine enhances inflammasome activation and inflammation in mice during *A. fumigatus* infection (58). Increased inflammasome activity was reported in monocytes and macrophages from CGD patients (58, 62–64). Therefore, autophagy (or LAP) may protect hosts from collateral damage by inflammation through downregulating NLRP3 inflammasome, which is activated by fungal infections.

NFκB-Mediated Immune Responses

We have discussed autophagy-mediated suppression in immune responses (23, 29, 65). Yet, autophagy can enhance antifungal immune responses in early stages of *C. albicans* infection by using mice conditionally lacking *Atg7* in myeloid cells (*Atg7* CKO) (43) (**Table 1**; **Figure 1**). The *Atg7* CKO mice showed increased fungal burdens in infected sites as a result of reduced neutrophil recruitment. This was due to reduced production of neutrophil-chemoattractants (CXCL1 and CXCL2) by tissue-resident macrophages at the site of infection in the absence of *Atg7* (43) (**Figure 1**). Unstimulated tissue-resident macrophages express high levels of A20, an NFκB inhibitor (66–69). After detection of *C. albicans*, autophagy in tissue-resident macrophages sequesters A20 and frees NFκB activation from A20-mediated inhibition. Autophagy-adaptor protein p62 was demonstrated to interact with A20 to carry A20 into LC3-associated cargos (43) (**Figure 1**). Indeed, the lack of ATG7 or p62 increases the levels of A20, causing reduction in NFκB activity and chemokine production by tissue-resident macrophages (43) (**Figure 1**). In summary, autophagy appears to function in balancing between inducing antifungal immunity and controlling excessive inflammation.

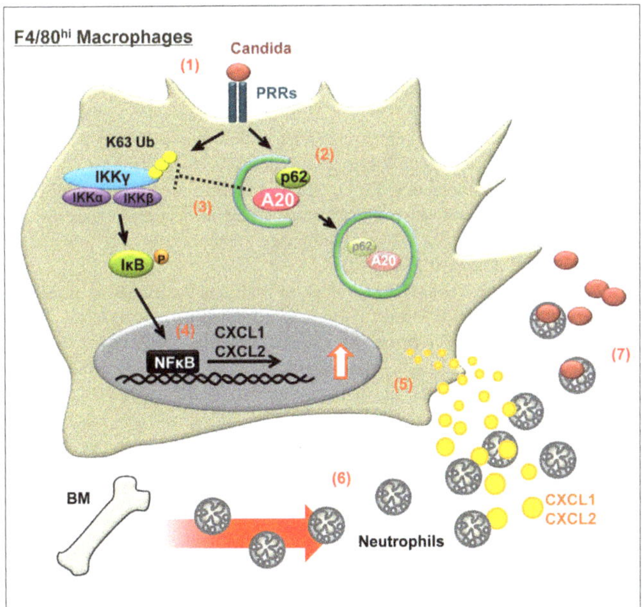

FIGURE 1 | Schematic illustration of a mechanism in which autophagy enhances antifungal immune responses by sequestering A20. Numbers in the figure indicate steps from fungal detection to killing by hosts and correspond to the following events: (1) detection of *C. albicans* by PRRs, such as dectin-1 and TLR2, expressed on F4/80hi tissue-resident macrophages; (2) autophagy induction and sequestration of A20 in autophagosomes, and A20 delivery to autophagosomes by p62; (3) IKKγ ubiquitination by sequestering A20; (4) enhancement of NFκB activity; (5) increased chemokine production; (6) recruitment of neutrophils to infected sites (i.e., where the responding tissue-resident macrophages are located); and (7) Killing of fungi by neutrophils.

IMPACTS OF AUTOPHAGY ON ADAPTIVE IMMUNITY

Limited amount of autophagy information is available on adaptive immune responses against fungal pathogens, but the involvement of autophagy in shaping adaptive immunity to protect hosts against fungal infections has been suggested. Autophagy enhances survival and functions of T cells (70–72) and B cells (73, 74) in various pathological conditions. Autophagic machinery including ATG5 plays an important role in processing and presenting extracellular microbial antigens in dendritic cells (75). Importantly, LC3-associated cargoes are involved in presenting fungal antigen from *S. cerevisiae* (47). Thus, autophagy appears to control antifungal adaptive immune responses via antigen presentation by DCs and other APCs. It is still not clear whether autophagy in T cells directly controls antifungal immunity.

CONCLUSION

Autophagy was initially described as a self-catabolic process, but it is now known to play a critical role in clearance of bacterial, viral, and parasitic pathogens. Although the role of autophagy and ATGs in host defense against fungi had not been made clear, recent studies demonstrated the involvement of autophagy and ATGs in modulating antifungal immunity. LC3-associated cargoes may include fungal spores; but the inclusion cannot be always seen, e.g., when live *Candida* is engulfed. Autophagy and LAP generally protect hosts from majority of fungal infections by inducing immune responses or by controlling excessive inflammation. There are, however, some exceptions that ATGs promote fungal infections. The outcomes of autophagy/LAP in shaping host immune responses appear to greatly vary depending on species of fungi. Interestingly, previous findings suggested that activation of autophagy/LAP by immunosuppressants, such as rapamycin or anakinra (IL-1R antagonist), may result in inducing host resistance against fungal infections. It might be possible that autophagy-related pathways are targeted for new antifungal therapeutics.

AUTHOR CONTRIBUTIONS

MK and MLS wrote the manuscript.

ACKNOWLEDGMENTS

The authors wish to thank support from NIH (R01-AI088100 and R21-105584) to MLS. We also thank past and present laboratory members, particularly Will Barclay and Gregory Swan for their editorial help to prepare the manuscript.

REFERENCES

1. Brown GD, Denning DW, Gow NA, Levitz SM, Netea MG, White TC. Hidden killers: human fungal infections. *Sci Transl Med* (2012) **4**:165rv13. doi:10.1126/scitranslmed.3004404
2. Shimodaira K, Okubo Y, Nakayama H, Wakayama M, Shinozaki M, Ishiwatari T, et al. Trends in the prevalence of invasive fungal infections from an analysis of annual records of autopsy cases of Toho University. *Mycoses* (2012) **55**:435–43. doi:10.1111/j.1439-0507.2012.02169.x
3. Cornely OA. *Aspergillus* to Zygomycetes: causes, risk factors, prevention, and treatment of invasive fungal infections. *Infection* (2008) **36**:296–313. doi:10.1007/s15010-008-7357-z
4. Bassiri-Jahromi S, Doostkam A. Fungal infection and increased mortality in patients with chronic granulomatous disease. *J Mycol Med* (2012) **22**:52–7. doi:10.1016/j.mycmed.2011.12.079
5. Marodi L, Schreiber S, Anderson DC, MacDermott RP, Korchak HM, Johnston RB Jr. Enhancement of macrophage candidacidal activity by interferon-gamma. Increased phagocytosis, killing, and calcium signal mediated by a decreased number of mannose receptors. *J Clin Invest* (1993) **91**:2596–601. doi:10.1172/JCI116498
6. Conti HR, Shen F, Nayyar N, Stocum E, Sun JN, Lindemann MJ, et al. Th17 cells and IL-17 receptor signaling are essential for mucosal host defense against oral candidiasis. *J Exp Med* (2009) **206**:299–311. doi:10.1084/jem.20081463
7. Opata MM, Hollifield ML, Lund FE, Randall TD, Dunn R, Garvy BA, et al. B lymphocytes are required during the early priming of CD4+ T cells for clearance of pneumocystis infection in mice. *J Immunol* (2015) **195**:611–20. doi:10.4049/jimmunol.1500112
8. Rohatgi S, Pirofski LA. Molecular characterization of the early B cell response to pulmonary *Cryptococcus neoformans* infection. *J Immunol* (2012) **189**:5820–30. doi:10.4049/jimmunol.1201514
9. Subramaniam KS, Datta K, Quintero E, Manix C, Marks MS, Pirofski LA. The absence of serum IgM enhances the susceptibility of mice to pulmonary challenge with *Cryptococcus neoformans*. *J Immunol* (2010) **184**:5755–67. doi:10.4049/jimmunol.0901638
10. Yuan R, Clynes R, Oh J, Ravetch JV, Scharff MD. Antibody-mediated modulation of *Cryptococcus neoformans* infection is dependent on distinct Fc receptor functions and IgG subclasses. *J Exp Med* (1998) **187**:641–8. doi:10.1084/jem.187.4.641
11. Nakahira K, Choi AM. Autophagy: a potential therapeutic target in lung diseases. *Am J Physiol Lung Cell Mol Physiol* (2013) **305**:L93–107. doi:10.1152/ajplung.00072.2013
12. Cuervo AM. Chaperone-mediated autophagy: Dice's 'wild' idea about lysosomal selectivity. *Nat Rev Mol Cell Biol* (2011) **12**:535–41. doi:10.1038/nrm3150
13. Marino G, Niso-Santano M, Baehrecke EH, Kroemer G. Self-consumption: the interplay of autophagy and apoptosis. *Nat Rev Mol Cell Biol* (2014) **15**:81–94. doi:10.1038/nrm3735
14. Mizushima N, Komatsu M. Autophagy: renovation of cells and tissues. *Cell* (2011) **147**:728–41. doi:10.1016/j.cell.2011.10.026
15. Choi AM, Ryter SW, Levine B. Autophagy in human health and disease. *N Engl J Med* (2013) **368**:651–62. doi:10.1056/NEJMra1205406
16. Salem M, Ammitzboell M, Nys K, Seidelin JB, Nielsen OH. ATG16L1: a multifunctional susceptibility factor in Crohn disease. *Autophagy* (2015) **11**:585–94. doi:10.1080/15548627.2015.1017187
17. Parkes M, Barrett JC, Prescott NJ, Tremelling M, Anderson CA, Fisher SA, et al. Sequence variants in the autophagy gene IRGM and multiple other replicating loci contribute to Crohn's disease susceptibility. *Nat Genet* (2007) **39**:830–2. doi:10.1038/ng2061
18. Che N, Li S, Gao T, Zhang Z, Han Y, Zhang X, et al. Identification of a novel IRGM promoter single nucleotide polymorphism associated with tuberculosis. *Clin Chim Acta* (2010) **411**:1645–9. doi:10.1016/j.cca.2010.06.009
19. Bahari G, Hashemi M, Taheri M, Naderi M, Eskandari-Nasab E, Atabaki M. Association of IRGM polymorphisms and susceptibility to pulmonary tuberculosis in Zahedan, Southeast Iran. *ScientificWorldJournal* (2012) **2012**:950801. doi:10.1100/2012/950801
20. Pierdominici M, Vomero M, Barbati C, Colasanti T, Maselli A, Vacirca D, et al. Role of autophagy in immunity and autoimmunity, with a special focus on systemic lupus erythematosus. *FASEB J* (2012) **26**:1400–12. doi:10.1096/fj.11-194175
21. Liang C, Feng P, Ku B, Dotan I, Canaani D, Oh BH, et al. Autophagic and tumour suppressor activity of a novel Beclin1-binding protein UVRAG. *Nat Cell Biol* (2006) **8**:688–99. doi:10.1038/ncb1426
22. Kimmey JM, Huynh JP, Weiss LA, Park S, Kambal A, Debnath J, et al. Unique role for ATG5 in neutrophil-mediated immunopathology during M. tuberculosis infection. *Nature* (2015) **528**:565–9. doi:10.1038/nature16451
23. Shi CS, Shenderov K, Huang NN, Kabat J, Abu-Asab M, Fitzgerald KA, et al. Activation of autophagy by inflammatory signals limits IL-1beta production by targeting ubiquitinated inflammasomes for destruction. *Nat Immunol* (2012) **13**:255–63. doi:10.1038/ni.2215
24. Zhou R, Yazdi AS, Menu P, Tschopp J. A role for mitochondria in NLRP3 inflammasome activation. *Nature* (2011) **469**:221–5. doi:10.1038/nature09663
25. Remijsen Q, Vanden Berghe T, Wirawan E, Asselbergh B, Parthoens E, De Rycke R, et al. Neutrophil extracellular trap cell death requires both autophagy and superoxide generation. *Cell Res* (2011) **21**:290–304. doi:10.1038/cr.2010.150
26. Tang S, Zhang Y, Yin SW, Gao XJ, Shi WW, Wang Y, et al. Neutrophil extracellular trap formation is associated with autophagy-related signalling in ANCA-associated vasculitis. *Clin Exp Immunol* (2015) **180**:408–18. doi:10.1111/cei.12589
27. Itakura A, McCarty OJ. Pivotal role for the mTOR pathway in the formation of neutrophil extracellular traps via regulation of autophagy. *Am J Physiol Cell Physiol* (2013) **305**:C348–54. doi:10.1152/ajpcell.00108.2013
28. Buffen K, Oosting M, Quintin J, Ng A, Kleinnijenhuis J, Kumar V, et al. Autophagy controls BCG-induced trained immunity and the response to intravesical BCG therapy for bladder cancer. *PLoS Pathog* (2014) **10**:e1004485. doi:10.1371/journal.ppat.1004485
29. Paul S, Kashyap AK, Jia W, He YW, Schaefer BC. Selective autophagy of the adaptor protein Bcl10 modulates T cell receptor activation of NF-kappaB. *Immunity* (2012) **36**:947–58. doi:10.1016/j.immuni.2012.04.008
30. Chen M, Kodali S, Jang A, Kuai L, Wang J. Requirement for autophagy in the long-term persistence but not initial formation of memory B cells. *J Immunol* (2015) **194**:2607–15. doi:10.4049/jimmunol.1403001
31. Chen M, Hong MJ, Sun H, Wang L, Shi X, Gilbert BE, et al. Essential role for autophagy in the maintenance of immunological memory against influenza infection. *Nat Med* (2014) **20**:503–10. doi:10.1038/nm.3521
32. Sanjuan MA, Dillon CP, Tait SW, Moshiach S, Dorsey F, Connell S, et al. Toll-like receptor signalling in macrophages links the autophagy pathway to phagocytosis. *Nature* (2007) **450**:1253–7. doi:10.1038/nature06421
33. Martinez J, Malireddi RK, Lu Q, Cunha LD, Pelletier S, Gingras S, et al. Molecular characterization of LC3-associated phagocytosis reveals distinct roles for Rubicon, NOX2 and autophagy proteins. *Nat Cell Biol* (2015) **17**:893–906. doi:10.1038/ncb3192
34. Vergne I, Deretic V. The role of PI3P phosphatases in the regulation of autophagy. *FEBS Lett* (2010) **584**:1313–8. doi:10.1016/j.febslet.2010.02.054
35. Brooks CR, Yeung MY, Brooks YS, Chen H, Ichimura T, Henderson JM, et al. KIM-1-/TIM-1-mediated phagocytosis links ATG5-/ULK1-dependent clearance of apoptotic cells to antigen presentation. *EMBO J* (2015) **34**:2441–64. doi:10.15252/embj.201489838
36. Kuballa P, Nolte WM, Castoreno AB, Xavier RJ. Autophagy and the immune system. *Annu Rev Immunol* (2012) **30**:611–46. doi:10.1146/annurev-immunol-020711-074948
37. Deretic V, Saitoh T, Akira S. Autophagy in infection, inflammation and immunity. *Nat Rev Immunol* (2013) **13**:722–37. doi:10.1038/nri3532
38. Martinez J, Almendinger J, Oberst A, Ness R, Dillon CP, Fitzgerald P, et al. Microtubule-associated protein 1 light chain 3 alpha (LC3)-associated phagocytosis is required for the efficient clearance of dead cells. *Proc Natl Acad Sci U S A* (2011) **108**:17396–401. doi:10.1073/pnas.1113421108
39. Jung CH, Jun CB, Ro SH, Kim YM, Otto NM, Cao J, et al. ULK-Atg13-FIP200 complexes mediate mTOR signaling to the autophagy machinery. *Mol Biol Cell* (2009) **20**:1992–2003. doi:10.1091/mbc.E08-12-1249
40. Florey O, Kim SE, Sandoval CP, Haynes CM, Overholtzer M. Autophagy machinery mediates macroendocytic processing and entotic cell death by targeting single membranes. *Nat Cell Biol* (2011) **13**:1335–43. doi:10.1038/ncb2363
41. Lenz HD, Haller E, Melzer E, Kober K, Wurster K, Stahl M, et al. Autophagy differentially controls plant basal immunity to biotrophic and necrotrophic pathogens. *Plant J* (2011) **66**:818–30. doi:10.1111/j.1365-313X.2011.04546.x

42. Harris J. Autophagy and cytokines. *Cytokine* (2011) **56**:140–4. doi:10.1016/j.cyto.2011.08.022
43. Kanayama M, Inoue M, Danzaki K, Hammer G, He YW, Shinohara ML. Autophagy enhances NFkappaB activity in specific tissue macrophages by sequestering A20 to boost antifungal immunity. *Nat Commun* (2015) **6**:5779. doi:10.1038/ncomms6779
44. Smeekens SP, Malireddi RK, Plantinga TS, Buffen K, Oosting M, Joosten LA, et al. Autophagy is redundant for the host defense against systemic *Candida albicans* infections. *Eur J Clin Microbiol Infect Dis* (2014) **33**:711–22. doi:10.1007/s10096-013-2002-x
45. Tam JM, Mansour MK, Khan NS, Seward M, Puranam S, Tanne A, et al. Dectin-1-dependent LC3 recruitment to phagosomes enhances fungicidal activity in macrophages. *J Infect Dis* (2014) **210**(11):1844–54. doi:10.1093/infdis/jiu290
46. Brothers KM, Gratacap RL, Barker SE, Newman ZR, Norum A, Wheeler RT. NADPH oxidase-driven phagocyte recruitment controls *Candida albicans* filamentous growth and prevents mortality. *PLoS Pathog* (2013) **9**:e1003634. doi:10.1371/journal.ppat.1003634
47. Ma J, Becker C, Lowell CA, Underhill DM. Dectin-1-triggered recruitment of light chain 3 protein to phagosomes facilitates major histocompatibility complex class II presentation of fungal-derived antigens. *J Biol Chem* (2012) **287**:34149–56. doi:10.1074/jbc.M112.382812
48. Gow NA, Netea MG, Munro CA, Ferwerda G, Bates S, Mora-Montes HM, et al. Immune recognition of *Candida albicans* beta-glucan by dectin-1. *J Infect Dis* (2007) **196**:1565–71. doi:10.1086/523110
49. Gantner BN, Simmons RM, Underhill DM. Dectin-1 mediates macrophage recognition of *Candida albicans* yeast but not filaments. *EMBO J* (2005) **24**:1277–86. doi:10.1038/sj.emboj.7600594
50. Nicola AM, Albuquerque P, Martinez LR, Dal-Rosso RA, Saylor C, De Jesus M, et al. Macrophage autophagy in immunity to *Cryptococcus neoformans* and *Candida albicans*. *Infect Immun* (2012) **80**:3065–76. doi:10.1128/IAI.00358-12
51. Saijo S, Fujikado N, Furuta T, Chung SH, Kotaki H, Seki K, et al. Dectin-1 is required for host defense against Pneumocystis carinii but not against *Candida albicans*. *Nat Immunol* (2007) **8**:39–46. doi:10.1038/ni1425
52. Taylor PR, Tsoni SV, Willment JA, Dennehy KM, Rosas M, Findon H, et al. Dectin-1 is required for beta-glucan recognition and control of fungal infection. *Nat Immunol* (2007) **8**:31–8. doi:10.1038/ni1408
53. Rosentul DC, Plantinga TS, Farcas M, Oosting M, Hamza OJ, Scott WK, et al. Role of autophagy genetic variants for the risk of Candida infections. *Med Mycol* (2014) **52**:333–41. doi:10.1093/mmy/myt035
54. Qin QM, Luo J, Lin X, Pei J, Li L, Ficht TA, et al. Functional analysis of host factors that mediate the intracellular lifestyle of *Cryptococcus neoformans*. *PLoS Pathog* (2011) **7**:e1002078. doi:10.1371/journal.ppat.1002078
55. Kyrmizi I, Gresnigt MS, Akoumianaki T, Samonis G, Sidiropoulos P, Boumpas D, et al. Corticosteroids block autophagy protein recruitment in *Aspergillus fumigatus* phagosomes via targeting dectin-1/Syk kinase signaling. *J Immunol* (2013) **191**:1287–99. doi:10.4049/jimmunol.1300132
56. Falcone EL, Holland SM. Invasive fungal infection in chronic granulomatous disease: insights into pathogenesis and management. *Curr Opin Infect Dis* (2012) **25**:658–69. doi:10.1097/QCO.0b013e328358b0a4
57. Palmer LB, Greenberg HE, Schiff MJ. Corticosteroid treatment as a risk factor for invasive aspergillosis in patients with lung disease. *Thorax* (1991) **46**:15–20. doi:10.1136/thx.46.1.15
58. de Luca A, Smeekens SP, Casagrande A, Iannitti R, Conway KL, Gresnigt MS, et al. IL-1 receptor blockade restores autophagy and reduces inflammation in chronic granulomatous disease in mice and in humans. *Proc Natl Acad Sci U S A* (2014) **111**:3526–31. doi:10.1073/pnas.1322831111
59. Gross O, Poeck H, Bscheider M, Dostert C, Hannesschlager N, Endres S, et al. Syk kinase signalling couples to the Nlrp3 inflammasome for anti-fungal host defence. *Nature* (2009) **459**:433–6. doi:10.1038/nature07965
60. Karki R, Man SM, Malireddi RK, Gurung P, Vogel P, Lamkanfi M, et al. Concerted activation of the AIM2 and NLRP3 inflammasomes orchestrates host protection against *Aspergillus* infection. *Cell Host Microbe* (2015) **17**:357–68. doi:10.1016/j.chom.2015.01.006
61. van de Veerdonk FL, Joosten LA, Shaw PJ, Smeekens SP, Malireddi RK, van der Meer JW, et al. The inflammasome drives protective Th1 and Th17 cellular responses in disseminated candidiasis. *Eur J Immunol* (2011) **41**:2260–8. doi:10.1002/eji.201041226
62. Rieber N, Hector A, Kuijpers T, Roos D, Hartl D. Current concepts of hyper-inflammation in chronic granulomatous disease. *Clin Dev Immunol* (2012) **2012**:252460. doi:10.1155/2012/252460
63. Meissner F, Seger RA, Moshous D, Fischer A, Reichenbach J, Zychlinsky A. Inflammasome activation in NADPH oxidase defective mononuclear phagocytes from patients with chronic granulomatous disease. *Blood* (2010) **116**:1570–3. doi:10.1182/blood-2010-01-264218
64. van de Veerdonk FL, Smeekens SP, Joosten LA, Kullberg BJ, Dinarello CA, van der Meer JW, et al. Reactive oxygen species-independent activation of the IL-1beta inflammasome in cells from patients with chronic granulomatous disease. *Proc Natl Acad Sci U S A* (2010) **107**:3030–3. doi:10.1073/pnas.0914795107
65. Kanayama M, He YW, Shinohara ML. The lung is protected from spontaneous inflammation by autophagy in myeloid cells. *J Immunol* (2015) **194**:5465–71. doi:10.4049/jimmunol.1403249
66. Hammer GE, Turer EE, Taylor KE, Fang CJ, Advincula R, Oshima S, et al. Expression of A20 by dendritic cells preserves immune homeostasis and prevents colitis and spondyloarthritis. *Nat Immunol* (2011) **12**:1184–93. doi:10.1038/ni.2135
67. Boone DL, Turer EE, Lee EG, Ahmad RC, Wheeler MT, Tsui C, et al. The ubiquitin-modifying enzyme A20 is required for termination of Toll-like receptor responses. *Nat Immunol* (2004) **5**:1052–60. doi:10.1038/ni1110
68. Lee EG, Boone DL, Chai S, Libby SL, Chien M, Lodolce JP, et al. Failure to regulate TNF-induced NF-kappaB and cell death responses in A20-deficient mice. *Science* (2000) **289**:2350–4. doi:10.1126/science.289.5488.2350
69. Shi CS, Kehrl JH. TRAF6 and A20 regulate lysine 63-linked ubiquitination of Beclin-1 to control TLR4-induced autophagy. *Sci Signal* (2010) **3**:ra42. doi:10.1126/scisignal.2000751
70. Matsuzawa Y, Oshima S, Takahara M, Maeyashiki C, Nemoto Y, Kobayashi M, et al. TNFAIP3 promotes survival of CD4 T cells by restricting MTOR and promoting autophagy. *Autophagy* (2015) **11**:1052–62. doi:10.1080/15548627.2015.1055439
71. Schlie K, Westerback A, DeVorkin L, Hughson LR, Brandon JM, MacPherson S, et al. Survival of effector CD8+ T cells during influenza infection is dependent on autophagy. *J Immunol* (2015) **194**:4277–86. doi:10.4049/jimmunol.1402571
72. Pua HH, Dzhagalov I, Chuck M, Mizushima N, He YW. A critical role for the autophagy gene Atg5 in T cell survival and proliferation. *J Exp Med* (2007) **204**:25–31. doi:10.1084/jem.20061303
73. Weindel CG, Richey LJ, Bolland S, Mehta AJ, Kearney JF, Huber BT. B cell autophagy mediates TLR7-dependent autoimmunity and inflammation. *Autophagy* (2015) **11**:1010–24. doi:10.1080/15548627.2015.1052206
74. Eriksen AB, Torgersen ML, Holm KL, Abrahamsen G, Spurkland A, Moskaug JO, et al. Retinoic acid-induced IgG production in TLR-activated human primary B cells involves ULK1-mediated autophagy. *Autophagy* (2015) **11**:460–71. doi:10.1080/15548627.2015.1009797
75. Lee HK, Mattei LM, Steinberg BE, Alberts P, Lee YH, Chervonsky A, et al. In vivo requirement for Atg5 in antigen presentation by dendritic cells. *Immunity* (2010) **32**:227–39. doi:10.1016/j.immuni.2009.12.006

4

The Mitogen-Activated Protein Kinase (MAPK) Pathway: Role in Immune Evasion by Trypanosomatids

*Mercedes Soares-Silva[1], Flavia F. Diniz[1], Gabriela N. Gomes[1] and Diana Bahia[1,2]**

[1] Departamento de Biologia Geral, Instituto de Ciências Biológicas, Universidade Federal de Minas Gerais, Minas Gerais, Brazil, [2] Departamento de Microbiologia, Imunologia e Parasitologia, Escola Paulista de Medicina, Universidade Federal de São Paulo, São Paulo, Brazil

Edited by:
Amy Rasley,
Lawrence Livermore National Laboratory, USA

Reviewed by:
Hridayesh Prakash,
University of Hyderabad, India
Diego A. Vargas-Inchaustegui,
National Cancer Institute, USA

***Correspondence:**
Diana Bahia
dianabahia@hotmail.com

Leishmania spp. and *Trypanosoma cruzi* are the causative agents of leishmaniasis and Chagas disease, respectively, two neglected tropical diseases that affect about 25 million people worldwide. These parasites belong to the family Trypanosomatidae, and are both obligate intracellular parasites that manipulate host signaling pathways and the innate immune system to establish infection. Mitogen-activated protein kinases (MAPKs) are serine and threonine protein kinases that are highly conserved in eukaryotes, and are involved in signal transduction pathways that modulate physiological and pathophysiological cell responses. This mini-review highlights existing knowledge concerning the mechanisms that *Leishmania* spp. and *T. cruzi* have evolved to target the host's MAPK signaling pathways and highjack the immune response, and, in this manner, promote parasite maintenance in the host.

Keywords: MAP kinase, *Trypanosoma cruzi*, *Leishmania*, immune evasion of parasites, cellular signaling

INTRODUCTION

Leishmania spp. and *Trypanosoma cruzi* are protozoan parasites of the Trypanosomatida order (Kent, 1980) and Trypanosomatidae family (Doflein, 1901). They are the etiological agents of leishmaniasis and Chagas disease, respectively, and are transmitted by the bite of infected sandflies (leishmaniasis) or through triatomine bug feces (Chagas disease). Both *Leishmania* spp. and *T. cruzi* have complex life cycles comprising diverse developing forms that alternate between the insect vector and the vertebrate host. *Leishmania* spp. promastigotes and amastigotes preferentially infect phagocytic cells of vertebrates, while *T. cruzi* metacyclic trypomastigotes, blood trypomastigotes and amastigotes are able to infect both phagocytic and non-phagocytic cells (Tanowitz et al., 1992; Alexander et al., 1999; Ferreira et al., 2012).

Although the persistence of *Leishmania* spp. and *T. cruzi* within a host depends on several factors, the manipulation of host signal transduction pathways involved in the modulation of the immune response is probably one of the most commonly used mechanisms by parasites. In this mini-review, we will focus on the mechanisms that *Leishmania* spp. and *T. cruzi* use to subvert mitogen-activated protein kinase (MAPK) signaling pathways—more specifically, extracellular-signal-regulated kinase (ERK), and p38 MAPK—that are highly relevant in the context of the regulation of the immune response against intracellular parasites.

MAPK PATHWAYS

Mitogen-activated protein kinases are protein kinases that phosphorylate their own dual serine and threonine residues (autophosphorylation), or those found on their substrates, to activate or de-activate their target (Johnson and Lapadat, 2002; Peti and Page, 2013). Accordingly, MAPKs regulate important cellular processes such as proliferation, stress responses, apoptosis and immune defense (Dong et al., 2002; Liu et al., 2007; Arthur and Ley, 2013). MAPKs are ubiquitously expressed and evolutionarily conserved in eukaryotes (Kyriakis and Avruch, 2001; Kyriakis and Avruch, 2012; Peti and Page, 2013). The activation of a MAPK cascade occurs in a module of consecutive phosphorylations, i.e., after a previous stimulus, each MAPK is phosphorylated by an upstream MAPKs. A MAPK module comprises a MAP3K that activates a MAP2K, which then, in turn, activates a MAPK (Pimienta and Pascual, 2007; Turjanski, Vaqué and Gutkind, 2007; Johnson, 2011; Kyriakis and Avruch, 2012; Peti and Page, 2013). MAPK phosphorylation events can be inactivated by MAPK protein phosphatases (MKPs) that dephosphorylate both phosphothreonine and phosphotyrosine residues on MAPKs (Liu et al., 2007; Pimienta and Pascual, 2007; Zhang and Dong, 2007).

There are three well-known MAPK pathways in mammalian cells (**Figure 1**): the ERK1/2, the c-JUN N-terminal kinase 1, 2 and 3 (JNK1/2/3), and the p38 MAPK α, β, δ, and γ pathways. ERK, JNK, and p38 isoforms are grouped according to their activation motif, structure and function (Owens and Keyse, 2007; Raman et al., 2007; Zhang and Dong, 2007). ERK1/2 is activated in response to growth factors, hormones and proinflammatory stimuli, while JNK1/2/3 and p38 MAPK α, β, δ, and γ are activated by cellular and environmental stresses, in addition to proinflammatory stimuli (Owens and Keyse, 2007; Kyriakis and Avruch, 2012; **Figure 1**).

ERK AND P38 MAPK PATHWAYS

The classical activation of ERK1 and ERK2 isoforms is initiated by the binding of a ligand to a receptor tyrosine kinase (RTK) at the plasma membrane (PM), followed by activation of the small G-protein, Ras. In turn, Ras recruits and activates the serine/threonine protein kinase, Raf, a MAP3K, which activates the MAP2K, MEK, that, in turn, phosphorylates the MAPK, ERK1/2, at both threonine and tyrosine residues within the TEY motif (Kolch, 2000; Chambard et al., 2007; Shaul and Seger, 2007; Knight and Irving, 2014). The Ras/Raf/MEK/ERK1/2 pathway can be deactivated by dual-specificity MAPK phosphatases (MKPs). For example, MKP2/4 dephosphorylates ERK1/2, but can also deactivate other MAPKs while MKP3 and MKP-X are specific to ERK (Owens and Keyse, 2007). The tyrosine phosphatase, SHP2, also acts on this signaling pathway by activating the G-protein, Ras (Zhang et al., 2004; Matozaki et al., 2009; **Figure 1A**).

Both stress and cytokines activate p38 MAPK isoforms that play an important role in inflammatory responses (Johnson and Lapadat, 2002; Yang et al., 2014), despite each isoform being encoded by different genes and showing different tissue expression patterns (Cuadrado and Nebreda, 2010). As with ERK isoforms, p38 MAPKs are also sequentially activated. A canonical activation occurs when, in response to stress or cytokines, a MAP3K, such as MEKK1, ASK1, or TAK1, is activated by TRAF [TNF (tumor necrosis factor) receptor-associated factor] 2/3/6 or by Rho proteins. In turn, the MAP3K phosphorylates a MAP2K, either MKK3 or MKK6, that then phosphorylates the TGY motif of p38 isoforms (Cuenda and Rousseau, 2007; Cuadrado and Nebreda, 2010; **Figure 1B**). The p38 MAPKs α, β, δ and γ, are dephosphorylated by several dual-specificity protein phosphatases, such as MPK2/4, that can also deactivate ERK. The phosphatases, MPK5/7, can also dephosphorylate JNK and p38, while MPK1 exhibits a higher specificity for p38 (Owens and Keyse, 2007; Salojin and Oravecz, 2007).

MAPKs AND THE IMMUNE SYSTEM

Many pathogens target host intracellular signaling pathways, including MAPK pathways, to inhibit immune responses (Roy and Mocarski, 2007; Arthur and Ley, 2013). The immune response is one of several key functions regulated by MAPKs, with the production of immunomodulatory cytokines, such as TNFα, interleukin (IL)-1, IL-10, and IL-12, induced by the activation of p38 MAPK, JNK, and ERK pathways (Dong et al., 2002; Arthur and Ley, 2013). The regulatory cytokines, IL-12 and IL-10, produced by specialized dendritic and macrophage cells, play an important role in the coordination of the immune response since IL-12 modulates the development of a Th1 response, which protects the host against intracellular parasites, and IL-10 promotes a Th2 response, which provides protection against extracellular infectious agents (Romagnani, 2006).

The production of IL-12 is regulated by p38 MAPK and, consequently, is involved in the induction of a Th1 response (Jackson et al., 2010; Arthur and Ley, 2013). Upon p38 MAPK activation and IL-12 production, Th cell differentiation is guided into a Th1-type cell that releases pro-inflammatory cytokines such as IL-2, IFN-γ, and TNF-α/β (Romagnani, 2006). Such cytokines mediate the immune response by acting to kill intracellular pathogens such as the protozoans, *Leishmania* spp. and *T. cruzi* (Beiting, 2014).

Conversely, the ERK1/2 pathway modulates the production of IL-10 (Chang et al., 2012) that induces Th cell differentiation into a Th2-type. In this manner, Th2 cells regulate the host's humoral immune response by releasing anti-inflammatory cytokines, such as IL-4, IL-5, IL-9, and IL-13, that are involved in allergic reactions and the elimination of extracellular pathogens (Mosmann et al., 2009). Furthermore, IL-10 can also act as a negative regulator of inflammation to prevent tissue damage (Haddad et al., 2003; Romagnani, 2006; Guilliams et al., 2009; Bosschaerts et al., 2010).

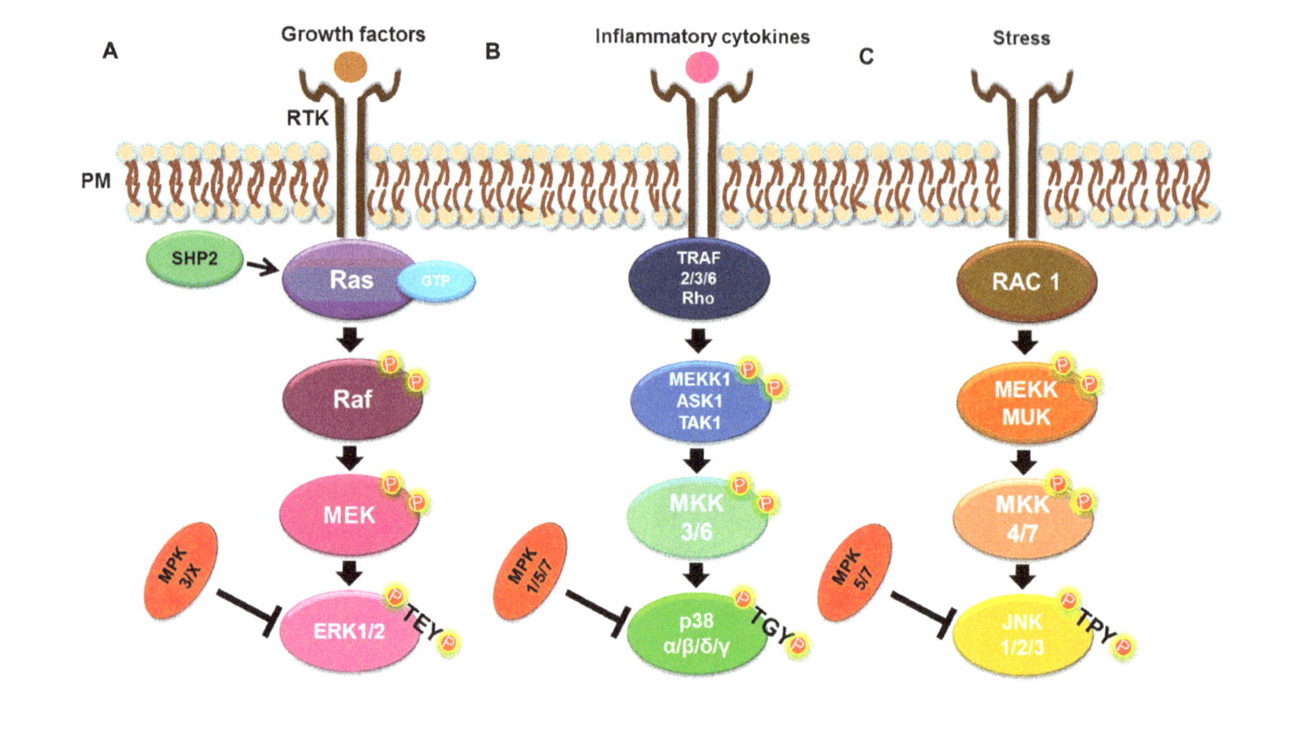

FIGURE 1 | Simplified MAPK signaling pathways. (A) ERK1/2 pathway. **(B)** p38 α, β, δ, and γ pathways. **(C)** JNK 1, 2, and 3 pathways. See text for details.

HOST IMMUNE RESPONSE SUBVERSION BY PROTOZOAN PARASITES

Protozoan parasites use many varied tactics to avoid and/or subvert the host's immune response, with the adoption of an intracellular lifestyle one the first mechanisms (Sacks and Sher, 2002). Other successful strategies employed by such pathogens to circumvent the immune system include: the expression of specific parasite antigens on the surface of infected cells to prevent recognition by immune cells (Sacks and Sher, 2002), the subversion of T cell responses by interfering with cytokine production (Engwerda et al., 2014), and the avoidance of direct killing by the complement system (Ouaissi and Ouaissi, 2005).

To protect an organism against protozoan intracellular infection, its immune system needs to identify and eliminate the parasite, but, at the same time, also needs to be balanced in order to minimize or avoid self-inflicted tissue damage (Dent, 2002). Phagocytosis is the primary mechanism used by the immune system in its response to intracellular parasites. This mechanism, mainly promoted by macrophages, depends on parasite recognition and is enhanced by opsonization and the complement system (Stafford et al., 2002). *Leishmania* spp. and *T. cruzi* escape from the host's complement system using different strategies. *T. cruzi* produces glycoproteins that inhibit C3b/C4b (gp160), factor B (gp58) and molecules that accelerate the decay of the C3 pathway. In contrast, *Leishmania* spp. adapt their membrane to prevent the insertion of C5b-C9. This species also produces a surface metalloproteinase, gp63, and a lipophosphoglycan (LPG) that cleave C3b, abrogating complement-mediated lysis (Sacks and Sher, 2002; Ouaissi and Ouaissi, 2005). *Leishmania* gp63 can also activate the complement system, leading to parasite opsonization and increased uptake by host macrophages in a highly advantageous mechanism that allows both forms of this parasite (promastigote and amastigote) to replicate inside host macrophages (Stafford et al., 2002; Gómez and Olivier, 2010).

Although phagocytosis is the first mechanism activated in immune protection, the induction of a Th1-type response is the most effective reaction against intracellular protozoans. The promotion of such an inflammatory response leads to the successful elimination of these parasites due to the release and intense activity of pro-inflammatory cytokines and mediators (Jankovic et al., 2001; Stafford et al., 2002; Guilliams et al., 2009; Bosschaerts et al., 2010; Beiting, 2014; Engwerda et al., 2014). As outlined previously, the generation of Th1 and Th2 responses is regulated by IL-12 and IL-10 cytokines, respectively, which are modulated by the host's intracellular MAPK signaling pathways (Dong et al., 2002; Arthur and Ley, 2013). So, by acting on the MAPK signaling pathway, intracellular protozoan parasites can switch the production of regulatory cytokines from IL-12 to IL-10 to prevent the formation of an inflammatory response (Jankovic et al., 2001; Stafford et al., 2002; Zambrano-Villa et al., 2002; Guilliams et al., 2009). In this way, *T. cruzi* and *Leishmania* spp. impair the development of a Th1 response to favor a Th2-type response. The mechanisms used by these parasites to deregulate the immune response will be briefly reviewed below.

Leishmania spp. SUBVERT THE MAPK PATHWAY BY ACTIVATING ERK1/2 TO INCREASE IL-10 AND DOWN-REGULATE IL-12

Leishmania spp. parasites use several strategies to survive inside host cells after infection. The methods employed by these parasites to subvert the host's immune defense systems include: (i) an intracellular stage in their life cycle, allowing them protection against humoral anti-leishmanicidal products; (ii) the suppression of the synthesis of reactive oxygen intermediates (ROI) or reactive nitrogen intermediates (RNI); (iii) the inhibition of antigen presentation by repressing the gene expression of major histocompatibility complex (MHC) class II, interfering with antigen loading, or by sequestering/highjacking the MHC II molecule or antigen; (iv) the subversion of host cellular signaling pathways such as STAT, PI3K/AKT, and MAPK; and (v) the modulation of host cytokines to avoid T cell differentiation and thus prevent the formation of a Th1-type immune response (Bogdan and Röllinghoff, 1998; Olivier et al., 2005; Liese et al., 2008; Martinez and Petersen, 2014).

To undermine the production of regulatory inflammatory cytokines and prevent the formation of a Th1-type immune response, *Leishmania* parasites target the MAPK signaling pathway, which is responsible for regulating the production of IL-12 (p38) and IL-10 (ERK 1/2) in macrophages and dendritic cells. By this means, *Leishmania* spp. promote switching from IL-12 to IL-10 production, consequently altering the formation of a Th1 response to Th2, and leading to parasite prevalence in the host (Ghalib et al., 1995; Olivier et al., 2005; Bhardwaj et al., 2010; Shadab and Ali, 2011).

Rub et al. (2009) have shown that *Leishmania* spp. often act upon the CD40/MAPK pathway. Expressed mainly on macrophages and dendritic cells, CD40 is an important co-stimulatory molecule involved in the differentiation of Th cells to a Th1-type, reflecting how CD40 induces the production of IL-12 (O'Sullivan and Thomas, 2003) by the activation of MAPK pathway members (Bhardwaj et al., 2010). In their study, Rub et al. (2009) showed that *L. major* triggered cholesterol depletion, and, in doing so, prevented CD40 reallocation to skew the CD40 signaling pathway from p38 and IL-12 to ERK1/2 and IL-10 production. As a consequence, this event led to an increased IL-10 production. Taken together, these events favored a parasite burden and confirmed that the CD40/MAPK pathway was important for *L. major* subversion of the host's immune response. However, the specific mechanism surrounding augmented IL-10 production remained unknown until Srivastava et al. (2011) and Khan et al. (2014) noted the participation of phosphatases in this process. Srivastava et al. (2011) demonstrated that in *L. major* infection, the phosphatases, MKP-1 and MKP-3, were differentially expressed. *L. major* induced the upregulation of MKP-1 (p38 high affinity phosphatase) and downregulation of MKP-3 (ERK1/2 high affinity phosphatase) to skew CD40 signaling toward the ERK1/2 pathway, favoring infection. Khan et al. (2014) demonstrated that the phosphotyrosine phosphatase, SHP-1, functioned in a similar manner and acted on Syk and Lyn (ERK1/2 and p38 MAPK activators, respectively). In a CD40 dose-dependent manner, SHP-1 modulates CD40-induced phosphorylation of p38 MAPK and ERK1/2 to favor ERK1/2-dependent IL-10 expression and parasite survival (**Figure 2**). Xin et al. (2008) and Boggiatto et al. (2009, 2014) also demonstrated the importance of the CD40/MAPK signaling pathway in *L. amazonensis* infection. *L. amazonensis* upregulated ERK1/2 in dendritic cells, increased IL-10 production and prevented the expression of CD40 and IL-12p40 (one of the subunits of IL-12), leading to the limited activation of dendritic cells and a deficient Th1-type response.

Extracellular-signal-regulated kinase 1/2 activation by *Leishmania* spp., associated with parasite persistence and IL-12 down-regulation, was also seen for *L. amazonensis* (Xin et al., 2008; Martinez and Petersen, 2014) and for *L. donovani*. Infection by *L. donovani* also led to the suppression of p38 MAPK and increased IL-10 production (Chandra and Naik, 2008; Shadab and Ali, 2011). Interestingly, treatment of *L. donovani*-infected macrophages with an immunoprophylactic glycolipid, arabinosylated lipoarabinomannan (Ara-LAM) isolated from *Mycobacterium smegmatis*, activated p38 MAPK with a concomitant abrogation of ERK1/2 phosphorylation. Furthermore, the production of IL-12 (Bhattacharya et al., 2011) and IFNγ responsiveness were restored (Chowdhury et al., 2015). Exploitation of host phosphatases can also be seen in *L. donovani* infection: Nandan and Reiner (2005) showed that *L. donovani* activates the phosphotyrosine phosphatase, SHP-1 that correlates with parasite survival.

Gp63 and LPG can also modulate the host MAPK signaling pathway and the production of cytokines. Gp63 acts on both p38 and the phosphotyrosine phosphatase SHP-1. To subvert MAPK signaling, gp63 of *L. major* promastigotes leads to a p38 inhibition in fibroblasts by mediating the proteolysis of TAK-1-binding protein-1 (TAB1), a p38 regulator (Halle et al., 2009). Gp63 of both *L. major* and *L. mexicana* amastigotes and promastigotes activates SHP-1 in macrophages in a cleavage-dependent manner leading to p38 down-regulation (Gomez et al., 2009; Abu-Dayyeh et al., 2010). Gp63-mediated inactivation of p38 could consequently inhibit IL-12 production. Conversely, the LPG of *L. braziliensis* and *L. infantum* also activates ERK1/2, but abrogates not only IL-12 production, but also that of IL-10 (Ibraim et al., 2013). Further studies are needed to understand the mechanisms by which LPG modulates MAPK and IL-12 production in an IL-10-independent manner. *L. mexicana* LPG can also manipulate the MAPK pathway and inhibit IL-12 but this seems to be due to the impairment of NFκB translocation caused by LPG (Cameron et al., 2004; Argueta-Donohué et al., 2008; **Table 1A**).

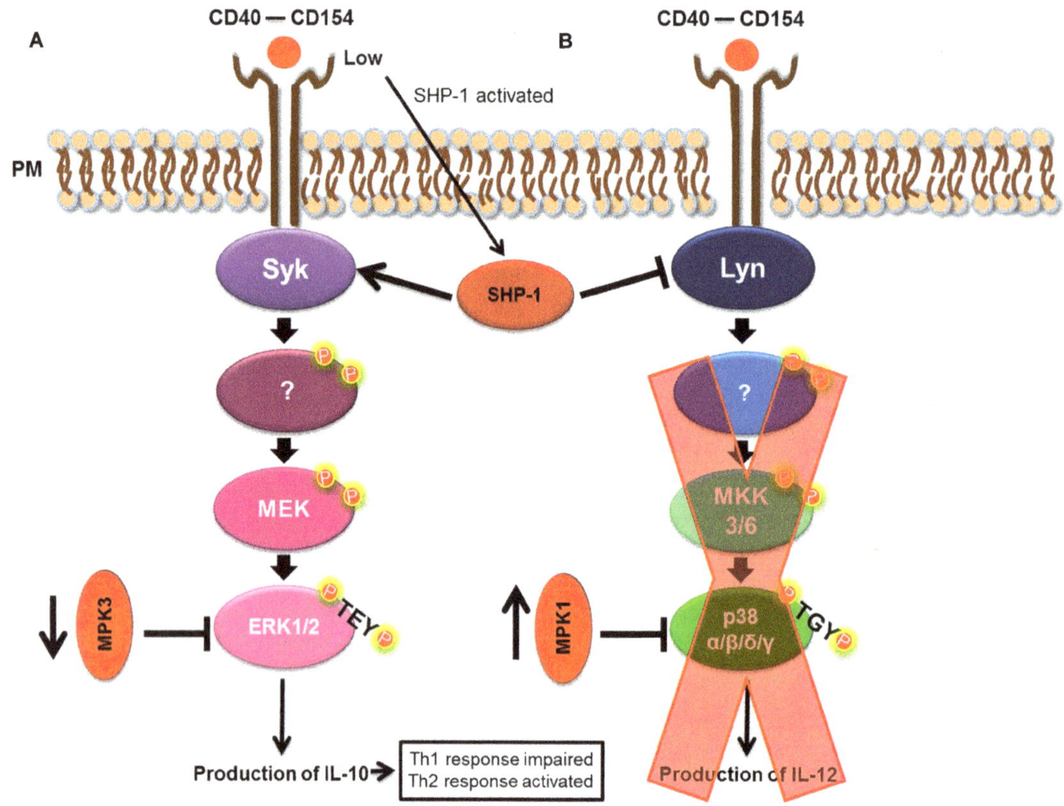

FIGURE 2 | *Leishmania* spp. model of immune modulation targeting the p38/ERK1/2 MAPK pathway. (A) CD40/ERK1/2 pathway. **(B)** CD40/p38 MAPK pathway. *Leishmania* spp. skew the CD40/MAPK signaling pathway, from p38 to ERK1/2, to favor IL-10 production.

Trypanosoma cruzi TRIGGERS MOLECULES TO REGULATE THE MAPK PATHWAY AND CYTOKINE PRODUCTION

The mechanisms by which *T. cruzi* evades the host's immune response by acting on the MAPK pathway have been poorly studied and are therefore incompletely understood. However, it is well established that upon cell invasion, *T. cruzi* begins to subvert signaling pathways and to use host molecules to favor its entry and survival inside host cells. For instance, *T. cruzi* extracellular amastigotes (EAs) recruit both host protein kinase D1 (PKD1) and cortactin to induce PKD1 autophosphorylation and cortactin activation by ERK, leading to the recruitment of host actin that allows parasite entry into HeLa cells (Bonfim-Melo et al., 2015). *T. cruzi*, as well as *Leishmania* spp., is able to induce: (i) ERK1/2, but not p38 MAPK, activation in macrophages and dendritic cells (Mukherjee et al., 2004); and (ii) increased IL-10 and decreased IL-12 production (Poncini et al., 2008). These effects impair the formation of an efficient Th1 inflammatory response (Alba Soto et al., 2003, 2010) to allow parasite evasion of the host immune response.

Some *T. cruzi* molecules are released and activate Toll-like receptors (TLRs), such as TLR2, TLR4, or TLR9, in dendritic cells and macrophages (Tarleton, 2007). This leads to the activation of p38 MAPK and the production of IL-12, favoring an inflammatory response (Ropert et al., 2001; Terrazas et al., 2011). This model is supported by the observations of Terrazas et al. (2011) who showed that dendritic cells exposed to *T. cruzi* antigens (TcAgs) and TLR ligands induced p38 phosphorylation that was dependent on TcAg-macrophage migration inhibitory factor (MIF) synergism. This led to the enhancement of IL-12 production, thus promoting a Th1-type response (Terrazas et al., 2011).

However, despite the activation of a pro-inflammatory immune response by some parasite molecules, it is well known that several other molecules of such pathogens act against the host's infected cells and signaling pathways, subverting the host's immune response against the parasite (Ouaissi et al., 1995; Hovsepian et al., 2011; Castillo et al., 2013; Ruiz Díaz et al., 2015).

One of the proteins released by *T. cruzi* that disrupts the host's immune response is Tc52. A protein of 52 kDa, Tc52 is composed of two homologous domains sharing significant homology with glutathione *S*-transferases (Schöneck et al., 1994), and exhibits both immunomodulatory and virulence roles (Ouaissi et al., 1995, 2002a). When localized in the cytoplasm, a 28 kDa peptide fragment derived from the C-terminal portion of Tc52 (Borges et al., 2003) induces the Tc52-mediated suppression of T cell proliferation, and exerts mitogen-dependent cytokine and chemokine-like activities. Thus, this peptide modulates genes that

TABLE 1 | Trypanosomatids-released proteins and their action on MAPK pathway in macrophages and T cells.

Protein	MAPK Target on macrophages	Mechanism	Target on T cells	Reference
(A) *Leishmania* spp. proteins				
gp63	p38 MAPK	Cleavage-dependent activation of SHP-1 leading to p38 inactivation and, presumably, inhibition of IL-12	IL-12	Gomez et al., 2009; Abu-Dayyeh et al., 2010
LPG	ERK	Activation of ERK abrogating both IL-10 and IL-12 production	IL-10 and IL-12	Feng et al., 1999; Ibraim et al., 2013
(B) *Trypanosoma cruzi* proteins				
Protein	**MAPK Target on macrophages**	**Mechanism**	**Target on T cells**	**Reference**
Tc52	p38 or ERK	Mitogen-dependent modulation of genes that encode IL-10 and IL-12 leading to increased IL-10 secretion and inhibition of IL-12	IL-10 and IL-12	Ouaissi et al., 2002b; Borges et al., 2003; Ouaissi and Ouaissi, 2005
AgC10	p38	Inhibition of p38 and IL-12	IL-12	De Diego et al., 1997; Alcaide and Fresno, 2004
GPIs and GPI-anchored mucins	ERK	Activation of ERK1/2 upon treatment with GPIs and GPI-anchored mucins associated to IL-12 decrease	IL-12	Ropert et al., 2001
TS	ERK	TS activates ERK and stimulates IL-10 secretion	IL-10	Chuenkova and Pereira, 2001; Ruiz Díaz et al., 2015

encode IL-10 and IL-12, leading to increased IL-10 secretion and thereby inhibiting IL-12 (Ouaissi et al., 2002b; Ouaissi and Ouaissi, 2005). Moreover, the events outlined above are probably mediated by MAPKs.

Other proteins related to immune modulation in *T. cruzi* infection are the glycosylphosphatidylinositol (GPI)-anchored mucins and *trans*-sialidases (TS). AgC10, a GPI-anchored mucin of 40–50 kDa, inhibits TNF and IL-12 secretion in a p38 MAPK inhibition-dependent manner, impairing the formation of a Th1 response (De Diego et al., 1997; Alcaide and Fresno, 2004). Ropert et al. (2001) reported that ERK1/2 activation was associated with a decrease in IL-12 in macrophages treated with *T. cruzi* GPI and GPI-mucins, corroborating the participation of these proteins in modulating the host's immune response. However, they also showed that GPIs and GPI-anchored mucins could activate p38 MAPK later than ERK1/2, thus increasing IL-12 synthesis and generating an opposing effect in the regulation of the immune response by promoting a Th1 response (Ropert et al., 2001). *T. cruzi* TS have been linked to ERK1/2 activation (Chuenkova and Pereira, 2001). Recently, Ruiz Díaz et al. (2015) confirmed the role of TS in IL-10-stimulated secretion, leading to an imbalance of the Th1 cell response toward an Th2 phenotype (**Table 1B**). However, despite current knowledge concerning the strategies used by *T. cruzi* to subvert the host's immune response, the precise mechanisms by which this occurs remain unknown.

CONCLUSION

The MAPK signaling pathway, responsible for regulating the production of Th1-and Th2-type responses, is targeted by trypanosomatids to modulate the host's immune response in order to favor parasite replication and survival. The mechanisms whereby *Leishmania* spp. skew the MAPK signaling pathway to subvert cytokine production and switch a Th1 to a Th2 response are well known compared to those for *T. cruzi*. *Leishmania* parasites often target the CD40/MAPK pathway, activating ERK1/2 to increase and decrease IL-10 and IL-12 production, respectively. In comparison, the strategy used by *T. cruzi* is to trigger molecules to subvert MAPK ERK1/2 and p38 pathways and thus modulate cytokine production. Further studies are required to increase our understanding of the intriguing mechanisms by which *T. cruzi* manipulates the host's immune response.

AUTHOR CONTRIBUTIONS

MS-S, FD and GG contributed equally to the writing of the review. DB conceived and wrote the review.

ACKNOWLEDGMENTS

The authors are indebted to Prof. Dr. Erika Suzuki Toledo (Escola Paulista de Medicina, Universidade Federal de São Paulo) for her critical revision of the manuscript and valuable comments. DB is the recipient of a CNPq fellowship. MS is the recipient of a FAPEMIG fellowship. The authors wish to gratefully acknowledge the financial support of FAPEMIG (APQ-01674-14) and FAPESP (07/50551-2). The authors would also like to thank BioMed Proofreading (http://www.biomedproofreading.com/).

REFERENCES

Abu-Dayyeh, I., Hassani, K., Westra, E. R., Mottram, J. C., and Olivier, M. (2010). Comparative study of the ability of *Leishmania mexicana Promastigotes* and amastigotes to alter macrophage signaling and functions. *Infect. Immun.* 78, 2438–2445. doi: 10.1128/IAI 00812–819,

Alba Soto, C. D., Mirkin, G. A., Solana, M. E., and González Cappa, S. M. (2003). *Trypanosoma cruzi* infection modulates in vivo expression of major histocompatibility complex class II molecules on antigen-presenting cells and T-cell stimulatory activity of dendritic cells in a strain-dependent manner. *Infect. Immun.* 71, 1194–1199. doi: 10.1128/IAI.71.3.1194-1199.2003

Alba Soto, C. D., Solana, M. E., Poncini, C. V., Pino-Martinez, A. M., Tekiel, V., and González-Cappa, S. M. (2010). Dendritic cells devoid of IL-10 induce protective immunity against the protozoan parasite *Trypanosoma cruzi. Vaccine* 28, 7407–7413. doi: 10.1016/j.vaccine.2010.08.105

Alcaide, P., and Fresno, M. (2004). AgC10, a mucin from *Trypanosoma cruzi*, destabilizes TNF and cyclooxygenase-2 mRNA by inhibiting mitogen-activated protein kinase p38. *Eur. J. Immunol.* 34, 1695–1704. doi: 10.1002/eji.200324660

Alexander, J., Satoskar, A. R., and Russell, D. G. (1999). Leishmania species: models of intracellular parasitism. *J. Cell Sci.* 112, 2993.

Argueta-Donohué, J., Carrillo, N., Valdés-Reyes, L., Zentella, A., Aguirre-García, M., Becker, I., et al. (2008). *Leishmania mexicana*: participation of NF-κB in the differential production of IL-12 in dendritic cells and monocytes induced by lipophosphoglycan (LPG). *Exp. Parasitol.* 120, 1–9. doi: 10.1016/j.exppara.2008.04.002

Arthur, J. S. C., and Ley, S. C. (2013). Mitogen-activated protein kinases in innate immunity. *Nat. Rev. Immunol.* 13, 679–692. doi: 10.1038/nri3495

Beiting, D. P. (2014). Protozoan parasites and type I interferons: a cold case reopened. *Trends Parasitol.* 30, 491–498. doi: 10.1016/j.pt.2014.07.007

Bhardwaj, S., Srivastava, N., Sudan, R., and Saha, B. (2010). *Leishmania interferes* with host cell signaling to devise a survival strategy. *J. Biomed. Biotechnol.* 2010, 1–13. doi: 10.1155/2010/109189

Bhattacharya, P., Gupta, G., Majumder, S., Adhikari, A., Banerjee, S., Halder, K., et al. (2011). Arabinosylated lipoarabinomannan skews Th2 phenotype towards thl during leishmania infection by chromatin modification: involvement of MAPK signaling. *PLoS ONE* 6:e24141. doi: 10.1371/journal.pone.0024141

Bogdan, C., and Röllinghoff, M. (1998). The immune response to Leishmania: mechanisms of parasite control and evasion. *Int. J. Parasitol.* 28, 121–134. doi: 10.1016/S0020-7519(97)00169-0

Boggiatto, P. M., Jie, F., Ghosh, M., Gibson-Corley, K. N., Ramer-Tait, A. E., Jones, D. E., et al. (2009). Altered dendritic cell phenotype in response to leishmania amazonensis amastigote infection is mediated by MAP kinase. *ERK. Am. J. Pathol.* 174, 1818–1826. doi: 10.2353/ajpath.2009.080905

Boggiatto, P. M., Martinez, P. A., Pullikuth, A., Jones, D. E., Bellaire, B., Catling, A., et al. (2014). Targeted extracellular signal-regulated kinase activation mediated by *Leishmania amazonensis* requires MP1 scaffold. *Microbes Infect.* 16, 328–336. doi: 10.1016/j.micinf.2013.12.006

Bonfim-Melo, A., Zanetti, B. F., Ferreira, É. R., Vandoninck, S., Han, S. W., Van Lint, J., et al. (2015). *T rypanosoma cruzi* extracellular amastigotes trigger the protein kinase D1-cortactin-actin pathway during cell invasion. *Cell. Microbiol.* 17, 1797–1810. doi: 10.1111/cmi.12472

Borges, M., Da Silva, A. C., Sereno, D., and Ouaissi, A. (2003). Peptide-based analysis of the amino acid sequence important to the immunoregulatory function of *Trypanosoma cruzi* Tc52 virulence factor. *Immunology* 109, 147–155. doi: 10.1046/j.1365-2567.2003.01637.x

Bosschaerts, T., Guilliams, M., Stijlemans, B., Morias, Y., Engel, D., Tacke, F., et al. (2010). Tip-DC development during parasitic infection is regulated by IL-10 and requires CCL2/CCR2, IFN-γ and MyD88 signaling. *PLoS Pathog.* 6:e1001045. doi: 10.1371/journal.ppat.1001045

Cameron, P., McGachy, A., Anderson, M., Paul, A., Coombs, G. H., Mottram, J. C., et al. (2004). Inhibition of lipopolysaccharide-induced macrophage IL-12 production by *Leishmania mexicana* amastigotes: the role of cysteine peptidases and the NF-kappaB signaling pathway. *J. Immunol.* 173, 3297–3304.

Castillo, C., Villarroel, A., Duaso, J., Galanti, N., Cabrera, G., Maya, J. D., et al. (2013). Phospholipase C gamma and ERK1/2 Mitogen Activated Kinase Pathways are differentially modulated by *Trypanosoma cruzi* during tissue invasion in human placenta. *Exp. Parasitol.* 133, 12–17. doi: 10.1016/j.exppara.2012.10.012

Chambard, J. C., Lefloch, R., Pouysségur, J., and Lenormand, P. (2007). ERK implication in cell cycle regulation. *Biochim. Biophys. Acta* 1773, 1299–1310. doi: 10.1016/j.bbamcr.2006.11.010

Chandra, D., and Naik, S. (2008). *Leishmania donovani* infection down-regulates TLR2-stimulated IL-12p40 and activates IL-10 in cells of macrophage/monocytic lineage by modulating MAPK pathways through a contact-dependent mechanism. *Clin. Exp. Immunol.* 154, 224–234. doi: 10.1111/j.1365-2249.2008.03741.x

Chang, C.-F., D'Souza, W. N., Ch'en, I. L., Pages, G., Pouyssegur, J., and Hedrick, S. M. (2012). Polar opposites: Erk direction of CD4 T cell subsets. *J. Immunol.* 189, 721–731. doi: 10.4049/jimmunol.1103015

Chowdhury, B. P., Bandyopadhyay, S., Das, S., Majumder, S., Jha, M. K., Majumdar, S. B., et al. (2015). The host-protective effect of arabinosylated lipoarabinomannan against *Leishmania donovani* infection is associated with restoration of IFN-γ responsiveness. *PLoS ONE* 10:e0117247. doi: 10.1371/journal.pone.0117247

Chuenkova, M. V., and Pereira, M. A. (2001). The *T. cruzi* trans-sialidase induces PC12 cell differentiation via MAPK/ERK pathway. *Neuroreport* 4, 3715–3718. doi: 10.1097/00001756-200112040-00022

Cuadrado, A., and Nebreda, A. R. (2010). Mechanisms and functions of p38 MAPK signalling. *Biochem. J.* 429, 403–417. doi: 10.1042/BJ20100323

Cuenda, A., and Rousseau, S. (2007). p38 MAP-kinases pathway regulation, function and role in human diseases. *Biochim. Biophys. Acta* 1773, 1358–1375. doi: 10.1016/j.bbamcr.2007.03.010

De Diego, J., Punzón, C., Duarte, M., and Fresno, M. (1997). Alteration of macrophage function by a *Trypanosoma cruzi* membrane mucin. *J. Immunol.* 15, 4983–4989.

Dent, L. A. (2002). For better or worse: common determinants influencing health and disease in parasitic infections, asthma and reproductive biology. *J. Reproduct. Immunol.* 57, 255–272. doi: 10.1016/S0165-0378(02)00017-7

Dong, C., Davis, R. J., and Flavell, R. A. (2002). MAP kinases in the immune response. *Annu. Rev. Immunol.* 20, 55–72. doi: 10.1146/annurev.immunol.20.091301.131133

Engwerda, C. R., Ng, S. S., and Bunn, P. T. (2014). The regulation of CD^{4+} T Cell responses during protozoan infections. *Front. Immunol.* 5:498. doi: 10.3389/fimmu.2014.00498

Feng, G., Goodridge, H. S., Harnett, M. M., Wei, X., Nikolaev, A. V., Higson, A. P., et al. (1999). Extracellular signal-related kinase (ERK) and p38 mitogen-activated protein (MAP) kinases differentially regulate the lipopolysaccharide-mediated induction of inducible nitric oxide synthase and IL-12 in macrophages: Leishmania phosphoglycans subvert macrophage IL-12 production by targeting ERK MAP kinase. *J. Immunol.* 163, 6403–6412.

Ferreira, É. R., Bonfim-Melo, A., Mortara, R. A., and Bahia, D. (2012). *Trypanosoma cruzi* extracellular amastigotes and host cell signaling: more pieces to the puzzle. *Front. Immunol.* 3:3. doi: 10.3389/fimmu.2012.00363

Ghalib, H. W., Whittle, J. A., Kubin, M., Hashim, F. A., El-Hassan, A. M., Grabstein, K. H., et al. (1995). IL-12 enhances Th1-type responses in human *Leishmania donovani* infections. *J. Immunol.* 154, 4623–4629.

Gomez, M. A., Contreras, I., Hallé, M., Tremblay, M. L., Mcmaster, R. W., and Olivier, M. (2009). Leishmania GP63 alters host signaling through cleavage-activated protein tyrosine phosphatases. *Sci Signal* 29, ra582. doi: 10.1126/scisignal.2000213

Gómez, M. A., and Olivier, M. (2010). Proteases and phosphatases during Leishmania-macrophage interaction: paving the road for pathogenesis. *Virulence* 1, 314–318. doi: 10.4161/viru.1.4.12194

Guilliams, M., Movahedi, K., Bosschaerts, T., VandenDriessche, T., Chuah, M. K., Herin, M., et al. (2009). IL-10 Dampens TNF/Inducible nitric oxide synthase-producing dendritic cell-mediated pathogenicity during parasitic infection. *J. Immunol.* 182, 1107–1118. doi: 10.4049/jimmunol.182.2.1107

Haddad, J. J., Saadé, N. E., and Safieh-Garabedian, B. (2003). Interleukin-10 and the regulation of mitogen-activated protein kinases: ae these signalling modules targets for the anti-inflammatory action of this cytokine? *Cell. Signal.* 15, 255–267. doi: 10.1016/S0898-6568(02)00075-X

Halle, M., Gomez, M. A., Stuible, M., Shimizu, H., McMaster, W. R., Olivier, M., et al. (2009). The leishmania surface protease GP63 cleaves multiple intracellular proteins and actively participates in p38 mitogen-activated protein kinase inactivation. *J. Biol. Chem.* 284, 6893–6908. doi: 10.1074/jbc.M805861200

Hovsepian, E., Mirkin, G. A., Penas, F., Manzano, A., Bartrons, R., and Goren, N. B. (2011). Modulation of inflammatory response and parasitism by 15-Deoxy-Δ12,14 prostaglandin J2 in *Trypanosoma cruzi*-infected cardiomyocytes. *Int. J. Parasitol.* 41, 553–562. doi: 10.1016/j.ijpara.2010.12.002

Ibraim, I. C., de Assis, R. R., Pessoa, N. L., Campos, M. A., Melo, M. N., Turco, S. J., et al. (2013). Two biochemically distinct lipophosphoglycans from *Leishmania braziliensis* and *Leishmania infantum* trigger different innate immune responses in murine macrophages. *Parasit. Vect.* 6, 54. doi: 10.1186/1756-3305-6-54

Jackson, A. M., Mulcahy, L. A., Porte, J., Franks, H. A., Refaee, M., El Wang, Q., et al. (2010). Role of mitogen-activated protein kinase and PI3K pathways in the regulation of IL-12-family cytokines in dendritic cells and the generation of T H-responses. *Eur. Cytok. Netw.* 21, 319–328. doi: 10.1684/ecn.2010.0219

Jankovic, D., Sher, A., and Yap, G. (2001). Th1/Th2 effector choice in parasitic infection: decision making by committee. *Curr. Opin. Immunol.* 13, 403–409. doi: 10.1016/S0952-7915(00)00234-X

Johnson, G. L. (2011). Defining MAPK interactomes. *ACS Chem. Biol.* 6, 18–20. doi: 10.1021/cb100384z

Johnson, G. L., and Lapadat, R. (2002). Mitogen-activated protein kinase pathways mediated by ERK, JNK, and p38 protein kinases. *Science* 298, 1911–1912. doi: 10.1126/science.1072682

Khan, T. H., Srivastava, N., Srivastava, A., Sareen, A., Mathur, R. K., Chande, A. G., et al. (2014). SHP-1 plays a crucial role in cd40 signaling reciprocity. *J. Immunol.* 193, 3644–3653. doi: 10.4049/jimmunol.1400620

Knight, T., and Irving, J. A. E. (2014). Ras/Raf/MEK/ERK Pathway activation in childhood acute lymphoblastic leukemia and its therapeutic targeting. *Front. Oncol.* 4:160. doi: 10.3389/fonc.2014.00160

Kolch, W. (2000). Meaningful relationships: the regulation of the Ras/Raf/MEK/ERK pathway by protein interactions. *Biochem. J.* 351(Pt 2), 289–305. doi: 10.1042/bj3510289

Kyriakis, J. M., and Avruch, J. (2001). Mammalian mitogen-activated protein kinase signal transduction pathways activated by stress and inflammation. *Physiol. Rev.* 81, 807–869.

Kyriakis, J. M., and Avruch, J. (2012). Mammalian MAPK signal transduction pathways activated by stress and inflammation: A 10-Year Update. *Physiol. Rev.* 92, 689–737. doi: 10.1152/physrev.00028.2011

Liese, J., Schleicher, U., and Bogdan, C. (2008). The innate immune response against *Leishmania parasites*. *Immunobiology* 213, 377–387. doi: 10.1016/j.imbio.2007.12.005

Liu, Y., Shepherd, E. G., and Nelin, L. D. (2007). MAPK phosphatases–regulating the immune response. *Nat. Rev. Immunol.* 7, 202–212. doi: 10.1038/nri2035

Martinez, P. A., and Petersen, C. A. (2014). Chronic infection by *Leishmania amazonensis* mediated through MAPK ERK mechanisms. *Immunol. Res.* 59, 153–165. doi: 10.1007/s12026-014-8535-y

Matozaki, T., Murata, Y., Saito, Y., Okazawa, H., and Ohnishi, H. (2009). Protein tyrosine phosphatase SHP-2: a roto-oncogene product that promotes Ras activation. *Cancer Sci.* 100, 1786–1793. doi: 10.1111/j.1349-7006.2009.01257.x

Mosmann, T. R., Kobie, J. J., Lee, F. E.-H., and Quataert, S. A. (2009). T helper cytokine patterns: defined subsets, random expression, and external modulation. *Immunol. Res.* 45, 173–184. doi: 10.1007/s12026-009-8098-5

Mukherjee, S., Huang, H., Petkova, S. B., Albanese, C., Pestell, R. G., Braunstein, V. L., et al. (2004). *Trypanosoma cruzi* infection activates extracellular signal-regulated kinase in cultured endothelial and smooth muscle cells. *Society* 72, 5274–5282.

Nandan, D., and Reiner, N. E. (2005). *Leishmania donovani* engages in regulatory interference by targeting macrophage protein tyrosine phosphatase SHP-1. *Clin. Immunol.* 114, 266–277. doi: 10.1016/j.clim.2004.07.017

Olivier, M., Gregory, D. J., and Forget, G. (2005). Subversion mechanisms by which leishmania parasites can escape the host immune response: a signaling point of view. *Clin. Microbiol. Rev.* 18, 293–305. doi: 10.1128/CMR.18.2.293-305.2005

O'Sullivan, B., and Thomas, R. (2003). CD40 and dendritic cell function. *Crit. Rev. Immunol.* 23, 83–107. doi: 10.1615/CritRevImmunol.v23.i12.50

Ouaissi, A., Guevara-Espinoza, A., Chabe, F., Gomez-Corvera, R., and Taibi, A. (1995). A novel and basic mechanism of immunosuppression in Chagas' disease: *Trypanosoma cruzi* releases in vitro and in vivo a protein which induces T cell unresponsiveness through specific interaction with cysteine and glutathione. *Immunol. Lett.* 48, 221–224. doi: 10.1016/0165-2478(95)02463-8

Ouaissi, A., Guilvard, E., Delneste, Y., Caron, G., Magistrelli, G., Herbault, N., et al. (2002a). The *Trypanosoma cruzi* Tc52-released protein induces human dendritic cell maturation, signals via toll-like receptor 2, and confers protection against lethal infection. *J. Immunol.* 168, 6366–6374. doi: 10.4049/jimmunol.168.12.6366

Ouaissi, A., Ouaissi, M., and Sereno, D. (2002b). Glutathione S-transferases and related proteins from pathogenic human parasites behave as immunomodulatory factors. *Immunol. Lett.* 81, 159–164. doi: 10.1016/S0165-2478(02)00035-4

Ouaissi, A., and Ouaissi, M. (2005). Molecular basis of *Trypanosoma cruzi* and *Leishmania interaction* with their host(s): exploitation of immune and defense mechanisms by the parasite leading to persistence and chronicity, features reminiscent of immune system evasion strategies in cancer dis. *Arch. Immunol. Ther. Exp.* 53, 102–114.

Owens, D. M., and Keyse, S. M. (2007). Differential regulation of MAP kinase signalling by dual-specificity protein phosphatases. *Oncogene* 26, 3203–3213. doi: 10.1038/sj.onc.1210412

Peti, W., and Page, R. (2013). Molecular basis of MAP kinase regulation. *Protein Sci.* 22, 1698–1710. doi: 10.1002/pro.2374

Pimienta, G., and Pascual, J. (2007). Canonical and alternative MAPK signaling. *Cell Cycle* 6, 2628–2632. doi: 10.4161/cc.6.21.4930

Poncini, C. V., Soto, C. D. A., Batalla, E., Solana, M. E., and Gonzalez Cappa, S. M. (2008). *Trypanosoma cruzi* induces regulatory dendritic cells in vitro. *Infect. Immun.* 76, 2633–2641. doi: 10.1128/IAI.01298-07

Raman, M., Chen, W., and Cobb, M. H. (2007). Differential regulation and properties of MAPKs. *Oncogene* 26, 3100–3112. doi: 10.1038/sj.onc.1210392

Romagnani, S. (2006). Regulation of the T cell response. *Clin. Exp. Allergy* 36, 1357–1366. doi: 10.1111/j.1365-2222.2006.02606.x

Ropert, C., Almeida, I. C., Closel, M., Travassos, L. R., Ferguson, M. A. J., Cohen, P., et al. (2001). Requirement of mitogen-activated protein kinases and I B phosphorylation for induction of proinflammatory cytokines synthesis by macrophages indicates functional similarity of receptors triggered by glycosylphosphatidylinositol anchors from parasitic prot. *J. Immunol.* 166, 3423–3431. doi: 10.4049/jimmunol.166.5.3423

Roy, C. R., and Mocarski, E. S. (2007). Pathogen subversion of cell-intrinsic innate immunity. *Nat. Immunol.* 8, 1179–1187. doi: 10.1038/ni1528

Rub, A., Dey, R., Jadhav, M., Kamat, R., Chakkaramakkil, S., Majumdar, S., et al. (2009). Cholesterol depletion associated with Leishmania major infection alters macrophage CD40 signalosome composition and effector function. *Nat. Immunol.* 10, 273–280. doi: 10.1038/ni.1705

Ruiz Díaz, P., Mucci, J., Meira, M. A., Bogliotti, Y., Musikant, D., Leguizamón, M. S., et al. (2015). *Trypanosoma cruzi* trans -sialidase prevents elicitation of th1 cell response via interleukin 10 and downregulates th1 effector cells. *Infect. Immun.* 83, 2099–2108. doi: 10.1128/IAI.00031-15

Sacks, D., and Sher, A. (2002). Evasion of innate immunity by parasitic protozoa. *Nat. Rev. Immunol.* 3, 1041–1047. doi: 10.1038/ni1102-1041

Salojin, K., and Oravecz, T. (2007). Regulation of innate immunity by MAPK dual-specificity phosphatases: knockout models reveal new tricks of old genes. *J. Leukocyte Biol.* 81, 860–869. doi: 10.1189/jlb.1006639

Schöneck, R., Plumas-Marty, B., Taibi, A., Billaut-Mulot, O., Loyens, M., Gras-Masse, H., et al. (1994). *Trypanosoma cruzi* cDNA encodes a tandemly repeated domain structure characteristic of small stress proteins and glutathione S-transferases. *Biol. Cell* 80, 1–10. doi: 10.1016/0248-4900(94)90011-6

Shadab, M., and Ali, N. (2011). Evasion of host defence by *Leishmania donovani*: subversion of signaling pathways. *Mol. Biol. Int.* 2011:343961. doi: 10.4061/2011/343961

Shaul, Y. D., and Seger, R. (2007). The MEK/ERK cascade: From signaling specificity to diverse functions. *Biochim. Biophys. Acta* 1773, 1213–1226. doi: 10.1016/j.bbamcr.2006.10.005

Srivastava, N., Sudan, R., and Saha, B. (2011). CD40-modulated dual-specificity phosphatases MAPK phosphatase (MKP)-1 and MKP-3 reciprocally regulate leishmania major infection. *J. Immunol.* 186, 5863–5872. doi: 10.4049/jimmunol.1003957

Stafford, J. L., Neumann, N. F., and Belosevic, M. (2002). Macrophage-mediated innate host defense against protozoan parasites. *Crit. Rev. Microbiol.* 28, 187–248. doi: 10.1080/1040-840291046731

Tanowitz, H. B., Kirchhoff, L. V., Simon, D., Morris, S. A., Weiss, L. M., and Wittner, M. (1992). Chagas' disease. *Clin. Microbiol. Rev.* 5, 400–419.

Tarleton, R. L. (2007). Immune system recognition of *Trypanosoma cruzi*. *Curr. Opin. Immunol.* 19, 430–434. doi: 10.1016/j.coi.2007.06.003

Terrazas, C. A., Huitron, E., Vazquez, A., Juarez, I., Camacho, G. M., Calleja, E. A., et al. (2011). MIF synergizes with *Trypanosoma cruzi* antigens to promote efficient dendritic cell maturation and IL-12 production via p38 MAPK. *Int. J. Biol. Sci.* 7, 1298–1310. doi: 10.7150/ijbs.7.1298

Turjanski, A. G., Vaqué, J. P., and Gutkind, J. S. (2007). MAP kinases and the control of nuclear events. *Oncogene* 26, 3240–3253. doi: 10.1038/sj.onc.1210415

Xin, L., Li, K., and Soong, L. (2008). Down-regulation of dendritic cell signaling pathways by *Leishmania amazonensis* amastigotes. *Mol. Immunol.* 45, 3371–3382. doi: 10.1016/j.molimm.2008.04.018

Yang, Y., Kim, S. C., Yu, T., Yi, Y., Rhee, M. H., Sung, G., and Cho, J. Y. (2014). Functional roles of p38 mitogen-activated protein kinase in macrophage-mediated inflammatory responses, 2014. *Med. Inflamm.* 2014:352371.

Zambrano-Villa, S., Rosales-Borjas, D., Carrero, J. C., and Ortiz-Ortiz, L. (2002). How protozoan parasites evade the immune response. *Trends Parasitol.* 18, 272–278. doi: 10.1016/S1471-4922(02)02289-4

Zhang, S. Q., Yang, W., Kontaridis, M. I., Bivona, T. G., Wen, G., Araki, T., et al. (2004). Shp2 regulates src family kinase activity and ras/erk activation by controlling csk recruitment. *Mol. Cell* 13, 341–355. doi: 10.1111/j.1582-4934.2008.00333.x

Zhang, Y., and Dong, C. (2007). Regulatory mechanisms of mitogen-activated kinase signaling. *Cell. Mol. Life Sci.* 64, 2771–2789. doi: 10.1007/s00018-007-7012-3

5

Interactions between *Trypanosoma cruzi* Secreted Proteins and Host Cell Signaling Pathways

Renata Watanabe Costa*[1]*, Jose F. da Silveira*[1] *and Diana Bahia*[1,2]*

[1] Departamento de Microbiologia, Imunologia e Parasitologia, Escola Paulista de Medicina, Universidade Federal de São Paulo, São Paulo, Brazil, [2] Departamento de Biologia Geral, Instituto de Ciências Biológicas, Universidade Federal de Minas Gerais, Minas Gerais, Brazil

Edited by:
Amy Rasley, Lawrence Livermore National Laboratory, USA

Reviewed by:
Adam Cunningham, University of Birmingham, UK
Julio Aliberti, Cincinnati Children's Hospital Medical Center, USA

***Correspondence:**
Diana Bahia
dianabahia@hotmail.com

Chagas disease is one of the prevalent neglected tropical diseases, affecting at least 6–7 million individuals in Latin America. It is caused by the protozoan parasite *Trypanosoma cruzi*, which is transmitted to vertebrate hosts by blood-sucking insects. After infection, the parasite invades and multiplies in the myocardium, leading to acute myocarditis that kills around 5% of untreated individuals. *T. cruzi* secretes proteins that manipulate multiple host cell signaling pathways to promote host cell invasion. The primary secreted lysosomal peptidase in *T. cruzi* is cruzipain, which has been shown to modulate the host immune response. Cruzipain hinders macrophage activation during the early stages of infection by interrupting the NF-kB P65 mediated signaling pathway. This allows the parasite to survive and replicate, and may contribute to the spread of infection in acute Chagas disease. Another secreted protein P21, which is expressed in all of the developmental stages of *T. cruzi*, has been shown to modulate host phagocytosis signaling pathways. The parasite also secretes soluble factors that exert effects on host extracellular matrix, such as proteolytic degradation of collagens. Finally, secreted phospholipase A from *T. cruzi* contributes to lipid modifications on host cells and concomitantly activates the PKC signaling pathway. Here, we present a brief review of the interaction between secreted proteins from *T. cruzi* and the host cells, emphasizing the manipulation of host signaling pathways during invasion.

Keywords: *T. cruzi*, secretome, secreted proteins, virulence factor, host parasite interaction, host cell signaling, host cell invasion, Chagas disease

INTRODUCTION

Trypanosoma cruzi: Life Cycle and Chagas Disease

Chagas disease is caused by the protozoan parasite *Trypanosoma cruzi*, affects 6–7 million individuals, primarily in Latin America, and is associated with negative economic impacts in developing countries (http://www.who.int/mediacentre/factsheets/fs340/en/). *T. cruzi* is transmitted to vertebrate hosts by the triatomine vector *Triatoma infestans*. Despite its high incidence and economic costs, Chagas disease remains a neglected tropical disease; it does not have an effective pharmacological treatment and there are minimal investments in finding a cure for Chagas disease (Clayton, 2010; Souza et al., 2010). The life cycle of *T. cruzi* has four developmental phases that occur in the hematophagous insect vector and bloodstream and tissues of mammalian hosts (Souza et al., 2010). The epimastigote (EPI) is a non-infectious replicative form found in the vector's digestive tract. The EPI differentiates into the metacyclic trypomastigote (MT), which is transmitted to mammals through the insect's feces during a blood meal or by the oral route. The MT invade mammalian host cells where they transform into an amastigote (AMA) that replicates intracellularly. After a multiple rounds of replication, the AMAs differentiate back into trypomastigotes (TCTs), which are released into the extracellular milieu when the host cell is disrupted. TCTs can invade neighboring host cells or be released into the blood stream where they can infect other tissues or be ingested by a feeding insect. Once the host has been infected, the parasite can invade and multiply in the myocardium, leading to acute myocarditis, which kills around 5% of untreated individuals (Ponce et al., 2013).

Similar to other intracellular protozoa, *T. cruzi* is an intracellular parasite that invades different types of cells to evade the host immune system (Guiñazú et al., 2007). Intracellular parasites have complex lifecycles that involve several developmental stages, and usually contain multiple secreted proteins that can manipulate host cell signaling pathways to promote parasite adhesion, recognition, and invasion (Burleigh and Woolsey, 2002). The complex interplay between proteins secreted by *T. cruzi* that affect the host cell environment or contribute to immune evasion likely influences the outcome of infection. Understanding the role of secreted proteins during *T. cruzi* infection is critical to deepen the knowledge of the pathogenesis of Chagas disease (McConville et al., 2002).

T. cruzi Secretome

In eukaryotes, secreted proteins typically contain an N-terminal signal peptide that directs them to the classical endoplasmic reticulum (ER)/Golgi-dependent secretion pathway. Secretory proteins that do not contain the signal peptide are secreted outside the plasma membrane using non-classical secretory pathways including, membrane-bound extracellular vesicles (EVs), such as exosomes and ectosomes (Nickel and Seedorf, 2008; Simpson and Mathivanan, 2012). Only a small fraction (~9%) of the proteins in the *T. cruzi* secretome contain an N-terminal signal peptide suggesting that they are secreted by classical pathways (Bayer-Santos et al., 2013), the remaining proteins are likely secreted by non-classical pathways (Torrecilhas et al., 2009, 2012; Bayer-Santos et al., 2013; Marcilla et al., 2014).

Secretion or shedding of EVs by *T. cruzi* can occur spontaneously or be induced by nutritional or chemical stress (da Silveira et al., 1979; Torrecilhas et al., 2009, 2012; Bayer-Santos et al., 2013; Marcilla et al., 2014). A considerable number of the *T. cruzi* secreted/excreted proteins have been characterized at the structural and functional levels. Some of the secreted *T. cruzi* proteins, such as the *trans*-sialidase (TS) glycoproteins (TS/SAPA, Tc85, gp82, gp90, CRP, TESA), mucin-associated surface proteins (MASP), cruzipain, gp63, mucins, and serine-, alanine-, and proline-rich proteins (SAP), are associated with the plasma membrane via a glycosylphosphatidylinositol (GPI) anchor (Torrecilhas et al., 2009, 2012; Bayer-Santos et al., 2013; Marcilla et al., 2014). Several of these proteins (e.g., TS/SAPA, CRP, mucins) are also spontaneously shed from the parasite surface in a soluble form that lacks the GPI anchor, possibly due to cleavage by an endogenous phospholipase C (Affranchino et al., 1989; de Almeida and Heise, 1993; Bartholomeu et al., 2009; de Pablos et al., 2011; Cánepa et al., 2012a). Others (e.g., Tc85) are shed with the GPI anchor linked to membrane vesicles (Zingales et al., 1985; Abuin et al., 1996a).

Trypomastigotes and AMAs release EVs containing virulence factors involved in: (i) host cell invasion and intracellular parasite development, such as the TS and TS-like proteins (Zingales et al., 1985; Gonçalves et al., 1991; Schenkman et al., 1991; Abuin et al., 1996a,b; Torrecilhas et al., 2009, 2012; Cánepa et al., 2012a; Maeda et al., 2012; Bayer-Santos et al., 2013; Marcilla et al., 2014; Mattos et al., 2014), peptidyl prolyl *cis-trans*-isomerase (Moro et al., 1995), oligopeptidases and proteases (Meirelles et al., 1992; Caler et al., 1998; Scharfstein et al., 2000; Cuevas et al., 2003; Bastos et al., 2005; Doyle et al., 2011; Maeda et al., 2014), phospolipases A1 and C (Andrews et al., 1988; Rosenberg et al., 1991; Furuya et al., 2000; Okura et al., 2005; Belaunzarán et al., 2007, 2013; Castillo et al., 2013); mucins and mucin-like proteins (Gazzinelli et al., 1991; de Diego et al., 1997; Kierszenbaum et al., 1999; Pereira-Chioccola et al., 2000; Kierszenbaum et al., 2002; Alcaide and Fresno, 2004; Cánepa et al., 2012b), MASP (de Pablos et al., 2011; Cánepa et al., 2012a), SAP (Baida et al., 2006; Zanforlin et al., 2013), P21 AMA specific proteins (da Silva et al., 2009), surface membrane proteins (TcSMP; Martins et al., 2015); (ii) immune evasion (Andrews et al., 1990; Norris et al., 1991; Norris and Schrimpf, 1994; Ouaissi et al., 1995; Reina-San-Martin et al., 2000; Martin et al., 2006; Mott et al., 2011; Kulkarni et al., 2013; Nogueira et al., 2015); and (iii) increased heart parasitism, inflammation, and arrhythmia that contribute to the pathogenesis of Chagas disease (Torrecilhas et al., 2009; Nogueira et al., 2015; Rodriguez-Angulo et al., 2015). In addition, some of the secreted/excreted proteins are diagnostic markers for Chagas disease (Jazin et al., 1995; Umezawa et al., 1996; Agusti et al., 2000; Bernabó et al., 2013). This mini review will focus on specific molecules secreted by *T. cruzi* that have already been identified as interfering with host cell signaling and that ultimately play a role in the ability of *T. cruzi* to evade the immune system.

T. cruzi Cruzipain: A Role in Evading the Host Immune Response and Promoting Survival in Cardiomyocytes

To facilitate their entry into non-phagocytic cells, infectious TCTs employ an arsenal of surface glycoproteins, secreted proteases, and signaling agonists to actively manipulate multiple host cell signaling pathways (Burleigh and Woolsey, 2002). Several studies using synthetic irreversible cysteine peptidase inhibitors have demonstrated that *T. cruzi* infectivity, host immune evasion, and intracellular growth depend on the activity of cruzipain (Meirelles et al., 1992; Waghabi et al., 2005; McKerrow et al., 2008). To facilitate entry into non-phagocytic cells like endothelial cells and cardiomyocytes, cruzipain acts on a cell-bound kininogen to generate bradykinin, which upon recognition by the B2 bradykinin receptor, triggers the Ca^{2+}

mobilization required for parasite internalization (Scharfstein et al., 2000; Guiñazú et al., 2007; Maeda et al., 2014).

Murine macrophages stimulated with cruzipain up-regulate arginase activity and increase production of IL-10 and TGF-β, thereby increasing *T. cruzi* survival (Stempin et al., 2002). TGF-β in particular can suppress some of the microbicidal functions of macrophages and is one way that parasites create a favorable cellular microenvironment to gain a survival advantage (Gantt et al., 2003; Waghabi et al., 2005). Previous studies have demonstrated that forms of *T. cruzi* are able to activate latent TGF-β (Waghabi et al., 2005). Treatment of macrophages with increasing doses of cruzipain promoted the activation of TGF-β in a dose-dependent manner, confirming that this peptidase is capable of activating latent TGF-β in the absence of any other host or parasite factors (Ferrão et al., 2015). In addition, transgenic EPIs overexpressing chagasin, a natural cruzipain inhibitor, were significantly less able to activate latent TGF-β when compared to wild type parasites (Santos et al., 2005; Ferrão et al., 2015).

The role of cruzipain in cell entry and TGF-β production suggest that it may function during the early events of macrophage infection to facilitate parasite survival and replication. Taken together, the data suggests that cruzipain is a potential pharmaceutical target as it may have an essential role in the pathogenesis of Chagas disease (Guiñazú et al., 2007; Doyle et al., 2011). Based on this evidence, cruzipain inhibitors are considered promising anti-*T. cruzi* chemotherapeutic agents (Ndao et al., 2014; Branquinha et al., 2015). Irreversible cruzipain inhibitors, such as the prototype molecule K777 (also known as K11777) have been efficacious in experimental models of *T. cruzi* infection (Engel et al., 1998; Barr et al., 2005; Doyle et al., 2011).

In parallel to the immunological findings, cruzipain promotes cardiomyocyte survival via the PI3K and MEK1-dependent signaling pathways (Aoki et al., 2004, 2006). Cardiomyocytes were pretreated with PI3K or MAPK inhibitors and grown in the presence or absence of cruzipain. Cardiomyocyte apoptosis was decreased after cruzipain treatment, but this protective effect was reduced by incubation with PI3K and MEK1 inhibitors, which had no effect on cruzipain-mediated cardiomyocyte survival in the absence of cruzipain. These findings suggest the survival effects of cruzipain are regulated by effector proteins downstream of PI3K and MEK1. Moreover, *T. cruzi* infection as well as cruzipain itself mediates the phosphorylation of ERK1/2 and Akt, and cruzipain inhibits proteolytic cleavage of caspase 3 via PI3K and MEK1-dependent signaling pathways (Fujio et al., 2000; Aoki et al., 2006). Together the data strongly suggest cruzipain mediates survival in part via anti-apoptotic PI3K/MEK1 signaling. Another study has shown that the anti-apoptotic effect of cruzipain is also mediated in part by arginase activity and Bcl-2 expression (Aoki et al., 2004). Thus, cruzipain activates at least two signaling pathways leading to enhanced cardiomyocyte survival. Parallel activation of these signal transduction pathways may represent a cellular strategy to amplify survival signals in the target cell. Elucidating the pro-survival pathways may lead to a better understanding of the parasite–host relationship and may provide useful targets for the treatment of Chagas disease (Aoki et al., 2004, 2006).

T. cruzi Phospholipase A1: A Role in Activating Host Protein Kinase C (PKC) Throughout Infection

Phospholipases play a critical role in some physiological processes including the generation of signaling lipids that are relevant to disease (Dennis, 2015). In the case of *T. cruzi*, phospholipid degrading enzymes are associated with the inflammatory responses elicited by degenerating AMA nests in the tissues of patients with Chagas disease (Wainszelbaum et al., 2001).

Throughout its life cycle, *T. cruzi* has to adapt to different environments through morphological and functional changes that involve complex networks of enzymatic pathways, including phospholipases. *T. cruzi* Phospholipase A1 (Tc-PLA1) is secreted by the parasite into the extracellular medium and shows remarkably higher membrane-bound activity in infectious AMAs and TCTs (Wainszelbaum et al., 2001; Belaunzarán et al., 2007). In VERO cells, treatment with Tc-PLA1 and PMA (phorbol 12-myristate 13-acetate), a known PKC activator, demonstrated that Tc-PLA1 is involved in host cell lipid modifications leading to PKC activation (34). Tc-PLA1 significantly modified the host cell lipid profile by generating secondary lipid messengers (DG, FFA, and LPC) and concomitant PKC activation. PKC has been implicated in increased parasite invasion, suggesting that Tc-PLA1 is involved in the early events of parasite–host cell interaction preceding parasite invasion (Belaunzarán et al., 2007). Specific anti- Tc-PLA1 antibodies can bind to the surface of the parasite and neutralize Tc-PLA1 activity, preventing parasite invasion. This suggests that Tc-PLA1 is an emerging virulence factor for *T. cruzi* and emphasizes the promise of Tc-PLA1 as a potential therapeutic target (Belaunzarán et al., 2007). Taking these findings into consideration, Tc-PLA1-mediated host cell PKC activation could modulate Ca^{2+} release from intracellular stores thereby contributing to parasite invasion. Ca^{2+} mobilization, in both host cell and parasite, is required during the internalization process (Villalta et al., 1999; Yoshida, 2006; Souza et al., 2010; Maeda et al., 2012). In addition, *T. cruzi* infective stages partially incorporated and metabolized LPC, therefore the remaining extracellular LPC might exert a toxic effect on the host cell, reinforcing the involvement of Tc-PLA1 in the pathogenesis. In this concern, it has been described that LPC inhibits nitric oxide production by *T. cruzi* stimulated macrophages (**Figure 1**), and thus interferes with the vertebrate host immune system (Belaunzarán et al., 2007).

Secreted *T. cruzi* Cyclophilin Inactivates the Lytic Vector Defense

Trypanosoma cruzi not only has to interact with the mammalian host but also with its insect vector (*Triatoma infestans*), and many of these interactions are still unknown. Innate immune cationic antimicrobial peptides (CAMPs) are expressed by a wide variety of insects to prevent microbial colonization and infection. Several CAMPs have been identified from the saliva, hemolymph, and intestinal tract of reduviid insects (Lopez et al., 2003). Kulkarni et al. (2013) studied the interactions between CAMPs and *T. cruzi*, and found a unique parasite-driven pathway that modified host CAMPs. Parasites exposed to cyclophilin-trialysin

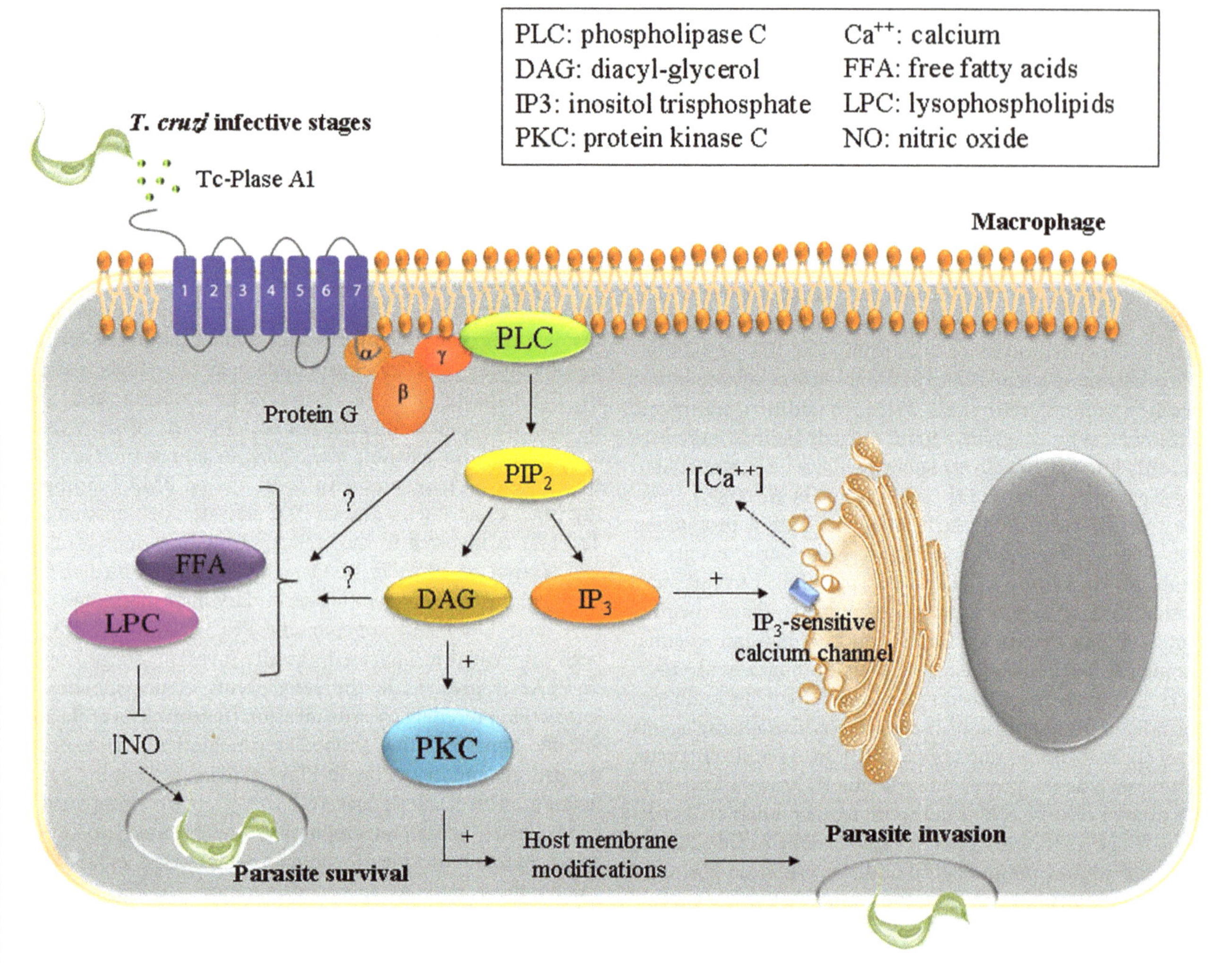

FIGURE 1 | To ensure successful invasion of the host cell *T. cruzi* has developed a multi-step process with redundant mechanisms involving diverse host and parasite molecules. The enzyme Plase A1 (Tc-Plase A1) is secreted by *T. cruzi* and is widely present during the infectious life stages. Tc-Plase A1 may contribute to lipid modifications on host cells and concomitantly activates the PKC signaling pathway. This suggests that Tc-Plase A1 is involved in the early events of the parasite–host cell interaction and precedes parasite invasion.

have enhanced binding and invasion in myoblasts pre-grown leading to higher infectivity. They found that secreted parasite cyclophilin, a peptidyl-prolyl isomerase involved in protein folding (Kulkarni et al., 2013; Carraro et al., 2015), binds to and inactivates trialysin via its proline residue. Replicating insect-stage parasites secrete cyclophilin 19 as they migrate through the reduviid gastrointestinal tract. Cyclophilin 19 binds to and isomerizes the CAMP peptide neutralizing its anti-parasitic activity. The cyclophilin-trialysin complex then synergistically acts on the parasites to activate calcineurin phosphatase signaling, which drives metabolic activation and ATP production leading to enhanced infectivity. This parasite pathway is a mechanism of CAMP recognition, evasion, and adaptation mediated through calcineurin intracellular signaling (Kulkarni et al., 2013; Carraro et al., 2015). These findings also represent one of the few descriptions of specific stimuli that enhance infectivity of *T. cruzi* and indicate a defined host molecule-based environmental sensing mechanism in this group of organisms (Kulkarni et al., 2013).

T. cruzi Soluble Factors: Effects on the Host Extracellular Matrix

Trypomastigotes trigger rapid changes in the host cell signaling pathways during their early interactions with mammalian host cells to facilitate the process of parasite entry into non-professional phagocytic cells (Giddings et al., 2006; Yoshida, 2008). However, *T. cruzi* also affects the host cell downstream of the invasion process. Transcriptional profiling of *T. cruzi*-infected fibroblasts showed that the earliest detectable changes triggered by infectious TCTs involved downregulation of a small subset of genes including members of the CCN family (cyr61 and ctgf/ccn2) that play critical roles in cardiovascular development (angiogenesis), injury repair, fibrotic disease, and extracellular matrix (ECM) homeostasis (Chen and Lau, 2009;

Mott et al., 2011). Connective tissue growth factor (CTGF/CCN2) promotes cell proliferation and cooperates with TGF-ß to promote myofibroblast differentiation and enhanced ECM synthesis. Mott et al. (2011) showed that *T. cruzi* may release a factor that inhibits TGF-ß-mediated expression of CTGF/CCN2. The expression of CTGF/CCN2 is also controlled by the ETS family of transcriptional factors, which are regulated through MAP kinase signaling. *T. cruzi*-dependent abrogation of CTGF/CCN2 expression in human dermal fibroblasts is associated with inhibition of both basal and agonist-induced activation of MAP kinase signaling (Mott et al., 2011). *T. cruzi*-mediated down-regulation of CTGF expression requires *de novo* host cell protein synthesis, indicating that the ability of *T. cruzi* to interfere with the host fibrogenic response is a complex process requiring input from multiple host cell signaling pathways (Unnikrishnan and Burleigh, 2004; Mott et al., 2011).

Regarding the impact of *T. cruzi* secreted factors on TGF-ß-induced fibroblast gene expression, a discrete subset of agonist-inducible fibroblast genes are sensitive to factors secreted/released by *T. cruzi*. A study reports that the group of TGFß-inducible genes that exhibit the highest sensitivity to a *T. cruzi* secreted/released fraction are MAP kinase-regulated genes that function in wound repair, ECM remodeling, and host response pathways. Inhibition of ECM synthesis because of these secreted parasite factors would facilitate dissemination from early sites of infection (Mott et al., 2011).

Secreted *T. cruzi* P21 Enhances Host Phagocytosis

P21 is a secreted protein expressed in all of the developmental stages in the *T. cruzi* lifecycle that may play an important role in parasite internalization (da Silva et al., 2009). Rodrigues et al. (2012) engineered a recombinant protein based on P21 (P21-His6) and then assessed its ability to upregulate phagocytosis in macrophages and alter host cell signaling. P21-His6 upregulated phagocytosis in macrophages in a manner dependent on CXCR4-binding and actin polymerization, and triggered the PI3K signaling pathway (Rodrigues et al., 2012). P21-His6 required PI3K signaling independent of AKT for its function (Vasudevan et al., 2009; Lee et al., 2011). PI3K-dependent signal transduction through the Rho-family GTPases occurs during FcR-mediated phagocytosis and that PI3K-dependent deactivation of Cdc42 is necessary for phagocytosis. Moreover, the activities of PI3K and Cdc42 are linked: FcR-activated Cdc42 stimulates PI3K, which increases concentrations of PI(3,4,5)P3 in phagocytic cups, allowing the PI(3,4,5)P3-dependent deactivation of Cdc42 that is necessary to complete phagocytosis (Beemiller et al., 2010). In addition, previous work has provided evidence of PI3K activation in non-professional phagocytic cells during *T. cruzi* cell invasion (Woolsey et al., 2003; Rodrigues et al., 2012). In sum, P21 serves as part of the host cell invasion machinery by triggering actin polymerization on the host cell through interactions with the CXCR4 chemokine receptor on the cell membrane, and favoring its own phagocytosis into the host (dos Santos et al., 2014).

T. cruzi MASP: A Role in Evading Host Immune Cells

The annotation of the *T. cruzi* genome revealed a new multigene family composed of approximately 1,300 genes, which became known as MASPs because they were clustered with genes encoding mucins and other surface protein families (El-Sayed et al., 2005). MASP proteins are GPI-anchored glycoproteins expressed on the surface of the circulating infectious forms of the parasite that can be secreted into the extracellular medium (Bartholomeu et al., 2009; dos Santos et al., 2012; Serna et al., 2014). MASP is the second largest gene family (1377 genes and 433 pseudogenes), representing approximately 6% of the T. cruzi genome (Serna et al., 2014).

dos Santos et al. (2012) using antibody recognition of several MASP peptides observed the interaction of these proteins with the host immune system during acute *T. cruzi* infection. The MASP family may play a role in promoting the polyclonal lymphocyte activation that leads to hypergammaglobulinemia and the delayed specific humoral immune response, characteristic of the acute phase of Chagas disease. Polyclonal B-cell activation might diffuse the immune response, preventing the development of a specific and neutralizing response against the parasite and its complete elimination. Additionally, MASP peptides could possibly mediate both specific T-cell dependent and non-specific T-cell independent immune responses. This hypothesis is partially supported by the differential recognition of MASPs by immunoglobulin (Ig) M and IgG and the difference in the antibody affinity levels against each of the synthetic peptides. All of these phenomena are suggestive of an immune evasion mechanism (Reina-San-Martin et al., 2000; Minoprio, 2001; Gao et al., 2002; dos Santos et al., 2012).

CONCLUDING REMARKS

Host cell invasion and parasite internalization are important steps in the evolution of parasitism by several pathogens. These processes present at least two important advantages: protection against the host immune response and access to a microenvironment rich in metabolic products (Barrias et al., 2012). Substantial progress has been made in understanding the roles of secreted proteins in infection and invasion by pathogenic *T. cruzi*. Host cell intracellular signaling can combat the infection; but it can also favor parasite entry. Parasites hijack the host immune response, phagocytosis, ECM, and anti-parasitic proteins for their own survival, replication, and immune evasion purposes. The complex networks are interconnected and require extensive study to identify intracellular rearrangements that facilitate parasite internalization; the tools in use today include bioinformatics, novel molecular level studies, and new experimental drugs. A multidisciplinary approach to understanding parasite host interaction will be critical to better understand *T. cruzi* physiopathology, diagnosis, and treatment.

AUTHOR CONTRIBUTIONS

All authors listed conceived and wrote the manuscript.

ACKNOWLEDGMENTS

The authors wish to thank the support of FAPESP (07/50551-2, 11/51475-3) and CAPES. JFS and DB are recipients of CNPq fellowships. Authors also thank BioMed Proofreading (http://www.biomedproofreading.com).

REFERENCES

Abuin, G., Colli, W., and Alves, M. J. (1996a). Turnover and shedding of the Tc-85 surface glycoprotein of *Trypanosoma cruzi* trypomastigotes. *Braz. J. Med. Biol. Res.* 29, 335–341.

Abuin, G., Couto, A. S., de Lederkremer, R. M., Casal, O. L., Galli, C., Colli, W., et al. (1996b). *Trypanosoma cruzi*: the Tc-85 surface glycoprotein shed by trypomastigotes bears a modified glycosylphosphatidylinositol anchor. *Exp. Parasitol.* 82, 290–297. doi: 10.1006/expr.1996.0036

Affranchino, J. L., Ibanez, C. F., Luquetti, A. O., Rassi, A., Reyes, M. B., Macina, R. A., et al. (1989). Identification of a *Trypanosoma cruzi* antigen that is shed during the acute phase of chagas' disease. *Mol. Biochem. Parasitol.* 34, 221–228. doi: 10.1016/0166-6851(89)90050-9

Agusti, R., Couto, A. S., Alves, M. J., Colli, W., and Lederkremer, R. M. (2000). Lipids shed into the culture medium by trypomastigotes of *Trypanosoma cruzi*. *Mem. Inst. Oswaldo Cruz* 97:e102.

Alcaide, P., and Fresno, M. (2004). The *Trypanosoma cruzi* membrane mucin AgC10 inhibits T cell activation and IL-2 transcription through L-selectin. *Int. Immunol.* 16, 1365–1375. doi: 10.1093/intimm/dxh138

Andrews, N. W., Abrams, C. K., Slatin, S. L., and Griffiths, G. (1990). A *T. cruzi*-secreted protein immunologically related to the complement component C9: evidence for membrane pore-forming activity at low pH. *Cell* 61, 1277–1287. doi: 10.1016/0092-8674(90)90692-8

Andrews, N. W., Robbins, E. S., Ley, V., Hing, K. S., and Nussenzweig, V. (1988). Developmental regulated phospholipase-C mediated release of the major surface glycoprotein of amastigotes of *Trypanosoma cruzi*. *J. Exp. Med.* 167, 300–314. doi: 10.1084/jem.167.2.300

Aoki, M. P., Cano, R. C., Pellegrini, A. V., Tanos, T., Guiñazú, N. L., Coso, O. A., et al. (2006). Different signaling pathways are involved in cardiomyocyte survival induced by a *Trypanosoma cruzi* glycoprotein. *Microbes Infect.* 8, 1723–1731. doi: 10.1016/j.micinf.2006.02.010

Aoki, M. P., Guiñazú, N. L., Pellegrini, A. V., Gotoh, T., Masih, D. T., and Gea, S. (2004). Cruzipain, a major *Trypanosoma cruzi* antigen, promotes arginase-2 expression and survival of neonatal mousecardiomyocytes. *Am. J. Physiol. Cell Physiol.* 286, C206–C212. doi: 10.1152/ajpcell.00282.2003

Baida, R. C., Santos, M. R., Carmo, M. S., Yoshida, N., Ferreira, D., Ferreira, A. T., et al. (2006). Molecular characterization of serine-, alanine-, and proline-rich proteins of *Trypanosoma cruzi* and their possible role in host cell infection. *Infect. Immun.* 74, 1537–1546. doi: 10.1128/IAI.74.3.1537-1546.2006

Barr, S. C., Warner, K. L., Kornreic, B. G., Piscitelli, J., Wolfe, A., Benet, L., et al. (2005). A cysteine protease inhibitor protects dogs from cardiac damage during infection by *Trypanosoma cruzi*. *Antimicrob. Agents Chemother.* 49, 5160–5161. doi: 10.1128/AAC.49.12.5160-5161.2005

Barrias, E. S., Reignault, L. C., De Souza, W., and Carvalho, T. M. (2012). *Trypanosoma cruzi* uses macropinocytosis as an additional entry pathway into mammalian host cell. *Microbes Infect.* 14, 1340–1351. doi: 10.1016/j.micinf.2012.08.003

Bartholomeu, D. C., Cerqueira, G. C., Leao, A. C., daRocha, W. D., Pais, F. S., Macedo, C., et al. (2009). Genomic organization and expression profile of the mucin-associated surface protein (masp) family of the human pathogen *Trypanosoma cruzi*. *Nucleic Acids Res.* 37, 3407–3417. doi: 10.1093/nar/gkp172

Bastos, I. M., Grellier, P., Martins, N. F., Cadavid-Restrepo, G., de Souza-Ault, M. R., Augustyns, K., et al. (2005). Molecular, functional and structural properties of the prolyl oligopeptidase of *Trypanosoma cruzi* (POP Tc80), which is required for parasite entry into mammalian cells. *Biochem. J.* 388(Pt 1), 29–38. doi: 10.1042/BJ20041049

Bayer-Santos, E., Aguilar-Bonavides, C., Rodrigues, S. P., Cordero, E. M., Marques, A. F., Varela-Ramirez, A., et al. (2013). Proteomic analysis of *Trypanosoma cruzi* secretome: characterization of two populations of extracellular vesicles and soluble proteins. *J. Proteome Res.* 12, 883–897. doi: 10.1021/pr300947g

Beemiller, P., Zhang, Y., Mohan, S., Levinsohn, E., Gaeta, I., Hoppe, A. D., et al. (2010). A Cdc42 activation cycle coordinated by PI 3-kinase during fc receptor-mediated phagocytosis. *Mol. Biol. Cell* 21, 470–480. doi: 10.1091/mbc.E08-05-0494

Belaunzarán, M. L., Wainszelbaum, M. J., Lammel, E. M., Gimenez, G., Aloise, M. M., Florin-Christensen, J., et al. (2007). Phospholipase A1 from *Trypanosoma cruzi* infective stages generates lipid messengers that activate host cell protein kinase c. *Parasitology* 134, 491–502. doi: 10.1017/S0031182006001740

Belaunzarán, M. L., Wilkowsky, S. E., Lammel, E. M., Giménez, G., Bott, E., Barbieri, M. A., et al. (2013). Phospholipase A1: a novel virulence factor in *Trypanosoma cruzi*. *Mol. Biochem. Parasitol.* 187, 77–86. doi: 10.1016/j.molbiopara.2012.12.004

Bernabó, G., Levy, G., Ziliani, M., Caeiro, L. D., Sánchez, D. O., and Tekiel, V. (2013). TcTASV-C, a protein family in *Trypanosoma cruzi* that is predominantly trypomastigote-stage specific and secreted to the medium. *PLoS ONE* 8:e71192. doi: 10.1371/journal.pone.0071192

Branquinha, M. H., Oliveira, S. S., Sangenito, L. S., Sodre, C. L., Kneipp, L. F., d'Avila-Levy, C. M., et al. (2015). Cruzipain: an update on its potential as chemotherapy target against the human pathogen *Trypanosoma cruzi*. *Curr. Med. Chem.* 22, 2225–2235. doi: 10.2174/0929867322666150521091652

Burleigh, B. A., and Woolsey, A. M. (2002). Cell signaling and *Trypanosoma cruzi* invasion. *Cell Microbiol.* 4, 701–711. doi: 10.1046/j.1462-5822.2002.00226.x

Caler, E. V., Vaena de Avalos, S., Haynes, P. A., Andrews, N. W., and Burleigh, B. A. (1998). Oligopeptidase B-dependent signaling mediates host cell invasion by *Trypanosoma cruzi*. *EMBO J.* 17, 4975–4986. doi: 10.1093/emboj/17.17.4975

Cánepa, G. E., Degese, M. S., Budu, A., Garcia, C. R. S., and Buscaglia, C. A. (2012a). Involvement of TSSA (trypomastigote small surface antigen) in *Trypanosoma cruzi* invasion of mammalian cells. *Biochem. J.* 444, 211–218. doi: 10.1042/BJ20120074

Cánepa, G. E., Mesías, A. C., Yu, H., Chen, X., and Buscaglia, C. A. (2012b). Structural features affecting trafficking, processing, and secretion of *Trypanosoma cruzi* mucins. *J. Biol. Chem.* 287, 26365–26376. doi: 10.1074/jbc.M112.354696

Carraro, R., Iribarne, F., and Paulino, M. (2015). Analysis of cyclosporin A and a set of analogs as inhibitors of a *T. cruzi* cyclophilin by docking and molecular dynamics. *J. Biomol. Struct. Dyn.* 5, 1–15. doi: 10.1080/07391102.2015.1038584

Castillo, C., Villarroel, A., Duaso, J., Galanti, N., Cabrera, G., Maya, J. D., et al. (2013). Phospholipase C gamma and ERK1/2 mitogen activated kinase pathways are differentially modulated by *Trypanosoma cruzi* during tissue invasion in human placenta. *Exp. Parasitol.* 133, 12–17. doi: 10.1016/j.exppara.2012.10.012

Chen, C. C., and Lau, L. F. (2009). Functions and mechanisms of action of CCN matricellular proteins. *Int. J. Biochem. Cell Biol.* 41, 771–783. doi: 10.1016/j.biocel.2008.07.025

Clayton, J. (2010). Chagas disease 101. *Nature* 465, S4–S5. doi: 10.1038/nature09220

Cuevas, I. C., Cazzulo, J. J., and Sanchez, D. O. (2003). Gp63 homologues in *Trypanosoma cruzi*: surface antigens with metalloprotease activity and a possible role in host cell infection. *Infect. Immun.* 71, 5739–5749. doi: 10.1128/IAI.71.10.5739-5749.2003

de Almeida, M. L., and Heise, N. (1993). Proteins anchored via glycosylphosphatidylinositol and solubilizing phospholipases in *Trypanosoma cruzi*. *Biol. Res.* 26, 285–312.

de Diego, J., Punzón, C., Duarte, M., and Fresno, M. (1997). Alteration of macrophage function by a *Trypanosoma cruzi* membrane mucin. *J. Immunol.* 159, 4983–4989.

de Pablos, L. M., Gonzalez, G. G., Solano Parada, J., Seco Hidalgo, V., Diaz Lozano, I. M., Gomez Samblas, M. M., et al. (2011). Differential expression and characterization of a member of the mucin-associated surface protein family secreted by *Trypanosoma cruzi*. *Infect. Immun.* 79, 3993–4001. doi: 10.1128/IAI.05329-11

da Silva, C. V., Kawashita, S. Y., Probst, C. M., Dallagiovanna, B., Cruz, M. C., da Silva, E. A., et al. (2009). Characterization of a 21kDa protein from *Trypanosoma cruzi* associated with mammalian cell invasion. *Microbes Infect.* 11, 563–570. doi: 10.1016/j.micinf.2009.03.007

da Silveira, J. F., Abrahamsohn, P. A., and Colli, W. (1979). Plasma membrane vesicles isolated from epimastigote forms of *Trypanosoma cruzi*. *Biochim. Biophys. Acta* 550, 222–232. doi: 10.1016/0005-2736(79) 90209-8

Dennis, E. A. (2015). Introduction to thematic review series: phospholipases: central role in lipid signaling and disease. *J. Lipid Res.* 56, 1245–1247. doi: 10.1194/jlr.E061101

dos Santos, M. A., Teixeira, F. B., Moreira, H. H., Rodrigues, A. A., Machado, F. C., Clemente, T. M., et al. (2014). A successful strategy for the recovering of active P21, an insoluble recombinant protein of *Trypanosoma cruzi*. *Sci. Rep.* 4:4259. doi: 10.1038/srep04259

dos Santos, S. L., Freitas, L. M., Lobo, F. P., Rodrigues-Luiz, G. F., Mendes, T. A., Oliveira, A. C., et al. (2012). The MASP family of *Trypanosoma cruzi*: changes in gene expression and antigenic profile during the acute phase of experimental infection. *PLoS Negl. Trop. Dis.* 6:e1779. doi: 10.1371/journal.pntd.0001779

Doyle, P. S., Zhou, Y. M., Hsieh, I., Greenbaum, D. C., McKerrow, J. H., and Engel, J. C. (2011). The *Trypanosoma cruzi* protease cruzain mediates immune evasion. *PLoS Pathog.* 7:e1002139. doi: 10.1371/journal.ppat.1002139

El-Sayed, N. M., Myler, P. J., Bartholomeu, D. C., Nilsson, D., Aggarwal, G., Tran, A. N., et al. (2005). The genome sequence of *Trypanosoma cruzi*, etiologic agent of chagas disease. *Science* 309, 409–415. doi: 10.1126/science.1112631

Engel, J. C., Doyle, P. S., Hsieh, I., and McKerrow, J. H. (1998). Cysteine protease inhibitors cure an experimental *Trypanosoma cruzi* infection. *J. Exp. Med.* 188, 725–734. doi: 10.1084/jem.188.4.725

Ferrão, P. M., d'Avila-Levy, C. M., Araujo-Jorge, T. C., Degrave, W. M., Gonçalves Ada, S., Garzoni, L. R., et al. (2015). Cruzipain activates latent TGF-β from host cells during *T. cruzi invasion*. *PLoS ONE* 10:e0124832. doi: 10.1371/journal.pone.0124832

Fujio, Y., Nguyen, T., Wencker, D., Kitsis, R. N., and Walsh, K. (2000). Akt promotes survival of cardiomyocytes in vitro and protects against ischemia-reperfusion injury in mouse heart. *Circulation* 101, 660–667. doi: 10.1161/01.CIR.101.6.660

Furuya, T., Kashuba, C., Docampo, R., and Moreno, S. N. (2000). A novel phosphatidylinositol-phospholipase C of *Trypanosoma cruzi* that is lipid modified and activated during trypomastigote to amastigote differentiation. *J. Biol. Chem.* 275, 6428–6438. doi: 10.1074/jbc.275.9.6428

Gantt, K. R., Schultz-Cherry, S., Rodriguez, N., Jeronimo, S. M. B., Nascimento, E. T., Goldman, T. L., et al. (2003). Activation of TGF-beta by *Leishmania chagasi*: importance for parasite survival in macrophages. *J. Immunol.* 170, 2613–2620. doi: 10.4049/jimmunol.170.5.2613

Gao, W., Wortis, H. H., and Pereira, M. A. (2002). The *Trypanosoma cruzi* trans-sialidase is a T cell-independent B cell mitogen and an inducer of non-specific Ig secretion. *Int. Immunol.* 14, 299–308. doi: 10.1093/intimm/14.3.299

Gazzinelli, R. T., Pereira, M. E. S., Romanha, A., Gazzinelli, G., and Brener, Z. (1991). Direct lysis of *Trypanosoma cruzi*: a novel effector mechanism of protection mediated by human anti-gal antibodies. *Parasite Immunol.* 13, 345–356. doi: 10.1111/j.1365-3024.1991.tb00288.x

Giddings, O. K., Eickhoff, C. S., Smith, T. J., Bryant, L. A., and Hoft, D. F. (2006). Anatomical route of invasion and protective mucosal immunity in *Trypanosoma cruzi* conjunctival infection. *Infect. Immun.* 74, 5549–5560. doi: 10.1128/IAI.00319-06

Gonçalves, M. F., Umezawa, E. S., Katzin, A. M., de Souza, W., Alves, M. J., Zingales, B., et al. (1991). *Trypanosoma cruzi*: shedding of surface antigens as membrane vesicles. *Exp. Parasitol.* 72, 43–53. doi: 10.1016/0014-4894(91)90119-H

Guiñazú, N., Pellegrini, A., Carrera-Silva, E. A., Aoki, M. P., Cabanillas, A. M., Gìronés, N., et al. (2007). Immunisation with a major *Trypanosoma cruzi* antigen promotes pro-inflammatory cytokines, nitric oxide production and increases TLR2 expression. *Int. J. Parasitol.* 37, 1243–1254. doi: 10.1016/j.ijpara.2007.03.010

Jazin, E. E., Bontempi, E. J., Sanchez, D. O., Aslund, L., Henriksson, J., Frasch, A. C., et al. (1995). *Trypanosoma cruzi* exoantigen is a member of a 160 kDa gene family. *Parasitology* 110(Pt 1), 61–69. doi: 10.1017/S0031182000081051

Kierszenbaum, F., de Diego, J. L., Fresno, M., and Sztein, M. B. (1999). Inhibitory effects of the *Trypanosoma cruzi* membrane glycoprotein AGC10 on the expression of IL-2 receptor chains and secretion of cytokines by subpopulations of activated human T lymphocytes. *Eur. J. Immunol.* 29, 1684–1691. doi: 10.1002/(SICI)1521-4141(199905)29:05<1684::AID-IMMU1684>3.0.CO;2-7

Kierszenbaum, F., Fresno, M., and Sztein, M. B. (2002). The *Trypanosoma cruzi* membrane glycoprotein AGC10 inhibits human lymphocyte activation by a mechanism preceding translation of both, interleukin-2 and its high-affinity receptor subunits. *Mol. Biochem. Parasitol.* 125, 91–101. doi: 10.1016/S0166-6851(02)00217-7

Kulkarni, M. M., Karafova, A., Kamysz, W., Schenkman, S., Pelle, R., and McGwire, B. S. (2013). Secreted trypanosome cyclophilin inactivates lytic insect defense peptides and induces parasite calcineurin activation and infectivity. *J. Biol. Chem.* 288, 8772–8784. doi: 10.1074/jbc.M112.421057

Lee, E. J., Kim, N., Kang, K. H., and Kim, J. W. (2011). Phosphorylation/inactivation of PTEN by Akt-independent PI3K signaling in retinal pigment epithelium. *Biochem. Biophys. Res. Commun.* 414, 384–389. doi: 10.1016/j.bbrc.2011.09.083

Lopez, L., Morales, G., Ursic, R., Wolff, M., and Lowenberger, C. (2003). Isolation and characterization of a novel insect defensin from *Rhodnius prolixus*, a vector of chagas disease. *Insect Biochem. Mol. Biol.* 33, 439–447. doi: 10.1016/S0965-1748(03)00008-0

Maeda, F. Y., Cortez, C., Izidoro, M. A., Juliano, L., and Yoshida, N. (2014). Fibronectin-degrading activity of *Trypanosoma cruzi* cysteine proteinase plays a role in host cell invasion. *Infect. Immun.* 82, 5166–5174. doi: 10.1128/IAI.02022-14

Maeda, F. Y., Cortez, C., and Yoshida, N. (2012). Cell signaling during *Trypanosoma cruzi* invasion. *Front. Immunol.* 28:361. doi:10.3389/fimmu.2012.00361

Marcilla, A., Martin-Jaular, L., Trelis, M., Menezes- Neto, A., Osuna, A., Bernal, D., et al. (2014). Extracellular vesicles in parasitic diseases. *J. Extracell. Vesicles* 3:25040. doi: 10.3402/jev.v3.25040

Martin, D. L., Weatherly, D. B., Laucella, S. A., Cabinian, M. A., Crim, M. T., Sullivan, S., et al. (2006). CD8+ T-cell responses to *Trypanosoma cruzi* are highly focused on strain-variant trans-sialidase epitopes. *PLoS Pathog.* 2:e77. doi: 10.1371/journal.ppat.0020077

Martins, N. O., Souza, R. T., Cordero, E. M., Maldonado, D. C., Cortez, C., Marini, M. M., et al. (2015). Molecular characterization of a novel family of *Trypanosoma cruzi* surface membrane proteins (TcSMP) involved in mammalian host cell invasion. *PLoS Negl. Trop. Dis.* 9:e0004216. doi: 10.1371/journal.pntd.0004216

Mattos, E. C., Tonelli, R. R., Colli, W., and Alves, M. J. (2014). The Gp85 surface glycoproteins from *Trypanosoma cruzi*. *Subcell. Biochem.* 74, 151–180. doi: 10.1007/978-94-007-7305-9_7

McConville, M. J., Mullin, K. A., Ilgoutz, S. C., and Teasdale, R. D. (2002). Secretory pathway of trypanosomatid parasites. *Microb. Mol. Biol. Rev.* 66, 122–154. doi: 10.1128/MMBR.66.1.122-154.2002

McKerrow, J. H., Rosenthal, P. J., Swenerton, R., and Doyle, P. (2008). Development of protease inhibitors for protozoan infections. *Curr. Opin. Infect. Dis.* 21, 668–672. doi: 10.1097/QCO.0b013e328315cca9

Meirelles, M. N., Juliano, L., Carmona, E., Silva, S. G., Costa, E. M., Murta, A. C., et al. (1992). Inhibitors of the major cysteinyl proteinase (gp57/51) impair host cell invasion and arrest the intracellular development of *Trypanosoma cruzi* in vitro. *Mol. Biochem. Parasitol.* 52, 175–184. doi: 10.1016/0166-6851(92)90050-T

Minoprio, P. (2001). Parasite polyclonal activators: new targets for vaccination approaches? *Int. J. Parasitol.* 31, 588–591. doi: 10.1016/S0020-7519(01)00171-0

Moro, A., Ruiz-Cabello, F., Fernandez-Cano, A., Stock, R. P., and Gonzalez, A. (1995). Secretion by *Trypanosoma cruzi* of a peptidyl-prolyl cis-trans isomerase involved in cell infection. *EMBO J.* 14, 2483–2490.

Mott, A., Costales, J. A., and Burleigh, B. A. (2011). A soluble factor from *Trypanosoma cruzi* inhibits transforming growth factor-ss-induced MAP kinase activation and gene expression in dermal fibroblasts. *PLoS ONE* 6:e23482. doi: 10.1371/journal.pone.0023482

Ndao, M., Beaulieu, C., Black, W. C., Isabel, E., Vasquez-Camargo, F., Nath-Chowdhury, M., et al. (2014). Reversible cysteine protease inhibitors show promise for a chagas disease cure. *Antimicrob. Agents Chemother.* 58, 1167–1178. doi: 10.1128/AAC.01855-13

Nickel, W., and Seedorf, M. (2008). Unconventional mechanisms of protein transport to the cell surface of eukaryotic cells. *Annu. Rev. Cell Dev. Biol.* 24, 287–308. doi: 10.1146/annurev.cellbio.24.110707.175320

Nogueira, P. M., Ribeiro, K., Silveira, A. C., Campos, J. H., Martins-Filho, O. A., Bela, S. R., et al. (2015). Vesicles from different *Trypanosoma cruzi* strains trigger differential innate and chronic immune responses. *J. Extracell. Vesicles* 4:28734. doi: 10.3402/jev.v4.28734

Norris, K. A., Bradt, B., Cooper, N. R., and So, M. (1991). Characterization of a *Trypanosoma cruzi* C3 binding protein with functional and genetic similarities to the human complement regulatory protein, decay-accelerating factor. *J. Immunol.* 147, 2240–2247.

Norris, K. A., and Schrimpf, J. E. (1994). Biochemical analysis of the membrane and soluble forms of the complement regulatory protein of *Trypanosoma cruzi*. *Infect. Immun.* 62, 236–243.

Okura, M., Fang, J., Salto, M. L., Singer, R. S., Docampo, R., and Moreno, S. N. (2005). A lipid-modified phosphoinositide-specific phospholipase C (TcPI-PLC) is involved in differentiation of trypomastigotes to amastigotes of *Trypanosoma cruzi*. *J. Biol. Chem.* 280, 16235–16243. doi: 10.1074/jbc.M414535200

Ouaissi, A., Guevara-Espinoza, A., Chabe, F., Gomez-Corvera, R., and Taibi, A. (1995). A novel and basic mechanism of immunosuppression in Chagas' disease: *Trypanosoma cruzi* releases in vitro and in vivo a protein which induces T cell unresponsiveness through specific interaction with cysteine and glutathione. *Immunol. Lett.* 48, 221–224. doi: 10.1016/0165-2478(95)02463-8

Pereira-Chioccola, V. L., Acosta-Serrano, A., Correia de Almeida, I., Ferguson, M. A., Souto-Padron, T., Rodrigues, M. M., et al. (2000). Mucin-like molecules form a negatively charged coat that protects *Trypanosoma cruzi* trypomastigotes from killing by human anti-α-galactosyl antibodies. *J. Cell Sci.* 113, 1299–1307.

Ponce, N. E., Carrera-Silva, E. A., Pellegrini, A. V., Cazorla, S. I., Malchiodi, E. L., Lima, A. P., et al. (2013). *Trypanosoma cruzi*, the causative agent of Chagas disease, modulates interleukin-6-induced STAT3 phosphorylation via gp130 cleavage in different host cells. *Biochim. Biophys. Acta (BBA)* 1832, 485–494. doi: 10.1016/j.bbadis.2012.12.003

Reina-San-Martin, B., Degrave, W., Rougeot, C., Cosson, A., Chamond, N., Cordeiro-Da-Silva, A., et al. (2000). A B-cell mitogen from a pathogenic trypanosome is a eukaryotic proline racemase. *Nat. Med.* 6, 890–897. doi: 10.1038/78651

Rodrigues, A. A., Clemente, T. M., Dos Santos, M. A., Machado, F. C., Gomes, R. G., Moreira, H. H., et al. (2012). A recombinant protein based on *Trypanosoma cruzi* P21 enhances phagocytosis. *PLoS ONE* 7:e51384. doi: 10.1371/journal.pone.0051384

Rodriguez-Angulo, H. O., Toro-Mendoza, J., Marques, J. A., Concepcion, J. L., Bonfante-Cabarcas, R., Higuerey, Y., et al. (2015). Evidence of reversible bradycardia and arrhythmias caused by immunogenic proteins secreted by *T. cruzi* in isolated rat hearts. *PLoS Negl. Trop. Dis.* 9:e0003512. doi: 10.1371/journal.pntd.0003512

Rosenberg, I., Prioli, R. P., Ortega-Barria, E., and Pereira, M. E. (1991). Stage-specific phospholipase C-mediated release of *Trypanosoma cruzi* neuraminidase. *Mol. Biochem. Parasitol.* 46, 303–305. doi: 10.1016/0166-6851(91)90054-A

Santos, C. C., Sant'Anna, C., Terres, A., Cunha-e-Silva, N. L., Scharfstein, J., and de Lima, A. P. (2005). Chagasin, the endogenous cysteine-protease inhibitor of *Trypanosoma cruzi*, modulates parasite differentiation and invasion of mammalian cells. *J. Cell Sci.* 118, 901–915. doi: 10.1242/jcs.01677

Scharfstein, J., Schmitz, V., Morandi, V., Capella, M. M. A., Lima, A. P. C. A., Morrot, A., et al. (2000). Host cell invasion by *Trypanosoma cruzi* is potentiated by activation of bradykinin B2 receptors. *J. Exp. Med.* 192, 1289–1299. doi: 10.1084/jem.192.9.1289

Schenkman, S., Jiang, M.-S., Hart, G. W., and Nussenzweig, V. (1991). A novel cell surface trans-sialidase of *Trypanosoma cruzi* generates a stage-specific epitope required for invasion of mammalian cells. *Cell* 65, 1117–1125. doi: 10.1016/0092-8674(91)90008-M

Serna, C., Lara, J. A., Rodrigues, S. P., Marques, A. F., Almeida, I. C., and Maldonado, R. A. (2014). A synthetic peptide from *Trypanosoma cruzi* mucin-like associated surface protein as candidate for a vaccine against Chagas disease. *Vaccine* 32, 3525–3532. doi: 10.1016/j.vaccine.2014.04.026

Simpson, R. J., and Mathivanan, S. (2012). Extracellular microvesicles: the need for internationally recognised nomenclature and stringent purification criteria. *J. Proteomics Bioinform.* 5:1. doi: 10.4172/jpb.10000e10

Souza, W., Carvalho, T. M. U., and Barria, E. S. (2010). Review on *Trypanosoma cruzi*: host cell interaction. *Int. J Cell Biol.* 2010:295394. doi: 10.1155/2010/295394

Stempin, C., Giordanengo, L., Gea, S., and Cerba, F. (2002). Alternative activation and increase of *Trypanosoma cruzi* survival in murine macrophages stimulated by cruzipain, a parasite antigen. *J. Leukoc. Biol.* 72, 727–734.

Torrecilhas, A. C., Schumacher, R. I., Alves, M. J., and Colli, W. (2012). Vesicles as carriers of virulence factors in parasitic protozoan diseases. *Microbes Infect.* 14, 1465–1474. doi: 10.1016/j.micinf.2012.07.008

Torrecilhas, A. C., Tonelli, R. R., Pavanelli, W. R., da Silva, J. S., Schumacher, R. I., de Souza, W., et al. (2009). *Trypanosoma cruzi*: parasite shed vesicles increase heart parasitism and generate an intense inflammatory response. *Microbes Infect.* 11, 29–39. doi: 10.1016/j.micinf.2008.10.003

Umezawa, E. S., Shikanai-Yasuda, M. A., and Stolf, A. M. (1996). Changes in isotype composition and antigen recognition of anti-*Trypanosoma cruzi* antibodies from acute to chronic chagas disease. *J. Clin. Lab. Anal.* 10, 407–413. doi: 10.1002/(SICI)1098-2825(1996)10:6<407::AID-JCLA16>3.3.CO;2-9

Unnikrishnan, M., and Burleigh, B. A. (2004). Inhibition of host connective tissue growth factor expression: a novel *Trypanosoma cruzi*-mediated response. *FASEB J.* 18, 1625–1635. doi: 10.1096/fj.04-1554com

Vasudevan, K. M., Barbie, D. A., Davies, M. A., Rabinovsky, R., McNear, C. J., Kim, J. J., et al. (2009). AKT-independent signaling downstream of oncogenic PIK3CA mutations in human cancer. *Cancer Cell* 16, 21–32. doi: 10.1016/j.ccr.2009.04.012

Villalta, F., Zhang, Y., Bibb, K. E., Pratap, S., Burns, J. M. Jr., and Lima, M. F. (1999). Signal transduction in human macrophages by gp83 ligand of *Trypanosoma cruzi*: trypomastigote gp83 ligand up-regulates trypanosome entry through protein kinase C activation. *Mol. Cell. Biol. Res. Commun.* 2, 64–70. doi: 10.1006/mcbr.1999.0150

Waghabi, M. C., Keramidas, M., Bailly, S., Degrave, W., Mendonça-Lima, L., Soeiro, M. N., et al. (2005). Uptake of host cell transforming growth factor-beta by *Trypanosoma cruzi* amastigotes in cardiomyocytes: potential role in parasite cycle completion. *Am. J. Pathol.* 167, 993–1003. doi: 10.1016/S0002-9440(10)61189-3

Wainszelbaum, M., Isola, E., Wilkowsky, S., Cannata, J. J., Florin-Christensen, J., and Florin-Christensen, M. (2001). Lysosomal phospholipase A1 in *Trypanosoma cruzi*: an enzyme with a possible role in the pathogenesis of Chagas disease. *Biochem. J.* 355, 765–770. doi: 10.1042/bj3550765

Woolsey, A. M., Sunwoo, L., Petersen, C. A., Brachmann, S. M., Cantley, L. C., and Burleigh, B. A. (2003). Novel PI 3-kinase-dependent mechanisms of trypanosome invasion and vacuole maturation. *J. Cell Sci.* 116(Pt 17), 3611–3622. doi: 10.1242/jcs.00666

Yoshida, N. (2006). Molecular basis of mammalian cell invasion by *Trypanosoma cruzi*. *An. Acad. Bras. Cienc.* 78, 87–111. doi: 10.1590/S0001-37652006000100010

Yoshida, N. (2008). *Trypanosoma cruzi* infection by oral route: how the interplay between parasite and host components modulates infectivity. *Parasitol. Int.* 57, 105–109. doi: 10.1016/j.parint.2007.12.008

Zanforlin, T., Bayer-Santos, E., Cortez, C., Almeida, I. C., Yoshida, N., and da Silveira, J. F. (2013). Molecular characterization of *Trypanosoma cruzi* SAP proteins with host-cell lysosome exocytosis-inducing activity required for parasite invasion. *PLoS ONE* 8:e83864. doi: 10.1371/journal.pone.0083864

Zingales, B., Katzin, A. M., Arruda, M. V., and Colli, W. (1985). Correlation of tunicamycin-sensitive surface glycoproteins from *Trypanosoma cruzi* with parasite interiorization into mammalian cells. *Mol. Biochem. Parasitol.* 16, 21–34. doi: 10.1016/0166-6851(85)90046-5

6

mTOR Regulation of Lymphoid Cells in Immunity to Pathogens

*Rachael Keating and Maureen Ann McGargill**

Department of Immunology, St. Jude Children's Research Hospital, Memphis, TN, USA

Edited by:
Diana Bahia,
Universidade Federal de Minas Gerais, Brazil

Reviewed by:
Hridayesh Prakash,
University of Hyderabad, India
Katharina F. Kubatzky,
University Hospital Heidelberg, Germany

***Correspondence:**
Maureen Ann McGargill
maureen.mcgargill@stjude.org

Immunity to pathogens exists as a fine balance between promoting activation and expansion of effector cells, while simultaneously limiting normal and aberrant responses. These seemingly opposing functions are kept in check by immune regulators. The mechanistic target of rapamycin (mTOR) is a serine/threonine kinase that senses nutrient availability and, in turn, regulates cell metabolism, growth, and survival accordingly. mTOR plays a pivotal role in facilitating immune defense against invading pathogens by regulating the differentiation, activation, and effector functions of lymphoid cells. Here, we focus on the emerging and sometimes contradictory roles of mTOR in orchestrating lymphoid cell-mediated host immune responses to pathogens. A thorough understanding of how mTOR impacts lymphoid cells in pathogen defense will provide the necessary base for developing therapeutic interventions for infectious diseases.

Keywords: mTOR, immune regulation, pathogen, metabolism, immune differentiation

The mechanistic target of rapamycin (mTOR) is an evolutionarily conserved serine/threonine kinase that is ubiquitously expressed in immune cells. mTOR integrates multiple environmental signals to regulate diverse cellular processes including protein translation, cell growth, proliferation, metabolism, migration, and survival. Accordingly, mTOR plays a key role in multiple components of both myeloid and lymphoid cell differentiation, activation, and acquisition of effector functions. Through regulation of these key mechanisms, mTOR has an essential role in generating and regulating immune cells to combat pathogens.

The emerging view describes mTOR not as a single trigger for a linear cascade of events, but rather as a multifunctional orchestrator of diverse immune responses. In this review, we detail the current understanding of mTOR as a multifaceted regulator of immunity to pathogens through its impact on lymphoid cells. Specifically, we describe mTOR regulation of natural killer (NK) cells, invariant natural killer T (iNKT) cells, CD8 and CD4 T cells, and B cells during immunity to pathogens. The regulation of myeloid cells has been extensively reviewed recently (1). We also explore how mTOR inhibition may be utilized to enhance immunity to pathogens and discuss implications for vaccine design.

mTOR COMPLEXES AND SIGNALING CASCADES

Mechanistic target of rapamycin functions as two signaling complexes in mammalian cells: mTOR complex 1 (mTORC1) and mTOR complex 2 (mTORC2) (**Figure 1**). mTORC1 includes a scaffolding protein, regulatory-associated protein of mTOR (Raptor), DEP-containing mTOR interacting protein (Deptor), mammalian lethal with Sec13 protein (mLST8), and the Proline-Rich AKT substrate (PRAS40). Similarly, mTORC2 also comprises mTOR, Deptor, and mLST8, with the addition of the scaffold protein Raptor-independent Companion of TOR (Rictor), the Protein observed with Rictor (Proctor), and the mammalian stress activated protein kinase-interacting protein 1 (mSIN1).

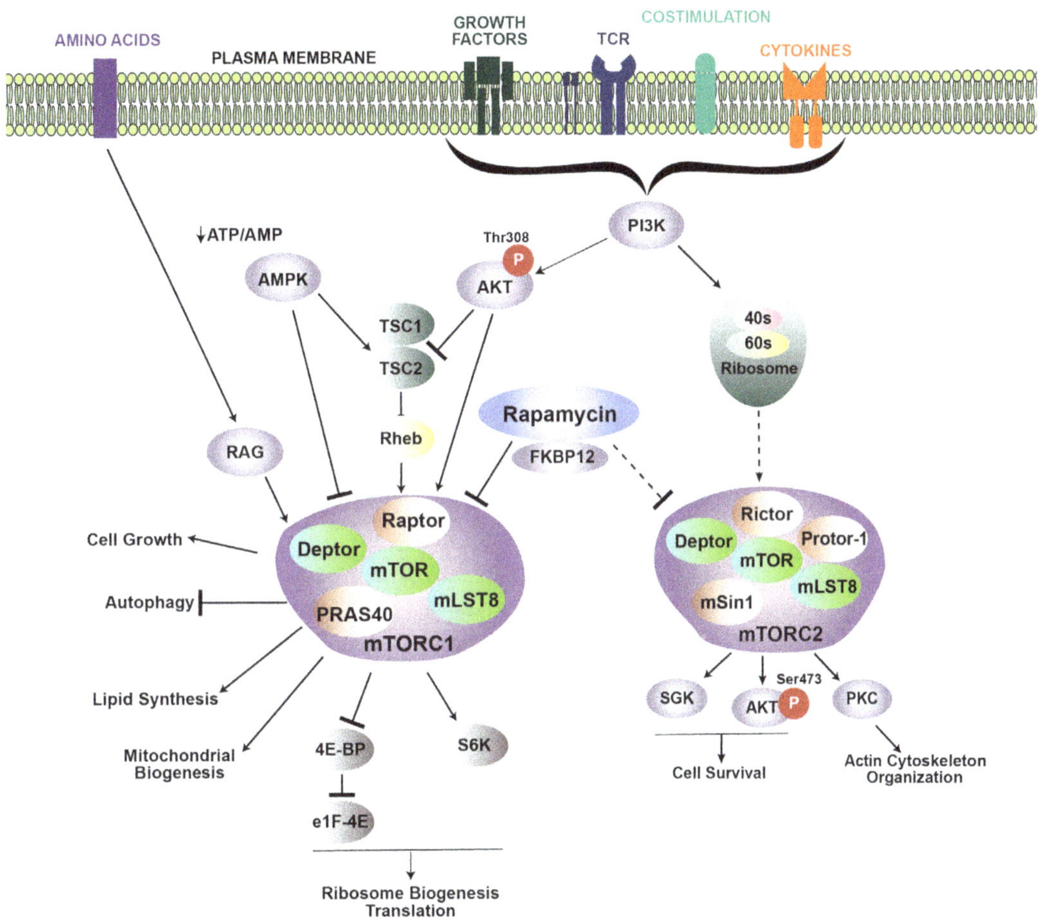

FIGURE 1 | Signaling cascades promoting mTOR activation. mTOR regulates multiple cellular processes via two distinct complexes: mTORC1 and mTORC2. Growth factors, TCR engagement, costimulation, and cytokines all contribute to PI3K activation, which leads to the recruitment of Akt to the membrane, where it is phosphorylated at position threonine 308. Activated Akt then phosphorylates the TSC1/TSC2 inhibitory complex, which releases Rheb and induces accumulation of Rheb-GTP to promote mTORC1 activity. Rapamycin inhibits mTOR by binding to the intracellular partner, FKBP12, which directly inhibits mTORC1. Upstream regulation of mTORC2 is less known, but is downstream of PI3K activation. Association with ribosomes regulates mTORC2 activation. mTORC2 is typically regarded as insensitive to rapamycin; however, prolonged rapamycin treatment can reduce mTORC2 activity. Arrows and bars represent activation and inhibition, respectively. Dashed lines indicate that the exact mechanism is unknown.

mTOR complex 1 is activated in response to various extracellular stimuli including nutrients, growth factors, stress, cytokines, and antigen receptor signaling. When nutrients and these stimuli are readily available, mTORC1 activity is high, and energy-demanding cellular processes such as translation and ribosomal biogenesis are promoted. Extracellular stimuli activate mTORC1 by triggering a signaling cascade through PI3 kinase (PI3K) and the protein kinase, Akt. Akt then phosphorylates the mTORC1 repression factor, consisting of tuberous sclerosis complex 1 (TSC1) and TSC2. Phosphorylation of the TSC complex prevents it from inhibiting Rheb, which is essential for mTORC1 activation. Additionally, mTORC1 can also be activated in a TSC-independent process, whereby Akt blocks inhibition of mTORC1 by phosphorylating PRAS40 to stabilize the mTOR–Raptor interaction (2, 3). Activation of mTORC1 leads to phosphorylation of p70–S6 kinase (S6K) and the eukaryotic initiation factor 4E-binding protein (4E-BP1), which helps regulate translation (4–6). mTORC1 signaling also increases the rate of glycolysis by inducing the expression of HIF-1α and c-Myc. Conversely, mTORC1 activity is inhibited when nutrient-associated cues are lacking. For example, in response to a decrease in the ATP/ADP ratio, the AMP-activated protein kinase (AMPK) becomes activated, and in turn, inhibits mTOR activity by phosphorylating TSC2 or Raptor.

Activation of mTORC2 is not as well defined as mTORC1. Similar to mTORC1, activation of mTORC2 occurs downstream of PI3K. However, activation of mTORC1 occurs downstream of Akt following phosphorylation at threonine 308 (T308), while mTORC2 acts upstream of Akt by phosphorylating serine 473 of Akt (S473). In contrast to mTORC1 activation, mTORC2 activation occurs independent of protein synthesis, but instead, relies on association with the ribosome (7). Activation of Akt by mTORC2 also leads to phosphorylation and inhibition of FOXO1 and FOXO3 (8). In addition to Akt, mTORC2 also activates serum glucocorticoid-regulated kinase 1 (SGK1) and protein kinase C alpha (PKCα). Signaling via mTORC2 is important for reorganization of the cytoskeleton through activation of RhoA GTPase and promoting cell survival by upregulating anti-apoptotic proteins (7, 9).

Much of the role and function of mTOR has been ascertained with rapamycin, a small molecule drug derived from *Streptomyces hygroscopicus*. Rapamycin binds a 12-kDa Fk506-binding protein (FKBP12) to form the rapmycin–FKBP12 complex, which binds mTOR and inhibits formation of the mTORC1 complex (10). Prolonged rapamycin exposure also inhibits mTORC2 (11). However, mTORC2 is typically regarded as insensitive to rapamycin. Use of knock-out mice, in particular, mice with a conditional deletion of *Raptor*, *Rictor*, *Rheb*, or *Tsc*, which are critical components of mTORC1 and mTORC2 complexes, have also greatly advanced our understanding of mTOR signaling pathways and function (**Table 1**).

Several pathways regulate activation of mTORC1 and mTORC2. When energy levels are low, mTORC1 is inactivated, and FOXO transcribes *Rictor*, promoting formation of mTORC2, which phosphorylates Akt S473 (12). Transcription of *Rictor* and subsequent phosphorylation of Akt S473 requires mTORC1 inhibition (13). Similarly, while mTORC1 activates protein synthesis and S6K, S6K activity can repress Rictor and mTORC2 function. In addition, recent studies highlight a positive feedback loop between Akt and mTORC2 via SIN1 phosphorylation, whereby Akt is activated following PDK1 phosphorylation. Next, Akt phosphorylates SIN1, enhancing mTORC2 activity, which then promotes phosphorylation and complete activation of Akt (14).

Pathogens can also influence activation of the mTOR pathway. mTORC1 regulates translation by phosphorylating 4E-BP1, which releases it from the 5′ cap-binding protein, eukaryotic translation initiation factor 4E (eIF4E) allowing translation to proceed (4–6). Pathogens that are dependent on the host's cellular 5′ cap-dependent translation must therefore maintain mTOR activity, or bypass the need for mTOR-mediated phosphorylation of 4E-BP1 to enable the translation complex to form. Indicative of the former approach, human papillomavirus (HPV) uses two early proteins, E6 and E7, to activate mTOR signaling, which phosphorylates and inactivates 4E-BP1 to support viral cap-dependent protein synthesis (15, 16). Similarly, Epstein–Barr virus (EBV) activates cap-dependent translation using a viral protein, LMP2A, to activate mTORC1 (17). Adenovirus also uses viral proteins (e4-ORF1 and e4-ORF4) to mimic stimulatory signals and activate mTORC1 activity in the absence of nutrients or growth factors to maintain translation of viral proteins (18). Bacterial pathogens including *Listeria monocytogenes* (*L. monocytogenes*) and *Staphylococcus aureas* can also activate mTOR to promote IL-10 production and increase their survival in the host (19). Alternatively, some pathogens have evolved mechanisms to bypass mTORC1 activity. For example, human cytomegalovirus (HCMV) bypasses mTORC1 activity by directly phosphorylating 4E-BP1 and eIF4G to maintain the activity of the translation complex (20). In contrast, some pathogens such as *Leishmania major* have proteases that block mTOR activation, which suppresses the type 1 IFN response, allowing the pathogen to survive within cells (21). Hence, a pathogen's translation requirements and the ability to resolve these requirements will influence whether the pathogen tries to enhance, bypass, or suppress mTOR activity, and in turn, will influence the counter approach by the host immune response.

mTOR REGULATION OF AUTOPHAGY IN HOST DEFENSE

Mechanistic target of rapamycin regulates cell processes in response to nutrient availability. A key component of cellular control by mTOR is through regulation of autophagy, which is an essential process in all myeloid and lymphoid cells. Autophagy facilitates turnover of unnecessary or damaged cellular components. These cellular components are surrounded by a double-membraned vesicle, targeted to a lysome, degraded, and then recycled. This process allows cells to survive under stress. When energy sources are low, mTOR activity is low, biosynthesis is attenuated, and autophagy is upregulated to recycle nutrients, rather than synthesize new material. This prevents translational arrest and cell death. Conversely, when energy and nutrients are readily available, mTOR is active and signals downstream pathways to generate new cellular material to promote cell growth and proliferation, while suppressing autophagy. Basal autophagy levels are essential for homeostatic clearance of protein aggregates and damaged organelles (22). Basal autophagy is regulated independent of mTOR; however, mTOR suppresses autophagy induction above basal levels (23). Regulation of autophagy by mTOR provides an interface for both pathogen assault and host defense, as intracellular pathogens compete with the host for energy and resources.

Stimuli triggered by pathogen infection can induce autophagy above basal levels to destroy intracellular pathogens, while simultaneously increasing the cell surface presentation of microbial antigens to stimulate the immune response. For example, infection with the bacteria, *Shigella flexneri*, causes amino acid starvation and subsequent downregulation of mTOR to induce autophagy (24). The adaptation of the immune system to detect and respond to intracellular pathogens has simultaneously provoked evolution of some pathogens to circumvent autophagy induction. Indeed, HSV-1 and HSV-2 prevent induction of autophagy to evade immune defense mechanisms (22, 25, 26). Similarly, *L. monocytogenes* and *Salmonella* attempt to subvert induction of autophagy by reactivating mTOR to downregulate the immune response (24, 27). Therefore, these pathogens hijack and maintain basal levels of autophagy to exploit host energy supplies and nutrients for their own replication. In such situations, it is beneficial for host defense mechanisms to inhibit mTOR and induce autophagy above basal levels.

In contrast, autophagy induction can benefit some pathogens by supporting their replication. Multiple subtypes of influenza A virus induce autophagy and autophagic cell death by suppressing mTOR to promote replication (28, 29). Hence, autophagy inhibitors may limit influenza virus infection. Datan et al. also reported that influenza virus infection induced autophagy in

TABLE 1 | Overview of studies demonstrating the role of mTOR in lymphoid cells following pathogen infection.

Cell type	mTOR modification	Pathogen	Cellular and pathogen outcome	Conclusion about mTORs activity during infection	Reference
NK cells	Rapamycin	MCMV	Blocked proliferation, IFN-γ synthesis and granzyme B expression. Higher viral titer	Promotes proliferation, IFN-γ synthesis and granzyme B expression, and pathogen clearance	(33)
	mTOR$^{-/-}$ NK cells	MCMV	Blocked proliferation and granzyme B expression. Higher viral titer	Promotes proliferation and granzyme B expression and pathogen clearance	(31)
CD8 effector	T cell-specific *Tsc2* deletion to enhance mTORC1 activity	Vaccinia-OVA	Excessive generation of effector CD8^{+} T cells, unable to differentiate into memory cells. High cytolytic activity. Robust IFN-γ and TNF-α	mTORC1 promotes generation of effectors and mTORC1 suppression promotes memory formation	(57)
	T cell-specific *Rictor* deletion to inhibit mTORC2	Vaccinia-OVA	Unaltered CD8 differentiation and effector function	mTORC2 does not regulate effector cells	(57)
	T cell-specific *Rheb* deletion to inhibit mTORC1	Vaccinia-OVA	Reduced CD8 effector function. Decreased IFN-γ, TNF-α, and cytolytic function	mTORC1 enhances CD8 effector function	(57)
	Rapamycin	LCMV and *Listeria*	Impaired CD8 effector function and reduced pathogen clearance	mTORC1 promotes effector function and pathogen clearance	(59)
	Rapamycin	Influenza	Reduces IRF4 expression, which is required for effector CD8 T cell differentiation and expansion. Impaired viral clearance and host recovery	mTOR regulates IRF4 expression to impact during CD8 T cell differentiation to promote pathogen clearance	(61)
	Rapamycin	LCMV and LM-OVA	Impaired effector CD8 T cell number and function, Reduced IFN-γ, TNF-α, granzyme B, and cytolytic activity. Reduced pathogen clearance and survival	mTORC1 promotes effector function, pathogen clearance, and host survival	(59)
CD8 memory	Rapamycin, *mTOR,* and *Raptor* deletion	LCMV	Enhanced memory cell quantity, quality, and persistence	mTORC1 suppresses memory quality and quantity	(67)
	In vitro rapamycin treatment prior to cell transfer	LCMV-gp 33 peptide	Enhanced and long-lived memory cell formation	mTORC1 suppresses memory formation	(69)
	In vitro rapamycin treatment of WT and *TRAF6*$^{-/-}$ cells prior to cell transfer and infection	LM-OVA	Restored the ability to develop memory cells and increased the recall response in the absence of *TRAF6*	mTOR blocks memory development and recall responses	(68)
	T cell-specific *Tsc2* deletion to enhance mTORC1 activity	LM-OVA	Effector cells were unaltered. Differentiation of effector cells to memory cells was impaired. Recall response was reduced	Excessive mTORC1 activity inhibits memory formation and is regulated by Tsc1	(70)
	Rapamycin	LCMV and LM-OVA	Enhanced CD8 memory formation	mTOR suppress memory formation	(59)
	Rapamycin	Canary poxvirus	Long-term, low dose rapamycin blocked memory formation. Short-term, high dose rapamycin enhanced CD8 memory	Sustained, low level mTOR activity supports memory formation	(71)
	Rapamycin	Vaccinia virus	IL-12-dependent increase in memory CD8 T cells	IL-12 regulates the mTORC1 block in formation of memory CD8 T cells	(76)
	T cell-specific *Rictor* deletion to inhibit mTORC2	Vaccinia-OVA	Enhanced generation of memory CD8 T cells	mTORC2 limits memory cell formation	(57)
CD8-resident memory	Rapmycin shRNA silenced mTOR	Vesicular stomatitis virus (VSV) and VSV-OVA	Rapamycin increased the quantity of memory CD8 in the spleen but reduced resident memory cells in the intestinal mucosa and vaginal mucosa	mTOR enhanced formation of memory cells in the intestinal and vaginal mucosa	(84)
CD8 secondary expansion	Rapamycin	LCMV, Pichinde virus	IL-15-dependent, virus-induced cell cycling of memory CD8 cells was blocked	Inflammatory IL-15 activates the mTORC1-signaling pathway to support preexisting memory cells and enhance antiviral protection	(78)
CD8 T cell exhaustion	Rapamycin	Chronic LCMV	Abrogated therapeutic effects of blocking PD-1, leading to CD8 T cell exhaustion and failure to control chronic infection	During chronic infection persistent antigen impairs mTOR activation, allowing FOXO1 activity to increase and promote differentiation of terminally exhausted CTLs	(79)
Tfh cells	shRNA silenced *Rictor* or *Raptor*	LCMV	Raptor silencing favored Tfh development over Th1 development. Rictor silencing favored Th1 over Tfh development	IL-2-mediated mTORC1 activation promotes Th1 over Tfh development. mTORC2 activation favors Tfh over TH1 development	(118)

(Continued)

TABLE 1 | Continued

Cell type	mTOR modification	Pathogen	Cellular and pathogen outcome	Conclusion about mTORs activity during infection	Reference
B cells	Mice hypomorphic for *mTOR* and B cell-specific deletion of *mTOR*	*S. pneumoniae*	Decreased germinal centers, high-affinity antibodies, and SMH/CSR. Higher mortality in hypomorphic mTOR mice	mTOR is a critical immunoregulator, promoting germinal center formation through AID signaling to generate high affinity antibodies	(125)
	ATP-competitive mTOR kinase inhibitor (TOR-KIs)	*Salmonella*	Early (d14) IgM response was unaltered and IgG2c decreased. Late (d28) IgM increased and Tfh cell% increased with some evidence of increased GC B cells	Partial inhibition of mTOR activity increases protective IgM responses	(138)
	Rapamycin	Influenza vaccination and heterosubtypic challenge	Delayed germinal center formation, reduced class switching, increased survival. Increased viral clearance	mTOR supports antibody class switching and affinity maturation, which may impair viral clearance to heterosubtypic infection	(137)

apoptotic cells in the presence of mTORC1 and mTORC2 activity, indicating that alternate regulatory mechanisms may override suppression of autophagy by mTOR (30). The degree to which various pathogens support or inhibit mTOR activity is therefore a reflection of the extent to which autophagy benefits or hinders their own replication, and the degree to which they have evolved to counter host immune responses.

mTOR IN NK CELL ACTIVATION AND EFFECTOR FUNCTIONS

Natural killer cells are a subset of innate lymphoid cells that limit infection by intracellular pathogens and promote tumor immunosurveillance. The name "natural killer" reflects their capacity to kill target cells without prior antigenic stimulation. NK cells develop from common lymphoid progenitor cells in the bone marrow in a process dependent on proliferation and mTOR. Mice with a conditional deletion of *mTOR* in NK cells had a dramatic reduction in NK cell development and differentiation due to defects in proliferation (31). Similarly, transplant patients treated with rapamycin had reduced NK cell numbers (32).

Following development and maturation in the bone marrow, NK cells enter the periphery in a metabolically resting, quiescent state. In the periphery, exposure to IL-15 or viruses promotes NK cell activation, leading to an increase in metabolism, cytokine production, and acquisition of cytotoxic effector functions. High IL-15 concentrations are required to activate mTOR (31). Complete NK cell activation requires mTOR signaling, as rapamycin and NK cell-specific deletion of *mTOR* blocked proliferation and granzyme B expression in response to *in vitro* cytokine stimulation, *in vitro* polyI:C stimulation, and MCMV infection (31, 33, 34). Consequently, inhibition of mTOR and NK function resulted in higher viral titers following MCMV infection (31, 33). Interestingly, IFN-γ secretion was unimpaired following MCMV infection in mice with a NK cell-specific mTOR deletion, yet it was blocked in MCMV-infected mice treated with rapamycin (33). This difference could be due to the fact that rapamycin inhibits mTOR in cells other than NK cells, suggesting that mTOR signaling in other cell types may impact NK cell effector functions. Alternatively, rapamycin may inhibit the mTOR pathway to a different extent than genetic deletion of mTOR in NK cells.

mTOR complex 1 regulates NK cell effector function by enhancing glucose uptake and promoting aerobic glycolysis. Accordingly, directly limiting glycolysis inhibits IFN-γ production and granzyme B expression by NK cells (35). Cells infected with pathogens typically increase glucose uptake and glycolysis and therefore, limit the amount of glucose available to surrounding immune cells (36). The effector functions of NK cells may therefore be hindered by the availability of glucose following infection. During the initial phase of MCMV infection, proliferation of NK cells is IL-15 and mTOR dependent. However, NK cell proliferation subsequently becomes IL-15 and mTOR independent, at which point proliferation is driven by activating receptors on the NK cell, such as Ly49H, which recognizes viral ligands on infected cells, but does not activate mTOR (37, 38). The later, mTOR-independent phase of such immune responses may be an adaptation to maintain NK effector functions with diminishing glucose supplies. Regardless, this model illustrates that immune response kinetics influence the requirement for mTOR signaling in pathogen defense.

mTOR IN iNKT CELL DEVELOPMENT, ACTIVATION, AND EFFECTOR FUNCTIONS

Signaling through mTOR is also important in the development, activation, and effector function of invariant NKT cells (iNKT) cells. iNKT cells share features common with both NK cells and T cells. Similar to NK cells, iNKT cells rapidly produce large amounts of cytokines following activation, including IFN-γ, TNF-α, IL-4, and IL-17. However, unlike NK cells, iNKT cells express a T cell receptor (TCR) similar to conventional T cells, albeit at intermediate levels and with decreased diversity. In mice, the iNKT TCR consists of a Vα14Jα18 α-chain paired with Vβ2, Vβ7, or Vβ8.2 β chains. The iNKT cell TCR recognizes lipid antigens bound to the non-classical major histocompatibility complex (MHC) homolog, CD1d. These cells contribute to tumor immunity, autoimmunity, and immunity to pathogens.

Invariant natural killer T cells can be divided into subsets based on transcription factor expression and cytokine production. Expression of the transcription factors T-bet, GATA3, and RORγt define the NKT1, NKT2, and NKT17 subsets, respectively (39, 40). Mice with a conditional deletion of *Raptor* in T cells showed that mTORC1 is required for the development of the NKT1 subset, and to a lesser degree, NKT2 cells (41, 42). Furthermore, the remaining iNKT cells were functionally compromised, as they failed to produce IFN-γ and TNF-α following stimulation with α-Gal-Cer. These data indicate that mTORC1 is differentially required for NKT cell development and effector functions.

While mTORC1 was required for development of NKT1 and NKT2 cells, mice with abrogated mTORC2 signaling had a defect in NKT17 development (43). Moreover, development of NKT17 cells was enhanced in a *Rictor*-dependent manner in the absence of PTEN, further clarifying that differentiation of the NKT17 lineage is mTORC2 dependent, but mTORC1 independent. However, Prevot et al. found that mTORC2 was required for the development of NKT2, but not NKT1 or NKT17 cells (44). The reason for this inconsistency is not known as both studies used mice with a conditional deletion of *Rictor* in T cells for their studies. Possible factors that contribute to such variable findings include diverse microbiomes associated with the different facilities, use of different markers to identify iNKT cell subsets, and technical difficulties associated with the reagents available to analyze iNKT cell transcription factors. An intriguing study recently showed that in the presence of IL-10 and rapamycin, iNKT cells expressed Foxp3 and acquired properties associated with immunosuppressive regulatory cells, identifying yet another population of iNKT cells: Foxp3$^+$ iNKT cells (iNKTregs) (45).

Stimulation of iNKT cells with α-Gal-Cer activates mTOR signaling (46). Mice with an inducible deletion of *Raptor* in mature iNKT cells showed that mTORC1 signaling is required for optimal proliferation and IL-4, IFN-γ, and TNF-α synthesis following *in vitro* and *in vivo* stimulation with α-Gal-Cer (42). Moreover, *Raptor* deficiency inhibited iNKT-mediated autoimmune hepatitis, which arises in part, due to TNF-α produced by iNKT cells, further highlighting a role for mTOR in regulating iNKT cell effector function.

A role for iNKT cells in pathogen defense is clearly evident following their activation with artificial ligands during infection with different pathogens including malaria (47), HIV (48), influenza virus (49), and hepatitis B virus associated hepatic carcinoma (50). In addition, iNKT cells mediate pathogen defense against several strains of bacteria that express microbial glycolipids and diacylglycerols capable of binding CD1d molecules, such as *Streptococcus pneumoniae* (51, 52), *Novosphingobium aromaticivorans* (53), and *Borrelia burgdorferi* (52). At present, mTOR has not been implicated in pathogen responses mediated by iNKT cells. However, as mTOR is required for the development and function of iNKT cells, it is perhaps only a matter of time before a role for mTOR in regulating pathogen defense by iNKT cells is established.

mTOR IN CD8 T CELL IMMUNITY TO PATHOGENS

CD8$^+$ T cells play a pivotal role in host defense against pathogens by directly killing infected cells via release of cytotoxic granules containing granzymes and perforin, and indirectly through cytokine secretion. A critical role for mTOR is implicated in several stages of CD8 T cell-mediated immunity including activation, differentiation, migration, and memory formation.

Naive CD8$^+$ T cells undergo sequential stages of activation, clonal expansion, and differentiation to generate pathogen-specific effector CD8$^+$ T cells. Activation of naive CD8$^+$ T cells requires recognition of their cognate antigen presented by MHC class I molecules on the surface of antigen-presenting cells (APC). Engagement of the TCR with antigen and MHC, in the context of costimulation, induces PI3K signaling and subsequent activation of mTORC1 via TSC1-dependent and -independent processes. Activation of mTORC1 independent of TSC1 occurs through binding of the chaperone protein, Hsp90 to Raptor. The interaction of Hsp90 and Raptor promotes mTORC1 activation and prevents T cell anergy, as activation of T cells during a blockage of Hsp90 led to T cell tolerance (54).

One way in which mTOR regulates T cell activation is by blocking negative regulators of T cell activation. In particular, naive CD8$^+$ T cells are kept in a state of quiescence by several transcription factors including KLF4 and E74-like factor 4 (ELF4). Signaling through mTOR inhibits KLF4 and ELF4 and reverses quiescence following TCR engagement, allowing CD8$^+$ T cell activation and proliferation to proceed (55).

Following CD8$^+$ T cell activation, antigen-specific T cells undergo clonal expansion and differentiate into effector cells. Progression to effector CD8$^+$ T cells is coordinated with a switch from catabolism to anabolism and oxidative glycolysis. CD8$^+$ T cells drastically increase their glucose metabolism as they respond to pathogens and differentiate into effectors (56). Signaling through mTORC1 promotes glycolysis in effector CD8$^+$ T cells, and mTOR inhibition with rapamycin suppresses this process. Indeed, CD8$^+$ T cells deficient in mTORC1 signaling failed to develop into effector cells following Vaccinia infection (57, 58). Likewise, rapamycin treatment during crucial stages of infection impaired glycolysis and impeded CD8$^+$ effector function and pathogen clearance during infection with lymphocytic choriomeningitis virus (LCMV) and *L. monocytogenes* (59). Conversely, genetic deletion of *Tsc2* and enhanced mTORC1 activity led to excessive generation of CD8$^+$ effector cells. Importantly, constitutively activate mTORC1 produced terminally differentiated effector cells and impaired memory development (57). While mTORC1 activity is required to generate effective CD8$^+$ effector cells, mTORC2 appears to be dispensable as mice with a conditional deletion of *Rictor* exhibit normal proliferation and effector functions (57).

Effector differentiation of CD8$^+$ T cells is tightly regulated by several transcription factors. The transcription factor HIF-1α is required to sustain glucose metabolism and expression of perforin and granzymes in effector CD8$^+$ T cells (58). HIF-1α controls glycolysis by regulating expression of the glucose transporter,

Glut1. As CD8+ T cells differentiate into effectors, they increase expression of HIF-1α in an mTORC1- and IL-2-dependent manner. Therefore, inhibition of mTORC1 with rapamycin inhibits HIF-1α and Glut1 expression, glucose uptake, and glycolysis (58). Signaling through mTOR also promotes glycolysis via the oncogene c-MYC, which regulates metabolic reprograming in T cells following activation (60). Expression of c-MYC is upregulated in an mTOR-dependent manner following TCR stimulation and influences the expression of rate-limiting glycolytic enzymes (60).

Although the HIF-1α complex is required to sustain glycolytic metabolism in CD8+ T cells, it is not required to initiate the process nor is it essential for T cell proliferation (58). In contrast, the transcription factor interferon regulatory factor 4 (IRF4) was critical for proliferation and survival of CD8+ T cells responding to infection with influenza virus and *L. monocytogenes* (61, 62). IRF4 maintains aerobic glycolysis and effector functions in activated CD8+ T cells by promoting the expression of multiple glycolytic enzymes (62). IRF4 expression in CD8+ T cells is regulated by mTOR, and is proportional to the strength of TCR stimulation, such that strong TCR stimulation increases mTOR activity, which enhances IRF4 expression and CD8+ T cell expansion and effector differentiation. Conversely, rapamycin treatment reduced mTOR signaling and impaired IRF4 expression and CD8+ T cell differentiation, leading to impaired viral clearance and host recovery from infection (61).

SEMA4A is a class IV semaphorin that activates mTORC1 signaling and is also required for optimal CD8+ T cell activation and differentiation (63). SEMA4A-mediated mTOR signaling was not important in the early phases of CD8+ T cell activation, but was important for acquisition of effector functions. Interestingly, IRF4 expression was unaltered in SEM4A4-deficient mice, yet effector differentiation was impaired, indicating that IRF4-dependent signaling may be necessary, but not sufficient, for differentiation of CD8+ effector T cells. Furthermore, the SEMA4A-plexin B2 axis shifts mTOR-mediated signaling from mTORC2 to mTORC1 in CD8+ T cells (63).

After expansion in the lymph nodes, effector CD8+ T cells travel to the site of infection to kill infected cells. Migration from the lymph node to the site of infection is in part, due to downregulation of CD62L and CCR7, which is dependent on activated mTOR (64). Such regulation by mTOR is evident from maintenance of CD62L and CCR7 in cells treated with rapamycin, leading to the accumulation of these cells in the secondary lymphoid organs rather than non-lymphoid tissue (65).

Following the dramatic CD8+ T cell expansion, and subsequent clearance of the pathogen, T cells go through a contraction phase leaving a resolute population of memory cells. The transition from effector to memory cells is coordinated with a transition from high mTOR activity to low mTOR activity. In contrast to CD8+ effector T cells, which favor glucose over fatty acid metabolism for energy, memory cells are less glycolytic and rely on lipid metabolism to break down fatty acids, amino acids, and glucose interchangeably (66).

The role of mTOR in regulating the transition from effector to memory cells has been studied extensively. Treatment of mice with a low dose of rapamycin or genetic depletion of *Raptor*-enhanced memory CD8+ T cell formation by promoting the generation of memory precursor cells, indicating that high mTOR signaling suppresses memory CD8+ T cell differentiation (67, 68). Similarly, adoptive transfer of LCMV-specific T cells cultured with rapamycin showed an increase in oxidative phosphorylation and were long-lived, compared to non-treated cells (69). Conversely, increasing mTOR via *Tsc1* deletion in T cells led to a decrease in memory T cells following *L. monocytogenes* infection, suggesting that TSC1 promotes memory CD8+ T cell formation by regulating mTOR activity (70). While reducing mTOR signaling increases memory T cells, it comes at the expense of an optimal primary effector response, which occurs due to impaired glycolysis in the CD8+ T cells after activation (59).

Although a low dose of rapamycin can augment memory formation in response to certain pathogens, long-term blockage of mTORC1 with the same low dose of rapamycin abrogated formation of memory CD8+ T cells in response to canary poxvirus vaccination (71). In this study, a short duration of a high dose of rapamycin during the expansion phase enhanced memory CD8+ T cell responses. Several factors could contribute to these observed differences in generating memory cells with rapamycin, including the strength of TCR stimulation, the level of mTOR activity induced with each pathogen, differential requirements and timing of pathogen replication, and experimental differences in the location (blood versus lymph node and spleen) where the T cells were analyzed. A recent study indicated that signaling through IFN-γR, with low, but not high TCR signaling, promoted mTOR signaling and generation of short lived effector cells, while simultaneously blocking the formation of memory precursors (72). Given that TCR signaling differs between pathogens, pathogen dose, and even between epitopes of the same pathogen (73), it is not surprising that rapamycin has varied impacts on the CD8 response to different pathogens. In addition, *Rictor*-deficient mice showed enhanced generation of memory CD8+ T cells, which implicates mTORC2 in suppressing memory cell formation (57). As mTORC2 does not impact formation of effector cells, yet suppresses memory formation, inhibition of mTORC2, rather than mTORC1, may provide an alternate means to boost CD8 memory T cells.

Studies investigating CD8-mediated tumor immunity demonstrate that mTOR controls memory CD8+ T cell differentiation by regulating two transcription factors, T-bet and Eomesodermin (Eomes) (74, 75). *In vitro* inhibition of mTOR with a low dose of rapamycin reduced T-bet expression and enhanced Eomes expression in CD8+ T cells, leading to augmented memory responses following adoptive transfer *in vivo* (74). A recent study indicates that mTOR also controls vaccinia virus-specific CD8+ T cell differentiation by regulating T-bet (76). Therefore, it is likely that mTOR similarly regulates memory differentiation in response to pathogens through T-bet and Eomes.

The effector to memory cell transition is also coupled with the cytokine environment, which appears to be critical for optimal memory development (77). Activation of mTOR by IL-7 promotes T-bet expression and upregulates IL-2Rβ. This, in turn, allows IL-15, which also utilizes IL-2Rβ for signaling, to promote differentiation of effector cells to memory cells through the upregulation of Eomes (75). IL-15 appears to be particularly important in promoting optimal memory following LCMV infection by transiently inducing cell-cycle progression, independent of antigen re-encounter, via a mTORC1-dependent pathway. This induces more rapid cell division and more protective memory cells following encounter with viral antigen (78). Thus, mTOR signaling impacts multiple stages of T cell activation and memory formation, and the impact of mTOR signaling can have different outcomes based on the stage of cellular differentiation.

An additional role for mTOR in regulating T cell exhaustion in response to chronic infection has emerged (79). T cell exhaustion refers to a state of dysfunction marked by reduced effector functions due to the persistence of antigen and inflammation (80). Following chronic LCMV infection, Akt and mTOR signaling were impaired in $CD8^+$ effector T cells. This reduction in mTOR activity led to increased activity of the transcription factor, FOXO1 and subsequent upregulation of the inhibitory receptor, programed cell death protein 1 (PD-1), which promotes $CD8^+$ T cell exhaustion (79). Blockade of PD-1 increased mTOR activity and decreased $CD8^+$ T cell exhaustion. Furthermore, suppression of the mTOR pathway abrogated the therapeutic effects of PD-1 blockade, suggesting that mTOR activity is required to reverse T cell exhaustion by PD-1 blockade.

Recently, a population of long-lived, resident memory $CD8^+$ T cells (T_{RM}) that resides in peripheral tissues has been defined. T_{RM} cells are non-recirculating memory T cells located at barrier sites, including the skin and mucosal tissue (81, 82). Reactivation of T_{RM} stimulates innate immunity against antigenically unrelated pathogens, and could potentially enhance vaccine efficacy (83). While rapamycin promotes memory development in lymphoid tissues, it blocks formation of T_{RM} cells (84). Specifically, rapamycin treatment blocked formation of Vesicular stomatitis virus (VSV)-specific T_{RM} cells in the intestine and vaginal mucosa; however, lung T_{RM} cells were unaltered (84). Thus, mTOR appears to promote formation of mucosal resident memory cells in specific tissues following VSV infection. As such, promoting mTOR activation rather than inhibiting it may prove more beneficial in enhancing formation of T_{RM} cells.

In summary, signaling mediated by mTOR has diverse roles in $CD8^+$ T cell-mediated immunity to pathogens including activation of naive precursors, migration of pathogen-specific cells, development of effector functions, regulation of chronically stimulated effector cells, and generation of stable memory cell populations. Each of these stages poses a potential for immune intervention. However, with such diverse roles, therapeutic regulation of the mTOR pathway will require extensive analysis of the potential implications.

mTOR IN CD4 T CELL DIFFERENTIATION AND IMMUNITY TO PATHOGENS

$CD4^+$ T cells play both supportive and direct roles in host defense against pathogens. As helper cells they regulate the responses of innate immune cells, cytotoxic $CD8^+$ T cells, and antibody-forming B cells. In particular, CD4 T cells are absolutely required for generating long-lived, protective antibodies. As direct effectors, CD4 T cells can utilize perforin-dependent cytolysis of infected cells (85, 86). Similar to $CD8^+$ T cells, naive $CD4^+$ T cells are initially primed in secondary lymphoid organs following binding of their TCR to microbial peptides presented by APCs. $CD4^+$ TCR engagement, coupled with co-stimulation, leads to PI3K signaling, mTOR activation, and subsequent development into effector and memory populations (87).

Following antigen encounter, naive $CD4^+$ T cells can differentiate into several distinct subsets including T helper cells (Th1, Th2, and Th17), T follicular helper cells (Tfh), effector cells, and regulatory T cells (Treg). Each $CD4^+$ T cell lineage is associated with the expression of specific transcription factors: T-bet (Th1), Gata-3 (Th2), RORγt (Th17) Foxp3 (Treg), and Bcl-6 (Tfh) (88). mTOR is a key regulator of differentiation of $CD4^+$ T cells. Specifically, differentiation into Th1, Th2, and Th17 cell lineages was severely inhibited in *mTOR*-deficient mice, even in the presence of polarizing cytokines (87, 89). In the absence of mTOR signaling, impaired phosphorylation of STATs in response to cytokine stimuli blocked the induction of the lineage specific transcription factors (89, 90). Conversely, suppression of mTOR with rapamycin or genetic deletion of *mTOR* in T cells promoted differentiation to $FoxP3^+$ Treg cells, indicating that mTOR signaling blocks Treg development (89).

Subsequent studies using conditional deletion of *Raptor*, *Rheb*, or *Rictor* in T cells ascertained the contribution of mTORC1 and mTORC2 in T helper differentiation. Deficiency of *Rheb*, which is an activator of mTORC1 signaling, impaired differentiation of Th1 and Th17 cell lineages (91). However, *Raptor*-deficient T cells showed a defect in Th17, but not Th1 differentiation (92). More recently, it was shown that Th2 differentiation was also drastically compromised in the absence of *Raptor* (93). On the other hand, deficiency of *Rictor* and mTORC2 signaling impaired Th2 differentiation, but not Th1 or Th17 (91). However, a conflicting study suggested that mTORC2 signaling was also required for Th1 and Th2 differentiation (94). Furthermore, the loss of Th2 differentiation observed in *Rictor*-deficient T cells was comparatively minor compared to the loss observed following rapamycin treatment of *Rictor*-deficient T cells, indicating a more prominent role for mTORC1 in Th2 differentiation (93). Thus, it is clear that both mTORC1 and mTORC2 complexes play important roles in Th differentiation; however, the precise roles of each pathway in each of the subsets remain to be resolved.

Similar to $CD8^+$ T cells, differentiation of naive $CD4^+$ T cells to Th1, Th2, and Th17 effector cells represents a shift from oxidative phosphorylation to glycolysis. Th1, Th2, and Th17 cells express high levels of Glut1 and are highly glycolytic via mTOR signaling, which sustains their high energy consumption and supports their diverse effector functions (95). Development and maintenance of Th17 cells, in particular, is heavily reliant on glycolysis, which is stimulated by HIF-1α downstream of mTOR (96, 97). Therefore, signaling through mTOR contributes to resistance to a variety of pathogens via regulating transcription factors essential for developing T-helper subsets, and by maintaining glycolysis of these subsets following activation.

mTOR IN Treg CELL-MEDIATED IMMUNITY TO INFECTION

Regulatory T cells contribute most significantly to pathogen defense by suppressing T cell responses to limit immunopathology following infection (98). Foxp3$^+$ expression distinguishes Treg cells from conventional T cells and confers Treg cell function (99, 100). Inhibition of mTOR signaling through either genetic deficiency or rapamycin treatment promoted expansion of preexisting Treg cells (89, 101–103) and induced the Treg cell phenotype on conventional T cells and Th17 cells (96, 104–106). Thus, mTOR is critical for negatively regulating Foxp3 expression and Treg cell numbers.

The stability and function of Treg cells is influenced by inflammation. In particular, Foxp3 Treg cells can be reprogramed into Th1 and Th17 effectors in the gut or sites of parasitic infection. However, reprograming of Treg cells into Th1/Th17 effectors is blocked with rapamycin, which stabilizes Foxp3 expression *in vivo* (107). Thus, mTOR signaling promotes differentiation of Treg cells into T helper cells when needed. Indeed, both Th17 and Treg cells require TGF-β for their differentiation, and the degree of mTOR activation delineates the relative development of each cell type. High levels of mTOR activation promote Th17 cell differentiation (108), whereas low mTOR signaling promotes Treg accumulation and sensitizes Treg cells to TGF-β (87, 109).

Furthermore, mTOR orchestrates a metabolic checkpoint for the differentiation between Treg cells and Th17 cells. Conditions that induced Th17 differentiation led to an induction of HIF-1α, which was dependent on mTOR signaling (96). HIF-1α, in turn, increased the expression of glycolytic enzymes and Th17 development, while simultaneously dampening Treg cell development. Thus, blocking glycolysis and mTOR promotes Treg cell generation by downregulation of HIF-1α (96). Similarly, the ratio between Th1 and Treg cells is also regulated by sphingosine 1-phosphate (S1P), which signals through mTOR and attenuates activity of SMAD3 to antagonize TGF-β, and inhibit generation of Treg cells and promote Th1 development (110).

In general, mTOR signaling suppresses Treg cell differentiation in favor of T helper differentiation, which increases immunity to pathogens by supporting the antimicrobial effector functions of T helper cells. Inhibition of mTOR can support development of Treg cells to restrain and control immune responses to circumvent excessive immunopathology. However, *Raptor* deficiency specifically in Treg cells unexpectedly impaired the suppressive function of Tregs and resulted in a fatal inflammatory disease, suggesting that the mTORC1 complex is also important to maintain Treg homeostasis and function (111). These seemingly contradictory data again highlight the complexity of the mTOR complexes, and the need to further delineate this signaling pathway.

mTOR IN Tfh CELL IMMUNITY TO PATHOGENS

Tfh cells are a subset of differentiated CD4$^+$ T cells with a crucial role in initiating and maintaining germinal center reactions and high affinity isotype-class-switched antibody responses (112). Tfh cells express the transcription factor, Bcl6, which defines Tfh cells and promotes expression of the chemokine receptor, CXCR5 (113–115). In contrast to differentiation of Th1, Th2, and Th17 cells, which is supported by IL-2-mediated activation of mTOR, differentiation of Tfh cells is suppressed by IL-2 activation of mTOR. Th1 differentiation requires STAT5 activation and subsequent Blimp-1 expression, whereas Tfh cells develop when Blimp-1 synthesis is suppressed, enabling Bcl6 expression (116, 117). Indeed, Blimp-1 and Bcl6 expression are mutually exclusive, and over expression of either drives differentiation of Th1 or Tfh cells, respectively (113). During acute LCMV infection, Akt and mTOR signaling were essential for Blimp-1 and T-bet expression, which induced Th1 differentiation and countered Tfh development (118). Silencing of *Rictor* or *Raptor* demonstrated that mTORC1 suppresses Tfh cell development and induces Th1 cells, while mTORC2 may suppress Th1 cell development to permit preferential differentiation of Tfh cells in response to LCMV infection (118). Thus, commitment to either Th1 or Tfh lineages is discerned by the level of IL-2–STAT5–mTOR signaling, with increased signaling correlating with the Th1 transcription factors, T-bet and Blimp1, and lower IL-2–STAT5–mTOR levels with the Tfh transcription factor, Bcl6 (115, 116, 119–121).

A recent study demonstrated that during influenza virus infection, TGF-β opposes IL-2 to produce more Tfh cells and isotype-switched antibody responses rather than Th1 cells (122). TGF-β suppressed mTOR activation in T cells early during infection, which promoted Tfh cell differentiation by limiting IL-2 signaling. Inhibition of mTOR with rapamycin also promoted Tfh cell differentiation, suggesting that TGF-β restricts IL-2 responsiveness and insulates early Tfh progenitor cells from mTOR signaling to promote Tfh cell differentiation and isotype-switched antibody responses (122).

Bcl6 expression in Tfh cells downregulates genes associated with glycolysis, while T-bet in Th1 cells inhibits Bcl6-mediated repression of these genes to promote glycolysis (123). Accordingly, Tfh cells are less proliferative, and less glycolytic than Th1 cells, due to a lack of IL-2 signaling and a lower level of mTOR activation. Tfh cells therefore have a reduced metabolic capacity, similar to Treg cells, and also utilize oxidative phosphorylation for cellular maintenance (118).

Similar to the reciprocal relationship between Th17 and Treg cell activation, mTOR activation also plays a key role in regulating the balance between Th1 versus Tfh responses. While both Treg and Tfh cells are suppressed by mTOR activity and share metabolic similarities, Treg cells typically suppress and control immune responses, whereas Tfh cells typically promote immune responses, primarily germinal center formation and high affinity antibody responses.

mTOR IN B CELL-MEDIATED IMMUNITY TO PATHOGENS

B cells are responsible for producing pathogen-specific antibodies that block infection and control pathogen spread. Once generated, antibodies persist to provide long-lasting immunity. B cells develop in the bone marrow through several stages of maturation and differentiation that are influenced by mTOR signaling. Mice with a hypomorphic allele of mTOR have a partial block in B cell development in the bone marrow, and altered proportions of B cell subsets in the spleen (124). Conditional deletion of *mTOR* in B cells via CD19-cre did not abrogate development, but decreased mature, T2 transitional, and marginal zone B cells in the spleen (125). In addition, B cells lacking components of the mTORC2 complex had altered development. B cells deficient in *Sin1* accumulated at the pro-B cell stage in the bone marrow, and had a reduced capacity to become IgM^+ immature B cells due to decreased Akt activation (126). Conditional deletion of *Rictor*, mediated by Vav-cre, did not perturb B cell development in the bone marrow, but did reduce mature B cells in the spleen (127). Interestingly, conditional deletion of *Tsc1* in B cells, which renders mTOR constitutively active, also impaired B cell maturation and significantly reduced marginal zone B cells (128, 129). Together, these studies show that both a reduction in mTOR activity and constitutive activation of mTOR can negatively impact B cell development, suggesting that the level of mTOR activation in B cells is critical for optimal B cell development.

Following development in the bone marrow, B cells migrate to secondary lymphoid organs and mature into follicular or marginal zone B cells, and remain quiescent until stimulated. B cells exhibit basal levels of mTOR activity in response to nutrients without stimulation. mTOR activity levels vary across B cell types, with marginal zone B cells maintaining high levels of mTOR activity, and follicular B cells having lower levels of mTOR activity (130). Activation through the BCR, TLR, or CD40 induces mTOR signaling (131). Similar to T cells, initial B cell activation increases glucose uptake and glycolysis to promote clonal expansion, which is also dependent on the mTOR pathway. Rapamycin inhibited B cell proliferation in response to anti-CD40, LPS, BAFF, and the polyclonal activator, *S. aureus* (131–136). In addition, B cells expressing the hypomorphic allele of mTOR had decreased proliferation in response to anti-IgM, anti-CD40, and LPS, with a bigger reduction following anti-IgM and anti-CD40 than LPS (124). These data demonstrate that mTOR signaling is essential for B cell proliferation in response to multiple stimuli.

Upon activation, B cells can either differentiate rapidly into plasmablasts that secrete low affinity IgM antibodies, or form germinal centers within secondary lymphoid organs, which are important for the generation of memory B cells and high affinity, isotype-class-switched antibody responses. As described above, the formation of germinal centers is dependent on interaction with Tfh cells. In the germinal centers, B cells proliferate and undergo somatic hypermutation (SHM) and class switch recombination (CSR). SHM increases antibody diversity by introducing point mutations into heavy chain Ig genes. CSR increases diversity further by rearranging the Ig heavy chain genes to express a constant region of a specific Ig antibody class. Initiation of these processes relies on activation-induced cytidine deaminase (*Aicda*; AID), which creates mutations in DNA by deaminating cytidine residues to generate uracil, leading to base pair mis-matching. Mutated B cells then undergo affinity-driven selection in the germinal centers, which is necessary for the generation of high affinity antibodies and optimal protection against pathogens.

Mechanistic target of rapamycin signaling is critical for the formation of germinal centers and the production of high affinity antibodies. Mice with a hypomorphic allele of mTOR had decreased antibody responses to T-independent and T-dependent antigens, with a more pronounced decrease in IgG compared to IgM antibodies (124). Further analysis of mice with the hypomorphic mTOR allele following immunization with either the model antigen NP-CGG, or heat-killed Pn14 derived from *S. pneumoniae* showed decreased germinal center formation, SHM, CSR, and IgG antibody affinity maturation (125). Challenge of mice with the hypomorphic mTOR allele with live *S. pneumoniae* led to reduced antibody titers and survival, compared to wild-type mice (125). Likewise, mice with a B cell-specific deletion of *mTOR* also had reduced germinal center formation and high affinity antibody generation following immunization with NP-CGG (125). In both cases, decreased mTOR signaling caused a reduction in the RNA and protein levels of AID. Rescue experiments, increasing AID levels in B cells with low mTOR activity, restored CSR, indicating that mTOR regulates the generation of high affinity antibodies through AID signaling (125).

Consistent with these findings, treatment of mice with a low dose of rapamycin during influenza vaccination blocked germinal center formation, *Aicda* transcription, and consequently, CSR (137). Furthermore, AID induction was dependent on signaling via mTORC1, as *Aicda* transcription was not induced in *Raptor*-deficient B cells following stimulation. In addition, mTOR was critical for CSR independent of proliferation, as the dose of rapamycin used in these experiments was low enough to support proliferation, yet *Aicda* transcription and CSR were inhibited (137).

Given that high affinity, class-switched antibodies are required for optimal protection against most pathogens, it was surprising that treatment of mice with a low dose of rapamycin during influenza vaccination improved protection against subsequent infections with different influenza subtypes (137). Analysis of the rapamycin-treated mice revealed a block in germinal center formation and CSR, which generated a broader antibody response, rather than a highly selected and affinity-matured repertoire. The majority of antibodies generated during influenza vaccination are high affinity antibodies specific for the globular head of the influenza hemagglutinin (HA) molecule. However, the HA protein is also the most variable protein between different influenza strains. Therefore, highly specific antibodies that protect against one subtype of influenza often do not protect against other subtypes. By blocking germinal center formation, rapamycin hampered the generation of high affinity antibodies specific for the variable region, with the unexpected benefit of allowing the lower affinity antibodies that are specific for conserved portions of influenza to become more prevalent. These data suggest that altering the levels of mTOR activation can steer the immune response away from strain-specific responses to more cross-reactive responses, which would be beneficial in generating a universal influenza vaccine.

It was also reported that mice with a conditional deletion of *Tsc1* in B cells, which renders mTOR constitutively active, also had a defect in germinal center formation, and a reduction in T-dependent and T-independent antibody production (128). These data suggest that the level of mTOR activation in B cells is

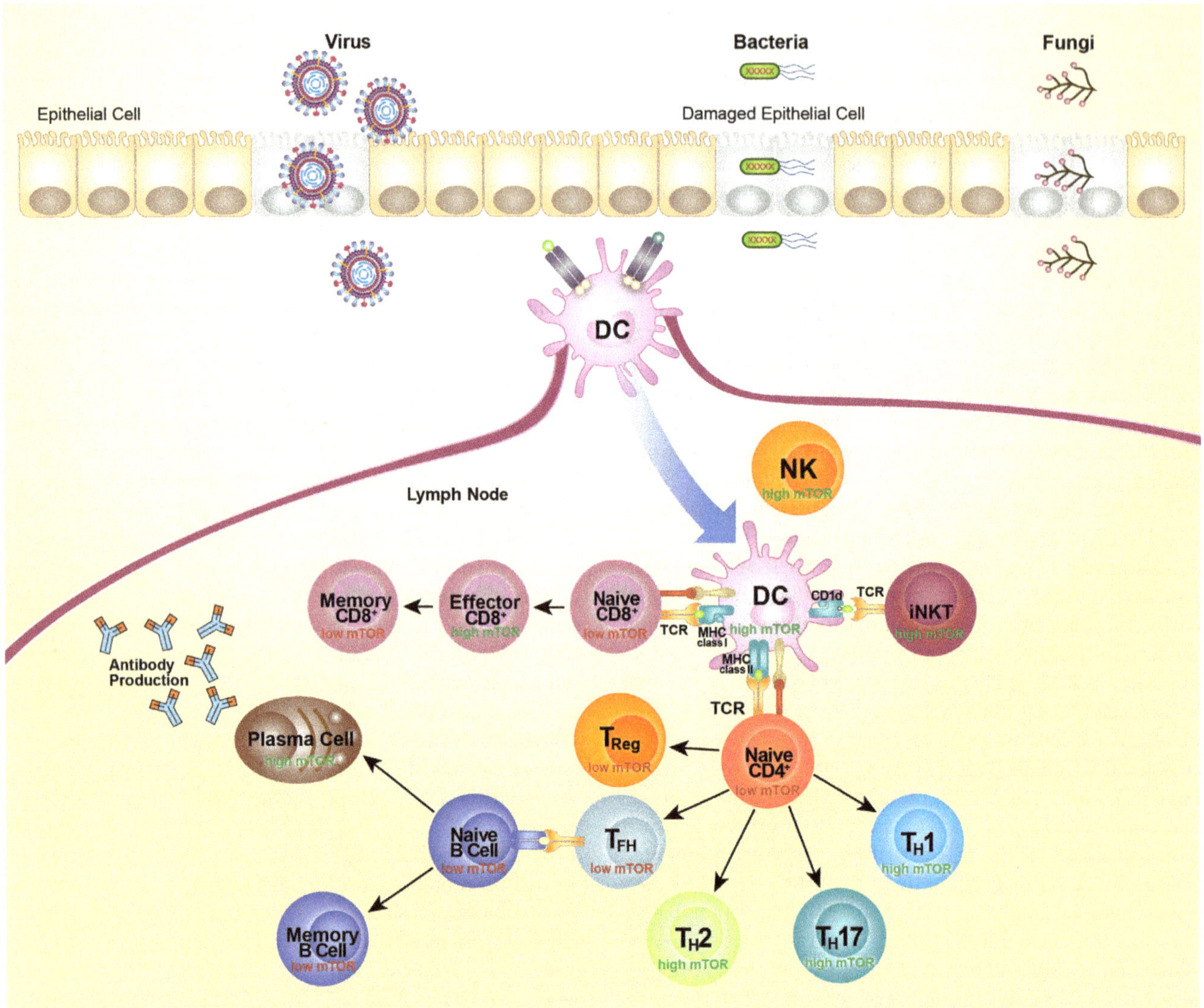

FIGURE 2 | mTOR activity differs across cell subsets. Recognition of microbial components via TLRs leads to activation and subsequent migration of DC from the site of infection to the draining lymph nodes. Activated DCs upregulate costimulatory molecules to become highly efficient antigen-presenting cells. TCR recognition of microbial antigens presented by DC leads to activation and differentiation of $CD4^+$ T cells, $CD8^+$ T cells, and iNKT cells. This TCR-mediated activation is coupled with activation of mTOR and acquisition of effector functions. High mTOR activity is typically associated with high glycolytic and metabolic activity, whereas low mTOR activity is typically associated with low glycolytic activity and a more quiescent state. Activation of naive $CD4^+$ T cells leads to an increase in mTOR activity and expression of transcription factors that promote the differentiation of Th1, Th2, and Th17 cells. Conversely, differentiation of $CD4^+$ T cells in the absence of mTOR signaling promotes development of Treg cells and Tfh cells. Treg cell differentiation is supported by IL-2 and Tfh cell differentiation is suppressed by IL-2. Mature Tfh cells initiate germinal center reactions and support generation of high affinity isotype-class-switched antibody responses. High mTOR activity also supports the generation of CD8 effectors cells, whereas a dampening of mTOR activity supports the transition to memory $CD8^+$ T cells. Cells are depicted in their activated state, unless specified as naive or memory.

critical for optimal germinal center formation. However, another group found that mice with a conditional deletion of *Tsc1* in B cells did not have defects in germinal center formation or high affinity antibody production (129). The reason for the discrepancy in these reports is not clear as both groups observed the same defects in B cell development in the absence of *Tsc1*.

Assessing the role of mTORC2 in antibody generation has also yielded seemingly conflicting data. In one study, inactivation of mTORC2, via *Rictor* deletion, impaired germinal center differentiation and antibody responses, demonstrating that mTORC2 supports antibody differentiation (127). However, in a subsequent study, partial mTORC2 inhibition through incomplete *Rictor* deletion increased CSR compared to wild-type mice, illustrating negative regulation of antibody responses by mTORC2 (138). Different levels of *Rictor* depletion or targeting B cells at different stages of development may have contributed to these discrepancies. The first study reported efficient deletion of *Rictor* in all hematopietic cells using Vav-cre, or a tamoxifen-inducible

Cre (127). Whereas Limon and colleagues obtained partial *Rictor* depletion with CD19-cre (138). In addition, Limon et al. used varying doses of a new class of ATP-competitive mTOR kinase inhibitors to show that partial mTORC2 deletion enhances CSR. Partial deletion of mTORC1 or complete deletion of both mTORC1 and mTORC2 resulted in decreased, rather than increased CSR. However, partial inhibition of both mTOR complexes with inhibitors, or partial deletion of *Rictor* enhanced CSR. The increase in CSR was dependent on FOXO transcription factors (138). These data again highlight that the level of mTOR activation is a critical determinant on the cellular response.

Tfh cells and B cells act in unison to promote germinal center formation, CSR, and high affinity antibody responses. Yet, mTORC1 signaling typically suppresses Tfh cell development and function, while promoting CSR in B cells. While such a scenario does not create a quandary for cell-specific mTOR signaling, inhibition of the mTOR in the whole organism through either deletion of mTOR or with rapamycin treatment is likely to have opposing affects on antibody formation via Tfh cells and B cells. Similar to the self-regulating mechanism influencing differentiation of Treg cells versus Th1 cells or Tfh versus Th17 cells, the balance between B cell and Tfh cell function may be regulated by as yet unknown mechanisms. Initial inhibition of mTOR could allow development and function of Tfh cells to promote germinal center formation, followed by a period of enhanced mTOR activity to allow B cell proliferation and CSR to proceed.

CONCLUSION

It is now clear that mTOR has a central role in coordinating the outcome of pathogen defense by modulating immunity mediated by lymphoid cells. However, many details of the host–pathogen interaction, and their implications, are still to be determined. The mTOR-signaling pathway has a key role in regulating cell metabolism and is therefore readily hijacked by pathogens seeking to acquire energy and control cell death for their own propagation. Survival of a pathogen often necessitates that mTOR pathways are either activated or inactivated, largely to satisfy particular replication requirements. In turn, the counter immune response also varies depending on the type and dose of the infecting pathogen. Moreover, mTOR signaling varies across responding cell types, cell states, and response kinetics. As mTOR senses the immune microenvironment to direct cellular activation and differentiation, activity of mTOR parallels metabolic activity. Highly glycolytic cells exhibit the highest mTOR activity, while lower mTOR activity is associated with cells more reliant on oxidative phosphorylation for their energy needs (**Figure 2**). Use of metabolic inhibitors, such as rapamycin, that target the mTOR pathway would preferentially promote responses by more metabolically inactive Treg, Tfh, and memory cells, but would likely do so at the expense of Th1, Th2, Th17, CD8 effector, and antibody responses. As most cellular responses are intimately linked, opposing outcomes would need to be weighed against each other. The outcome of therapeutic intervention is also likely to be highly variable depending on the nature and stage of infection. Moreover, the precise level of mTOR activation is critical for particular immune responses as complete inhibition of a pathway can generate very different outcomes than partial inhibition. Selectively targeting components of the mTOR pathway and metabolic programing may prove more effective in designing vaccine strategies. For example, suppression of mTORC2 has potential for promoting the generation of memory $CD8^+$ T cells, without negatively affecting the development of effector $CD8^+$ T cells. The impact of any therapeutic approach on the metabolic pathway must be thoroughly anticipated and tested as it is clear that there is still much to learn about utilizing the mTOR pathway for successful therapeutic intervention.

AUTHOR CONTRIBUTIONS

RK and MM researched the literature and wrote the review.

FUNDING

This work was funded by National Institutes of Health, 1R01AI11472801A1 and ALSAC.

REFERENCES

1. Weichhart T, Hengstschläger M, Linke M. Regulation of innate immune cell function by mTOR. *Nat Rev Immunol* (2015) **15**(10):599–614. doi:10.1038/nri3901
2. Dan HC, Ebbs A, Pasparakis M, Van Dyke T, Basseres DS, Baldwin AS. Akt-dependent activation of mTORC1 complex involves phosphorylation of mTOR (mammalian target of rapamycin) by IκB kinase α (IKKα). *J Biol Chem* (2014) **289**(36):25227–40. doi:10.1074/jbc.M114.554881
3. Vander Haar E, Lee S-I, Bandhakavi S, Griffin TJ, Kim D-H. Insulin signalling to mTOR mediated by the Akt/PKB substrate PRAS40. *Nat Cell Biol* (2007) **9**(3):316–23. doi:10.1038/ncb1547
4. Mamane Y, Petroulakis E, LeBacquer O, Sonenberg N. mTOR, translation initiation and cancer. *Oncogene* (2006) **25**(48):6416–22. doi:10.1038/sj.onc.1209888
5. Beugnet A, Tee AR, Taylor PM, Proud CG. Regulation of targets of mTOR (mammalian target of rapamycin) signalling by intracellular amino acid availability. *Biochem J* (2003) **372**(Pt 2):555–66. doi:10.1042/bj20021266
6. Hay N, Sonenberg N. Upstream and downstream of mTOR. *Genes Dev* (2004) **18**(16):1926–45. doi:10.1101/gad.1212704
7. Zinzalla V, Stracka D, Oppliger W, Hall MN. Activation of mTORC2 by association with the ribosome. *Cell* (2011) **144**(5):757–68. doi:10.1016/j.cell.2011.02.014
8. Guertin DA, Stevens DM, Thoreen CC, Burds AA, Kalaany NY, Moffat J, et al. Ablation in mice of the mTORC components raptor, rictor, or mLST8 reveals that mTORC2 is required for signaling to Akt-FOXO and PKCalpha, but not S6K1. *Dev Cell* (2006) **11**(6):859–71. doi:10.1016/j.devcel.2006.10.007
9. Goncharova EA, Goncharov DA, Li H, Pimtong W, Lu S, Khavin I, et al. mTORC2 is required for proliferation and survival of TSC2-null cells. *Mol Cell Biol* (2011) **31**(12):2484–98. doi:10.1128/MCB.01061-10
10. Abraham RT, Wiederrecht GJ. Immunopharmacology of rapamycin. *Annu Rev Immunol* (1996) **14**:483–510. doi:10.1146/annurev.immunol.14.1.483
11. Sarbassov DD, Ali SM, Sengupta S, Sheen J-H, Hsu PP, Bagley AF, et al. Prolonged rapamycin treatment inhibits mTORC2 assembly and Akt/PKB. *Mol Cell* (2006) **22**(2):159–68. doi:10.1016/j.molcel.2006.03.029

12. Sarbassov DD, Guertin DA, Ali SM, Sabatini DM. Phosphorylation and regulation of Akt/PKB by the rictor-mTOR complex. *Science* (2005) **307**(5712):1098–101. doi:10.1126/science.1106148
13. Breuleux M, Klopfenstein M, Stephan C, Doughty CA, Barys L, Maira S-M, et al. Increased AKT S473 phosphorylation after mTORC1 inhibition is rictor dependent and does not predict tumor cell response to PI3K/mTOR inhibition. *Mol Cancer Ther* (2009) **8**(4):742–53. doi:10.1158/1535-7163.MCT-08-0668
14. Yang G, Murashige DS, Humphrey SJ, James DE. A positive feedback loop between Akt and mTORC2 via SIN1 phosphorylation. *Cell Rep* (2015) **12**(6):937–43. doi:10.1016/j.celrep.2015.07.016
15. Pim D, Massimi P, Dilworth SM, Banks L. Activation of the protein kinase B pathway by the HPV-16 E7 oncoprotein occurs through a mechanism involving interaction with PP2A. *Oncogene* (2005) **24**(53):7830–8. doi:10.1038/sj.onc.1208935
16. Spangle JM, Münger K. The human papillomavirus type 16 E6 oncoprotein activates mTORC1 signaling and increases protein synthesis. *J Virol* (2010) **84**(18):9398–407. doi:10.1128/JVI.00974-10
17. Moody CA, Scott RS, Amirghahari N, Nathan C-A, Young LS, Dawson CW, et al. Modulation of the cell growth regulator mTOR by Epstein-Barr virus-encoded LMP2A. *J Virol* (2005) **79**(9):5499–506. doi:10.1128/JVI.79.9.5499-5506.2005
18. O'Shea C, Klupsch K, Choi S, Bagus B, Soria C, Shen J, et al. Adenoviral proteins mimic nutrient/growth signals to activate the mTOR pathway for viral replication. *EMBO J* (2005) **24**(6):1211–21. doi:10.1038/sj.emboj.7600597
19. Weichhart T, Costantino G, Poglitsch M, Rosner M, Zeyda M, Stuhlmeier KM, et al. The TSC-mTOR signaling pathway regulates the innate inflammatory response. *Immunity* (2008) **29**(4):565–77. doi:10.1016/j.immuni.2008.08.012
20. Kudchodkar SB, Yu Y, Maguire TG, Alwine JC. Human cytomegalovirus infection induces rapamycin-insensitive phosphorylation of downstream effectors of mTOR kinase. *J Virol* (2004) **78**(20):11030–9. doi:10.1128/JVI.78.20.11030-11039.2004
21. Jaramillo M, Gomez MA, Larsson O, Shio MT, Topisirovic I, Contreras I, et al. Leishmania repression of host translation through mTOR cleavage is required for parasite survival and infection. *Cell Host Microbe* (2011) **9**(4):331–41. doi:10.1016/j.chom.2011.03.008
22. Yakoub AM, Shukla D. Autophagy stimulation abrogates herpes simplex virus-1 infection. *Sci Rep* (2015) **5**:9730. doi:10.1038/srep09730
23. Kim YC, Guan K-L. mTOR: a pharmacologic target for autophagy regulation. *J Clin Invest* (2015) **125**(1):25–32. doi:10.1172/JCI73939
24. Tattoli I, Sorbara MT, Yang C, Tooze SA, Philpott DJ, Girardin SE. Listeria phospholipases subvert host autophagic defenses by stalling pre-autophagosomal structures. *EMBO J* (2013) **32**(23):3066–78. doi:10.1038/emboj.2013.234
25. Gutierrez MG, Master SS, Singh SB, Taylor GA, Colombo MI, Deretic V. Autophagy is a defense mechanism inhibiting BCG and *Mycobacterium tuberculosis* survival in infected macrophages. *Cell* (2004) **119**(6):753–66. doi:10.1016/j.cell.2004.11.038
26. Yakoub AM, Shukla D. Basal autophagy is required for herpes simplex virus-2 infection. *Sci Rep* (2015) **5**:12985. doi:10.1038/srep12985
27. Tattoli I, Sorbara MT, Vuckovic D, Ling A, Soares F, Carneiro LAM, et al. Amino acid starvation induced by invasive bacterial pathogens triggers an innate host defense program. *Cell Host Microbe* (2012) **11**(6):563–75. doi:10.1016/j.chom.2012.04.012
28. Zhu H, Han L, Shi X, Wang B, Huang H, Wang X, et al. Baicalin inhibits autophagy induced by influenza A virus H3N2. *Antiviral Res* (2015) **113**:62–70. doi:10.1016/j.antiviral.2014.11.003
29. Ma J, Sun Q, Mi R, Zhang H. Avian influenza A virus H5N1 causes autophagy-mediated cell death through suppression of mTOR signaling. *J Genet Genomics* (2011) **38**(11):533–7. doi:10.1016/j.jgg.2011.10.002
30. Datan E, Shirazian A, Benjamin S, Matassov D, Tinari A, Malorni W, et al. mTOR/p70S6K signaling distinguishes routine, maintenance-level autophagy from autophagic cell death during influenza A infection. *Virology* (2014) **45**(2–453):175–90. doi:10.1016/j.virol.2014.01.008
31. Marçais A, Cherfils-Vicini J, Viant C, Degouve S, Viel S, Fenis A, et al. The metabolic checkpoint kinase mTOR is essential for IL-15 signaling during the development and activation of NK cells. *Nat Immunol* (2014) **15**(8):749–57. doi:10.1038/ni.2936
32. Wai L-E, Fujiki M, Takeda S, Martinez OM, Krams SM. Rapamycin, but not cyclosporine or FK506, alters natural killer cell function. *Transplantation* (2008) **85**(1):145–9. doi:10.1097/01.tp.0000296817.28053.7b
33. Nandagopal N, Ali AK, Komal AK, Lee S-H. The critical role of IL-15-PI3K-mTOR pathway in natural killer cell effector functions. *Front Immunol* (2014) **5**:187. doi:10.3389/fimmu.2014.00187
34. Eissens DN, Van Der Meer A, Van Cranenbroek B, Preijers FWMB, Joosten I. Rapamycin and MPA, but not CsA, impair human NK cell cytotoxicity due to differential effects on NK cell phenotype. *Am J Transplant* (2010) **10**(9):1981–90. doi:10.1111/j.1600-6143.2010.03242.x
35. Donnelly RP, Loftus RM, Keating SE, Liou KT, Biron CA, Gardiner CM, et al. mTORC1-dependent metabolic reprogramming is a prerequisite for NK cell effector function. *J Immunol* (2014) **193**(9):4477–84. doi:10.4049/jimmunol.1401558
36. Ripoli M, D'Aprile A, Quarato G, Sarasin-Filipowicz M, Gouttenoire J, Scrima R, et al. Hepatitis C virus-linked mitochondrial dysfunction promotes hypoxia-inducible factor 1 alpha-mediated glycolytic adaptation. *J Virol* (2010) **84**(1):647–60. doi:10.1128/JVI.00769-09
37. Arase H, Mocarski ES, Campbell AE, Hill AB, Lanier LL. Direct recognition of cytomegalovirus by activating and inhibitory NK cell receptors. *Science* (2002) **296**(5571):1323–6. doi:10.1126/science.1070884
38. Sun JC, Ma A, Lanier LL. Cutting edge: IL-15-independent NK cell response to mouse cytomegalovirus infection. *J Immunol* (2009) **183**(5):2911–4. doi:10.4049/jimmunol.0901872
39. Lee YJ, Holzapfel KL, Zhu J, Jameson SC, Hogquist KA. Steady-state production of IL-4 modulates immunity in mouse strains and is determined by lineage diversity of iNKT cells. *Nat Immunol* (2013) **14**(11):1146–54. doi:10.1038/ni.2731
40. Constantinides MG, Bendelac A. Transcriptional regulation of the NKT cell lineage. *Curr Opin Immunol* (2013) **25**(2):161–7. doi:10.1016/j.coi.2013.01.003
41. Zhang L, Tschumi BO, Corgnac S, Rüegg MA, Hall MN, Mach J-P, et al. Mammalian target of rapamycin complex 1 orchestrates invariant NKT cell differentiation and effector function. *J Immunol* (2014) **193**(4):1759–65. doi:10.4049/jimmunol.1400769
42. Shin J, Wang S, Deng W, Wu J, Gao J, Zhong X-P. Mechanistic target of rapamycin complex 1 is critical for invariant natural killer T-cell development and effector function. *Proc Natl Acad Sci U S A* (2014) **111**(8):E776–83. doi:10.1073/pnas.1315435111
43. Wei J, Yang K, Chi H. Cutting edge: discrete functions of mTOR signaling in invariant NKT cell development and NKT17 fate decision. *J Immunol* (2014) **193**(9):4297–301. doi:10.4049/jimmunol.1402042
44. Prevot N, Pyaram K, Bischoff E, Sen JM, Powell JD, Chang C-H. Mammalian target of rapamycin complex 2 regulates invariant NKT cell development and function independent of promyelocytic leukemia zinc-finger. *J Immunol* (2015) **194**(1):223–30. doi:10.4049/jimmunol.1401985
45. Huijts CM, Schneiders FL, Garcia-Vallejo JJ, Verheul HM, de Gruijl TD, van der Vliet HJ. mTOR inhibition per se induces nuclear localization of FOXP3 and conversion of invariant NKT (iNKT) cells into immunosuppressive regulatory iNKT cells. *J Immunol* (2015) **195**(5):2038–45. doi:10.4049/jimmunol.1402710
46. Wu J, Shin J, Xie D, Wang H, Gao J, Zhong X-P. Tuberous sclerosis 1 promotes invariant NKT cell anergy and inhibits invariant NKT cell-mediated antitumor immunity. *J Immunol* (2014) **192**(6):2643–50. doi:10.4049/jimmunol.1302076
47. Gonzalez-Aseguinolaza G, Van Kaer L, Bergmann CC, Wilson JM, Schmieg J, Kronenberg M, et al. Natural killer T cell ligand alpha-galactosylceramide enhances protective immunity induced by malaria vaccines. *J Exp Med* (2002) **195**(5):617–24. doi:10.1084/jem.20011889
48. Huang Y, Chen A, Li X, Chen Z, Zhang W, Song Y, et al. Enhancement of HIV DNA vaccine immunogenicity by the NKT cell ligand, alpha-galactosylceramide. *Vaccine* (2008) **26**(15):1807–16. doi:10.1016/j.vaccine.2008.02.002
49. Youn H-J, Ko S-Y, Lee K-A, Ko H-J, Lee Y-S, Fujihashi K, et al. A single intranasal immunization with inactivated influenza virus and alpha-galactosylceramide induces long-term protective immunity without redirecting antigen to the central nervous system. *Vaccine* (2007) **25**(28):5189–98. doi:10.1016/j.vaccine.2007.04.081

50. Shibolet O, Alper R, Zlotogarov L, Thalenfeld B, Engelhardt D, Rabbani E, et al. NKT and CD8 lymphocytes mediate suppression of hepatocellular carcinoma growth via tumor antigen-pulsed dendritic cells. *Int J Cancer* (2003) **106**(2):236–43. doi:10.1002/ijc.11201
51. Kawakami K, Yamamoto N, Kinjo Y, Miyagi K, Nakasone C, Uezu K, et al. Critical role of Valpha14+ natural killer T cells in the innate phase of host protection against *Streptococcus pneumoniae* infection. *Eur J Immunol* (2003) **33**(12):3322–30. doi:10.1002/eji.200324254
52. Kinjo Y, Tupin E, Wu D, Fujio M, Garcia-Navarro R, Benhnia MR-E-I, et al. Natural killer T cells recognize diacylglycerol antigens from pathogenic bacteria. *Nat Immunol* (2006) **7**(9):978–86. doi:10.1038/ni1380
53. Mattner J, Savage PB, Leung P, Oertelt SS, Wang V, Trivedi O, et al. Liver autoimmunity triggered by microbial activation of natural killer T cells. *Cell Host Microbe* (2008) **3**(5):304–15. doi:10.1016/j.chom.2008.03.009
54. Delgoffe GM, Kole TP, Cotter RJ, Powell JD. Enhanced interaction between Hsp90 and raptor regulates mTOR signaling upon T cell activation. *Mol Immunol* (2009) **46**(13):2694–8. doi:10.1016/j.molimm.2009.05.185
55. Yamada T, Gierach K, Lee P-H, Wang X, Lacorazza HD. Cutting edge: expression of the transcription factor E74-like factor 4 is regulated by the mammalian target of rapamycin pathway in CD8+ T cells. *J Immunol* (2010) **185**(7):3824–8. doi:10.4049/jimmunol.1000718
56. Maciver NJ, Jacobs SR, Wieman HL, Wofford JA, Coloff JL, Rathmell JC. Glucose metabolism in lymphocytes is a regulated process with significant effects on immune cell function and survival. *J Leukoc Biol* (2008) **84**(4):949–57. doi:10.1189/jlb.0108024
57. Pollizzi KN, Patel CH, Sun I-H, Oh M-H, Waickman AT, Wen J, et al. mTORC1 and mTORC2 selectively regulate CD8⁺ T cell differentiation. *J Clin Invest* (2015) **125**(5):2090–108. doi:10.1172/JCI77746
58. Finlay DK, Rosenzweig E, Sinclair LV, Feijoo-Carnero C, Hukelmann JL, Rolf J, et al. PDK1 regulation of mTOR and hypoxia-inducible factor 1 integrate metabolism and migration of CD8+ T cells. *J Exp Med* (2012) **209**(13):2441–53. doi:10.1084/jem.20112607
59. Goldberg EL, Smithey MJ, Lutes LK, Uhrlaub JL, Nikolich-Žugich J. Immune memory-boosting dose of rapamycin impairs macrophage vesicle acidification and curtails glycolysis in effector CD8 cells, impairing defense against acute infections. *J Immunol* (2014) **193**(2):757–63. doi:10.4049/jimmunol.1400188
60. Wang R, Dillon CP, Shi LZ, Milasta S, Carter R, Finkelstein D, et al. The transcription factor Myc controls metabolic reprogramming upon T lymphocyte activation. *Immunity* (2011) **35**(6):871–82. doi:10.1016/j.immuni.2011.09.021
61. Yao S, Buzo BF, Pham D, Jiang L, Taparowsky EJ, Kaplan MH, et al. Interferon regulatory factor 4 sustains CD8(+) T cell expansion and effector differentiation. *Immunity* (2013) **39**(5):833–45. doi:10.1016/j.immuni.2013.10.007
62. Man K, Miasari M, Shi W, Xin A, Henstridge DC, Preston S, et al. The transcription factor IRF4 is essential for TCR affinity-mediated metabolic programming and clonal expansion of T cells. *Nat Immunol* (2013) **14**(11):1155–65. doi:10.1038/ni.2710
63. Ito D, Nojima S, Nishide M, Okuno T, Takamatsu H, Kang S, et al. mTOR complex signaling through the SEMA4A-Plexin B2 axis is required for optimal activation and differentiation of CD8+ T cells. *J Immunol* (2015) **195**(3):934–43. doi:10.4049/jimmunol.1403038
64. Finlay D, Cantrell D. Phosphoinositide 3-kinase and the mammalian target of rapamycin pathways control T cell migration. *Ann N Y Acad Sci* (2010) **1183**:149–57. doi:10.1111/j.1749-6632.2009.05134.x
65. Sinclair LV, Finlay D, Feijoo C, Cornish GH, Gray A, Ager A, et al. Phosphatidylinositol-3-OH kinase and nutrient-sensing mTOR pathways control T lymphocyte trafficking. *Nat Immunol* (2008) **9**(5):513–21. doi:10.1038/ni.1603
66. van der Windt GJW, Everts B, Chang C-H, Curtis JD, Freitas TC, Amiel E, et al. Mitochondrial respiratory capacity is a critical regulator of CD8+ T cell memory development. *Immunity* (2012) **36**(1):68–78. doi:10.1016/j.immuni.2011.12.007
67. Araki K, Turner AP, Shaffer VO, Gangappa S, Keller SA, Bachmann MF, et al. mTOR regulates memory CD8 T-cell differentiation. *Nature* (2009) **460**(7251):108–12. doi:10.1038/nature08155
68. Pearce EL, Walsh MC, Cejas PJ, Harms GM, Shen H, Wang L-S, et al. Enhancing CD8 T-cell memory by modulating fatty acid metabolism. *Nature* (2009) **460**(7251):103–7. doi:10.1038/nature08097
69. He S, Kato K, Jiang J, Wahl DR, Mineishi S, Fisher EM, et al. Characterization of the metabolic phenotype of rapamycin-treated CD8+ T cells with augmented ability to generate long-lasting memory cells. *PLoS One* (2011) **6**(5):e20107. doi:10.1371/journal.pone.0020107
70. Shrestha S, Yang K, Wei J, Karmaus PWF, Neale G, Chi H. Tsc1 promotes the differentiation of memory CD8+ T cells via orchestrating the transcriptional and metabolic programs. *Proc Natl Acad Sci U S A* (2014) **111**(41):14858–63. doi:10.1073/pnas.1404264111
71. Li Q, Rao R, Vazzana J, Goedegebuure P, Odunsi K, Gillanders W, et al. Regulating mammalian target of rapamycin to tune vaccination-induced CD8(+) T cell responses for tumor immunity. *J Immunol* (2012) **188**(7):3080–7. doi:10.4049/jimmunol.1103365
72. Stoycheva D, Deiser K, Stärck L, Nishanth G, Schlüter D, Uckert W, et al. IFN-γ regulates CD8+ memory T cell differentiation and survival in response to weak, but not strong, TCR signals. *J Immunol* (2015) **194**(2):553–9. doi:10.4049/jimmunol.1402058
73. Cukalac T, Chadderton J, Zeng W, Cullen JG, Kan WT, Doherty PC, et al. The influenza virus-specific CTL immunodominance hierarchy in mice is determined by the relative frequency of high-avidity T cells. *J Immunol* (2014) **192**(9):4061–8. doi:10.4049/jimmunol.1301403
74. Rao RR, Li Q, Odunsi K, Shrikant PA. The mTOR kinase determines effector versus memory CD8+ T cell fate by regulating the expression of transcription factors T-bet and Eomesodermin. *Immunity* (2010) **32**(1):67–78. doi:10.1016/j.immuni.2009.10.010
75. Li Q, Rao RR, Araki K, Pollizzi K, Odunsi K, Powell JD, et al. A central role for mTOR kinase in homeostatic proliferation induced CD8+ T cell memory and tumor immunity. *Immunity* (2011) **34**(4):541–53. doi:10.1016/j.immuni.2011.04.006
76. Garcia K, Sun Z, Mattson E, Li L, Smyth K, Xiao Z. IL-12 is required for mTOR regulation of memory CTLs during viral infection. *Genes Immun* (2014) **15**(6):413–23. doi:10.1038/gene.2014.33
77. Mehlhop-Williams ER, Bevan MJ. Memory CD8+ T cells exhibit increased antigen threshold requirements for recall proliferation. *J Exp Med* (2014) **211**(2):345–56. doi:10.1084/jem.20131271
78. Richer MJ, Pewe LL, Hancox LS, Hartwig SM, Varga SM, Harty JT. Inflammatory IL-15 is required for optimal memory T cell responses. *J Clin Invest* (2015) **125**(9):3477–90. doi:10.1172/JCI81261
79. Staron MM, Gray SM, Marshall HD, Parish IA, Chen JH, Perry CJ, et al. The transcription factor FoxO1 sustains expression of the inhibitory receptor PD-1 and survival of antiviral CD8(+) T cells during chronic infection. *Immunity* (2014) **41**(5):802–14. doi:10.1016/j.immuni.2014.10.013
80. Wherry EJ. T cell exhaustion. *Nat Immunol* (2011) **12**(6):492–9. doi:10.1038/ni.2035
81. Klonowski KD, Williams KJ, Marzo AL, Blair DA, Lingenheld EG, Lefrançois L. Dynamics of blood-borne CD8 memory T cell migration in vivo. *Immunity* (2004) **20**(5):551–62. doi:10.1016/S1074-7613(04)00103-7
82. Masopust D, Choo D, Vezys V, Wherry EJ, Duraiswamy J, Akondy R, et al. Dynamic T cell migration program provides resident memory within intestinal epithelium. *J Exp Med* (2010) **207**(3):553–64. doi:10.1084/jem.20090858
83. Schenkel JM, Fraser KA, Beura LK, Pauken KE, Vezys V, Masopust D. T cell memory. Resident memory CD8 T cells trigger protective innate and adaptive immune responses. *Science* (2014) **346**(6205):98–101. doi:10.1126/science.1254536
84. Sowell RT, Rogozinska M, Nelson CE, Vezys V, Marzo AL. Cutting edge: generation of effector cells that localize to mucosal tissues and form resident memory CD8 T cells is controlled by mTOR. *J Immunol* (2014) **193**(5):2067–71. doi:10.4049/jimmunol.1400074
85. Brown DM, Dilzer AM, Meents DL, Swain SL. CD4 T cell-mediated protection from lethal influenza: perforin and antibody-mediated mechanisms give a one-two punch. *J Immunol* (2006) **177**(5):2888–98. doi:10.4049/jimmunol.177.5.2888
86. McKinstry KK, Strutt TM, Kuang Y, Brown DM, Sell S, Dutton RW, et al. Memory CD4+ T cells protect against influenza through multiple synergizing mechanisms. *J Clin Invest* (2012) **122**(8):2847–56. doi:10.1172/JCI63689

87. Powell JD, Pollizzi KN, Heikamp EB, Horton MR. Regulation of immune responses by mTOR. *Annu Rev Immunol* (2012) **30**:39–68. doi:10.1146/annurev-immunol-020711-075024
88. Zhu J, Yamane H, Paul WE. Differentiation of effector CD4 T cell populations (*). *Annu Rev Immunol* (2010) **28**:445–89. doi:10.1146/annurev-immunol-030409-101212
89. Delgoffe GM, Kole TP, Zheng Y, Zarek PE, Matthews KL, Xiao B, et al. The mTOR kinase differentially regulates effector and regulatory T cell lineage commitment. *Immunity* (2009) **30**(6):832–44. doi:10.1016/j.immuni.2009.04.014
90. Powell JD, Delgoffe GM. The mammalian target of rapamycin: linking T cell differentiation, function, and metabolism. *Immunity* (2010) **33**(3):301–11. doi:10.1016/j.immuni.2010.09.002
91. Delgoffe GM, Pollizzi KN, Waickman AT, Heikamp E, Meyers DJ, Horton MR, et al. The kinase mTOR regulates the differentiation of helper T cells through the selective activation of signaling by mTORC1 and mTORC2. *Nat Immunol* (2011) **12**(4):295–303. doi:10.1038/ni.2005
92. Kurebayashi Y, Nagai S, Ikejiri A, Ohtani M, Ichiyama K, Baba Y, et al. PI3K-Akt-mTORC1-S6K1/2 axis controls Th17 differentiation by regulating Gfi1 expression and nuclear translocation of RORγ. *Cell Rep* (2012) **1**(4):360–73. doi:10.1016/j.celrep.2012.02.007
93. Yang K, Shrestha S, Zeng H, Karmaus PWF, Neale G, Vogel P, et al. T cell exit from quiescence and differentiation into Th2 cells depend on Raptor-mTORC1-mediated metabolic reprogramming. *Immunity* (2013) **39**(6):1043–56. doi:10.1016/j.immuni.2013.09.015
94. Lee K, Gudapati P, Dragovic S, Spencer C, Joyce S, Killeen N, et al. Mammalian target of rapamycin protein complex 2 regulates differentiation of Th1 and Th2 cell subsets via distinct signaling pathways. *Immunity* (2010) **32**(6):743–53. doi:10.1016/j.immuni.2010.06.002
95. Michalek RD, Gerriets VA, Jacobs SR, Macintyre AN, MacIver NJ, Mason EF, et al. Cutting edge: distinct glycolytic and lipid oxidative metabolic programs are essential for effector and regulatory CD4+ T cell subsets. *J Immunol* (2011) **186**(6):3299–303. doi:10.4049/jimmunol.1003613
96. Shi LZ, Wang R, Huang G, Vogel P, Neale G, Green DR, et al. HIF1alpha-dependent glycolytic pathway orchestrates a metabolic checkpoint for the differentiation of TH17 and Treg cells. *J Exp Med* (2011) **208**(7):1367–76. doi:10.1084/jem.20110278
97. Dang EV, Barbi J, Yang H-Y, Jinasena D, Yu H, Zheng Y, et al. Control of T(H)17/T(reg) balance by hypoxia-inducible factor 1. *Cell* (2011) **146**(5):772–84. doi:10.1016/j.cell.2011.07.033
98. Sanchez AM, Yang Y. The role of natural regulatory T cells in infection. *Immunol Res* (2011) **49**(1–3):124–34. doi:10.1007/s12026-010-8176-8
99. Fontenot JD, Rudensky AY. A well adapted regulatory contrivance: regulatory T cell development and the forkhead family transcription factor Foxp3. *Nat Immunol* (2005) **6**(4):331–7. doi:10.1038/ni1179
100. Sakaguchi S, Ono M, Setoguchi R, Yagi H, Hori S, Fehervari Z, et al. Foxp3+ CD25+ CD4+ natural regulatory T cells in dominant self-tolerance and autoimmune disease. *Immunol Rev* (2006) **212**:8–27. doi:10.1111/j.0105-2896.2006.00427.x
101. Battaglia M, Stabilini A, Roncarolo M-G. Rapamycin selectively expands CD4+CD25+FoxP3+ regulatory T cells. *Blood* (2005) **105**(12):4743–8. doi:10.1182/blood-2004-10-3932
102. Battaglia M, Stabilini A, Migliavacca B, Horejs-Hoeck J, Kaupper T, Roncarolo M-G. Rapamycin promotes expansion of functional CD4+CD25+FOXP3+ regulatory T cells of both healthy subjects and type 1 diabetic patients. *J Immunol* (2006) **177**(12):8338–47. doi:10.4049/jimmunol.177.12.8338
103. Haxhinasto S, Mathis D, Benoist C. The AKT-mTOR axis regulates de novo differentiation of CD4+Foxp3+ cells. *J Exp Med* (2008) **205**(3):565–74. doi:10.1084/jem.20071477
104. Gao W, Lu Y, El Essawy B, Oukka M, Kuchroo VK, Strom TB. Contrasting effects of cyclosporine and rapamycin in de novo generation of alloantigen-specific regulatory T cells. *Am J Transplant* (2007) **7**(7):1722–32. doi:10.1111/j.1600-6143.2007.01842.x
105. Valmori D, Tosello V, Souleimanian NE, Godefroy E, Scotto L, Wang Y, et al. Rapamycin-mediated enrichment of T cells with regulatory activity in stimulated CD4+ T cell cultures is not due to the selective expansion of naturally occurring regulatory T cells but to the induction of regulatory functions in conventional CD4+ T cells. *J Immunol* (2006) **177**(2):944–9. doi:10.4049/jimmunol.177.2.944
106. Kopf H, de la Rosa GM, Howard OMZ, Chen X. Rapamycin inhibits differentiation of Th17 cells and promotes generation of FoxP3+ T regulatory cells. *Int Immunopharmacol* (2007) **7**(13):1819–24. doi:10.1016/j.intimp.2007.08.027
107. Yurchenko E, Shio MT, Huang TC, Da Silva Martins M, Szyf M, Levings MK, et al. Inflammation-driven reprogramming of CD4+ Foxp3+ regulatory T cells into pathogenic Th1/Th17 T effectors is abrogated by mTOR inhibition in vivo. *PLoS One* (2012) **7**(4):e35572. doi:10.1371/journal.pone.0035572
108. Gulen MF, Kang Z, Bulek K, Youzhong W, Kim TW, Chen Y, et al. The receptor SIGIRR suppresses Th17 cell proliferation via inhibition of the interleukin-1 receptor pathway and mTOR kinase activation. *Immunity* (2010) **32**(1):54–66. doi:10.1016/j.immuni.2009.12.003
109. Gabryšová L, Christensen JR, Wu X, Kissenpfennig A, Malissen B, O'Garra A. Integrated T-cell receptor and costimulatory signals determine TGF-β-dependent differentiation and maintenance of Foxp3+ regulatory T cells. *Eur J Immunol* (2011) **41**(5):1242–8. doi:10.1002/eji.201041073
110. Liu G, Yang K, Burns S, Shrestha S, Chi H. The S1P(1)-mTOR axis directs the reciprocal differentiation of T(H)1 and T(reg) cells. *Nat Immunol* (2010) **11**(11):1047–56. doi:10.1038/ni.1939
111. Zeng H, Yang K, Cloer C, Neale G, Vogel P, Chi H. mTORC1 couples immune signals and metabolic programming to establish T(reg)-cell function. *Nature* (2013) **499**(7459):485–90. doi:10.1038/nature12297
112. Crotty S. Follicular helper CD4 T cells (TFH). *Annu Rev Immunol* (2011) **29**:621–63. doi:10.1146/annurev-immunol-031210-101400
113. Johnston RJ, Poholek AC, DiToro D, Yusuf I, Eto D, Barnett B, et al. Bcl6 and Blimp-1 are reciprocal and antagonistic regulators of T follicular helper cell differentiation. *Science* (2009) **325**(5943):1006–10. doi:10.1126/science.1175870
114. Poholek AC, Hansen K, Hernandez SG, Eto D, Chandele A, Weinstein JS, et al. In vivo regulation of Bcl6 and T follicular helper cell development. *J Immunol* (2010) **185**(1):313–26. doi:10.4049/jimmunol.0904023
115. Choi YS, Kageyama R, Eto D, Escobar TC, Johnston RJ, Monticelli L, et al. ICOS receptor instructs T follicular helper cell versus effector cell differentiation via induction of the transcriptional repressor Bcl6. *Immunity* (2011) **34**(6):932–46. doi:10.1016/j.immuni.2011.03.023
116. Choi YS, Yang JA, Yusuf I, Johnston RJ, Greenbaum J, Peters B, et al. Bcl6 expressing follicular helper CD4 T cells are fate committed early and have the capacity to form memory. *J Immunol* (2013) **190**(8):4014–26. doi:10.4049/jimmunol.1202963
117. Ray JP, Marshall HD, Laidlaw BJ, Staron MM, Kaech SM, Craft J. Transcription factor STAT3 and type I interferons are corepressive insulators for differentiation of follicular helper and T helper 1 cells. *Immunity* (2014) **40**(3):367–77. doi:10.1016/j.immuni.2014.02.005
118. Ray JP, Staron MM, Shyer JA, Ho P-C, Marshall HD, Gray SM, et al. The interleukin-2-mTORc1 kinase axis defines the signaling, differentiation, and metabolism of T helper 1 and follicular B helper T Cells. *Immunity* (2015) **43**(4):690–702. doi:10.1016/j.immuni.2015.08.017
119. Pepper M, Pagán AJ, Igyártó BZ, Taylor JJ, Jenkins MK. Opposing signals from the Bcl6 transcription factor and the interleukin-2 receptor generate T helper 1 central and effector memory cells. *Immunity* (2011) **35**(4):583–95. doi:10.1016/j.immuni.2011.09.009
120. Choi YS, Eto D, Yang JA, Lao C, Crotty S. Cutting edge: STAT1 is required for IL-6-mediated Bcl6 induction for early follicular helper cell differentiation. *J Immunol* (2013) **190**(7):3049–53. doi:10.4049/jimmunol.1203032
121. Ballesteros-Tato A, León B, Graf BA, Moquin A, Adams PS, Lund FE, et al. Interleukin-2 inhibits germinal center formation by limiting T follicular helper cell differentiation. *Immunity* (2012) **36**(5):847–56. doi:10.1016/j.immuni.2012.02.012
122. Marshall HD, Ray JP, Laidlaw BJ, Zhang N, Gawande D, Staron MM, et al. The transforming growth factor beta signaling pathway is critical for the formation of CD4 T follicular helper cells and isotype-switched antibody responses in the lung mucosa. *eLife* (2015) **4**:e04851. doi:10.7554/eLife.04851
123. Oestreich KJ, Read KA, Gilbertson SE, Hough KP, McDonald PW, Krishnamoorthy V, et al. Bcl-6 directly represses the gene program of the glycolysis pathway. *Nat Immunol* (2014) **15**(10):957–64. doi:10.1038/ni.2985

124. Zhang S, Readinger JA, DuBois W, Janka-Junttila M, Robinson R, Pruitt M, et al. Constitutive reductions in mTOR alter cell size, immune cell development, and antibody production. *Blood* (2011) **117**(4):1228–38. doi:10.1182/blood-2010-05-287821
125. Zhang S, Pruitt M, Tran D, Du Bois W, Zhang K, Patel R, et al. B cell-specific deficiencies in mTOR limit humoral immune responses. *J Immunol* (2013) **191**(4):1692–703. doi:10.4049/jimmunol.1201767
126. Lazorchak AS, Liu D, Facchinetti V, Di Lorenzo A, Sessa WC, Schatz DG, et al. Sin1-mTORC2 suppresses rag and il7r gene expression through Akt2 in B cells. *Mol Cell* (2010) **39**(3):433–43. doi:10.1016/j.molcel.2010.07.031
127. Lee K, Heffington L, Jellusova J, Nam KT, Raybuck A, Cho SH, et al. Requirement for Rictor in homeostasis and function of mature B lymphoid cells. *Blood* (2013) **122**(14):2369–79. doi:10.1182/blood-2013-01-477505
128. Benhamron S, Tirosh B. Direct activation of mTOR in B lymphocytes confers impairment in B-cell maturation andloss of marginal zone B cells. *Eur J Immunol* (2011) **41**(8):2390–6. doi:10.1002/eji.201041336
129. Ci X, Kuraoka M, Wang H, Carico Z, Hopper K, Shin J, et al. TSC1 promotes B cell maturation but is dispensable for germinal center formation. *PLoS One* (2015) **10**(5):e0127527. doi:10.1371/journal.pone.0127527
130. Donahue AC, Fruman DA. Distinct signaling mechanisms activate the target of rapamycin in response to different B-cell stimuli. *Eur J Immunol* (2007) **37**(10):2923–36. doi:10.1002/eji.200737281
131. Donahue AC, Fruman DA. Proliferation and survival of activated B cells requires sustained antigen receptor engagement and phosphoinositide 3-kinase activation. *J Immunol* (2003) **170**(12):5851–60.
132. Sakata A, Kuwahara K, Ohmura T, Inui S, Sakaguchi N. Involvement of a rapamycin-sensitive pathway in CD40-mediated activation of murine B cells in vitro. *Immunol Lett* (1999) **68**(2–3):301–9. doi:10.1016/S0165-2478(99)00053-X
133. Wicker LS, Boltz RC Jr, Matt V, Nichols EA, Peterson LB, Sigal NH. Suppression of B cell activation by cyclosporin A, FK506 and rapamycin. *Eur J Immunol* (1990) **20**(10):2277–83. doi:10.1002/eji.1830201017
134. Kay JE, Kromwel L, Doe SE, Denyer M. Inhibition of T and B lymphocyte proliferation by rapamycin. *Immunology* (1991) **72**(4):544–9.
135. Aagaard-Tillery KM, Jelinek DF. Inhibition of human B lymphocyte cell cycle progression and differentiation by rapamycin. *Cell Immunol* (1994) **156**(2):493–507. doi:10.1006/cimm.1994.1193
136. Zeng Q, Zhang H, Qin J, Xu Z, Gui L, Liu B, et al. Rapamycin inhibits BAFF-stimulated cell proliferation and survival by suppressing mTOR-mediated PP2A-Erk1/2 signaling pathway in normal and neoplastic B-lymphoid cells. *Cell Mol Life Sci* (2015) **72**(24):4867–84. doi:10.1007/s00018-015-1976-1
137. Keating R, Hertz T, Wehenkel M, Harris TL, Edwards BA, McClaren JL, et al. The kinase mTOR modulates the antibody response to provide cross-protective immunity to lethal infection with influenza virus. *Nat Immunol* (2013) **14**(12):1266–76. doi:10.1038/ni.2741
138. Limon JJ, So L, Jellbauer S, Chiu H, Corado J, Sykes SM, et al. mTOR kinase inhibitors promote antibody class switching via mTORC2 inhibition. *Proc Natl Acad Sci U S A* (2014) **111**(47):E5076–85. doi:10.1073/pnas.1407104111

7

Spatio-Temporal Regulation of PKC Isoforms Imparts Signaling Specificity

Arkajyoti Mukherjee, Sayoni Roy, Bhaskar Saha and Debasri Mukherjee**

National Centre for Cell Science, Pune, India

Keywords: PKC isoforms, spatio-temporal regulation, subcellular distribution, PKC-signaling module, intracellular pertubators

Edited by:
Diana Bahia,
Universidade Federal de Minas Gerais, Brazil

Reviewed by:
Michael Freeley,
Trinity College Dublin, Ireland

****Correspondence:***
Bhaskar Saha
sahab@nccs.res.in;
Debasri Mukherjee
mukherjee.debasri@gmail.com

INTRODUCTION

Nishizuka and colleagues discovered protein kinase C (PKC) as a calcium-dependent, lipid cofactor-sensitive protein kinase (1). Initially known as the only physiological effector for tumor-promoting phorbol esters, this ubiquitous enzyme eventually took center stage in cellular signaling. Part of the AGC kinase branch of the kinome, PKC is a family of serine/threonine kinases comprising 11 isoforms encoded by 9 genes and grouped into 4 classes – classical (cPKCs-α, βI, βII, γ), novel (nPKCs-δ, ϵ, η, θ), atypical (aPKCs-ζ, ι/λ), and PKCμ (a form between novel and atypical isoforms) (**Box 1**). Their considerable structural homology, overlapping substrate specificities, and biochemical properties indicated at least partial enzymatic redundancy and rendered the task of identifying isoform-specific functions challenging. The initial phase of PKC research correlated the unique structural features of the isoforms with their functions (**Box 1**) (2). But, with their structural overlaps and hugely varying functions in different models, specificity of the isoforms became a confounding puzzle demanding stringent isoform-specific regulation to avoid functional redundancy. To define this stringency, subsequent PKC research focused on the upstream and downstream regulatory mechanisms (3, 4), highlighting the importance of subcellular distribution as a function of time (5). As simultaneous activation of all PKC isoforms would be energetically and spatially conflicting for decoding the message received by the cell surface receptor, cell type, and stimulus-specific selective and sequential activation deemed justified (6). The recent phase of PKC research propounds a PKC-signaling module (7, 8) wherein an inter-isoform network regulates the PKC isoforms' activity (**Box 2; Figure 1B**). Here, we propose activator- and cofactor-specific sequential activation of PKC isoforms in a spatio-temporal model wherein selective subcellular compartmentalization quantitatively determines the isoform-specific effector functions.

STRUCTURAL BIAS TO PKC FUNCTION?

Structural reconfigurations of the PKC isoforms play a central role in orchestrating their spatial distribution and activation (15). C1 domain ligands – diacylglycerol (DAG) and phosphatidylserine (PS) – recruit PKC by altering the surface properties of the domain to favor membrane penetration. Studies on GFP–PKC with the fluorescent phorbol ester analog sapintoxin-D showed differential subcellular localization of C1 ligands determining subcellular targeting of PKCs (16). The nPKC C1B domain has a 100-fold higher affinity for DAG compared to cPKCs due to an invariant tryptophan residue at position 22 as opposed to a tyrosine in case of cPKCs resulting in rapid plasma membrane localization for cPKCs versus slower and sustained Golgi localization for nPKCs (17). The C2 domain functions as a Ca^{2+}-regulated membrane anchor in cPKCs (6, 15) while regulating protein–protein interactions and spatial distribution through phosphotyrosine-binding modules in

BOX 1 | Structural features of PKC isoforms defining functional specificity.

PKC serve as a paradigm for the reversible regulation of membrane localization by the concerted action of two membrane-targeting modules.

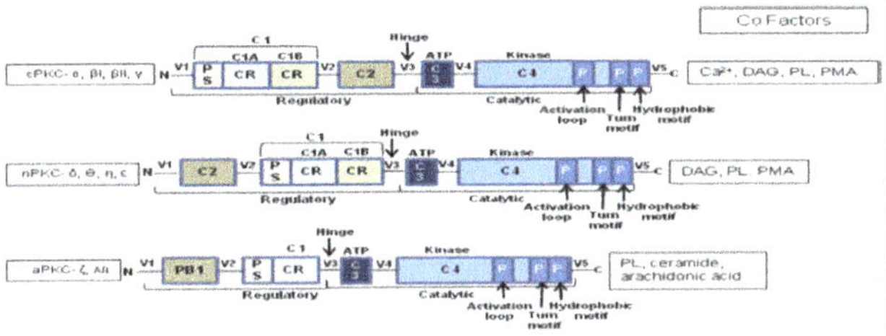

C1: a cysteine-rich region of approximately 50 residues present in all PKC isozymes. In cPKCs and nPKCs, it is present as a tandem repeat C1A and C1B. aPKCs contain a single copy of the domain termed atypical because it does not bind phorbol esters. The domain contains two pulled apart β-sheet forming the ligand-binding pocket. Two zinc atoms are coordinated by two histidines and six cysteines at opposite ends of the primary sequence stabilizing the domain. In aPKCs, one face of the ligand-binding pocket is compromised, so that the module cannot bind phorbol esters or DAG.

Ligand binding dramatically alters the surface properties of the module (C1B). The ligand caps the hydrophilic ligand-binding pocket, so that the top-third of the C1 domain presents a continuous hydrophobic surface thus achieving membrane targeting by simply altering membrane properties.

C2: an independent membrane-targeting module that binds calcium in cPKCs, but not in nPKCs. C2 domain is a β-strand rich globular domain with loops formed by sequences at opposite ends of the primary structure. Two topological variants exist – Type I for domains that follow the C1 domain (cPKCs) and Type II for domains that precede C1 (nPKCs). In the calcium-responsive C2 domains of cPKCs, the pocket is lined by multiple aspartic acid residues that coordinate two to three calcium ions that act as a bridge between the C2 domain and the phospholipid head groups of the membrane.

CR, cysteine rich; PS, pseudosubstrate; DAG, diacylglycerol; PMA, phorbol-12-myristate 13-acetate; PB1, Phox–Bem 1; C, constant regions; V, variable regions.

BOX 2 | Chronological epochs of PKC research.

1977–1987: introducing PKC

- Discovery of PKC by Nishizuka and colleagues (1).
- Structural analysis of the PKC isoforms.
- Takai et al. showed PKC to be reversibly activated by Ca^{2+}/phospholipid and DAG (9).
- Identification of PKC as a target for phorbol ester class of tumor promoters by Castagna et al. (10).
- Discovery of the pseudosubstrate region by House and Kemp (2).
- *1988–1997: the decade of elucidation of PKC structure–function relationship*
- Elucidation of the structural basis of PKC function.
- Identification of receptors for activated C-kinases by Mochly-Rosen and coworkers (11).
- First report by Nishizuka about the role of PKC in cellular signaling (3).
- In 1994, Dekker and Parker for the first time raised the question of specificity of PKC isoforms (4).
- Cloning of first RACK by Mochly-Rosen et al. (12).
- Role of anchoring protein in localization of PKCs (5).
- For the first time, the role of binding proteins in PKC isoform-specific functions was suggested.

1998–2007: the decade of studies on the spatio-temporal regulation of PKC isoforms

- Identification of PKC anchoring proteins as a means for isozyme selectivity by Mochly-Rosen and colleagues in 1998 (13).
- Detection of substrates that interact with C-kinases (STICKs) through overlay assay by Jaken (14).
- Compartmentalization of PKC through binding proteins and substrates.
- Studies on the temporal kinetics of PKC function.
- Elucidation of cell and stimulus-specific actions of PKC isozymes.

2008–present: inter-PKC regulation and PKC-signaling module

This is the decade where the concept of inter-PKC regulation in calibrating receptor triggered effector functions is gaining popularity leading to the possible build-up of a PKC-signaling network in space-time coordinates.

nPKCs (18, 19). The aPKCs are not responsive to either DAG or Ca^{2+} and instead possess a Phox-Bem (PB)1 domain that facilitates interactions with scaffolding proteins leading to constitutive activation (20). Phosphorylation plays a key role in determining the cellular levels of PKC rendering them catalytically competent and protecting them from degradation (17). Classical and novel PKCs are constitutively phosphorylated at three conserved residues: the activation loop – the first rate-limiting step governed by phospho-inositide-dependent kinase (PDK)-1 that aligns residues within the active site for catalysis, followed by phosphorylation at the turn motif and autophosphorylation of the hydrophobic loop. For cPKCs and nPKCs, these phosphorylation events require mTORC2. Phosphorylation at the hydrophobic motif controls the stability of the enzyme promoting degradation on dephosphorylation (17). A phosphomimetic glutamic acid prevents phosphorylation at the hydrophobic motif of aPKCs. Although catalytically competent, these phosphorylated PKCs remain inactive in the cytosol due to binding of the pseudosubstrate to its kinase domain (**Figure 1A**) until appropriate cofactor interaction provides the necessary energy to expel the pseudosubstrate and activate PKC. So, the key to functional specificity *in vivo* may lie in the isoforms' primary structure and the cofactors governed conformational plasticity. Yet, their promiscuity *in vitro* suggests involvement of factors beyond the primary structure leading to the studies on the spatio-temporal regulation of these isozymes.

PKC ISOFORM FUNCTIONS IN SPACE-TIME COORDINATES

Functional specificity of the PKC isoforms depends on the proximity to their substrates in specific intracellular compartments effected through multiple binding proteins that also determine the precise duration and amplitude of PKC activity thus providing a mechanism for integrating PKC-mediated signaling with other cellular activities. PKCβI interacts with Bruton's tyrosine kinase to positively regulate JNK signaling and cytokine gene expression in mast cells (6, 21). Partition defective-3 (PAR3) protein interacts with PKCζ to activate pathways leading to embryonic polarity and asymmetric cell division (6, 22), whereas PAR4 inactivates PKCζ leading to apoptosis (6, 23). Ras-related nuclear protein-binding protein-9/10 integrates PKCδ and γ signals to

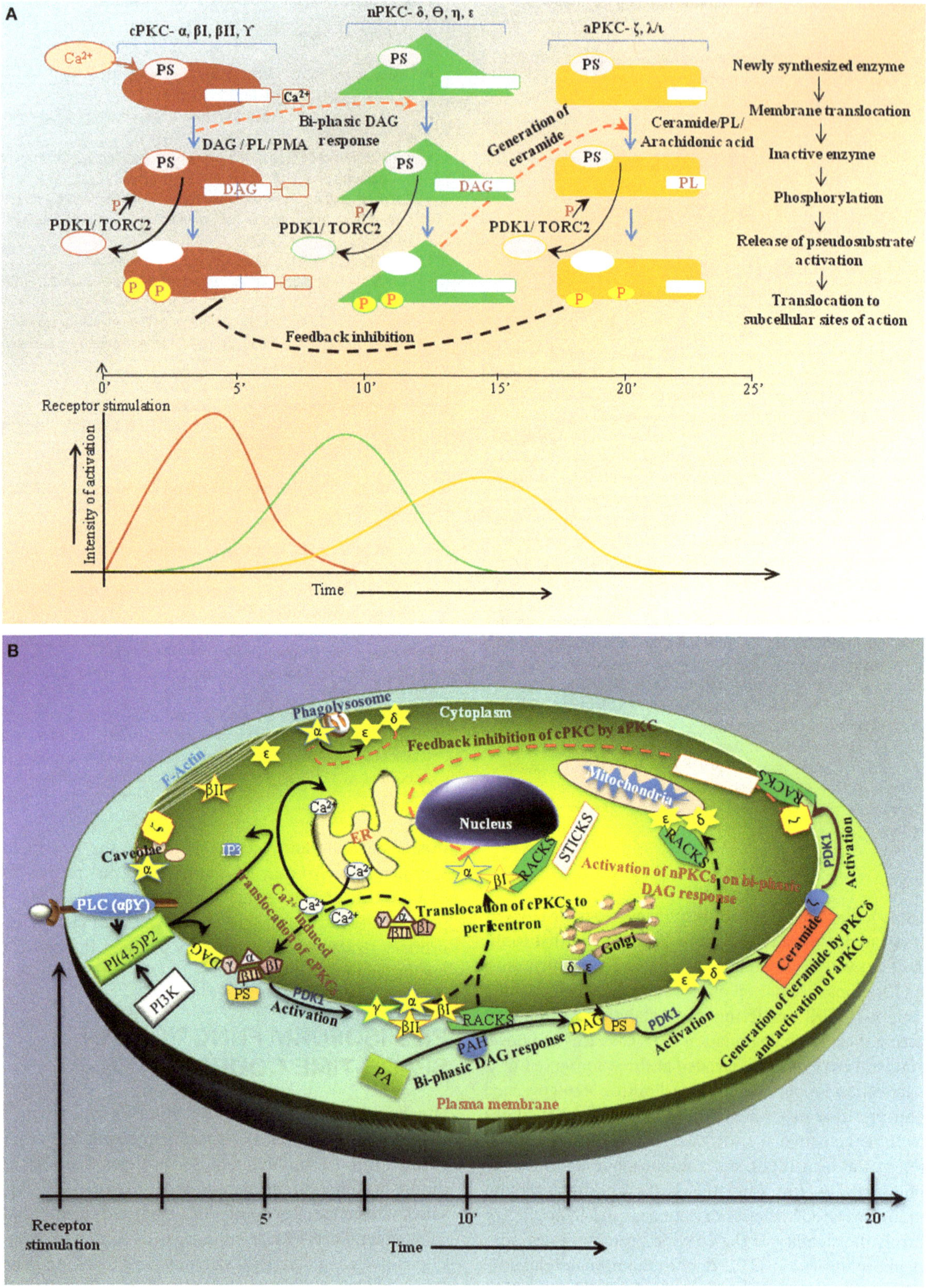

FIGURE 1 | **Continued**

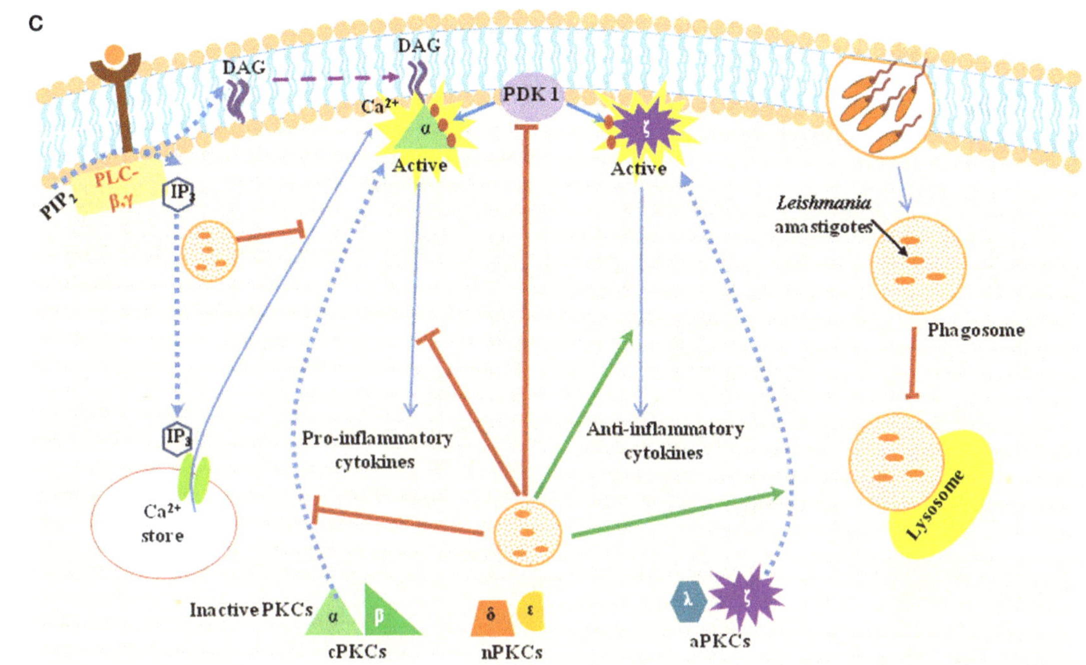

FIGURE 1 | **(A)** Hypothetical figure showing sequential activation of the PKC isoforms in response to receptor-stimulus coupling. Analyses from various studies indicate that in cases where a single stimulus activates multiple isoforms of different classes the classical PKC isoforms are usually activated first and to the greatest degree followed by the novel and atypical forms. **(B)** Kinetics of activation and subcellular distribution of PKC isoforms. The various classes of PKC isoforms are localized at different subcellular sites both prior and post activation. They are translocated to the membrane for activation in response to their respective cofactors. Post activation, the isoforms are translocated to specific subcellular locations by their binding partners, which bring them close to their respective substrates. DAG, diacylglycerol; IP3, inositol tri-phosphate; PA, phosphatidic acid; PAH, phosphatidic acid hydrolase; PI3K, phosphatidyl inositol-3-kinase; PLC, phospholipase C; PKC, protein kinase C; PhosIns(4,5)P2, phosphatidyl-inositol-4,5-bis phosphate; RACK, receptors for activated C-kinases; STICK, subtrates that interact with C-kinases. **(C)** Modulation of PKC isoforms by *Leishmania* infection. The figure shows modulation of PKC isoforms by *Leishmania* as it enters a host cell, for instance, macrophages, by preventing phagosome maturation. It interferes with the translocation of isoforms to membranes and inhibits PDK-1. It inhibits Ca^{2+} efflux to prevent DAG-mediated activation of cPKCs α and β, which are involved in proinflammatory cytokine production thus resulting in disease progression concomitantly enhancing PKCδ- and ζ-mediated production of anti-inflammatory cytokines to suppress host immune response conducing in parasite survival.

dictate efficient regulation of dopaminergic D1 receptor signaling (24). In CHO cells, initial phorbol 12-myristate 13-acetate (PMA) treatment translocates and colocalizes receptor for activated C-kinases (RACK) 1 and PKCβII to the cell periphery that later move to the perinuclear area (25). Thus, RACK–PKC complex can move from one cellular site to another likely resulting in different molecular events at each site depending on the available substrate. PKCϵ interacts specifically with filamentous actin through a binding site located between its two cysteine-rich regions to enhance glutamate exocytosis from nerve terminals (26). Notwithstanding the considerable stimulus and cellular variability in their subcellular distribution, certain PKC isoforms exhibit unique localizations. PKCθ is redox dependently recruited to the plasma membrane of naive T-cells (27). The scaffolding protein A-kinase anchor protein (AKAP) 450 (25) associates with nascent PKCϵ within Golgi/centrosome membranes and dissociates on maturation of PKCϵ (17). PKCζ shows a remarkable range of functions reflecting its multiple cellular locations and interacting partners (28). Interactions with binding partners sometimes also affect PKC pharmacological profile like AKAP-79 protecting PKC from certain ATP competitive inhibitors and altering the susceptibility of PDK-1 to ATP analog inhibitors (29). These observations indicate that the structural features determine the isoforms' translocation and activation pattern, whereas the binding proteins contribute plasticity and specificity.

As multiple PKC isoforms may be activated by a single stimulus, it is logical to assume that differential activation dynamics can impose specificity. Following receptor ligation, the isoforms exhibited differences in activation kinetics and subcellular localization. The kinetics of Ca^{2+}-induced translocation triggered by the C2 domain is faster than that triggered by the C1 domain (30) (**Figure 1A**) indicating a possible sequential activation pattern (31) guided by stimulus specificities in many cell types including macrophages and T-cells (8, 31). Indeed, studies in chick muscle cells have shown short-term stimulation to initially trigger PKCα/β membrane translocation followed by PKCδ to a lesser degree (32). PKCδ-activated acid sphingomyelinase cleaves sphingomyelin releasing ceramide that activates PKCζ and inactivates the cPKCs (15). These observations support the "sequential

PKC isoform activation" hypothesis and an inter-isoform regulation (**Figure 1A**).

INTER-ISOFORM REGULATION AND THEIR SITES OF PERTURBATION

Coordination between the concomitantly expressed and activated PKC isoforms within a temporal framework seems to be regulated by transphosphorylation of the isoforms (33). Following receptor stimulation, classical isoforms are acutely activated as the release of calcium is fast and transient. Calcium sequestration into the endoplasmic reticulum is also rapid (**Figure 1B**), so deactivation of the classical isoforms likely occurs earlier than the others. Studies have noted a feedback inhibition of cPKCs on activation of novel and atypical isoforms (8, 15). PKCδ and ϵ have opposing effects in multiple pathological conditions, including cardiac ischemia, cancer, apoptosis, and cell proliferation (34). During endosome formation in phagocytic cells, the initial calcium burst activates the PKCα leading to initiation of phagosome formation (35). This is sequentially followed by translocation and activation of PKCδ and ϵ that play a role in phagosome maturation and lysosomal fusion (36). PKCα can phosphorylate PKCϵ, so it is possible that the effects of PKCα are shared by PKCϵ (37). Ceramide generated by PKCδ activates PKCζ that inhibits PKCα/βII at the perinuclear space (15, 38). PKCζ has constitutive kinase activity that can activate other PKC isoforms. Sequential activation of PKC isoforms thus explains the dynamic modulation of cellular responsiveness dictated by the strength of stimulus (**Figure 1B**). Following TCR activation, PKCα acts upstream of PKCθ to activate NFκB (39). Inhibition of PKCθ abrogates the PKCα response indicating presence of a feedback loop between the isoforms or, as proposed here, a sequential activation of PKCα and PKCθ. Conversely, PKCθ and PKCβ seem to have physiologically redundant roles in TCR/CD28-dependent NFκB and NFAT transactivation in primary mouse $CD3^+$ T cells (40). In intestinal epithelium, PKCα downregulates while PKCϵ upregulates cyclin D1 thus contributing to the opposing effects of these isoforms in tumor progression (41). nPKC isoforms are sequentially recruited to the immunological synapse with PKCϵ and η being recruited first followed by PKCθ (42). Opposing effects of PKCη and PKCθ on relative numbers of $CD4^+$ and $CD8^+$ T cells have been observed in mice (43). PKCα and PKCβ cooperate functionally in CD3-induced *de novo* IL2 mRNA transcriptional transactivation in primary mouse T cells independently of the actions of PKCθ (44). PKCϵ acts upstream of PKCα in the signal transduction of ischemic preconditioning of human myocardium (45). These inter-PKC regulations through feedback loops and sequential activation constitute a functioning PKC module. Mechanistically, an adaptor with dynamically controlled multiple scaffolds may connect one PKC isoform to the next.

DIFFERENTIAL PKC ISOFORM REGULATION IN INFECTION

With so many available isoforms, PKC is an ideal candidate for intracellular perturbators (**Figure 1C**). Reciprocal action of PKC isoforms has been observed in many infection and disease models. Histone deamination in neutrophils during pathogen infection or chronic inflammation is activated by PKCζ while PKCα inhibits it (28). In bone marrow-derived mast cells (BMMs), PKCα and θ positively regulate IL6 and TNFα production against filarial nematode *Acanthocheilonema viteae* infection, whereas PKCβ and ϵ act as negative regulators (46). cPKC activation seems to be associated with proinflammation as evident by the activation of these isoforms on coincubation with IFNγ (47). In *Mycobacterium tuberculosis*-infected macrophages, PKCα upregulates proinflammatory response in conjugation with TLR2 on pretreatment with arabinosylated lipoarabinomannan (48). While PKCα/β mediates CD40-induced p38MAPK phosphorylation and IL-12 expression, PKCδ and ζ inhibit it reciprocally by enhancing ERK1/2 phosphorylation and IL-10 production (8). PKCα degrades periphagosomal F-actin required for phagosomal maturation (49). This is key to the survival or elimination of the pathogens, which are either phagocytosed or internalized *via* receptor-mediated endocytosis. Infection might cause impairment of Ca^{2+}-host signal transduction, which in turn may affect classical PKC isoforms. *Leishmania donovani* infection or recombinant IL-10 treatment of macrophages inhibits both the activity and expression of the cPKC isoforms (50). *Leishmania major* also impairs PKCα, βI, βII isoforms while enhancing PKCδ and ζ isoforms in macrophages (8, 37). Increased generation of membrane ceramide (51) and concomitant cholesterol extrusion (52) may cause the inhibition of PKCα/β and activation of PKCζ in macrophages during *Leishmania* infection (8). Comparison between DAG and ceramide elucidates the different kinetics of aPKC isoforms from the other two. Ceramide activates the alternate signaling pathways leading to anti-inflammatory responses. PKCζ has been shown to be involved in the activation of arginase I, the enzyme responsible for inhibition of iNOS and inflammatory responses toward parasites (53). Intracellular pathogens can interfere with any of these mechanisms in order to tune the PKC-mediated signaling pathways according to their convenience (**Figure 1C**). Being least specific in regards to cofactor and activator requirements, nPKCs play a dual role in inflammation (8, 36, 47, 54). So, the PKC isoforms and the inter-isoform regulation might serve as targets for cellular signaling modulation for therapeutic intervention against pathophysiological conditions.

CONCLUSION

From the evidences gathered, the existence of inter-PKC regulation and a PKC-signaling module seem a logical plausibility. The signaling specificity is generated by the combinatorial assemblies and spatio-temporal dynamics of the PKC isoforms allowing calibration and kinetic modulation of the pathways of the receptor-regulated cellular responsiveness. Although the extreme diversity of PKC responses based on cellular and stimulus differences and the lack of appropriate tools and specific inhibitors pose a major hurdle in building of a generic PKC-signaling map, the analyses provide a conceptual framework placing all PKC isoforms in a single space-time network and a novel principle for devising therapeutic strategies against pathophysiological conditions.

AUTHOR CONTRIBUTIONS

AM and SR wrote the manuscript and built the figures. DM and BS critically reviewed the manuscript and made the necessary edits. The final manuscript was a result of the joint efforts of all the authors.

ACKNOWLEDGMENTS

The authors graciously acknowledge Arathi Nayer for her critical suggestions in drawing the figures.

FUNDING

DM gratefully acknowledges DST- Inspire Faculty program for providing the funds for publishing this opinion.

REFERENCES

1. Takai Y, Kishimoto A, Inoue M, Nishizuka Y. Studies on cyclic nucleotide-independent protein kinase and its pro-enzyme in mammalian tissues. I. Purification and characterization of an active enzyme from bovine cerebellum. *J Biol Chem* (1977) **252**:7603–9.
2. House C, Kemp BE. Protein kinase C contains a pseudosubstrate prototope in its regulatory domain. *Science* (1987) **238**(4834):1726–8. doi:10.1126/science.3686012
3. Nishizuka Y. Intracellular signalling by hydrolysis of phospholipids and activation of protein kinase C. *Science* (1992) **258**:607–14. doi:10.1126/science.1411571
4. Dekker LV, Parker PJ. Protein kinase C – a question of specificity. *Trends Biochem Sci* (1994) **19**:73–7. doi:10.1016/0968-0004(94)90038-8
5. Mochly-Rosen D. Localization of protein kinases by anchoring proteins: a theme in signal transduction. *Science* (1995) **268**:247–51. doi:10.1126/science.7716516
6. Newton AC. Protein kinase C: structural and spatial regulation by phosphorylation, cofactors and macromolecular interactions. *Chem Rev* (2001) **101**:2353–64. doi:10.1021/cr0002801
7. Freeley M, Kelleher D, Long A. Regulation of protein kinase C function by phosphorylation on conserved and non-conserved sites. *Cell Signal* (2011) **23**:753–62. doi:10.1016/j.cellsig.2010.10.013
8. Sudan R, Srivastava N, Pandey SP, Majumdar S, Saha B. Reciprocal regulation of protein kinase C isoforms results in differential cellular responsiveness. *J Immunol* (2012) **188**:2328–37. doi:10.4049/jimmunol.1101678
9. Takai Y, Kishimoto A, Kikkawa U, Mori T, Nishizuka Y. Unsaturated diacylglycerol as a possible messenger for the activation of calcium-activated phospholipid-dependent protein-kinase system. *Biochem Biophys Res Commun* (1979) **91**:1218–24. doi:10.1016/0006-291X(79)91197-5
10. Castagna MY, Takai Y, Kaibuchi K, Kikkawa SU, Nishizuka Y. Direct activation of calcium-activated phospholipid-dependent protein kinase by tumor promoting phorbol esters. *J Biol Chem* (1982) **257**:7847–51.
11. Mochly-Rosen D, Khaner H, Lopez J. Identification of intracellular receptor proteins for activated protein kinase C. *Proc Natl Acad Sci U S A* (1991) **88**:3997–4000. doi:10.1073/pnas.88.9.3997
12. Ron D, Chen CH, Caldwell J, Jamieson L, Orr E, Mochly-Rosen D. Cloning of an intracellular receptor for protein kinase C: a homolog of the β-subunit of G-proteins. *Proc Natl Acad Sci U S A* (1994) **91**:839–43. doi:10.1073/pnas.91.3.839
13. Mochly-Rosen D, Khaner H, Lopez J. Identification of intracellular receptor proteins for activated protein kinase C. *Proc Natl Acad Sci* (1991) **88**:3997–4001.
14. Jaken S. An overlay assay for detecting protein kinase C binding proteins and substrates. *Meth Mol Biol* (2003) **233**:359–68. doi:10.1385/1-59259-397-6:359
15. Steinberg SF. Structural basis of protein kinase C isoform function. *Physiol Rev* (2008) **88**:1341–78. doi:10.1152/physrev.00034.2007
16. Cho W. Membrane targeting by C1 and C2 domains. *J Biol Chem* (2001) **276**(35):32407–10. doi:10.1074/jbc.R100007200
17. Antal CE, Newton AC. Protein kinase C signalling in health and disease. *Biochem Soc Trans* (2014) **42**:1477–83. doi:10.1042/BST20140172
18. Benes CH, Wu N, Elia AEH, Dharia T, Cantley LC, Soltoff SP. The C2 domain of PKCδ is a phosphotyrosine binding domain. *Cell* (2005) **121**:271–80. doi:10.1016/j.cell.2005.02.019
19. Stahelin RV, Kong KF, Raha S, Tian W, Melowic HR, Ward KE, et al. Protein kinase Cθ C2 domain is a phosphotyrosine binding module that plays a key role in its activation. *J Biol Chem* (2012) **287**(36):30518–28. doi:10.1074/jbc.M112.391557
20. Tsai LCI, Xie I, Dore K, Xie L, Del Rio JC, King CC, et al. Zeta inhibitory peptide disrupts electrostatic interactions that maintain atypical protein kinase C in its active conformation on the scaffold p62. *J Biol Chem* (2015) **290**(36):21845–56. doi:10.1074/jbc.M115.676221
21. Kawakami Y, Kitaura J, Hartman SE, Lowell CA, Siragaman RP, Kawakami T. Regulation of protein kinase C βI by two protein-tyrosine kinases, Btk and Syk. *Proc Natl Acad Sci U S A* (2000) **97**(13):7423–8. doi:10.1073/pnas.120175097
22. Tabuse Y, Izumi Y, Piano F, Kemphues KJ, Miwa J, Ohno S. Atypical protein kinase C cooperates with PAR-3 to establish embryonic polarity in *Caenorhabditis elegans*. *Development* (1998) **125**(18):3607–14.
23. Diaz-Meco MT, Municio MM, Frutos S, Sanchez P, Lozano J, Sanz L, et al. The product of Par-4, a gene induced during apoptosis, interacts selectively with the atypical isoforms of protein kinase C. *Cell* (1996) **86**(5):777–86. doi:10.1016/S0092-8674(00)80152-X
24. Rex EB, Rankin ML, Yang Y, Lu Q, Gerfen CR, Jose PA, et al. Identification of RanBP 9/10 as interacting partners for protein kinase C (PKC) γ/δ and the D1 dopamine receptor: regulation of PKC-mediated receptor phosphorylation. *Mol Pharmacol* (2010) **78**:69–80. doi:10.1124/mol.110.063727
25. Schechtman D, Mochly-Rosen D. Adaptor proteins in protein kinase C-mediated signal transduction. *Oncogene* (2001) **20**:6339–47. doi:10.1038/sj.onc.1204778
26. Prekeris R, Mayhew MW, Cooper JB, Terrian DM. Identification and localization of an actin-binding motif that is unique to the epsilon isoforms of protein kinase C and participates in the regulation of synaptic function. *J Cell Biol* (1996) **132**(1,2):77–90. doi:10.1083/jcb.132.1.77
27. Rode von Essen M, Kongsbak M, Levring TB, Hansen AK, Boding L, Holst Lauritsen JP, et al. PKC-θ exists in an oxidized inactive form in naive human T cells. *Eur J Immunol* (2013) **43**:1659–66. doi:10.1002/eji.201243140
28. Neeli I, Radic M. Opposition between PKC isoforms regulates histone deamination and neutrophil extracellular chromatin release. *Front Immunol* (2013) **4**:38. doi:10.3389/fimmu.2013.00038
29. Hoshi N, Langeberg LK, Gould CM, Newton AC, Scott JD. Interaction with AKAP79 modifies the cellular pharmacology of PKC. *Mol Cell* (2010) **37**(4):541–50. doi:10.1016/j.molcel.2010.01.014
30. Oancea E, Meyer T. Protein kinase C as a molecular machine for decoding calcium and di-acyl glycerol signals. *Cell* (1998) **95**(3):307–18. doi:10.1016/S0092-8674(00)81763-8
31. Urtreger AJ, Kazanietz MG, Bal de Kier Joffe ED. Contribution of individual PKC isoforms to breast cancer progression. *Life* (2012) **64**(1):18–26. doi:10.1002/iub.574
32. Marinissen MJ, Capiati D, Boland R. 1, 25 $(OH)_2$-vitamin D_3 affects the subcellular distribution of protein kinase C isoenzymes in muscle cells. *Cell Signal* (1998) **10**(2):91–100. doi:10.1016/S0898-6568(97)00096-X
33. Dutil EM, Keranen LM, Depaoli-Roach AA, Newton AC. *In vivo* regulation of protein kinase C by transphosphorylation followed by autophosphorylation. *J Biol Chem* (1994) **269**(47):29359–62.
34. Murriel CL, Mochly-Rosen D. Opposing roles of δ and ε PKC in cardiac ischemia and reperfusion: targeting the apoptotic machinery. *Arch Biochem Biophys* (2003) **420**:246–54. doi:10.1016/j.abb.2003.08.038

35. Holm A, Tejle K, Gunnarsson T, Magnusson KE, Descoteaux A, Rasmusson B. Role of protein kinase C α for uptake of unopsonized prey and phagosomal maturation in macrophages. *Biochem Biophys Res Commun* (2003) **302**(4):653–8. doi:10.1016/S0006-291X(03)00231-6
36. Pingel S, Wang ZE, Locksley RM. Distribution of protein kinase C isoforms after infection of macrophages with *Leishmania major*. *Infect Immun* (1998) **66**(4):1795–9.
37. Loegering DJ, Lennartz MR. Protein kinase C and toll-like receptor signaling. *Enz Res* (2011) **2011**:1–7. doi:10.4061/2011/537821
38. White WO, Seibenhener ML, Wooten MW. Phosphorylation of tyrosine 256 facilitates nuclear import of atypical protein kinase C. *J Cell Biochem* (2002) **85**:42–53. doi:10.1002/jcb.10101
39. Trushin SA, Pennington KN, Carmona EM, Asin S, Savoy DN, Billadeau DD, et al. Protein kinase Cα (PKCα) acts upstream of PKCθ to activate IκB kinase and NF-κB in T lymphocytes. *Mol Cell Biol* (2003) **23**(19):7068–81. doi:10.1128/MCB.23.19.7068-7081.2003
40. Thuille N, Wachowicz K, Hermann-Kleiter N, Kaminski S, Fresser F, Lutz-Nicoladoni C, et al. PKC θ/β and CYLD are antagonistic partners in the NFκB and NFAT transactivation pathways in primary mouse $CD3^+$ T lymphocytes. *PLoS One* (2013) **8**(1):e53709. doi:10.1371/journal.pone.0053709
41. Pysz MA, Hao F, Hizli AA, Lum MA, Swetzig WM, Black AR, et al. Differential regulation of cyclin D1 expression by protein kinase C α and ε signaling in intestinal epithelial cells. *J Biol Chem* (2014) **289**(32):22268–83. doi:10.1074/jbc.M114.571554
42. Basu R, Chen Y, Quann EJ, Huse M. The variable hinge region of novel PKCs determines localization to distinct regions of the immunological synapse. *PLoS One* (2014) **9**(4):e95531. doi:10.1371/journal.pone.0095531
43. Fu G, Hu J, Neiderberger-Magnenat N, Rybakin V, Casas J, Yachi PP, et al. Protein kinase C η is required for T-cell activation and homeostatic proliferation. *Sci Signal* (2011) **4**(202):ra84. doi:10.1126/scisignal.2002058
44. Lutz-Nicoladoni C, Thuille N, Wachowicz K, Gruber T, Leitges M, Baier G. PKCα and PKCβ cooperate functionally in CD3-induced de novo IL2 mRNA transcription. *Immunol Lett* (2013) **151**(1–2):31–8. doi:10.1016/j.imlet.2013.02.002
45. Hassouna A, Matata BM, Galiñanes M. PKC-ε is upstream and PKC-α is downstream of mitoK_{ATP} channels in the signal transduction pathway of ischemic preconditioning of human myocardium. *Am J Physiol Cell Physiol* (2004) **287**:C1418–25. doi:10.1152/ajpcell.00144.2004
46. Bell KS, Al-Riyami L, Lumb FE, Britton GJ, Poole AW, Williams CM, et al. The role of individual protein kinase C isoforms in mouse mast cell function and their targeting by the immuno-modulatory parasitic worm product, ES-62. *Immunol Lett* (2015) **168**:31–40. doi:10.1016/j.imlet.2015.09.001
47. Castrillo A, Pennington DJ, Otto F, Parker PJ, Owen MJ, Boscá L. Protein kinase C ε is required for macrophage activation and defense against bacterial infection. *J Exp Med* (2001) **194**(9):1231–42. doi:10.1084/jem.194.9.1231
48. Das S, Bhattacharjee O, Goswami A, Pal NK, Majumdar S. Arabinosylated lipoarabinomannan (Ara-Lam) mediated intracellular mechanisms against tuberculosis infection: involvement of PKC mediated signaling. *Tuberculosis* (2015) **95**:208–16. doi:10.1016/j.tube.2014.11.007
49. Holm A, Tejle K, Magnusson KE, Descoteaux A, Rasmusson B. *Leishmania donovani* lipiphosphoglycan causes periphagosomal actin accumulation: correlation with impaired translocation of PKC α and defective phagosome maturation. *Cell Microbiol* (2001) **3**(7):439–47. doi:10.1046/j.1462-5822.2001.00127.x
50. Bhattacharyya S, Ghosh S, Jhonson PL, Bhattacharya SK, Majumdar S. Immunomodulatory role of interleukin-10 in visceral leishmaniasis: defective activation of protein kinase C-mediated signal transduction events. *Infect Immun* (2001) **69**(3):1499–507. doi:10.1128/IAI.69.3.1499-1507.2001
51. Majumder S, Dey R, Bhattacharjee S, Rub A, Gupta G, Bhattacharyya Majumdar S, et al. *Leishmania*-induced biphasic ceramide generation in macrophages is crucial for uptake and survival of the parasite. *J Infect Dis* (2012) **205**(10):1607–16. doi:10.1093/infdis/jis229
52. Rub A, Dey R, Jadhav M, Kamat R, Chakkaramakil S, Majumdar S, et al. Cholesterol depletion associated with *Leishmania major* infection alters macrophage CD40 signalosome composition and effector function. *Nat Immunol* (2009) **10**(3):273–80. doi:10.1038/ni.1705
53. Dey R, Majumder N, Bhattacharjee S, Bhattacharyya Majumdar S, Banerjee R, Ganguly S, et al. *Leishmania donovani*-induced ceramide as the key mediator of Akt dephosphorylation in murine macrophages: role of protein kinase C ζ and phosphatase. *Infect Immun* (2007) **75**(2):2136–42. doi:10.1128/IAI.01589-06
54. Yang CS, Lee JS, Song SH, Hur GM, Lee SJ, Tanaka S. Protein kinase C ζ plays an essential role for *Mycobacterium tuberculosis*-induced extracellular signal regulated kinase 1/2 activation in monocytes/macrophages via toll-like receptor 2. *Cell Microbiol* (2007) **9**(2):382–96. doi:10.1111/j.1462-5822.2006.00797.x

8

Close Encounters of Lymphoid Cells and Bacteria

Aranzazu Cruz-Adalia and Esteban Veiga**

Department of Molecular and Cellular Biology, Centro Nacional de Biotecnología, Consejo Superior de Investigaciones, Científicas (CNB-CSIC), Madrid, Spain

Edited by:
Abhay Satoskar, Ohio State University, USA

Reviewed by:
Paul Fisch, University Medical Center Freiburg, Germany
Danuta Radzioch, McGill University, Canada

***Correspondence:**
Aranzazu Cruz-Adalia
acruz@cnb.csic.es;
Esteban Veiga
eveiga@cnb.csic.es

During infections, the first reaction of the host against microbial pathogens is carried out by innate immune cells, which recognize conserved structures on pathogens, called pathogen-associated molecular patterns. Afterward, some of these innate cells can phagocytose and destroy the pathogens, secreting cytokines that would modulate the immune response to the challenge. This rapid response is normally followed by the adaptive immunity, more specific and essential for a complete pathogen clearance in many cases. Some innate immune cells, usually named antigen-presenting cells, such as macrophages or dendritic cells, are able to process internalized invaders and present their antigens to lymphocytes, triggering the adaptive immune response. Nevertheless, the traditional boundary of separated roles between innate and adaptive immunity has been blurred by several studies, showing that very specialized populations of lymphocytes (cells of the adaptive immunity) behave similarly to cells of the innate immunity. These "innate-like" lymphocytes include γδ T cells, invariant NKT cells, B-1 cells, mucosal-associated invariant T cells, marginal zone B cells, and innate response activator cells, and together with the newly described innate lymphoid cells are able to rapidly respond to bacterial infections. Strikingly, our recent data suggest that conventional $CD4^+$ T cells, the paradigm of cells of the adaptive immunity, also present innate-like behavior, capturing bacteria in a process called transinfection. Transinfected $CD4^+$ T cells digest internalized bacteria like professional phagocytes and secrete large amounts of proinflammatory cytokines, protecting for further bacterial challenges. In the present review, we will focus on the data showing such innate-like behavior of lymphocytes following bacteria encounter.

Keywords: innate-like lymphocytes, conventional T cells, unconventional T cells, gamma delta T cells, B cells, bacteria–lymphocyte interactions

INTRODUCTION

Classically, the immune system is classified into innate and adaptive immunity. During pathogen challenges, the first host defense involves the innate immune system that provides an immediate response. Innate immune systems are widely spread in nature and can be found in all plants and animals (1). Cells of the innate immune system include mast cells, eosinophils, basophils, natural killers (NKs), and phagocytes [macrophages, neutrophils, and dendritic cells (DCs)]. These cells recognize conserved structures, shared by different pathogens, called pathogen-associated molecular patterns (PAMPs) by their pattern-recognition receptors (PRRs). Afterward, they eliminate pathogens, either by combating through contact or by engulfing them. Some of these phagocytes are also antigen-presenting cells (APCs), such as DCs, and after engulfment pathogens, they are able to process and present the invader antigens to activate the cells of the adaptive immunity, the lymphocytes. Albeit both innate and adaptive immunity can distinguish between self and non-self molecules, adaptive immunity is defined by its capacity to specifically recognize a large amount of different antigens, defined by any non-self substance that can be recognized by the immune system. Theoretically, more than 10^{13} different antigens could be recognized by the adaptive immunity (2). This highly specific adaptive response takes more time to occur and generates memory, i.e., a second exposition to the same antigen results in faster and more potent response.

B and T cells are the major types of lymphocytes within the adaptive response and need to be activated by professional APCs during the antigen presentation. Regarding T cell activation, processed antigens are presented into the major histocompatibility complex (MHC) molecules on the membrane of APCs. There are two main subtypes of T cells: helper T cells ($CD4^+$) and cytotoxic T cells ($CD8^+$). Activation of $CD4^+$ T cells occurs by the recognition of antigens coupled to class II MHC molecules (MHC-II) by T cell receptor (TCR). Typically, MHC-II molecules expose antigens degraded into lysosomal compartments (i.e., "foreign" antigens). On the other hand, $CD8^+$ T cells are activated by the detection of antigens coupled to class I MHC (MHC-I). MHC-I molecules present antigens from the cytoplasm (i.e., self-antigens, or viral antigens), but DCs, are also able to present foreign antigens, degraded the lysosome, in their MHC-I by a process called cross-presentation, which is of the major relevance in antibacterial and antitumor fight. B cells express B cell receptors (BCRs) that recognize soluble molecules from pathogens with no need for antigen processing. It has been shown, however, that B cell activation requires presentation by professional APCs *in vivo* (3). This presentation does not require MHC molecules.

Antigen presentation by APCs triggers activation and differentiation of naïve lymphocytes to effector cells. B cells suffer immunoglobulin isotype switching and somatic hypermutation, which increase the affinity of the antibodies, and T cells develop distinct effector functions (for example, the secretion of a different array of cytokines or cytolytic activity). This textbook view of the innate and adaptive immunity role separation is being blurred by the discovery of lymphoid cells behaving in an innate-like manner (4, 5). Similarly, there exists an increasing body of evidences showing that cells of the innate immunity present adaptive-like behavior developing memory-like characteristics, termed "trained immunity." Trained monocytes respond more efficiently to a second exposition of the same (and different) challenges (6).

In this review, we will focus on the innate-like role of lymphoid cells. These innate-like lymphocytes include specialized populations of lymphocytes, i.e., unconventional (γδ) T cells, invariant NKT cells (iNKT), mucosal-associated invariant T (MAIT) cells, B-1 cells, marginal zone (MZ) B cells, innate response activator (IRA) B cells, and the innate lymphoid cells (ILCs) (4). Surprisingly, we have recently shown that conventional αβ $CD4^+$ T cells, paradigm of adaptive immune cells, are able to capture bacteria from DCs in a process called transinfection and contribute to the early immune response (7). Here, we discuss in some detail the innate-like functions performed by different types of lymphocytes during bacteria encounter.

γδ T Cells

These T cells, expressing the unconventional γδ TCR, were discovered from the accidental identification of the TCRγ chain (8). γδ TCRs and αβ TCRs have qualitatively distinct modes of antigen recognition; γδ TCRs are not restricted to the recognition of peptides bound to MHC molecules (9). Unlike conventional αβ T cells, cytokine stimulation, or bacterial contact, is sufficient for activation γδ T cells, making these cells rapid and potent mediators of inflammation.

They are much less abundant than classical αβ T cells (1–4%) in thymus and lymphoid organs of adult mice, but they are in highest abundance in mucosal sites, being ~20–40% of the intestinal intraepithelial T cells, ~10–20% of total T cells in the reproductive tracks, and ~50–70% of skin dermal T cells (10).

In humans, the population of peripheral blood γδ T cells is increased in response to infections (11). Initial characterization of human γδ T cells suggested that antigens recognized by γδ T cells were small, non-peptide compounds that contained critical phosphate residues (12). The mainstream γδ T cells in human peripheral blood express the TCR Vγ9Vδ2, and they can recognize (E)-4-hydroxy-3-methyl-but-2enyl pyrophosphate (HMBPP), which are usually referred as phosphoantigens, derived from various bacteria (13). Moreover, human $V\gamma 2V\delta 2^+$ T cells can expand 2- to 10-fold during infections and recognize primary alkylamines derived from microbes, releasing interleukine-2 (IL-2) (14). Lysates or culture supernatants from many bacteria (including mycobacteria, other Gram-negative and Gram-positive cocci, protozoal parasites, and even plants extracts) stimulate $V\gamma 2V\delta 2^+$ T cells (15). Thus, human peripheral blood γδ T cells can respond to specific antigens from bacteria [e.g., *Mycobacterium tuberculosis* (16) and *Listeria monocytogenes* (17)]. Non-peptidic mycobacterial ligands in human $V\gamma 9V\delta 2^+$ T cells induce massive tumor necrosis factor (TNF) production (18). Moreover, $V\gamma 2V\delta 2^+$ T cells respond to non-peptide bacterial antigens predominantly producing Th1 cytokines such as interferon-γ (IFN-γ), although few of them (<5%) also produce IL-4 (15). It has been reported that *Helicobacter pylori* can directly interact with human peripheral γδ T cells *in vitro*, upregulating the activation molecule CD69, TNF-α, IFN-γ, and chemokines, such as MIP-1β and RANTES, favoring an inflammatory environment (19).

In mice, γδ T cells expand dramatically after challenge with *Mycobacteria*, *Listeria*, and *Salmonella* spp. (20–22), rapidly producing cytokines. They are able to produce IFN-γ after *L. monocytogenes* infection and IL-4 in response to *Nippostrongylus brasiliensis* (23). Moreover, it has also been reported that γδ T cells are the major IL-17 producers during infections (24, 25).

Toll-Like Receptors in γδ T Cells

Recognition of bacterial PAMPs by innate immune cells is driven mainly by TLRs, a type of PRRs that recognize bacterial patterns including peptidoglycan and bacterial lipopeptides (TLR1, 2, and 6), lipopolysaccharide (LPS) (TLR4), or flagellin (TLR5), others recognize nucleic acid including double-stranded viral RNA (TLR3), single-stranded RNA (TLR7 and 8), or unmethylated bacterial DNA (TLR9). Both human and mouse $\gamma\delta$ T cells express functional TLRs. There are several studies showing TLR2 expression on $\gamma\delta$ T cells in mice (26, 27) and humans (28), supporting a role in early responses to bacterial infections.

In mice, it has been shown that CCR6$^+$IL-17-producing $\gamma\delta$ T cells, but not other $\gamma\delta$ T cells, express TLR1 and TLR2, as well as dectin-1, and could directly interact with certain pathogens. Toll-like receptor (TLR) stimulation in synergy with IL-23 results in cell expansion, IL-17 production, and further recruitment of neutrophils *in vivo* (29). In addition, the expression of TLR1, TLR2, TLR-6, TLR-9, and even TLR-4 by mice $\gamma\delta$ T cells has been confirmed (30). Furthermore, it has been shown that IL-23 stimulation of splenic $\gamma\delta$, but not $\alpha\beta$, T cells leads to enhanced TLR1, -2, and -4 mRNA expression. TLR2 agonist Pam3CSK4, but not IL-23, stimulates splenic $\gamma\delta$ T cell expansion *in vitro* (30). However, TLR agonist Pam3CSK4 and other pathogen products alone do not stimulate dermal $\gamma\delta$ T cell proliferation, which require IL-23 (31). Additionally, TLR agonists Pam3CSK4 (TLR2), Gardiquimod (TLR7), and CpG (TLR9), but not LPS (TLR4) or dectin-1 ligand curdlan, stimulate dermal $\gamma\delta$ T cells to produce IL-17, which is enhanced in the presence of IL-23 (31). It has also been reported that TLR4 is involved in the production of IL-17 and IFN-γ by $\gamma\delta$ T cells during experimental autoimmune encephalomyelitis (EAE) induction (32).

In humans, the two major $\gamma\delta$ T cell subsets, Vδ1 and Vδ2, express TLR1, TLR2, and TLR3 (33). Indeed, it has been demonstrated that human $\gamma\delta$ T cells isolated from blood express high levels of TLR2, and its engagement promotes the release of IFNγ (28). Furthermore, the expression of TLR3 on human $\gamma\delta$ cells has been verified by flow cytometry and confocal microscopy (34). Human $\gamma\delta$ T cells secrete IFN-γ and upregulate CD69 after stimulation *via* TCR in the presence of poly (I:C), a TLR-3 agonist, without APC engagement (34).

In brief, $\gamma\delta$ T cells rapidly respond after bacteria encounter secreting cytokines regulating the immune response, similarly to cells of the innate immunity (**Figure 1A**). Interestingly, human $\gamma\delta$ T cells, after activation, express molecules typically found in APCs, involved in antigen presentation, such as MHC-II and CD86 (33, 35), and it has been shown that they are able to present soluble antigens activating conventional T cells. Furthermore, some reports have also shown that human $\gamma\delta$ T cells are able to phagocytose-opsonized beads and bacteria (36), presenting bacterial antigens on MHC class II (*in vitro*), highlighting the innate immune role of $\gamma\delta$ T cells. However, it remains to be elucidated whether the $\gamma\delta$ T cell-mediated antigen presentation occurs *in vivo* during the course of bacterial infections and the relevance of such antigen presentation. Moreover, it is not clear if they could elicit a memory response.

Invariant NKT

Invariant NKT cells express an invariant TCRα chain and recognize lipid and glycolipid antigens presented by CD1d, a non-polymorphic MHC class I molecule (37). They also express several receptors, such as NK1.1 (in some mouse strains), and members of the Ly49 family that are typical of the NK cell lineage. iNKT cells are most abundant in liver, thymus, spleen, and bone marrow, but are also found in lymph nodes, peripheral blood, adipose tissue, skin, and mucosal surfaces of intestine and lungs.

It has been described that iNKT cells participate in the response to microbial pathogens in mice (38, 39) (**Figure 1B**). Upon activation, they produce cytokines such as IL-4 and IFN-γ. iNKTs can be activated by TCR stimulation with microbial antigens presented by CD1d (direct activation) or with endogenous antigens and/or cytokines (indirect activation) produced by APCs. The glycosphingolipid (GSL) α-galactosyl ceramide (α GalCer) was the first antigen identified recognized by mouse [Vα14 (40)] and human [Vα24 (41)] iNKT cells. Recently, it has been described a novel GSL antigen for iNKT cells, DB06-1, which induces preferentially IFN-γ *in vivo* in mice (42).

It has been reported that mouse iNKTs recognize cell wall GSL expressed by *Sphingomonas* spp., which are Gram-negative bacteria and have abundant GSLs, similar to α GalCer (43). Mice deficient of Vα14 iNKT cells present reduced spirochete clearance and are more susceptible to chronic inflammation following *Borrelia burgdorferi* infection (44). Furthermore, iNKTs recognize glycolipids expressed by *Helicobacter pilori* (45) and diacylglycerol-containing glycolipids from *Streptococcus pneumoniae* and group B *Streptococcus* (46). After infection of mice with *S. pneumoniae*, Vα14 iNKT cells produce IFN-γ, through TCR engagement (46). Recently, it has been shown that respiratory infection with *Francisella tularensis*, Gram-negative facultative intracellular bacteria that cause lethal pulmonary tularemia, activates iNKT cells which produce IFN-γ and propagates a sepsis-like proinflammatory response (47).

On the other hand, it has been demonstrated in both humans and mice that the responses by iNKT to some bacteria, such as *Salmonella typhimurium*, is due to indirect recognition of endogenous lysosomal GSL expressed by activated DCs combined with TLR activation (43, 48).

TLRs in iNKT Cells

It has been found that mouse iNKT cells activated by TCR resulted in increased expression of TLRs (49). In this regard, TLR4 engagement is required for production of IL-4 to further stimulate B-1 cells (50). The expression of TLR3 and 9 has been confirmed at protein level, and it has been shown that TLR signaling enhances iNKT activation. TLR stimulation of iNKT cells leads to IFN-γ, IL-4, and TNF-α production (51). On the other hand, although human iNKT cells express all TLRs, except from TLR8, they do not respond directly to TLR ligands (52).

MAIT Cells

Mucosal-associated invariant T cells express a semi-invariant TCRα chain that recognizes small molecules, pterin analogs, and riboflavin metabolites, presented by the non-polymorphic MHC class I-related molecule, MR1 (53). MR1 is highly conserved (90% gene sequence identity between mouse and human), allowing for considerable species cross-reactivity of MAIT cells. Human MAIT cells develop effector capacity before exiting the thymus, in contrast to conventional T cells that remain naïve until antigen stimulation in the periphery (54, 55). Human MAIT cells are

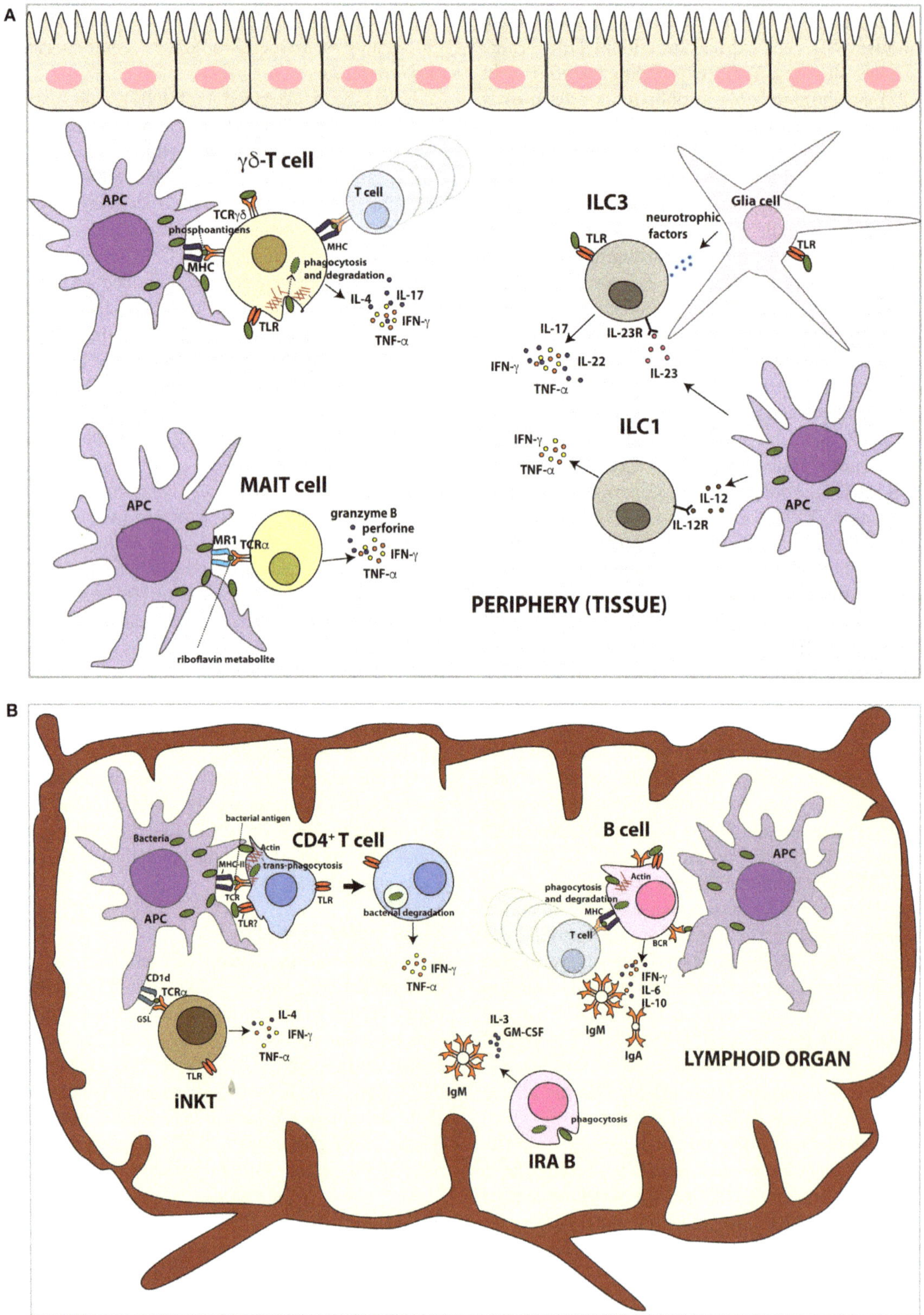

FIGURE 1 | Innate-like functions by different lymphoid cells after bacteria encounter. (A) Lymphoid populations in peripheral tissues. γδ T cells recognize non-peptide phosphoantigens derived from bacteria and induce production of TNF-α, IFN-γ, IL-4, and IL-17. Moreover, human γδ T cells are able to phagocytose bacteria and present bacterial antigens on MHC class II to activate T cells *in vitro*. TLR stimulation of γδ T cells also results in IL-17 and IFN-γ production. MAIT cells recognize microbial riboflavin metabolites presented by MR1 and produce TNF-α and IFN-γ. Furthermore, human MAIT cells destroy infected cells by secretion of

(Continued)

FIGURE 1 | Continued

cytotoxic granzyme and perforin. ILC1 cells promote defense against intracellular bacteria by TNF-α and IFN-γ. ILC3 cells respond to extracellular bacteria, inducing mainly IL-17 and IL-22. TLR stimulation of ILC3 cells also results in secretion of proinflammatory cytokines such as TNF-α and IFN-γ production. Mouse ILC3 cells can form a functional unit with glial cells in the epithelial gut sensing the environment in TLR–MyD88-dependent manner and control the immune response *via* IL-22 secretion. **(B)** Lymphoid populations in lymphoid organs. iNKT cells recognize bacterial glycosphingolipid (GSL) presented by CD1d and produce IFN-γ. Additionally, TLR engagement of iNKT cells leads to IFN-γ, IL-4, and TNF-α production. B cells capture soluble antigens or antigens exposed by APCs. On the other hand, they can phagocytose bacteria and present bacterial antigens to T lymphocytes. After bacteria encounter, B cells secrete large amount of IgM and/or IgA and proinflammatory cytokines. The newly recently described IRA B cells population protects against microbial sepsis, secreting IgM, IL-3, and GM-CSF. Human IRA B cells also phagocyte bacteria at least *in vitro*. Conventional CD4+ T cells can trans-phagocyte bacteria from infected dendritic cells (APCs) and secrete proinflammatory cytokines.

found principally in lungs, liver, and blood (56, 57), but they are less abundant in common laboratory mouse strains. Recently, it has been discovered that MAIT cells are more frequent in inbred CAST/EiJ mice than in C57BL/6 (58).

The role of MAIT cells in the control of bacterial infection was observed due to the absence of peripheral MAIT cells in germ-free mice and its expansion after microbial colonization (59). Both human and mouse MAIT cells can recognize bacterial (and fungi) infected cells in an MR1-dependet manner (56, 60). Human MAIT cells produce proinflammatory cytokines, e.g., IFN-γ and TNF in response to infection with *Mycobacterium smegmatis*, *Escherichia coli*, *Salmonella enterica*, or *Staphylococcus aureus* (56). However, not all microorganisms tested can activate MAIT cells. Several bacteria, such as *Enterococcus faecalis*, group A *Streptococcus*, and *L. monocytogenes*, do not stimulate human and mouse MAIT cells, neither do viruses (56, 60). Indeed, only the microorganisms that can synthetize riboflavin metabolites which bind MR1 are able to activate MAIT cells (i.e., most bacteria and some fungi) (53, 61). In agreement, MAIT cells accumulation in lungs of mice infected with *S. typhimurium* depends on microbial riboflavin synthesis (62). However, despite the fact the viruses do not produce riboflavin, viral infections (e.g., HIV) can reduce the numbers of peripheral MAIT cells (by mechanisms that are not totally understood), therefore increasing the susceptibility to opportunistic infections of bacteria and fungi (63). Nevertheless, an open question remains to be elucidated is how MAIT cells are able to recognize specific pathogens. Recently, using MR1 tetramers, it has been identified populations of MR1-restricted cells which assist to have different antigen recognition in humans (64) and mice (65).

In addition to the secretion of inflammatory cytokines, human MAIT cells achieve antibacterial immunity destroying infected cells by secretion of cytotoxic granzyme and perforin (66, 67). The role of MAIT cells during an infection *in vivo* has been demonstrated using MR1-deficient mice. These mice infected with *Klebsiella pneumoniae* develop higher bacterial burden, hypothermia, and have increased mortality in the first 4 days of infection compared with infected WT mice (68). In other infection model, using *Mycobacterium* bacillus Calmette–Guérin (BCG), MR-1 deficient mice also show higher bacterial burden in the lung compared to the WT mice (69). In both models, protection by mouse MAIT cells occurs within the first days of the infection, suggesting that they act as innate lymphocytes (**Figure 1A**).

Finally, it remains unresolved whether MAIT cells expressed TLRs influencing its activation or function upon stimulation.

B Lymphocytes

It is well known that B-lymphocytes, components of the adaptive response and responsible for humoral immunity, are also APCs. They can capture soluble antigens or antigens exposed by macrophages, DCs, and follicular (FO) DCs (70). There are several subtypes of B cell lymphocytes that include B-1 (B-1a and B-1b) and conventional B-2 cells that comprise two populations designated as MZ and FO B cells.

B-1 Cells

B-1 cells are not considered part of the adaptive immune system, as they do not develop into memory cells. B-1 cells have been identified in both human and mouse, but due to the logistical difficulties in isolating B-1 cells of humans, the vast majority of the studies have been performed in mouse models (71). B-1 resides principally in the peritoneal and pleural cavities but is in a minor fraction in lymph nodes and spleen (72). B-1 cells in the peritoneal cavity express CD11b (Mac-1) and are subdivided based on the expression of CD5.

B-1 cells play a relevant role in innate immunity by their contribution in the first line of defense against bacterial infection. B-1 cells alter their normal migration patterns (73), accumulating rapidly in the omentum, lymph nodes, and spleen, following activation by stimuli, such as IL-10, IL-5 (74), TLR agonists, such as LPSs (73, 75), or even whole bacteria, such as *S. pneumoniae* (76) or *Borrelia hermsii* (77). The exit of these cells from peritoneal cavity in response of LPS or bacteria is controlled by myeloid differentiation primary response protein 88 (MyD88), a key adaptor for TLRs signaling, that downregulates the expression of integrins and CD9, thereby promoting cell migration (73). After migration, B-1 cells differentiate and secrete rapidly large amount of IgM and/or IgA (73, 75). They are able to produce antibodies in response to T-cell-independent type 2 antigens (mainly repetitive structures from encapsulated bacteria) along with MZ B cells (76).

Regarding the phagocytic capacity of B-1 cells after bacteria encounter, there are several reports showing that B-1 cells are able to phagocyte *S. aureus*, *E. coli*, and polystyrene fluorescent microspheres (78, 79). Upon phagocytosis, B-1 cells kill internalized bacteria *via* phagolysosomes and present bacterial antigens in MHC-II molecules (78, 79). *Salmonella* spp. has been shown to be degraded through both proteasomal and lysosomal processing, resulting in MHC-I antigen presentation (80). On the other hand, not all bacteria are killed; some *Salmonella* can survive within B cells, using therefore B cells as Trojan horses to disseminate through the infected host (81), similarly to what has been observed in myeloid cells (82). It has been also described that B-1 cells undergo differentiation to acquire a mononuclear phagocyte phenotype *in vitro* (B-1CDP), and they are able to phagocytose *Coxiella burnetii* and kill them more effectively than peritoneal macrophages and bone marrow-derived macrophages (BMMf) (83). Moreover, it has been demonstrated that B-1 cell differentiation into phagocytes occurs also *in vivo* (84). These results revealed that mammalian B-1 cells have phagocytic and microbicidal abilities to strengthen the innate nature of these cells (**Figure 1B**). In agreement with the innate behavior of B1 cells, a recent report shows that during aging, mouse B1 cells express high levels of the costimulatory molecule CD86 and become potent activators of $CD8^+$ T cells, a role deserved for specialized populations of APCs (85).

Conventional B-2 Cells

The current consensus is that B-1 cells are phagocytic, whereas the phagocytic abilities of mouse conventional B cells and the mechanisms for bacterial uptake are less clear (79). BCR-blocking antibodies do not alter the internalization of bacteria, indicating that BCR is not involved in bacteria entry. In contrast to these data, it has been demonstrated that mouse liver B cells (both B-1 and B-2 cells) actively phagocytose and kill bacteria, such as *E. coli*, in a complement-dependent manner (86). Accordingly, it has been shown that splenic mouse B cells can internalize opsonized *Brucella abortus* (87), but *B. abortus* can survive inside B cells. In addition, it has also been reported that *S. typhimurium* are able to infect and survive within both mouse splenic B-1 and B-2 cell subpopulations. *Salmonella* infection stimulates expression of PD-L1 on mouse B cells, suggesting that PD-1/PD-L1 pathway may be involved in turning off the cytotoxic effector response during persistent infection (80). Furthermore, it has been reported that human primary B cells are able to internalize *S. typhimurium* (88). This process is BCR mediated and leads to efficient antigen loading into MHC-II, inducing $CD4^+$ T cell help to boost *Salmonella*-specific antibody production. *Salmonella*-specific B cells that phagocytose *Salmonella* upon BCR ligation reactivate human memory $CD8^+$ T cells *via* cross-presentation (89). Additionally, it has been demonstrated that both human peripheral blood and mouse splenic B-lymphocytes serve as a niche for intracellular *Salmonella* promoting systemic spreading of infection (81).

Conventional B-2 cells are divided into two populations designated as MZ and FO B cell. MZ B cells are a special population of mostly non-recirculating B cells enriched primarily in the MZ of the spleen. They are one of the first cells that take contact with blood-borne pathogens, supporting the first line of host defense. Pathogens trapped in the MZ activate MZ B cells, which maturate to plasma cells secreting IgM or to APCs. It has been reported that MZ B cells capture, process, and present antigens to T cells more efficiently than FO B cells both *in vitro* (90) and *in vivo* (91). MZ B cells appear, therefore, as excellent APCs, which, together with lymphoid DCs, play essential roles in the initial steps of *in vivo* T-cell activation. Consequently, they participate in T-cell-dependent (TD) immune response through the capture and import of blood-borne antigens to FO areas of the spleen (91).

Mice depleted of MZ B cells and infected with *B. burgdorferi* show elevated pathogen burden and reduced levels of *B. burgdorferi*-specific IgG and IgM, correlated with diminished splenic $CD4^+$ T-cell responses (92). Similarly, these mice show an increased susceptibility to *S. aureus* (93). The clearance of the bacteria *L. monocytogenes* depends on the interactions between marginal zone macrophages (MZ M) and MZ B cells (94). MZ M bind pathogens and capture antigens through various PRRs, including scavenger receptors and C-type lectin receptors (CLRs) (95), and then, they expose native antigens and establish direct cell–cell contact for the activation of MZ B cells (96) that are required for potent responses (94).

TLRs in B Cells

B cells can interact with bacteria *via* BCRs or TLRs. The expression and functionality of TLRs in B cells has been well characterized during last years. Both mouse and human B cells express a variety of TLRs (TLR1–10) (97, 98), but mouse TLR10 is not functional.

Bacterial proteins can regulate the expression of TLR in mouse B cells, such as *Shigella dysenteriae* porin, which increases the levels of TLR2, -4, and Myd88 on peritoneal B-1 cells (99). Stimulation of TLR in B cells can modify many effectors functions, and the effects depend on the development phase of B cell. TLR4 and 9 engagements at the immature and transitional B cell stage promote proliferation and survival (100). The proliferation of peritoneal B-1 cells in response to TLR stimulation is lower than splenic B-2 cells (101). On the other hand, TLR stimulation of mature B cells promotes proinflammatory cytokines production (102, 103); MZ B cells produce IL-6 and IL-10, FO B cells secrete IFN-γ and IL-6 (98), and peritoneal B-1 cells produce high levels of IL-10, limiting the clearance of *B. hermsii* infection (101). Moreover, many surface proteins are expressed in response to TLR signaling in B cells such as the receptors for B cell-activating factor belonging to the TNF family (BAFF), an important B cell survival factor in the periphery, and APRIL (97, 104).

Toll-like receptor signaling in B cells can also result in differentiation into plasma cells or influence class switching and affinity maturation (105). TLR agonists stimulate the proliferation of mouse MZ B cells and their phenotypic maturation process, increasing MHC-II, CD40, and CD86 molecules. Depending on TLR agonist, they also secrete a distinct cytokine profile (106). TLR agonists also activate MZ B cells *in vivo* and promote the migration from the MZ, accelerating the Ag-specific IgM response (107). It is shown that p110δ activity mediates TLR-induced proliferation and antibody responses by MZ B cells (108). Recently, it has been described that TLR4 stimulation can promote activation-induced cell death (AICD) in MZ B cells, increasing FasL and Fas expression, regulating T-cell-independent B cell responses (109).

In humans, TLR ligands can promote the differentiation of transitional B cells into MZ-like B cells, and patients with defective TLR signaling have reduced numbers of MZ B cells (110).

Therefore, TLR signaling in B cells induces functional responses including cytokine, immunoglobulin production, antigen presentation, proliferation, and modulation of several surface receptors. These responses depend on the B cell development stage.

Innate Response Activator Cells

Innate response activator (IRA) B cells have recently been described as a B cell population that protects against microbial sepsis in mice. They are accumulated in the spleen in a mouse model of sepsis and in response to *E. coli* infection, indicating that IRA B cell expansion is an overall characteristic of the body's reaction to bacteria (111). IRA B cells are different phenotypically and functionally from other B cell populations. They contain large amounts of intracellular IgM and spontaneously secrete IgM, but not IgA or IgG1. In addition, they are able to secrete granulocyte macrophage colony-stimulating factor (GM-CSF) and IL-3 (**Figure 1B**). IRA B cells derived from B-1 cell precursors are activated by TLR stimuli, and they protect against septic shock by controlling neutrophil-dependent bacterial clearance (111).

B-1a cells migrate to the lung in response to microbial airway infection, producing IgM. This process is depended on IRA B cells, which controls IgM production *via* autocrine GM-CSF signaling, conferring a first-line defense against bacteria in the lungs (112).

Recently, it has been described IRA B cells in humans. They reside in tonsils, within FO areas, which are the first route of defense from infection of the upper respiratory tract and are able to phagocyte bacteria, such as *S. aureus*, at least *in vitro* (113).

Therefore, IRA B cells seem to play important roles in bacterial clearance, but further work is required to clarify its function *in vivo*, and it remains to be studied whether IRA B cells directly interact with infecting bacteria and the nature of such interactions.

Innate Lymphoid Cells

Innate lymphoid cells are a recently identified member of the lymphoid lineage, which are enriched at epithelial barriers, such as skin, intestine, and lung, where contacts with microorganisms normally occur (114, 115). These innate lymphocytes mediate immune responses against infections and regulate homeostasis and inflammation (116, 117). They neither express TCRs or BCRs nor respond in an antigen-specific manner.

Innate lymphoid cells are divided into three subsets: group 1 ILCs (ILC1s and NK cells), group 2 ILCs (ILC2s), and group 3 ILCs (ILC3s and LTi cells). This nomenclature was unified to classify these emerging cell populations, which had been called by different terms including NK-22 cells, LTi-like cells, natural helper cells, nuocytes, and innate helper cells (118). ILCs are crucial in the protective immunity against bacteria (ILC1s and ILC3s) (119–121), intracellular parasites (ILC1s) (122), fungi (ILC3) (123), and parasitic worms (ILC2s) (124, 125).

Group 1 ILCs

Group 1 ILCs consisted of ILC1 and NK cells that produce IFN-γ and TNF-α after stimuli (when stimulated by IL-12, IL-15, or IL-18) (126) and had the T-box transcription factor (T-bet) as a key transcription factor (122, 127, 128). NKs were first described as innate lymphocytes with cytotoxic activity (129) that kills target cells. However, ILC1s are barely cytotoxic and seems to emerge from ILC3s (130) and are accumulated in inflamed mucosa tissue (131).

ILC1 populations have an important role in promoting defense against intracellular pathogens (**Figure 1A**). They secrete IFN-γ and TNF-α in mice infected with oral pathogen *Toxoplasma gondii*, recruiting myeloid cells that cease infection (122).

However, recently, it has been demonstrated that ILC1s also are important in promoting immunity to extracellular bacteria such as *Clostridium difficile* (119).

The deficiency of IFN-γ or T-bet-expressing ILC1s in Rag1$^{-/-}$ mice increases susceptibility to *C. difficile* (119). Furthermore, it has been shown that Nfil3, an important transcription factor for the development of NKs and ILC1s, plays a role in the intestinal innate immune defense against acute bacterial infection with *Citrobacter rodentium* and *C. difficile* (132). Nfil3 deficiency results in more susceptibility to both intestinal pathogens but also corresponds to severely reduction of ILC3s and ILC2s, revealing a general requirement for this transcription factor in the development of all ILC lineages (132).

Group 2 ILCs

Group 2 ILCs, also referred as natural helper cells, noucytes, or innate helper 2 cells, are innate lymphocytes that produce IL-5 and IL-13 when stimulated with IL-25, IL-33, or thymic stromal lymphopoietin (TSLP) (133, 134). They were discovered after administration of IL-25 intranasally in Rag2$^{-/-}$ mice, which lack conventional B and T cells (135, 136). ILC2s have been identified in fat tissue, spleen, nasal tissue, lung, intestine, and skin (137). ILC2 populations protect against helminth such as *N. brasiliensis* secreting IL-13 after infection (124, 133). IL-13 is necessary for the elimination of the parasite from the gastrointestinal tract, and transferring ILC2s into IL-13-deficient mice shows that IL-13 production by ILC2s is sufficient to resolve helminth infection (125). Moreover, it has been reported that ILC2s can promote IL-13-mediated immunity to other parasites in mice (138).

Group 3 ILC: ILC3s

Three groups characterized ILC3s in intestine as ILCs almost simultaneously (120, 139, 140). ILC3 populations secrete IL-17A, IL-22, TNF-α, and GM-CSF when activated (140–142). The transcription factor RORγt is an important regulator of this population (140). ILC3s are referred as NCR22 cells, NKp46^{+} ILCs, ILC22s, and NKR-LTi cells in the literature. This family is described in mucosal tissues, particularly in the intestinal tract, where they mediate the balance between the immune system and the symbiotic microbiota (120). It has been recently shown that mouse ILC3 cells form a functional unit together with glial cells that sense the gut environment in a MyD88-dependent manner and control the immune response *via* IL-22 secretion (143) (**Figure 1A**). ILC3 populations rapidly respond to infection of mice either extracellular bacteria (120, 121, 144) or fungi (123). ILC3s produce IL-22 after *C. rodentium* challenge in mice (140, 145), which is essential for host protection (146). IL-22 stimulates intestinal epithelial cells (IECs) to produce antimicrobial peptides and mucus, limiting the replication, dissemination, and tissue damage induced by pathogenic bacteria (116).

Similarly, ILC3s located in the oral mucosa produce IL-17 and IL-22 promoting immunity in mice against the fungal pathogen *Candida albicans* (123, 147). IL-17 acting alone or synergistically with IL-22 induces the recruitment of neutrophils to the site of infection. Moreover, it is shown that ILC3s also regulate neutrophils in neonatal mice, important for resistance to sepsis with Gram-negative opportunistic bacteria (148).

The production of IFN-γ by T-bet-expressing ILC3 contributes to the protection of the epithelial barrier during against *S. typhimurium* infection in mice (149). On other hand, it has been described that the expression of IL-17 and IFN-γ from ILC3s has been involved to drive inflammation in *Helicobacter hepaticus*-induced colitis (150), a mouse model of colitis. However, depletion of IL-22-producing ILCs localized in intestinal tissue results in peripheral dissemination of commensal bacteria, such as *Alcaligenes* species, promoting systemic inflammation (121). Consequently, these data indicate that ILCs regulate selective containment of lymphoid-resident bacteria to prevent systemic inflammation associated with chronic diseases.

ILC1s together with ILC3s mediate the recovery from *C. difficile* infection in mice (119). Previously, it has been suggested that ILC3s could play a role in the infection of these extracellular bacteria because the deficiency of the transcription factor Nfil3 resulted in a reduction of ILC3s with an increased of susceptibility to *C. difficile* infection (132). Nonetheless, it has been also demonstrated that ILC3s mediate protection against *S. pneumoniae* in respiratory tract (151).

Group 3 ILCs: LTi Cells

They are closely related to ILC3s, but their relationship is still controversial (114). LTi cells were first described in fetal and neonatal lymph nodes (152, 153), where they also showed that were crucial for lymphoid organogenesis. They are able to produce IL-17A and IL-22 mediating immunity to enteric pathogens (154, 155).

TLRs in ILCs

Mouse splenic ILC3s can produce IL-17 and IL-22 *in vivo* after contact with TLR2 ligands (154). Indeed, it has been shown that human RORγt$^+$ ILCs (LTi-like ILC) express functional TLR2, and its stimulation with agonists induces IL-5, IL-13, and IL-22 expression in a nuclear factor κ B (NF-κB)-dependent manner (142). Recently, it has been reported that human ILCs isolated from duodenum biopsies express TLR2, 3, and 9, but only TLR3 agonists stimulate them to produce TNF-α and IFN-γ (156).

The expression of TLRs in ILC2s has yet to be identified. There is a report showing that TLRs stimulation of purified ILC2s does not induce IL-9 (157), but further studies must be done to verify the expression and functionality of TLRs in these cells.

Natural killer cells and NCR$^+$RORγt$^+$ ILCSs (ILC3s) may interact directly with bacteria through natural cytotoxicity receptors (NCRs), such as NKp44 and NKp46, which can be activated by components derived form commensal bacteria (158, 159).

Conventional T Cells

In addition to the specialized lymphocyte populations with innate functions described above, we have recently described that conventional CD4$^+$ T cells, the paradigm of the adaptive immunity, also play innate-like roles during bacterial infections, contrary to the current view of immunology (7). CD4$^+$ T cells of both mouse and human origin are able to internalize different bacteria (pathogenic and non-pathogenic) such as *L. monocytogenes*, *S. aureus*, *E. coli*, and *S. enterica* from infected DCs, in a process called transinfection. Bacteria play a passive role in this process, driven by T cells (7); therefore, it would be more appropriate to term it transphagocytosis. Transphagocytic (ti) CD4$^+$ T cells kill internalized bacteria in a manner reminiscent of innate immune cells and secrete proinflammatory Th-1 cytokines (IFN-γ, TNF-α, and IL-6) in a rapid innate-like response (**Figure 1B**). Furthermore, tiCD4$^+$ T cells protect against bacterial infections *in vivo*, highly reducing the bacterial load found in liver and spleen 24 and 48 h after infections, contributing to the early innate immune response (7). This route of bacterial capture by T cells could be used for some pathogenic bacteria to spread. In this regard, it has been shown that T cells can serve as reservoir of bacteria *in vivo* (160–162). Moreover, *Shigella flexneri* manipulates the migration capacity of infected T cells in a type III secretion system-dependent manner (161–163). Transphagocytosis depends on T cell cytoskeleton, but the molecular mechanisms of how T cells can capture bacteria remain largely unknown. T cells are unable to directly capture bacteria (7); transphagocytosis requires T cell/DC intimate contact, and it is enhanced by antigen recognition by the TCR. On the other hand, T cells are unable to uptake latex beads from DCs, indicating that bacterial PAMPs are also involved in the transphagocytic process and suggest a role of T cell TLRs in this recently discovered process of bacterial uptake by CD4$^+$ T cells.

TLRs in Conventional T Cells

The expression of almost all TLRs in CD4$^+$ T cells, which would recognize bacterial PAMPs, has been identified at the mRNA level in CD4$^+$ T cells (164, 165). However, it has been shown that activated mouse CD4$^+$ T cells express TLR-3 and TLR-9 but not TLR-2 and TLR-4. Stimulation of TLR3 and 9 enhances survival in a NF-κB activation and is associated with Bcl-xL upregulation, without increased proliferation (166). On the contrary, it has been shown that TLR2 engagement induces Th1 activation in the absence of TCR stimulation, activating cell proliferation, cell survival, and IFN-γ production. IL-2 or IL-12 significantly enhances TLR-2-mediated IFN-γ production through the augmented activation of MAPKs (167). Furthermore, it has been described that TLR2 stimulation by porin of *S. dysenteriae* directly promotes CD4$^+$ T cell survival and proliferation in mouse cells (168). Human-activated CD4$^+$ T cells express TLR2 and TLR4 mRNA, but only activated cells show quantifiable surface expression of either TLR by flow cytometry (169). TLR2 activation, but not TLR4, promotes proliferation and IFN-γ, IL-2, and TNF-α production in activated CD4$^+$ T cells, indicating its costimulatory nature. In memory CD4$^+$ T cells, TLR2 expression is constitutive, and its activation leads to proliferation and IFN-γ production (169). On the other hand, TLR2 stimulation promotes Th17 differentiation both *in vivo* and *in vitro*, inducing proliferation and IL-17 production (30, 170). TLR9 stimulation in mouse CD4$^+$ T cells induces NF-κB-dependent survival (166) and provides costimulation to T cells (171). TLR9 engagement, in combination with TCR activation, reduces irradiation-induced apoptosis in mouse CD4$^+$ T cells and increases the rate of DNA repair (172). TLR9 stimulation in human effector CD4$^+$ T cells promotes cell cycle entry (173).

TLR3 stimulation also induces NF-κB, MAPK, and the survival of CD4$^+$ T cells (166). On the other hand, TLR5 engagement in combination with TCR activation results in increased proliferation and production of IL-2 in human CD4$^+$ T cells (174). TLR5 and TLR7/8 act also as costimulators, upregulating proliferation and IFN-γ, IL-8, and IL-10, but not IL-4, production by human CD4$^+$ T cells (175). Moreover, engagement of TLR7 in human CD4$^+$ T cells prevents cell cycle entry and proinflammatory cytokines production, by increasing intracellular calcium concentrations, which leads to dephosphorylation of NFATc2 and its translocation to the cell nucleus; this activates an anergic gene expression program (176).

FIGURE 2 | **Innate behaviors by lymphoid cells**. Summary of different innate conducts by distinct populations of lymphoid cells.

The role of T cell TLRs in bacterial capture and their roles in the recently described innate-like functions deserve future investigations.

CONCLUSION AND FUTURE PERSPECTIVES

Innate, rapid responses-sensing bacteria involve complex networks of cells working in a cooperative way [e.g., ILC3-glia cells collaboration (143)]. These responses include bacteria recognition by cellular PRRs, cytokine secretion, bacteria capture and killing by phagocytosis, and antigen presentation (**Figure 2**). Besides classical innate immune cells, specialized populations of lymphocytes, i.e., gamma delta (γ/δ) T, iNKT, MAIT, B-1, MZ B, and IRA B cells, behave in an innate-like manner, rapidly responding upon bacteria encounter. Surprisingly, it has been demonstrated that conventional lymphocytes (both B and T cells) can internalize bacteria in an innate-like manner. CD4+ T cells can capture and kill bacteria by transphagocytosis from infected DCs. A similar way of bacteria capture from one infected cell to another has been also recently described for macrophages (177), and it is known from long as a mechanism from viral spread (i.e., HIV and hepatitis C virus) (178). The precise role of the CD4+ T cell-dependent bacterial clearance during infections *in vivo* remains to be determined, as the number of bacteria directly cleared by transphagocytosis seems to be low, suggesting other mechanisms for the reduction of bacterial load (i.e., cytokine release or antigen presentation). In agreement with this hypothesis, transphagocytic T cells secrete large amounts of proinflammatory cytokines, mounting a potent Th-1 response.

One of the hallmarks of the innate immunity is the antigen-presentation capacity of phagocytes; it has been proposed that gamma delta T cells are able to present antigen from degraded bacteria, and whether this occurs *in vivo*, and its role during infections, remains unknown. B1 cells (and B2) have the capacity of present antigens and this ability, in addition to play a major role during infections, has been used for years to study the molecular mechanisms of T cell activation occurring during the immunological synapse *in vitro*. Whether recently discovered transphagocytic T cells (7) are able to present antigens from engulfed and killed bacteria remains unsolved and deserve further investigations. Indeed, it has been demonstrated that human T cells can process and present soluble antigens to stimulate other T lymphocytes (179, 180).

A major issue of the lymphocyte's innate-like responses is bacteria recognition. It is not fully clear which cellular receptors are involved in this process. TLRs are membrane-bound PRR involved in the recognition of extracellular PAMPs, initially characterized in innate immune cells. The expression of several TLRs has been found in the different subsets of lymphocytes, even in conventional T and B cells. Therefore, TLRs seem to be the best candidates for innate-like recognition of bacteria by lymphocytes. Bacterial recognition by lymphocytes during innate like responses and the role that TLRs would play deserve future research. Due to the similarities in TLRs activation between bacterial PAMPs and danger signals found in malignant cells, the study of lymphocyte activation by bacteria could improve the immunotherapies against cancer.

AUTHOR CONTRIBUTIONS

Both EV and AC-A contribute equally to this work. Both are also co-corresponding authors.

ACKNOWLEDGMENTS

This work was supported by the grants from the Spanish Ministry of Science and Technology (MICINN; BFU2011-29450) to EV and Ministry of Economy and Competitiveness (MINECO; SAF2014-58895-JIN to AC-A and SAF2014-56716-REDT and BFU2014-59585-R to EV).

REFERENCES

1. Litman GW, Cannon JP, Dishaw LJ. Reconstructing immune phylogeny: new perspectives. *Nat Rev Immunol* (2005) 5:866–79. doi:10.1038/nri1712
2. Nikolich-Zugich J, Slifka MK, Messaoudi I. The many important facets of T-cell repertoire diversity. *Nat Rev Immunol* (2004) 4:123–32. doi:10.1038/nri1292
3. Carrasco YR, Batista FD. B cells acquire particulate antigen in a macrophage-rich area at the boundary between the follicle and the subcapsular sinus of the lymph node. *Immunity* (2007) 27:160–71. doi:10.1016/j.immuni.2007.06.007
4. Lanier LL. Shades of grey – the blurring view of innate and adaptive immunity. *Nat Rev Immunol* (2013) 13:73–4. doi:10.1038/nri3389
5. Fan X, Rudensky AY. Hallmarks of tissue-resident lymphocytes. *Cell* (2016) 164:1198–211. doi:10.1016/j.cell.2016.02.048
6. Netea MG, Latz E, Mills KHG, O'Neill LAJ. Innate immune memory: a paradigm shift in understanding host defense. *Nat Immunol* (2015) 16:675–9. doi:10.1038/ni.3178
7. Cruz-Adalia A, Ramirez-Santiago G, Calabia-Linares C, Torres-Torresano M, Feo L, Galán-Díez M, et al. T cells kill bacteria captured by transinfection from dendritic cells and confer protection in mice. *Cell Host Microbe* (2014) 15:611–22. doi:10.1016/j.chom.2014.04.006
8. Saito H, Kranz DM, Takagaki Y, Hayday AC, Eisen HN, Tonegawa S. Complete primary structure of a heterodimeric T-cell receptor deduced from cDNA sequences. *Nature* (1984) 309:757–62. doi:10.1038/309757a0
9. Vantourout P, Hayday A. Six-of-the-best: unique contributions of γδ T cells to immunology. *Nat Rev Immunol* (2013) 13:88–100. doi:10.1038/nri3384
10. Fay NS, Larson EC, Jameson JM. Chronic inflammation and γδ T cells. *Front Immunol* (2016) 7:210. doi:10.3389/fimmu.2016.00210
11. Hara T, Mizuno Y, Takaki K, Takada H, Akeda H, Aoki T, et al. Predominant activation and expansion of V gamma 9-bearing gamma delta T cells *in vivo* as well as *in vitro* in *Salmonella* infection. *J Clin Invest* (1992) 90:204–10. doi:10.1172/JCI115837
12. Morita CT, Beckman EM, Bukowski JF, Tanaka Y, Band H, Bloom BR, et al. Direct presentation of nonpeptide prenyl pyrophosphate antigens to human gamma delta T cells. *Immunity* (1995) 3:495–507. doi:10.1016/1074-7613(95)90178-7
13. Chen ZW. Multifunctional immune responses of HMBPP-specific Vγ2Vδ2 T cells in *M. tuberculosis* and other infections. *Cell Mol Immunol* (2013) 10:58–64. doi:10.1038/cmi.2012.46
14. Bukowski JF, Morita CT, Brenner MB. Human gamma delta T cells recognize alkylamines derived from microbes, edible plants, and tea: implications for innate immunity. *Immunity* (1999) 11:57–65. doi:10.1016/S1074-7613(00)80081-3

15. Morita C, Lee H, Leslie D, Tanaka Y, Bukowski J, Märker-Hermann E. Recognition of nonpeptide prenyl pyrophosphate antigens by human gammadelta T cells. *Microbes Infect* (1999) 1:175–86. doi:10.1016/S1286-4579(99)80032-X
16. Constant P, Davodeau F, Peyrat MA, Poquet Y, Puzo G, Bonneville M, et al. Stimulation of human gamma delta T cells by nonpeptidic mycobacterial ligands. *Science* (1994) 264:267–70. doi:10.1126/science.8146660
17. Guo Y, Ziegler HK, Safley SA, Niesel DW, Vaidya S, Klimpel GR. Human T-cell recognition of *Listeria monocytogenes*: recognition of listeriolysin O by TcR alpha beta+ and TcR gamma delta+ T cells. *Infect Immun* (1995) 63:2288–94.
18. Lang F, Peyrat MA, Constant P, Davodeau F, David-Ameline J, Poquet Y, et al. Early activation of human V gamma 9V delta 2 T cell broad cytotoxicity and TNF production by nonpeptidic mycobacterial ligands. *J Immunol* (1995) 154:5986–94.
19. Romi B, Soldaini E, Pancotto L, Castellino F, Del Giudice G, Schiavetti F. *Helicobacter pylori* induces activation of human peripheral γδ+ T lymphocytes. *PLoS One* (2011) 6:e19324. doi:10.1371/journal.pone.0019324
20. Hiromatsu K, Yoshikai Y, Matsuzaki G, Ohga S, Muramori K, Matsumoto K, et al. A protective role of gamma/delta T cells in primary infection with *Listeria monocytogenes* in mice. *J Exp Med* (1992) 175:49–56. doi:10.1084/jem.175.1.49
21. Emoto M, Danbara H, Yoshikai Y. Induction of gamma/delta T cells in murine salmonellosis by an avirulent but not by a virulent strain of *Salmonella choleraesuis*. *J Exp Med* (1992) 176:363–72. doi:10.1084/jem.176.2.363
22. Janis EM, Kaufmann SH, Schwartz RH, Pardoll DM. Activation of gamma delta T cells in the primary immune response to *Mycobacterium tuberculosis*. *Science* (1989) 244:713–6. doi:10.1126/science.2524098
23. Ferrick DA, Schrenzel MD, Mulvania T, Hsieh B, Ferlin WG, Lepper H. Differential production of interferon-gamma and interleukin-4 in response to Th1- and Th2-stimulating pathogens by gamma delta T cells *in vivo*. *Nature* (1995) 373:255–7. doi:10.1038/373255a0
24. Hirota K, Duarte JH, Veldhoen M, Hornsby E, Li Y, Cua DJ, et al. Fate mapping of IL-17-producing T cells in inflammatory responses. *Nat Immunol* (2011) 12:255–63. doi:10.1038/ni.1993
25. Sheridan BS, Romagnoli PA, Pham Q-M, Fu H-H, Alonzo F, Schubert W-D, et al. γδ T cells exhibit multifunctional and protective memory in intestinal tissues. *Immunity* (2013) 39:184–95. doi:10.1016/j.immuni.2013.06.015
26. Mokuno Y, Matsuguchi T, Takano M, Nishimura H, Washizu J, Ogawa T, et al. Expression of toll-like receptor 2 on gamma delta T cells bearing invariant V gamma 6/V delta 1 induced by *Escherichia coli* infection in mice. *J Immunol* (2000) 165:931–40. doi:10.4049/jimmunol.165.2.931
27. Schwacha MG, Daniel T. Up-regulation of cell surface toll-like receptors on circulating gammadelta T-cells following burn injury. *Cytokine* (2008) 44:328–34. doi:10.1016/j.cyto.2008.09.001
28. Deetz CO, Hebbeler AM, Propp NA, Cairo C, Tikhonov I, Pauza CD. Gamma interferon secretion by human Vgamma2Vdelta2 T cells after stimulation with antibody against the T-cell receptor plus the toll-like receptor 2 agonist Pam3Cys. *Infect Immun* (2006) 74:4505–11. doi:10.1128/IAI.00088-06
29. Martin B, Hirota K, Cua DJ, Stockinger B, Veldhoen M. Interleukin-17-producing gammadelta T cells selectively expand in response to pathogen products and environmental signals. *Immunity* (2009) 31:321–30. doi:10.1016/j.immuni.2009.06.020
30. Reynolds JM, Pappu BP, Peng J, Martinez GJ, Zhang Y, Chung Y, et al. Toll-like receptor 2 signaling in CD4(+) T lymphocytes promotes T helper 17 responses and regulates the pathogenesis of autoimmune disease. *Immunity* (2010) 32:692–702. doi:10.1016/j.immuni.2010.04.010
31. Cai Y, Shen X, Ding C, Qi C, Li K, Li X, et al. Pivotal role of dermal IL-17-producing γδ T cells in skin inflammation. *Immunity* (2011) 35:596–610. doi:10.1016/j.immuni.2011.08.001
32. Reynolds JM, Martinez GJ, Chung Y, Dong C. Toll-like receptor 4 signaling in T cells promotes autoimmune inflammation. *Proc Natl Acad Sci U S A* (2012) 109:13064–9. doi:10.1073/pnas.1120585109
33. Kress E, Hedges JF, Jutila MA. Distinct gene expression in human Vδ1 and Vδ2 γδ T cells following non-TCR agonist stimulation. *Mol Immunol* (2006) 43:2002–11. doi:10.1016/j.molimm.2005.11.011
34. Wesch D, Beetz S, Oberg H-H, Marget M, Krengel K, Kabelitz D. Direct costimulatory effect of TLR3 ligand poly(I:C) on human gamma delta T lymphocytes. *J Immunol* (2006) 176:1348–54. doi:10.4049/jimmunol.176.3.1348
35. Brandes M, Willimann K, Moser B. Professional antigen-presentation function by human gammadelta T cells. *Science* (2005) 309:264–8. doi:10.1126/science.1110267
36. Wu Y, Wu W, Wong WM, Ward E, Thrasher AJ, Goldblatt D, et al. Human T cells: a lymphoid lineage cell capable of professional phagocytosis. *J Immunol* (2009) 183:5622–9. doi:10.4049/jimmunol.0901772
37. Borg NA, Wun KS, Kjer-Nielsen L, Wilce MCJ, Pellicci DG, Koh R, et al. CD1d-lipid-antigen recognition by the semi-invariant NKT T-cell receptor. *Nature* (2007) 448:44–9. doi:10.1038/nature05907
38. Ranson T, Bregenholt S, Lehuen A, Gaillot O, Leite-de-Moraes MC, Herbelin A, et al. Invariant V alpha 14+ NKT cells participate in the early response to enteric *Listeria monocytogenes* infection. *J Immunol* (2005) 175:1137–44. doi:10.4049/jimmunol.175.2.1137
39. Sada-Ovalle I, Chiba A, Gonzales A, Brenner MB, Behar SM. Innate invariant NKT cells recognize *Mycobacterium tuberculosis*-infected macrophages, produce interferon-gamma, and kill intracellular bacteria. *PLoS Pathog* (2008) 4:e1000239. doi:10.1371/journal.ppat.1000239
40. Kawano T, Cui J, Koezuka Y, Toura I, Kaneko Y, Motoki K, et al. CD1d-restricted and TCR-mediated activation of valpha14 NKT cells by glycosylceramides. *Science* (1997) 278:1626–9. doi:10.1126/science.278.5343.1626
41. Nieda M, Nicol A, Koezuka Y, Kikuchi A, Takahashi T, Nakamura H, et al. Activation of human Valpha24NKT cells by alpha-glycosylceramide in a CD1d-restricted and Valpha24TCR-mediated manner. *Hum Immunol* (1999) 60:10–9. doi:10.1016/S0198-8859(98)00100-1
42. Birkholz AM, Girardi E, Wingender G, Khurana A, Wang J, Zhao M, et al. A novel glycolipid antigen for NKT cells that preferentially induces IFN-γ production. *J Immunol* (2015) 195:924–33. doi:10.4049/jimmunol.1500070
43. Mattner J, Debord KL, Ismail N, Goff RD, Cantu C, Zhou D, et al. Exogenous and endogenous glycolipid antigens activate NKT cells during microbial infections. *Nature* (2005) 434:525–9. doi:10.1038/nature03408
44. Tupin E, Benhnia MR-E-I, Kinjo Y, Patsey R, Lena CJ, Haller MC, et al. NKT cells prevent chronic joint inflammation after infection with *Borrelia burgdorferi*. *Proc Natl Acad Sci U S A* (2008) 105:19863–8. doi:10.1073/pnas.0810519105
45. Chang Y-J, Kim HY, Albacker LA, Lee HH, Baumgarth N, Akira S, et al. Influenza infection in suckling mice expands an NKT cell subset that protects against airway hyperreactivity. *J Clin Invest* (2011) 121:57–69. doi:10.1172/JCI44845
46. Kinjo Y, Illarionov P, Vela JL, Pei B, Girardi E, Li X, et al. Invariant natural killer T cells recognize glycolipids from pathogenic Gram-positive bacteria. *Nat Immunol* (2011) 12:966–74. doi:10.1038/ni.2096
47. Hill TM, Gilchuk P, Cicek BB, Osina MA, Boyd KL, Durrant DM, et al. Border patrol gone awry: lung NKT cell activation by *Francisella tularensis* exacerbates tularemia-like disease. *PLoS Pathog* (2015) 11:e1004975. doi:10.1371/journal.ppat.1004975
48. Brigl M, Bry L, Kent SC, Gumperz JE, Brenner MB. Mechanism of CD1d-restricted natural killer T cell activation during microbial infection. *Nat Immunol* (2003) 4:1230–7. doi:10.1038/ni1002
49. Kulkarni RR, Villanueva AI, Elawadli I, Jayanth P, Read LR, Haeryfar SMM, et al. Costimulatory activation of murine invariant natural killer T cells by toll-like receptor agonists. *Cell Immunol* (2012) 277:33–43. doi:10.1016/j.cellimm.2012.06.002
50. Askenase PW, Itakura A, Leite-de-Moraes MC, Lisbonne M, Roongapinun S, Goldstein DR, et al. TLR-dependent IL-4 production by invariant Valpha14+Jalpha18+ NKT cells to initiate contact sensitivity *in vivo*. *J Immunol* (2005) 175:6390–401. doi:10.4049/jimmunol.175.10.6390
51. Villanueva AI, Haeryfar SMM, Mallard BA, Kulkarni RR, Sharif S. Functions of invariant NK T cells are modulated by TLR ligands and IFN-α. *Innate Immun* (2015) 21:275–88. doi:10.1177/1753425914527327
52. Moreno M, Mol BM, Mensdorff-Pouilly von S, Verheijen RHM, de Jong EC, Blomberg von BME, et al. Differential indirect activation of human invariant natural killer T cells by toll-like receptor agonists. *Immunotherapy* (2009) 1:557–70. doi:10.2217/imt.09.30

53. Kjer-Nielsen L, Patel O, Corbett AJ, Le Nours J, Meehan B, Liu L, et al. MR1 presents microbial vitamin B metabolites to MAIT cells. *Nature* (2012) 491:717–23. doi:10.1038/nature11605
54. Gold MC, Ehlinger HD, Cook MS, Smyk-Pearson SK, Wille PT, Ungerleider RM, et al. Human innate *Mycobacterium tuberculosis*-reactive alphabeta TCR+ thymocytes. *PLoS Pathog* (2008) 4:e39. doi:10.1371/journal.ppat.0040039
55. Gold MC, Eid T, Smyk-Pearson S, Eberling Y, Swarbrick GM, Langley SM, et al. Human thymic MR1-restricted MAIT cells are innate pathogen-reactive effectors that adapt following thymic egress. *Mucosal Immunol* (2013) 6:35–44. doi:10.1038/mi.2012.45
56. Gold MC, Cerri S, Smyk-Pearson S, Cansler ME, Vogt TM, Delepine J, et al. Human mucosal associated invariant T cells detect bacterially infected cells. *PLoS Biol* (2010) 8:e1000407. doi:10.1371/journal.pbio.1000407
57. Jeffery HC, van Wilgenburg B, Kurioka A, Parekh K, Stirling K, Roberts S, et al. Biliary epithelium and liver B cells exposed to bacteria activate intrahepatic MAIT cells through MR1. *J Hepatol* (2016) 64:1118–27. doi:10.1016/j.jhep.2015.12.017
58. Cui Y, Franciszkiewicz K, Mburu YK, Mondot S, Le Bourhis L, Premel V, et al. Mucosal-associated invariant T cell-rich congenic mouse strain allows functional evaluation. *J Clin Invest* (2015) 125:4171–85. doi:10.1172/JCI82424
59. Treiner E, Duban L, Bahram S, Radosavljevic M, Wanner V, Tilloy F, et al. Selection of evolutionarily conserved mucosal-associated invariant T cells by MR1. *Nature* (2003) 422:164–9. doi:10.1038/nature01433
60. Le Bourhis L, Martin E, Péguillet I, Guihot A, Froux N, Coré M, et al. Antimicrobial activity of mucosal-associated invariant T cells. *Nat Immunol* (2010) 11:701–8. doi:10.1038/ni.1890
61. Soudais C, Samassa F, Sarkis M, Le Bourhis L, Bessoles S, Blanot D, et al. *In vitro* and *in vivo* analysis of the Gram-negative bacteria-derived riboflavin precursor derivatives activating mouse MAIT cells. *J Immunol* (2015) 194:4641–9. doi:10.4049/jimmunol.1403224
62. Chen Z, Wang H, D'Souza C, Sun S, Kostenko L, Eckle SBG, et al. Mucosal-associated invariant T-cell activation and accumulation after in vivo infection depends on microbial riboflavin synthesis and co-stimulatory signals. *Mucosal Immunol* (2016). doi:10.1038/mi.2016.39
63. Saeidi A, Ellegård R, Yong YK, Tan HY, Velu V, Ussher JE, et al. Functional role of mucosal-associated invariant T cells in HIV infection. *J Leukoc Biol* (2016) 100:305–14. doi:10.1189/jlb.4RU0216-084R
64. Gherardin NA, Keller AN, Woolley RE, Le Nours J, Ritchie DS, Neeson PJ, et al. Diversity of T cells restricted by the MHC class I-related molecule MR1 facilitates differential antigen recognition. *Immunity* (2016) 44:32–45. doi:10.1016/j.immuni.2015.12.005
65. Sakala IG, Kjer-Nielsen L, Eickhoff CS, Wang X, Blazevic A, Liu L, et al. Functional heterogeneity and antimycobacterial effects of mouse mucosal-associated invariant T cells specific for riboflavin metabolites. *J Immunol* (2015) 195:587–601. doi:10.4049/jimmunol.1402545
66. Le Bourhis L, Dusseaux M, Bohineust A, Bessoles S, Martin E, Premel V, et al. MAIT cells detect and efficiently lyse bacterially-infected epithelial cells. *PLoS Pathog* (2013) 9:e1003681. doi:10.1371/journal.ppat.1003681
67. Kurioka A, Ussher JE, Cosgrove C, Clough C, Fergusson JR, Smith K, et al. MAIT cells are licensed through granzyme exchange to kill bacterially sensitized targets. *Mucosal Immunol* (2015) 8:429–40. doi:10.1038/mi.2014.81
68. Georgel P, Radosavljevic M, Macquin C, Bahram S. The non-conventional MHC class I MR1 molecule controls infection by *Klebsiella pneumoniae* in mice. *Mol Immunol* (2011) 48:769–75. doi:10.1016/j.molimm.2010.12.002
69. Chua W-J, Truscott SM, Eickhoff CS, Blazevic A, Hoft DF, Hansen TH. Polyclonal mucosa-associated invariant T cells have unique innate functions in bacterial infection. *Infect Immun* (2012) 80:3256–67. doi:10.1128/IAI.00279-12
70. Batista FD, Iber D, Neuberger MS. B cells acquire antigen from target cells after synapse formation. *Nature* (2001) 411:489–94. doi:10.1038/35078099
71. Tangye SG. To B1 or not to B1: that really is still the question! *Blood* (2013) 121:5109–10. doi:10.1182/blood-2013-05-500074
72. Hayakawa K, Hardy RR, Herzenberg LA, Herzenberg LA. Progenitors for Ly-1 B cells are distinct from progenitors for other B cells. *J Exp Med* (1985) 161:1554–68. doi:10.1084/jem.161.6.1554
73. Ha S-A, Tsuji M, Suzuki K, Meek B, Yasuda N, Kaisho T, et al. Regulation of B1 cell migration by signals through toll-like receptors. *J Exp Med* (2006) 203:2541–50. doi:10.1084/jem.20061041
74. Nisitani S, Tsubata T, Murakami M, Honjo T. Administration of interleukin-5 or -10 activates peritoneal B-1 cells and induces autoimmune hemolytic anemia in anti-erythrocyte autoantibody-transgenic mice. *Eur J Immunol* (1995) 25:3047–52. doi:10.1002/eji.1830251110
75. Murakami M, Tsubata T, Shinkura R, Nisitani S, Okamoto M, Yoshioka H, et al. Oral administration of lipopolysaccharides activates B-1 cells in the peritoneal cavity and lamina propria of the gut and induces autoimmune symptoms in an autoantibody transgenic mouse. *J Exp Med* (1994) 180:111–21. doi:10.1084/jem.180.1.111
76. Martin F, Oliver AM, Kearney JF. Marginal zone and B1 B cells unite in the early response against T-independent blood-borne particulate antigens. *Immunity* (2001) 14:617–29. doi:10.1016/S1074-7613(01)00129-7
77. Alugupalli KR, Leong JM, Woodland RT, Muramatsu M, Honjo T, Gerstein RM. B1b lymphocytes confer T cell-independent long-lasting immunity. *Immunity* (2004) 21:379–90. doi:10.1016/j.immuni.2004.06.019
78. Gao J, Ma X, Gu W, Fu M, An J, Xing Y, et al. Novel functions of murine B1 cells: active phagocytic and microbicidal abilities. *Eur J Immunol* (2012) 42:982–92. doi:10.1002/eji.201141519
79. Parra D, Rieger AM, Li J, Zhang Y-A, Randall LM, Hunter CA, et al. Pivotal advance: peritoneal cavity B-1 B cells have phagocytic and microbicidal capacities and present phagocytosed antigen to CD4+ T cells. *J Leukoc Biol* (2012) 91:525–36. doi:10.1189/jlb.0711372
80. Lopez-Medina M, Pérez-López A, Alpuche-Aranda C, Ortiz-Navarrete V. *Salmonella* induces PD-L1 expression in B cells. *Immunol Lett* (2015) 167:131–40. doi:10.1016/j.imlet.2015.08.004
81. Souwer Y, Griekspoor A, de Wit J, Martinoli C, Zagato E, Janssen H, et al. Selective infection of antigen-specific B lymphocytes by *Salmonella* mediates bacterial survival and systemic spreading of infection. *PLoS One* (2012) 7:e50667. doi:10.1371/journal.pone.0050667
82. Cohen TS, Hilliard JJ, Jones-Nelson O, Keller AE, ODay T, Tkaczyk C, et al. *Staphylococcus aureus* toxin potentiates opportunistic bacterial lung infections. *Sci Transl Med* (2016) 8:ra31–329. doi:10.1126/scitranslmed.aad9922
83. Popi AF, Zamboni DS, Mortara RA, Mariano M. Microbicidal property of B1 cell derived mononuclear phagocyte. *Immunobiology* (2009) 214:664–73. doi:10.1016/j.imbio.2008.12.007
84. Popi AF, Osugui L, Perez KR, Longo-Maugéri IM, Mariano M. Could a B-1 cell derived phagocyte "be one" of the peritoneal macrophages during LPS-driven inflammation? *PLoS One* (2012) 7:e34570. doi:10.1371/journal.pone.0034570
85. Lee-Chang C, Bodogai M, Moritoh K, Chen X, Wersto R, Sen R, et al. Aging converts innate B1a cells into potent CD8+ T cell inducers. *J Immunol* (2016) 196:3385–97. doi:10.4049/jimmunol.1502034
86. Nakashima M, Kinoshita M, Nakashima H, Habu Y, Miyazaki H, Shono S, et al. Pivotal advance: characterization of mouse liver phagocytic B cells in innate immunity. *J Leukoc Biol* (2012) 91:537–46. doi:10.1189/jlb.0411214
87. Goenka R, Guirnalda PD, Black SJ, Baldwin CLB. Lymphocytes provide an infection niche for intracellular bacterium *Brucella abortus*. *J Iinfect Dis* (2012) 206:91–8. doi:10.1093/infdis/jis310
88. Souwer Y, Griekspoor A, Jorritsma T, de Wit J, Janssen H, Neefjes J, et al. B cell receptor-mediated internalization of *Salmonella*: a novel pathway for autonomous B cell activation and antibody production. *J Immunol* (2009) 182:7473–81. doi:10.4049/jimmunol.0802831
89. de Wit J, Souwer Y, Jorritsma T, Klaasse Bos H, Brinke ten A, Neefjes J, et al. Antigen-specific B cells reactivate an effective cytotoxic T cell response against phagocytosed *Salmonella* through cross-presentation. *PLoS One* (2010) 5:e13016. doi:10.1371/journal.pone.0013016.g005
90. Oliver AM, Martin F, Kearney JF. IgMhighCD21high lymphocytes enriched in the splenic marginal zone generate effector cells more rapidly than the bulk of follicular B cells. *J Immunol* (1999) 162:7198–207.
91. Attanavanich K, Kearney JF. Marginal zone, but not follicular B cells, are potent activators of naive CD4 T cells. *J Immunol* (2004) 172:803–11. doi:10.4049/jimmunol.172.2.803
92. Belperron AA, Dailey CM, Booth CJ, Bockenstedt LK. Marginal zone B-cell depletion impairs murine host defense against *Borrelia burgdorferi* infection. *Infect Immun* (2007) 75:3354–60. doi:10.1128/IAI.00422-07
93. Tanigaki K, Han H, Yamamoto N, Tashiro K, Ikegawa M, Kuroda K, et al. Notch-RBP-J signaling is involved in cell fate determination of marginal zone B cells. *Nat Immunol* (2002) 3:443–50. doi:10.1038/ni793

94. Aichele P, Zinke J, Grode L, Schwendener RA, Kaufmann SHE, Seiler P. Macrophages of the splenic marginal zone are essential for trapping of blood-borne particulate antigen but dispensable for induction of specific T cell responses. *J Immunol* (2003) 171:1148–55. doi:10.4049/jimmunol.171.3.1148
95. Kang Y-S, Do Y, Lee H-K, Park SH, Cheong C, Lynch RM, et al. A dominant complement fixation pathway for pneumococcal polysaccharides initiated by SIGN-R1 interacting with C1q. *Cell* (2006) 125:47–58. doi:10.1016/j.cell.2006.01.046
96. Bergtold A, Desai DD, Gavhane A, Clynes R. Cell surface recycling of internalized antigen permits dendritic cell priming of B cells. *Immunity* (2005) 23:503–14. doi:10.1016/j.immuni.2005.09.013
97. Treml LS, Carlesso G, Hoek KL, Stadanlick JE, Kambayashi T, Bram RJ, et al. TLR stimulation modifies BLyS receptor expression in follicular and marginal zone B cells. *J Immunol* (2007) 178:7531–9. doi:10.4049/jimmunol.178.12.7531
98. Barr TA, Brown S, Ryan G, Zhao J, Gray D. TLR-mediated stimulation of APC: distinct cytokine responses of B cells and dendritic cells. *Eur J Immunol* (2007) 37:3040–53. doi:10.1002/eji.200636483
99. Ray A, Karmakar P, Biswas T. Up-regulation of CD80-CD86 and IgA on mouse peritoneal B-1 cells by porin of *Shigella dysenteriae* is toll-like receptors 2 and 6 dependent. *Mol Immunol* (2004) 41:1167–75. doi:10.1016/j.molimm.2004.06.007
100. Wechsler-Reya RJ, Monroe JG. Lipopolysaccharide prevents apoptosis and induces responsiveness to antigen receptor cross-linking in immature B cells. *Immunology* (1996) 89:356–62. doi:10.1046/j.1365-2567.1996.d01-749.x
101. Sindhava V, Woodman ME, Stevenson B, Bondada S. Interleukin-10 mediated autoregulation of murine B-1 B-cells and its role in *Borrelia hermsii* infection. *PLoS One* (2010) 5:e11445. doi:10.1371/journal.pone.0011445
102. Agrawal S, Gupta S. TLR1/2, TLR7, and TLR9 signals directly activate human peripheral blood naive and memory B cell subsets to produce cytokines, chemokines, and hematopoietic growth factors. *J Clin Immunol* (2011) 31:89–98. doi:10.1007/s10875-010-9456-8
103. Silva-Barrios S, Smans M, Duerr CU, Qureshi ST, Fritz JH, Descoteaux A, et al. Innate immune B cell activation by *Leishmania donovani* exacerbates disease and mediates hypergammaglobulinemia. *Cell Rep* (2016) 15:2427–37. doi:10.1016/j.celrep.2016.05.028
104. Ng LG, Ng C-H, Woehl B, Sutherland APR, Huo J, Xu S, et al. BAFF costimulation of toll-like receptor-activated B-1 cells. *Eur J Immunol* (2006) 36:1837–46. doi:10.1002/eji.200635956
105. Bekeredjian-Ding I, Jego G. Toll-like receptors – sentries in the B-cell response. *Immunology* (2009) 128:311–23. doi:10.1111/j.1365-2567.2009.03173.x
106. Bialecki E, Paget C, Fontaine J, Capron M, Trottein F, Faveeuw C. Role of marginal zone B lymphocytes in invariant NKT cell activation. *J Immunol* (2009) 182:6105–13. doi:10.4049/jimmunol.0802273
107. Rubtsov AV, Swanson CL, Troy S, Strauch P, Pelanda R, Torres RM. TLR agonists promote marginal zone B cell activation and facilitate T-dependent IgM responses. *J Immunol* (2008) 180:3882–8. doi:10.4049/jimmunol.180.6.3882
108. Durand CA, Hartvigsen K, Fogelstrand L, Kim S, Iritani S, Vanhaesebroeck B, et al. Phosphoinositide 3-kinase p110 delta regulates natural antibody production, marginal zone and B-1 B cell function, and autoantibody responses. *J Immunol* (2009) 183:5673–84. doi:10.4049/jimmunol.0900432
109. Figgett WA, Fairfax K, Vincent FB, Le Page MA, Katik I, Deliyanti D, et al. The TACI receptor regulates T-cell-independent marginal zone B cell responses through innate activation-induced cell death. *Immunity* (2013) 39:573–83. doi:10.1016/j.immuni.2013.05.019
110. Weller S, Bonnet M, Delagreverie H, Israel L, Chrabieh M, Maródi L, et al. IgM+IgD+CD27+ B cells are markedly reduced in IRAK-4-, MyD88-, and TIRAP- but not UNC-93B-deficient patients. *Blood* (2012) 120:4992–5001. doi:10.1182/blood-2012-07-440776
111. Rauch PJ, Chudnovskiy A, Robbins CS, Weber GF, Etzrodt M, Hilgendorf I, et al. Innate response activator B cells protect against microbial sepsis. *Science* (2012) 335:597–601. doi:10.1126/science.1215173
112. Weber GF, Chousterman BG, Hilgendorf I, Robbins CS, Theurl I, Gerhardt LMS, et al. Pleural innate response activator B cells protect against pneumonia via a GM-CSF-IgM axis. *J Exp Med* (2014) 211:1243–56. doi:10.1084/jem.20131471
113. Chiappini N, Cantisani R, Pancotto L, Ruggiero P, Rosa D, Manetti A, et al. Innate response activator (IRA) B cells reside in human tonsils and internalize bacteria in vitro. *PLoS One* (2015) 10:e0129879. doi:10.1371/journal.pone.0129879
114. Walker JA, Barlow JL, McKenzie ANJ. Innate lymphoid cells – how did we miss them? *Nat Rev Immunol* (2013) 13:75–87. doi:10.1038/nri3349
115. Sonnenberg GF, Mjösberg J, Spits H, Artis D. Snapshot: innate lymphoid cells. *Immunity* (2013) 39:622.e–622.e. doi:10.1016/j.immuni.2013.08.021
116. Sonnenberg GF, Artis D. Innate lymphoid cells in the initiation, regulation and resolution of inflammation. *Nat Med* (2015) 21:698–708. doi:10.1038/nm.3892
117. Cording S, Medvedovic J, Aychek T, Eberl G. Innate lymphoid cells in defense, immunopathology and immunotherapy. *Nat Immunol* (2016) 17:755–7. doi:10.1038/ni.3448
118. Spits H, Artis D, Colonna M, Diefenbach A, Di Santo JP, Eberl G, et al. Innate lymphoid cells – a proposal for uniform nomenclature. *Nat Rev Immunol* (2013) 13:145–9. doi:10.1038/nri3365
119. Abt MC, Lewis BB, Caballero S, Xiong H, Carter RA, Sušac B, et al. Innate immune defenses mediated by two ILC subsets are critical for protection against acute *Clostridium difficile* infection. *Cell Host Microbe* (2015) 18:27–37. doi:10.1016/j.chom.2015.06.011
120. Sanos SL, Bui VL, Mortha A, Oberle K, Heners C, Johner C, et al. RORgammat and commensal microflora are required for the differentiation of mucosal interleukin 22-producing NKp46+ cells. *Nat Immunol* (2009) 10:83–91. doi:10.1038/ni.1684
121. Sonnenberg GF, Monticelli LA, Alenghat T, Fung TC, Hutnick NA, Kunisawa J, et al. Innate lymphoid cells promote anatomical containment of lymphoid-resident commensal bacteria. *Science* (2012) 336:1321–5. doi:10.1126/science.1222551
122. Klose CSN, Flach M, Möhle L, Rogell L, Hoyler T, Ebert K, et al. Differentiation of type 1 ILCs from a common progenitor to all helper-like innate lymphoid cell lineages. *Cell* (2014) 157:340–56. doi:10.1016/j.cell.2014.03.030
123. Gladiator A, Wangler N, Trautwein-Weidner K, LeibundGut-Landmann S. Cutting edge: IL-17-secreting innate lymphoid cells are essential for host defense against fungal infection. *J Immunol* (2013) 190:521–5. doi:10.4049/jimmunol.1202924
124. Moro K, Yamada T, Tanabe M, Takeuchi T, Ikawa T, Kawamoto H, et al. Innate production of T(H)2 cytokines by adipose tissue-associated c-Kit(+) Sca-1(+) lymphoid cells. *Nature* (2010) 463:540–4. doi:10.1038/nature08636
125. Neill DR, Wong SH, Bellosi A, Flynn RJ, Daly M, Langford TKA, et al. Nuocytes represent a new innate effector leukocyte that mediates type-2 immunity. *Nature* (2010) 464:1367–70. doi:10.1038/nature08900
126. Fuchs A, Vermi W, Lee JS, Lonardi S, Gilfillan S, Newberry RD, et al. Intraepithelial type 1 innate lymphoid cells are a unique subset of IL-12- and IL-15-responsive IFN-γ-producing cells. *Immunity* (2013) 38:769–81. doi:10.1016/j.immuni.2013.02.010
127. Robinette ML, Fuchs A, Cortez VS, Lee JS, Wang Y, Durum SK, et al. Transcriptional programs define molecular characteristics of innate lymphoid cell classes and subsets. *Nat Immunol* (2015) 16:306–17. doi:10.1038/ni.3094
128. Spits H, Bernink JH, Lanier L. NK cells and type 1 innate lymphoid cells: partners in host defense. *Nat Immunol* (2016) 17:758–64. doi:10.1038/ni.3482
129. Kiessling R, Klein E, Pross H, Wigzell H. "Natural" killer cells in the mouse. II. Cytotoxic cells with specificity for mouse Moloney leukemia cells. Characteristics of the killer cell. *Eur J Immunol* (1975) 5:117–21. doi:10.1002/eji.1830050209
130. Crellin NK, Trifari S, Kaplan CD, Cupedo T, Spits H. Human NKp44+IL-22+ cells and LTi-like cells constitute a stable RORC+ lineage distinct from conventional natural killer cells. *J Exp Med* (2010) 207:281–90. doi:10.1084/jem.20091509
131. Bernink JH, Peters CP, Munneke M, Velde te AA, Meijer SL, Weijer K, et al. Human type 1 innate lymphoid cells accumulate in inflamed mucosal tissues. *Nat Immunol* (2013) 14:221–9. doi:10.1038/ni.2534
132. Geiger TL, Abt MC, Gasteiger G, Firth MA, O'Connor MH, Geary CD, et al. Nfil3 is crucial for development of innate lymphoid cells and host protection against intestinal pathogens. *J Exp Med* (2014) 211:1723–31. doi:10.1084/jem.20140212
133. Fallon PG, Ballantyne SJ, Mangan NE, Barlow JL, Dasvarma A, Hewett DR, et al. Identification of an interleukin (IL)-25-dependent cell population that provides IL-4, IL-5, and IL-13 at the onset of helminth expulsion. *J Exp Med* (2006) 203:1105–16. doi:10.1084/jem.20051615

134. Mjösberg JM, Trifari S, Crellin NK, Peters CP, van Drunen CM, Piet B, et al. Human IL-25- and IL-33-responsive type 2 innate lymphoid cells are defined by expression of CRTH2 and CD161. *Nat Immunol* (2011) 12:1055–62. doi:10.1038/ni.2104
135. Fort MM, Cheung J, Yen D, Li J, Zurawski SM, Lo S, et al. IL-25 induces IL-4, IL-5, and IL-13 and Th2-associated pathologies in vivo. *Immunity* (2001) 15:985–95. doi:10.1016/S1074-7613(01)00243-6
136. Hurst SD, Muchamuel T, Gorman DM, Gilbert JM, Clifford T, Kwan S, et al. New IL-17 family members promote Th1 or Th2 responses in the lung: in vivo function of the novel cytokine IL-25. *J Immunol* (2002) 169:443–53. doi:10.4049/jimmunol.169.1.443
137. Bostick JW, Zhou L. Innate lymphoid cells in intestinal immunity and inflammation. *Cell Mol Life Sci* (2016) 73:237–52. doi:10.1007/s00018-015-2055-3
138. Spencer SP, Wilhelm C, Yang Q, Hall JA, Bouladoux N, Boyd A, et al. Adaptation of innate lymphoid cells to a micronutrient deficiency promotes type 2 barrier immunity. *Science* (2014) 343:432–7. doi:10.1126/science.1247606
139. Luci C, Reynders A, Ivanov II, Cognet C, Chiche L, Chasson L, et al. Influence of the transcription factor RORgammat on the development of NKp46+ cell populations in gut and skin. *Nat Immunol* (2009) 10:75–82. doi:10.1038/ni.1681
140. Satoh-Takayama N, Vosshenrich CAJ, Lesjean-Pottier S, Sawa S, Lochner M, Rattis F, et al. Microbial flora drives interleukin 22 production in intestinal NKp46+ cells that provide innate mucosal immune defense. *Immunity* (2008) 29:958–70. doi:10.1016/j.immuni.2008.11.001
141. Mortha A, Chudnovskiy A, Hashimoto D, Bogunovic M, Spencer SP, Belkaid Y, et al. Microbiota-dependent crosstalk between macrophages and ILC3 promotes intestinal homeostasis. *Science* (2014) 343:1249288–1249288. doi:10.1126/science.1249288
142. Crellin NK, Trifari S, Kaplan CD, Satoh-Takayama N, Di Santo JP, Spits H. Regulation of cytokine secretion in human CD127(+) LTi-like innate lymphoid cells by Toll-like receptor 2. *Immunity* (2010) 33:752–64. doi:10.1016/j.immuni.2010.10.012
143. Ibiza S, García-Cassani B, Ribeiro H, Carvalho T, Almeida L, Marques R, et al. Glial-cell-derived neuroregulators control type 3 innate lymphoid cells and gut defence. *Nature* (2016) 535:440–3. doi:10.1038/nature18644
144. Sawa S, Lochner M, Satoh-Takayama N, Dulauroy S, Bérard M, Kleinschek M, et al. RORγt+ innate lymphoid cells regulate intestinal homeostasis by integrating negative signals from the symbiotic microbiota. *Nat Immunol* (2011) 12:320–6. doi:10.1038/ni.2002
145. Lee JS, Cella M, McDonald KG, Garlanda C, Kennedy GD, Nukaya M, et al. AHR drives the development of gut ILC22 cells and postnatal lymphoid tissues via pathways dependent on and independent of Notch. *Nat Immunol* (2012) 13:144–51. doi:10.1038/ni.2187
146. Zheng Y, Valdez PA, Danilenko DM, Hu Y, Sa SM, Gong Q, et al. Interleukin-22 mediates early host defense against attaching and effacing bacterial pathogens. *Nat Med* (2008) 14:282–9. doi:10.1038/nm1720
147. Sonnenberg GF, Nair MG, Kirn TJ, Zaph C, Fouser LA, Artis D. Pathological versus protective functions of IL-22 in airway inflammation are regulated by IL-17A. *J Exp Med* (2010) 207:1293–305. doi:10.1084/jem.20092054
148. Deshmukh HS, Liu Y, Menkiti OR, Mei J, Dai N, O'Leary CE, et al. The microbiota regulates neutrophil homeostasis and host resistance to *Escherichia coli* K1 sepsis in neonatal mice. *Nat Med* (2014) 20:524–30. doi:10.1038/nm.3542
149. Klose CSN, Kiss EA, Schwierzeck V, Ebert K, Hoyler T, d'Hargues Y, et al. A T-bet gradient controls the fate and function of CCR6-RORγt+ innate lymphoid cells. *Nature* (2013) 494:261–5. doi:10.1038/nature11813
150. Buonocore S, Ahern PP, Uhlig HH, Ivanov II, Littman DR, Maloy KJ, et al. Innate lymphoid cells drive interleukin-23-dependent innate intestinal pathology. *Nature* (2010) 464:1371–5. doi:10.1038/nature08949
151. Van Maele L, Carnoy C, Cayet D, Ivanov S, Porte R, Deruy E, et al. Activation of Type 3 innate lymphoid cells and interleukin 22 secretion in the lungs during *Streptococcus pneumoniae* infection. *J Infect Dis* (2014) 210:493–503. doi:10.1093/infdis/jiu106
152. Mebius RE, Streeter PR, Michie S, Butcher EC, Weissman IL. A developmental switch in lymphocyte homing receptor and endothelial vascular addressin expression regulates lymphocyte homing and permits CD4+ CD3- cells to colonize lymph nodes. *Proc Natl Acad Sci U S A* (1996) 93:11019–24. doi:10.1073/pnas.93.20.11019
153. Mebius RE, Rennert P, Weissman IL. Developing lymph nodes collect CD4+CD3- LTbeta+ cells that can differentiate to APC, NK cells, and follicular cells but not T or B cells. *Immunity* (1997) 7:493–504. doi:10.1016/S1074-7613(00)80371-4
154. Takatori H, Kanno Y, Watford WT, Tato CM, Weiss G, Ivanov II, et al. Lymphoid tissue inducer-like cells are an innate source of IL-17 and IL-22. *J Exp Med* (2009) 206:35–41. doi:10.1084/jem.20072713
155. Sonnenberg GF, Monticelli LA, Elloso MM, Fouser LA, Artis D. CD4(+) lymphoid tissue-inducer cells promote innate immunity in the gut. *Immunity* (2011) 34:122–34. doi:10.1016/j.immuni.2010.12.009
156. Marafini I, Monteleone I, Di Fusco D, Cupi ML, Paoluzi OA, Colantoni A, et al. TNF-α producing innate lymphoid cells (ILCs) are increased in active celiac disease and contribute to promote intestinal atrophy in mice. *PLoS One* (2015) 10:e0126291. doi:10.1371/journal.pone.0126291
157. Wilhelm C, Hirota K, Stieglitz B, Van Snick J, Tolaini M, Lahl K, et al. An IL-9 fate reporter demonstrates the induction of an innate IL-9 response in lung inflammation. *Nat Immunol* (2011) 12:1071–7. doi:10.1038/ni.2133
158. Esin S, Batoni G, Counoupas C, Stringaro A, Brancatisano FL, Colone M, et al. Direct binding of human NK cell natural cytotoxicity receptor NKp44 to the surfaces of *Mycobacteria* and other bacteria. *Infect Immun* (2008) 76:1719–27. doi:10.1128/IAI.00870-07
159. Chaushu S, Wilensky A, Gur C, Shapira L, Elboim M, Halftek G, et al. Direct recognition of *Fusobacterium nucleatum* by the NK cell natural cytotoxicity receptor NKp46 aggravates periodontal disease. *PLoS Pathog* (2012) 8:e1002601. doi:10.1371/journal.ppat.1002601
160. McElroy DS, Ashley TJ, DŁOrazio SEF. Lymphocytes serve as a reservoir for *Listeria monocytogenes* growth during infection of mice. *Microb Pathog* (2009) 46:214–21. doi:10.1016/j.micpath.2009.01.003
161. Salgado-Pabon W, Celli S, Arena ET, Nothelfer K, Roux P, Sellge G, et al. *Shigella* impairs T lymphocyte dynamics in vivo. *Proc Natl Acad Sci U S A* (2013) 110:4458–63. doi:10.1073/pnas.1300981110
162. Telford JL, Baldari CT. *Shigella* targets T cells. *Cell Host Microbe* (2011) 9:253–4. doi:10.1016/j.chom.2011.04.003
163. Konradt C, Frigimelica E, Nothelfer K, Puhar A, Salgado-Pabon W, Di Bartolo V, et al. The *Shigella flexneri* type three secretion system effector IpgD inhibits T cell migration by manipulating host phosphoinositide metabolism. *Cell Host Microbe* (2011) 9:263–72. doi:10.1016/j.chom.2011.03.010
164. Zarember KA, Godowski PJ. Tissue expression of human toll-like receptors and differential regulation of toll-like receptor mRNAs in leukocytes in response to microbes, their products, and cytokines. *J Immunol* (2002) 168:554–61. doi:10.4049/jimmunol.168.2.554
165. Hornung V, Rothenfusser S, Britsch S, Krug A, Jahrsdörfer B, Giese T, et al. Quantitative expression of toll-like receptor 1-10 mRNA in cellular subsets of human peripheral blood mononuclear cells and sensitivity to CpG oligodeoxynucleotides. *J Immunol* (2002) 168:4531–7. doi:10.4049/jimmunol.168.9.4531
166. Gelman AE, Zhang J, Choi Y, Turka LA. Toll-like receptor ligands directly promote activated CD4+ T cell survival. *J Immunol* (2004) 172:6065–73. doi:10.4049/jimmunol.172.10.6065
167. Imanishi T, Hara H, Suzuki S, Suzuki N, Akira S, Saito T. Cutting edge: TLR2 directly triggers Th1 effector functions. *J Immunol* (2007) 178:6715–9. doi:10.4049/jimmunol.178.11.6715
168. Biswas A, Banerjee P, Biswas T. Porin of *Shigella dysenteriae* directly promotes toll-like receptor 2-mediated CD4+ T cell survival and effector function. *Mol Immunol* (2009) 46:3076–85. doi:10.1016/j.molimm.2009.06.006
169. Komai-Koma M, Jones L, Ogg GS, Xu D, Liew FY. TLR2 is expressed on activated T cells as a costimulatory receptor. *Proc Natl Acad Sci U S A* (2004) 101:3029–34. doi:10.1073/pnas.0400171101
170. Nyirenda MH, Sanvito L, Darlington PJ, O'Brien K, Zhang G-X, Constantinescu CS, et al. TLR2 stimulation drives human naive and effector regulatory T cells into a Th17-like phenotype with reduced suppressive function. *J Immunol* (2011) 187:2278–90. doi:10.4049/jimmunol.1003715
171. Bendigs S, Salzer U, Lipford GB, Wagner H, Heeg K. CpG-oligodeoxynucleotides co-stimulate primary T cells in the absence of antigen-presenting cells. *Eur J Immunol* (1999) 29:1209–18. doi:10.1002/(SICI)1521-4141(199904)29:04<1209::AID-IMMU1209>3.0.CO;2-J
172. Zheng L, Asprodites N, Keene AH, Rodriguez P, Brown KD, Davila E. TLR9 engagement on CD4 T lymphocytes represses gamma-radiation-induced

apoptosis through activation of checkpoint kinase response elements. *Blood* (2008) 111:2704–13. doi:10.1182/blood-2007-07-104141

173. Funderburg N, Luciano AA, Jiang W, Rodriguez B, Sieg SF, Lederman MM. Toll-like receptor ligands induce human T cell activation and death, a model for HIV pathogenesis. *PLoS One* (2008) 3:e1915. doi:10.1371/journal.pone.0001915
174. Crellin NK, Garcia RV, Hadisfar O, Allan SE, Steiner TS, Levings MK. Human CD4+ T cells express TLR5 and its ligand flagellin enhances the suppressive capacity and expression of FOXP3 in CD4+CD25+ T regulatory cells. *J Immunol* (2005) 175:8051–9. doi:10.4049/jimmunol.175.12.8051
175. Caron G, Duluc D, Frémaux I, Jeannin P, David C, Gascan H, et al. Direct stimulation of human T cells via TLR5 and TLR7/8: flagellin and R-848 up-regulate proliferation and IFN-gamma production by memory CD4+ T cells. *J Immunol* (2005) 175:1551–7. doi:10.4049/jimmunol.175.3.1551
176. Dominguez-Villar M, Gautron A-S, de Marcken M, Keller MJ, Hafler DA. TLR7 induces anergy in human CD4(+) T cells. *Nat Immunol* (2015) 16:118–28. doi:10.1038/ni.3036
177. Steele S, Radlinski L, Taft-Benz S, Brunton J, Kawula TH. Trogocytosis-associated cell to cell spread of intracellular bacterial pathogens. *Elife* (2016) 5:e1000211–3. doi:10.7554/eLife.10625
178. Izquierdo-Useros N, Naranjo-Gómez M, Erkizia I, Puertas MC, Borràs FE, Blanco J, et al. HIV and mature dendritic cells: trojan exosomes riding the trojan horse? *PLoS Pathog* (2010) 6:e1000740. doi:10.1371/journal.ppat.1000740.g004
179. Lanzavecchia A, Roosnek E, Gregory T, Berman P, Abrignani S. T cells can present antigens such as HIV gp120 targeted to their own surface molecules. *Nature* (1988) 334:530–2. doi:10.1038/334530a0
180. Barnaba V, Watts C, de Boer M, Lane P, Lanzavecchia A. Professional presentation of antigen by activated human T cells. *Eur J Immunol* (1994) 24:71–5. doi:10.1002/eji.1830240112

9

A Systematic Review of Immunological Studies of Erythema Nodosum Leprosum

*Anastasia Polycarpou**, Stephen L. Walker and Diana N. J. Lockwood*

Faculty of Infectious and Tropical Diseases, Clinical Research Department, London School of Hygiene and Tropical Medicine, London, UK

Edited by:
Abhay Satoskar,
Ohio State University at Columbus, USA

Reviewed by:
Abraham Aseffa,
Armauer Hansen Research Institute, Ethiopia
Paul Fisch,
University Medical Center Freiburg, Germany

****Correspondence:***
Anastasia Polycarpou
anastpoly@yahoo.gr

Erythema nodosum leprosum (ENL) is a painful inflammatory complication of leprosy occurring in 50% of lepromatous leprosy patients and 5–10% of borderline lepromatous patients. It is a significant cause of economic hardship, morbidity and mortality in leprosy patients. Our understanding of the causes of ENL is limited. We performed a systematic review of the published literature and critically evaluated the evidence for the role of neutrophils, immune complexes (ICs), T-cells, cytokines, and other immunological factors that could contribute to the development of ENL. Searches of the literature were performed in PubMed. Studies, independent of published date, using samples from patients with ENL were included. The search revealed more than 20,000 articles of which 146 eligible studies were included in this systematic review. The studies demonstrate that ENL may be associated with a neutrophilic infiltrate, but it is not clear whether it is an IC-mediated process or that the presence of ICs is an epiphenomenon. Increased levels of tumor necrosis factor-α and other pro-inflammatory cytokines support the role of this cytokine in the inflammatory phase of ENL but not necessarily the initiation. T-cell subsets appear to be important in ENL since multiple studies report an increased $CD4^+/CD8^+$ ratio in both skin and peripheral blood of patients with ENL. Microarray data have identified new molecules and whole pathophysiological pathways associated with ENL and provides new insights into the pathogenesis of ENL. Studies of ENL are often difficult to compare due to a lack of case definitions, treatment status, and timing of sampling as well as the use of different laboratory techniques. A standardized approach to some of these issues would be useful. ENL appears to be a complex interaction of various aspects of the immune system. Rigorous clinical descriptions of well-defined cohorts of patients and a systems biology approach using available technologies such as genomics, epigenomics, transcriptomics, and proteomics could yield greater understanding of the condition.

Keywords: erythema nodosum leprosum, leprosy, type 2 reaction, immunology, systematic review, TNF-α, neutrophils, immune complexes

INTRODUCTION

Leprosy is an infectious disease predominantly of skin and peripheral nerves, caused by the obligate, intracellular, acid-fast bacillus *Mycobacterium leprae*. The organism shows tropism for macrophages and Schwann cells (1). The pathology and clinical phenotype of leprosy is determined by the host immune response to *M. leprae* (2). Patients develop leprosy on a clinical spectrum ranging from tuberculoid leprosy through borderline forms to lepromatous leprosy (LL) of the Ridley–Jopling classification (2). Patients with tuberculoid leprosy have a strong cell-mediated immune response to *M. leprae* limiting the disease to a few well-defined skin lesions and/or peripheral nerves (3). Patients with LL have absent cellular immunity and high titers of antibodies against *M. leprae*, which are not effective in controlling the bacilli (4).

Multi-drug therapy (MDT) is highly effective for treating the infection (1). However, despite this, 30–40% of patients with leprosy undergo immune-mediated inflammatory episodes such as Type 1 reactions (T1R) and erythema nodosum leprosum (ENL or Type 2 reactions) (5).

ENL is a painful inflammatory complication occurring in 50% of LL patients and 5–10% of borderline lepromatous leprosy (BL) patients particularly those with a bacterial index above 4 (6), whereas T1R predominantly affect those with borderline tuberculoid leprosy (BT), mid-borderline, and BL leprosy. Individuals with ENL present crops of painful, erythematous skin nodules with systemic symptoms of fever and malaise (6). ENL is a multisystem disorder and other organ involvement includes iritis, arthritis, lymphadenitis, orchitis, and neuritis (6). The histology of ENL skin lesions often shows an intense perivascular infiltrate of neutrophils throughout the dermis and subcutis (7) and vasculitis with edema of the endothelium together with granulocyte infiltration of vessels walls (8–10). However, not all ENL skin biopsies show evidence of vasculitis (10–13).

ENL is usually treated with high-dose oral corticosteroids or thalidomide if it is available and affordable. High doses of clofazimine are also commonly used (6). Treatment often lasts for many months or years. Few patients experience a single episode of acute ENL with the majority experiencing recurrent or chronic disease (6, 14). Prolonged use of oral corticosteroids is associated with multiple adverse effects (6). Our group has demonstrated that ENL results in significant economic hardship, morbidity, and mortality in patients (15, 16).

ENL is often described as a neutrophilic immune-complex-mediated condition, while there is evidence that T-cells further complicate the immunopathology. Elevated levels of certain cytokines such as tumor necrosis factor (TNF)-α and other immunological factors have been associated with episodes of ENL.

We performed a systematic review of the published literature and critically evaluated the current evidence for the role of immunological factors that have been associated with the ENL. We created a flowchart showing our search strategy by identifying the studies to be included in this systematic review (**Figure 1**). We divided the systematic review into sections according to the immune parameter under investigation including neutrophils, immune complexes/complement, T-cells, and cytokines. Furthermore, we sought to identify possible methodological issues that might account for discrepancies between studies and to make recommendations for future immunological studies of ENL. The studies that we considered to have the most important findings are discussed in detail, while all the studies included in the review are summarized in the comprehensive tables.

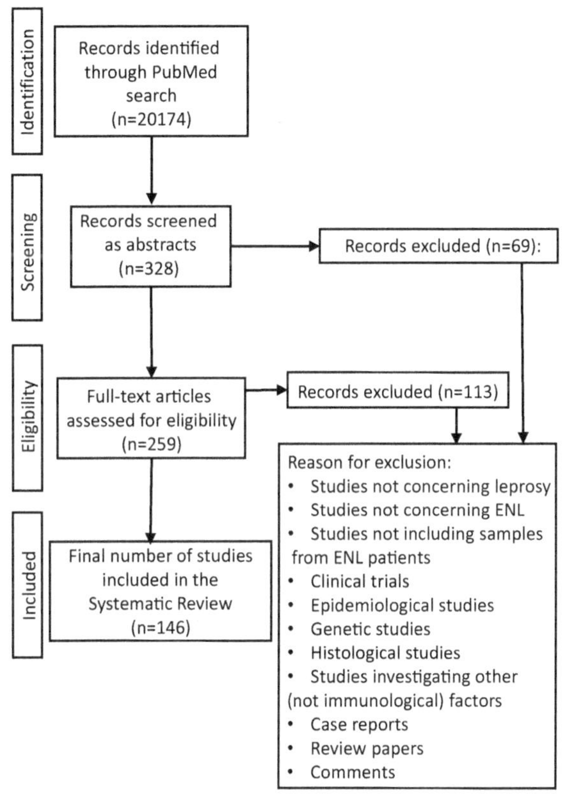

FIGURE 1 | **Flowchart.** Flowchart of included studies.

METHODOLOGY

The Preferred Reporting Items for Systematic review and Meta-Analysis Protocols (PRISMA-P) 2015 guideline was used to prepare this systematic review (17).

Searching

Searches of the literature were performed up to 31st October 2016 in PubMed by the first author. Keywords used were: Hansen* OR Type 2 OR Type II OR leprosy OR lepra*, AND reaction OR erythema nodosum leprosum OR ENL. The references included in each study were also checked for potentially relevant publications.

Inclusion Criteria

Immunological studies in PubMed, independent of published date, using samples from patients with ENL were included. Human samples including sera, peripheral blood mononuclear cells (PBMC), skin biopsies, or any other tissue were eligible for inclusion. Publications in languages other than English were translated.

An immunological study was defined as any study of the molecular and cellular components that comprise the immune system, including their function and interaction.

Results of Search

The search in PubMed revealed 95,771 records, which were narrowed down by using restrictions, species (humans), and search fields (title/abstracts), leading to 20,174 records (**Figure 1**). A total of 19,846 studies were excluded by title because they did not address leprosy or ENL. Others were excluded because they did not include samples from ENL patients or they were clinical trials, epidemiological studies, case reports, review papers, commentaries, histological studies, genetic studies, and investigations of non-immunological factors. The abstracts of the remaining 328 titles were reviewed and a further 69 studies were excluded due to the same considerations.

The 259 papers were obtained full text of which 113 were excluded for the reasons described above. When there was doubt about studies, the first and second author agreed on whether they should be included in the systematic review. Data were extracted from the 146 eligible studies. Of these 146 eligible studies, 5 studies investigated the role of neutrophils in ENL, 28 studies investigated the role of immune complexes and complement in ENL, 44 studies investigated the role of T-cells in ENL, and 49 studies investigated the role of cytokines in ENL, of which 30 investigated the role of TNF-α in ENL. Sixty-four studies investigated the role of other immunological factors in ENL.

Data Synthesis and Analysis

Data extraction from each study was conducted by the first author. Structured forms were designed for each of the five main sections of the systematic review: neutrophils, immune complexes and complement, T-cellular immunity, cytokines, and other immunological molecules or factors involved in the pathophysiology of ENL. Data were collected on the setting (study location and country of affiliation of the authors), study design and characteristics of the subjects (ENL case definition, study population included, number of patients with ENL, control subjects, timing of sampling, treatment for ENL and leprosy treatment), study measures and main findings reported by the study authors. A study could include multiple measures and therefore be part of more than one section of the systematic review.

WHAT IS THE ROLE OF NEUTROPHILS IN ENL?

Neutrophils are the predominant immune cell population in human blood and provide protection through phagocytosis, generation of neutrophil extracellular traps (NETs), and secretion of antimicrobial peptides (18). Recent evidence supports a role for neutrophils in the orchestration of adaptive immunity, engaged with lymphocytes and antigen-presenting cells (APCs) (19).

Neutrophils are considered to be the histological hallmark of ENL (7, 13). The histology of ENL skin lesions shows an intense perivascular infiltrate of neutrophils throughout the dermis and subcutis (7, 13). However, not all ENL lesions are characterized by the presence of neutrophils (12, 20–22) and the timing of biopsies appears crucial in detecting neutrophil infiltration (7, 23). A study of skin biopsies of ENL lesions within 72 h of onset showed a predominance of neutrophils in 30.4% of biopsies. Skin biopsies performed between 9 and 12 days showed neutrophils in 1.6% of specimens and increasing numbers of lymphocytes, plasma cells, and histiocytes (7). Neutrophils may precede the chemotaxis of lymphocytes into ENL lesions, but it is unclear why neutrophils are not always present in the initial stage of ENL.

The study by Lee et al. used DNA microarray and bioinformatic pathway analysis of gene expression profiles in skin biopsies obtained from six patients with ENL compared to seven LL controls (24). They identified 57 functional groups and 17 canonical pathways characteristic of ENL. Their striking finding was the "cell movement" functional pathway composed of 188 genes. From the list of genes of the "cell movement" pathway, 25 were identified to be involved specifically in neutrophil recruitment including the genes for P-selectin, E-selectin, and its ligands (24). Using immunohistochemistry, they showed that E-selectin was expressed in a vascular pattern and at higher levels in ENL skin lesions than in LL, although this was not quantified (24). They described an integrated pathway of TLR2/Fc Receptor activation triggering induction of interleukin (IL)-1β, which together with interferon (IFN)-γ, induced E-selectin expression on endothelial cells and neutrophil migration and adhesion to endothelial cells (24). Interestingly, thalidomide inhibited this neutrophil recruitment pathway (24).

A recent Brazilian study reported that surface CD64 (FcγRI) expression on circulating neutrophils increased significantly during ENL, while BL/LL patients without ENL had lower levels of CD64 (25). In addition, CD64 expression on neutrophils decreased after thalidomide treatment (25). Moreover, the higher levels of CD64 on circulating neutrophils were correlated with disease severity (25). This study demonstrated the potential of CD64 as an early biomarker for ENL and as a marker of severity (25). CD64 (FcγRI) is the high-affinity receptor for monomeric IgG1 and IgG3 (26). While resting neutrophils express low levels of CD64 (26), an increase of neutrophil CD64 surface expression is observed in certain Gram negative bacterial infections (27) and has been associated with the prognosis of disseminated intravascular coagulation during sepsis (28). The authors suggested that CD64 upregulation during ENL could be due to the presence of inflammatory cytokines such as IFN-γ and GM-CSF (29) or certain intracellular components of fragmented *M. leprae* bacilli following treatment with MDT (25). This was further supported by clinical studies showing that although ENL may also occur before initiation of treatment with MDT, the incidence of ENL is higher during treatment with MDT (5, 30).

Studies in the 70s tried to assess the polymorphonuclear leukocyte (PMN) functions in different forms of leprosy and ENL, investigating whether ENL is associated with PMN activation (31, 32). The nitro blue tetrazolium (NBT) test that measures PMN activation was increased in six patients described as LL with leprosy reactions compared with non-reactional leprosy patients (from across the leprosy spectrum) and healthy controls (31). In addition, LL patients with reactions had lower PMN activation when treated with steroids or thalidomide, although this was not significant (31). Another study found the resting NBT levels in different leprosy groups (tuberculoid, lepromatous, and patients with ENL) to be within normal limits (32). However, the sera from patients with ENL produced significantly increased levels of PMN activation as measured by the NBT test when incubated with PMN cells from healthy controls and patients with ENL

(32). This finding suggested that sera from ENL patients may lead to activation of neutrophils. However, when cell motility was studied as a marker of PMN activation using random migration, chemotaxis, and chemokinesis, all three were defective in lepromatous patients with or without complicating ENL (32).

Oliveira et al. reported the apoptotic rate of neutrophils to be greatly accelerated in ENL patients compared to BL/LL patients and healthy volunteers (33). Neutrophils isolated from leprosy patients (ENL and BL/LL) released TNF-α and IL-8, after stimulation with lipopolysaccharide (LPS) or *M. leprae* (33). Interestingly, *in vitro* TNF-α production by neutrophils was inhibited by thalidomide at both 3 and 6 h post-stimulation with LPS (33). This supports the role of neutrophils as effector cells actively producing pro-inflammatory cytokines and not only as migratory cells following chemoattractants.

There is little direct evidence of the actual role of neutrophils in ENL, despite the cell being the histological hallmark of ENL. There are multiple histological studies showing the presence of neutrophils in ENL lesions; however, only five studies investigated whether neutrophils actively take part in ENL as effector cells (**Table 1**). It remains unclear whether the neutrophil initiates ENL or is recruited to the site of the affected skin lesion under the action of chemokines such as IL-8 secreted by other cell types.

WHAT IS THE ROLE OF IMMUNE COMPLEXES IN ENL?

An IC or antigen-antibody complex is the result of binding of one or more antibody molecules with one or more antigen molecules (34). The ability of ICs to activate the complement system and to interact with a number of cells determines their biological properties (35). ICs activate complement pathways that opsonize or coat antigen–antibody complexes with large numbers of C3 molecules (36). Opsonization facilitates the clearance of ICs by the macrophage system (36). By maintaining complexes in solution, the complement allows clearance of ICs from their site of formation, minimizing local inflammatory consequences (36).

It was hypothesized that ENL is an IC-mediated disorder because it has some clinical features in common with the Arthus reaction, a type III hypersensitivity reaction that involves the deposition of ICs mainly in the vascular walls, serosa, and glomeruli and is characterized histologically by vasculitis with a polymorphonuclear cell infiltrate (37). The multisystem involvement of ENL resembling autoimmune diseases associated with ICs such as systemic lupus erythematosus (SLE), also lends credence to this theory.

TABLE 1 | Studies of neutrophils in ENL.

Reference; study site(s)	Study population	Timing of screening	MDT status	ENL treatment	Type of samples	Measures	Findings
Goihman-Yahr et al. (31); Venezuela	6 ENL, 32 BL/LL, 6 treated ENL, 9 indeterminate, 11 tuberculoid, 14 HC	ND	ND	Excluded patients on steroids except treated ENL	Peripheral blood neutrophils	Reduction of nitro blue tetrazolium (NBT)	Increased neutrophil activation in ENL
					Serum	Neutrophil response to endotoxin	Lower neutrophil activation after ENL treatment
					Plasma	Effect of adding sera and plasma from ENL to neutrophils of HC	Sera from ENL did not activate neutrophils from HC
Sher et al. (32); South Africa	8 ENL, 17 BT, 11 lepromatous, HC	ND	ND	ENL not receiving steroids or other anti-inflammatory drugs	Peripheral blood neutrophils	PMN leukocyte motility	Defect in random migration, chomotaxis, and chemokinesis in both ENL and lepromatous patients
					Serum	Reduction of nitro blue tetrazolium (NBT)	Reconstitution of PMN leukocytes from HC and ENL with sera from ENL led to increased neutrophil activation
Oliveira et al. (33); Brazil δ	10 BL/LL:6 ENL, 10 HC	ND	On MDT	ND	Peripheral blood neutrophils	Apoptosis DNA fragmentation extracted from neutrophils TNF-α and IL-8	Increased apoptosis in ENL Stimulated neutrophils secrete IL-8 and tumor necrosis factor (TNF)-α
Lee et al. (24); USA ε	6 ENL, 7 LL	ND	ND	ND	Skin	Microarrays and gene expression analysis Ability of HUVEC to bind neutrophils from HC	Genes involved in neutrophil recruitment identified Thalidomide diminished neutrophil binding to HUVECs stimulated with cytokines
Schmitz et al. (25); Brazil ε	62 leprosy: 22 ENL, 16 HC	ENL: before and 7 days after thalidomide	Patients before and after MDT	ENL: before and after thalidomide	Peripheral blood neutrophils	CD64 expression	CD64 upregulated on neutrophils during ENL Higher CD64 on neutrophils from severe ENL CD64 decreased after thalidomide

β, also in ***Table 2****; γ, also in* ***Table 3****; δ, also in* ***Table 4****; ε, also in* ***Table 5****.*

BB, mid-borderline leprosy; BL, borderline lepromatous leprosy; BT, borderline tuberculoid leprosy; ENL, erythema nodosum leprosum; HC, healthy controls; HUVEC, human umbilical vein endothelial cells; ICs, immune complexes; LL, lepromatous leprosy polar; ND, not described; PMN, polymorphonuclear; SLE, systemic lupus erythematosus; TB, tuberculosis; TT, tuberculoid leprosy polar.

Multiple studies have been performed investigating ICs in ENL. The widely cited study of Wemambu et al included 17 patients with ENL and six uncomplicated LL controls (37). Direct immunofluorescence demonstrated granular deposits of immunoglobulin and complement in a perivascular distribution in association with a polymorph infiltrate in the dermis of 10 out of 17 ENL lesions but not in any lesions of uncomplicated LL (37). However, such deposition is not conclusive evidence of ICs. The presence of soluble mycobacterial antigen was seen in ICs in only 3 out of 17 ENL lesions (37). The authors hypothesized that ENL results from the deposition of ICs in and around venules of the connective tissue septa of subcutaneous fat (37). The study was repeated using 38 patients with ENL and 13 LL controls and demonstrated the presence of immunoglobulin, complement, and mycobacterial antigen in less than half of the skin biopsies from patients with ENL and none of the LL control biopsies (22). Non-specific granular deposits of IgG were demonstrated along the collagen and elastic fibers in the dermis of all 25 patients with ENL in another study, not in any of the 10 LL patient controls (38). However, the deposits were not consistently seen in and around the blood vessels (38). Later studies in ENL suggest that these ICs are extravascular and hence ENL differs from the Arthus reaction (39, 40). These studies taken together provide evidence of an association of ICs and ENL but they do not necessarily support that ICs are the trigger leading to ENL.

Circulating ICs have been demonstrated in patients across the leprosy spectrum (41). The level of circulating ICs in the sera of leprosy patients have been measured in many studies using different immunological techniques (42–54) of which the most commonly used are C1q immunoassays (42, 43, 51). This highlights the fact that the use of different immunoassays to detect circulating ICs in studies may explain the contradictory results. The first study measuring ICs in sera of leprosy patients performed C1q immunoassays in samples from LL patients, tuberculoid leprosy patients, and healthy volunteers and showed that more than 70% of LL patients had demonstrable ICs (43). A subsequent study demonstrated increased occurrence of ICs in both the sera of ENL patients (80%) and uncomplicated LL patients (82%), indicating that the presence of circulating ICs is not a characteristic feature of ENL *per se* (46). Wager et al. analyzed sera from 135 leprosy patients using the platelet aggregation test (PAT) which had been previously suggested to be a sensitive detector of IgG complexes in other immunological and infectious diseases (55, 56) and concluded that PAT is a sensitive detector of IgG complexes peculiar to LL (44). No ICs were detected in the sera of leprosy patients using the C1q immunoassay (44).

Specific mycobacterial antigens (41) or antibodies against *M. leprae* antigens (50, 57) have been identified in the ICs derived from sera of lepromatous patients with or without ENL. Rojas et al. precipitated ICs from sera and detected antibodies against phenolic glycolipid-1 (PGL-1) (50) and major cytosolic protein of *M. leprae* (MCP-I). The finding that ICs are composed of anti-PGL-I and anti-MCP-I antibodies supports the concept that ENL is an IC-mediated disorder (50). However, the composition of circulating ICs of leprosy controls (combined BT and BL/LL) also showed high levels of anti-PGL-I antibodies (50) again suggesting that ICs are not specific to ENL.

Dupnik et al. used DNA microarrays to examine gene expression in PBMC isolated from patients with ENL and matched leprosy controls (58). Several components of the classical complement pathway showed increased expression in PBMC from patients with ENL: C1qA, B, and C and the complement receptors C3AR1 and C5AR1 (58). Increased intensity of fluorescent staining for C1q in skin lesions of ENL compared to BT and BL/LL controls was demonstrated (58). The finding of increased C1q deposition in the skin of ENL does not necessarily mean IC deposition has occurred (35). However, these data do support activation of the classical complement pathway in ENL, which may result from antigen–antibody formation.

Earlier studies in leprosy looked at the role of free complement in the sera of lepromatous patients (59). The serum C3 levels were decreased in patients with ENL, whereas they were elevated in LL controls (60). The low levels of C3 supported the concept that ENL is mediated by an antigen-antibody reaction and may be due to its utilization during the course of such antigen-antibody reactions. Similar decreased serum complement levels have been reported in other IC disorders such as acute glomerulonephritis (61–63) and acute systemic lupus erythematosus (SLE) (64, 65). It has been suggested that ENL is characterized by complement hypercatabolism because the level of the C3 breakdown product C3d in the sera was increased in 70% of the patients with ENL but in only 18% of patients with uncomplicated LL (46).

In other IC-associated diseases such as SLE, systemic vasculitides, and nephritis, defective complement-mediated solubilization of immune precipitates have been observed (48, 49, 66). Similarly, leprosy patients with ENL were shown to have markedly reduced solubilization levels that remained low for 3 months, whereas the C3d and circulating IC levels returned to baseline levels (48). Circulating ICs isolated from sera across the leprosy spectrum as PEG precipitates were shown to be efficient activators of the alternative complement pathway. In addition, PEG precipitates from BL/LL leprosy patients including those with ENL were shown to activate the classical complement pathway as well (52).

A Brazilian study of 46 patients with ENL investigated the association between the MHC class III complement proteins C2, BF, C4A, and C4B and leprosy (67). All patients who were homozygous for the silent C4B allele (C4B*Q0) and thus C4B-deficient had ENL (67). Increased frequency of ENL was also associated with those who were hemizygous for the C4B*Q0 allele. The relative risk of patients suffering from ENL carrying the C4B*Q0 allele was 5.3 compared with LL patients without C4B*Q0 (67). Interestingly, their findings suggested that C4B deficiency could play an important role in the abnormal immune response to *M. leprae* and to the lack of IC clearance, leading to ENL reactions (67). Hemizygous C4 deficiencies are associated with immune complex diseases such as SLE (68).

There is lack of evidence to support a causative role of ICs in ENL, which requires the deposition of ICs in tissues, the presence of bacterial antigens in these ICs, and the interaction of the ICs with the complement cascade and with phagocytic cells (35). Although there are 28 studies investigating the presence of ICs in the skin or circulating ICs in the sera of patients with ENL (**Table 2**), their role remains uncertain. It is unclear

TABLE 2 | Human studies on ENL investigating immune complexes and complement.

Reference; study site(s)	Study population	Timing of screening	MDT status	ENL treatment	Type of samples	Measures	Findings
de Azevedo and de Melo (59); Brazil	37 lepromatous, 33 tuberculoid, 18 "lepra reaction"	ND	ND	ND	Serum	Complement unit (K) Angular inclination (1/n)	Reduced complement activity in the reactional group
Wemambu et al. (37); United Kingdom and Malaysia	17 ENL, 6 lepromatous	ND	ND	ND	Skin Serum	Immunoglobulin Complement	Perivascular deposits of immunoglobulin and complement Mycobacterial antigen in some ENL skin lesions
Waters et al. (22); United Kingdom and Malaysia ε	38 lepromatous with ENL, 13 lepromatous	ND	ND	ND	Skin Serum	Immunoglobulin and complement Detection of mycobacterial antigen in the ICs	Immunoglobulin and complement perivascular in some ENL skin lesions Mycobacterial antigen present in ICs
Gelber et al. (69); Taiwan and USA	15 LL with ENL, 47 BT-LL	3 or more specimens time-span up to 4 months in BT and LL/ENL and up to 6 months to LL without ENL	ND	ND	Serum	C1q precipitin activity Complement levels C3 Cryoglobulins	Association of C1q precipitin activity with ENL
Bjorvatn et al. (46); Ethiopia and Switzerland	13 ENL, 7 LL, 6 tuberculoid, pulmonary TB, 30 HC	ND	All on dapsone or clofazimine	ENL patients received treatment for ENL	Serum	ICs with ^{125}I-C1q binding assay Complement	ICs increased in ENL and LL but also in tuberculoid leprosy Increased C3d level in most patients with ENL
Tung et al. (51); Ethiopia	22 BL/LL with ENL, 23TT-LL, 17 SLE	ND	19/23 non-ENL on dapsone	Untreated ENL	Serum	C1q ICs Raji test ICs	Circulating ICs in 67% of leprosy by C1q test Only 7% of this 67% showed ICs by the Raji test
Anthony et al. (60); India ε	25 LL with ENL, 10 LL without ENL	Active ENL lesions at the time of the biopsy	ND	ND	Skin Serum	Immunoglobulin deposits in skin Complement in sera	Immunoglobulin deposits in ENL skin but not in LL Decreased serum complement in ENL Elevated levels in LL
Wager et al. (44); Finland, Brazil, and Ethiopia	11 ENL, 112 leprosy, 61 LL, 7 tuberculoid, 28 SLE, 42 RA, 374 HC	ND	ND	ND	Serum	ICs with Platelet aggregation test (PAT) Other sero-immunological parameters	Higher PAT titers toward the lepromatous end of the spectrum No significant differences for ENL patients
Izumi et al. (70); Japan γ	12 ENL, 49 active lepromatous, 24 inactive lepromatous, 7 borderline, 6 tuberculoid, 9 HC	ND	ND	ND	Serum	C4, C3c, C3 activator	C3 activator and C3c concentrations higher in ENL compared with active lepromatous

(Continued)

TABLE 2 | Continued

Reference; study site(s)	Study population	Timing of screening	MDT status	ENL treatment	Type of samples	Measures	Findings
Harikrishan et al. (71); India ε	20 active LL, 15 ENL active and subsided, 20 HC	ENL: during the active and the subsided phase	ND	ND	Serum	Complement factor C3	Increased levels of C3 in LL and ENL Decrease in C3 during the "subsided phase" of reaction
Saha et al. (72); India	20 ENL, 15 HC	Initial sample first visit, subsequent on ENL clinical remission 4 weeks later	ND	Second sample: on antireactional treatment	Serum	Complement C1q, C3, C4	C3 level decreased during ENL, while increased after remission C3d increased during ENL, and remained elevated after clinical remission in most patients No significant difference in C1q or C4 during ENL
Valentijn et al. (54); Netherlands and Surinam	70 leprosy throughout the whole spectrum, 11 HC	ND	ND	ENL patients possibly on thalidomide treatment?	Serum	ICs Complement C1q, C3, C4	Elevated C3d significantly associated with ENL
Mshana et al. (73); Ethiopia	26 ENL, 20 BL/LL	Skin biopsies of ENL less than 12 h old	ND	ND	Skin	Immunoglobulin deposits Complement deposits Mycobacterial antigens	No ICs around blood vessels in ENL lesions IC formation common feature in LL Absence of immunoglobulin or C3 deposits in early ENL Extracellular antigens not seen
Ridley and Ridley (39); Malaysia, PNG, Ethiopia, and UK	20 ENL, 10 non-reactional leprosy	ND	ND	ND	Skin	Immunoglobulins IgG, IgM, IgA, IgE Complement C3, C4, C1q, C3d	ENL lesions had disintegration of macrophages and release of bacterial antigen combined first with IgM, later with IgG, present together with complement components of the classical pathway ICs were both extracellular and in neutrophils and macrophages ICs were extravascular
Ramanathan et al. (74); India	10 BT, 10 LL, 10 BT reactional, 30 LL reactional	ND	All patients on dapsone	Sampling before antireactional treatment	Serum	C3 and C4 ICs Isolated ICs for IgG, IgA, IgM, C3, C4 and antimycobacterial antibody	Increased C3d in both BT reactional (T1R) and LL reactional Circulating ICs in all reactional patients No antimycobacterial antibody in ICs from LL reactional patients
Saha et al. (75); India	20 ENL, 15 HC	Before and 4 weeks after starting treatment for ENL	ND	Second sample: on antireactional treatment	Serum	Quantitative analysis of composition of PEG precipitates (immunoglobulins, complement components, autoantibodies and acute phase proteins) and anticomplementary activity of PEG precipitates	Anticomplementary activity of PEG precipitates more in the lepromatous than in normal sera, independent of the presence of ENL

(Continued)

TABLE 2 | **Continued**

Reference; study site(s)	Study population	Timing of screening	MDT status	ENL treatment	Type of samples	Measures	Findings
Ramanathan et al. (48); India	32 LL in reaction, 10 BT, 10 BT in reaction, 10 LL uncomplicated, 15 HC	ND	BT and LL without reaction treatment for at least 2 years; all patients were on dapsone	ND	Serum	ICs by fluid phase ^{125}I-labeled conglutinin binding assay; serum C3d Complement-mediated solubilization of immune precipitates	Reduced solubilization of *in vitro* formed immune precipitates by the sera of ENL patients C3d, ICs, and solubilization levels correlated with the clinical course of reaction ICs and C3d decline after clinical subsidence of ENL
Sehgal et al. (76); India ε	21 T1R or ENL	ND	ND	ND	Serum	Complement C3	Lower level of C3 during ENL
Chakrabarty et al. (77); India	27 BB-BL-LL: 7 ENL, 4 T1R	Initial blood collected at the onset of reaction and subsequent 4 weeks after ENL remission	Patients on MDT	Second blood sample on antireactional treatment	Serum	Solubilization of preformed ICs (^{125}I-BSA-anti-BSA complexes)	The mean solubilizing capacity of the reaction patients' sera during the reaction was not significantly different from LL without ENL After clinical remission of the reaction, most patients showed no increase in the ICs solubilization
Rao and Rao (78); India ε	44 ENL, 39 BL/LL, 22 post-ENL	ENL: before starting treatment with anti-inflammatory drugs/steroids Post-ENL: ensuring that the patient had not taken anti-inflammatory drugs/steroids for at least 3 or 7 days	20 BL/LL untreated and 19 BL/LL treated with dapsone less than a year	Untreated first sample and second post-ENL sample after discontinuation of antireactional treatment	Serum	C3 and C4 levels IgG, IgA, IgM, C3, and C4 levels in the ICs ICs by PEG method	C3 and C4 levels were not significantly different in ENL compared to BL/LL and post-ENL C3 and C4 levels in the ICs reduced insignificantly in ENL than BL/LL and post-ENL IgG, IgA, and IgM in ICs showed no significant differences from LL to ENL and post-ENL
Sehgal et al. (79); India	17 ENL	Before antireactional treatment and 1 week after the clinical subsidence of ENL	On MDT	First sample untreated and second sample on prednisolone in 10 patients	Serum	Complement components: classic pathway: C1q and C4 Alternative pathway: C3, C3d, Factor B	No significant change in classical pathway in ENL reaction C3 elevated, C3d decreased and increase of Factor B after ENL
Jayapal 1989 (47); India	37 leprosy: 9 ENL, 6 bacterial endocarditis, syphilis, SLE, HC	ND	All leprosy patients on dapsone	ENL patients on clofazimine, prednisolone, antihistamine and chloroquine	Serum	ICs with PEG method	ICs higher in ENL than in LL

(*Continued*)

TABLE 2 | Continued

Reference; study site(s)	Study population	Timing of screening	MDT status	ENL treatment	Type of samples	Measures	Findings
Sehgal et al. (81); India	18 T1R, 17 ENL, non-reactional controls	During and after reaction	On MDT	ND	Serum	Complement components: classic pathway: C1q and C4 Alternative pathway: C3, C3d, Factor B	Classic pathway: no significant change in C1q and C4 during ENL Alternative pathway: increase in C3d during ENL; decrease of C3 during ENL; reduction of Factor B during ENL; elevation of C3 and Factor B after ENL
Tyagi et al. (52); India	20 BL/LL with ENL, 20 TT/BT, 20 BT with reaction, 20 BL/LL; 15 HC	ND	ND	ND	Serum	ICs by PEG precipitation Mycobacterial ICs in PEG precipitates; CH50 assay and AH50 assay (complement consumption)	PEG ICs from BL/LL and ENL higher IgG and IgM antimycobacterial antibodies than TT/BT, BT reactional (T1R) and HC No significant functional differences between the PEG ICs from reactional and non-reactional leprosy
Ramanathan et al. (49); India ε	26 BL/LL: 11 ENL, 24 HC	Before initiation of treatment and 2-monthly intervals	Untreated and then on MDT	Treated but after sampling	Serum	ICs by PEG method C3d Complement-induced IC solubilization	High levels of ICs in both LL and ENL Lower levels of complement-induced IC solubilization in ENL Highest levels of ICs and C3d at the time of ENL
Scollard et al. (82); Thailand ε	4 cured leprosy, 10 non-reactional leprosy BT/BL/LL, 8 ENL patients (5 LL/3 BL), 3 T1R, 4 HC	ND	ND	ND	Blisters induced over representative skin lesion Serum	ICs	ICs in ENL similar to that of active leprosy (either lepromatous or tuberculoid) Higher ICs in blisters than in matching sera
Rojas et al (50); Brazil ε	19 ENL, 10 BL/LL, 13 family contacts; 15 healthy non-contacts	ND	Both untreated and patients on MDT for 1–72 months	ND	Serum	ICs; anti-PGL-I IgM in IC precipitated from sera; anti-10-kDa hsp IgG in IC precipitated from sera	ENL highest levels of ICs compared with all other groups IgM anti-PGL-I and IgG anti-MCP-1 heat shock protein antibodies constituents of ICs in ENL
Dupnik et al. (58); Brazil δ, ε	11 ENL, 11 T1R, 19 non-reactional leprosy, additional 6 ENL, 11T1R, 11 HC	ND	3 ENL pre-treatment, 2 ENL on treatment and 6 ENL post-treatment; Leprosy controls matched for stage of treatment	Excluded patients who had received corticosteroids within 7 days or thalidomide within 28 days of enrollment	PBMC Skin	Microarray and qPCR for transcriptional profile of PBMC; IHC for C1q in skin lesions	Complement and coagulation pathway common in ENL and T1R Transcripts uniquely increased in ENL included complement receptors C3AR1 and C5AR1 C1q staining higher in both ENL and T1R compared with non-reactional leprosy

*α, also in **Table 1**; γ, also in **Table 3**; δ, also in **Table 4**; ε, also in **Table 5**.*

BB, mid-borderline leprosy; BL, borderline lepromatous leprosy; BT, borderline tuberculoid leprosy; ENL, erythema nodosum leprosum; HC, healthy controls; ICs, immune complexes; LL, lepromatous leprosy polar; ND, not described; PEG, polyethylene glycol; PNG, Papua New Guinea; RA, rheumatoid arthritis; SLE, systemic lupus erythematosus; TB, tuberculosis; TT, tuberculoid leprosy polar; WHO, World Health Organization.

whether they are involved in the pathogenesis of ENL or simply an epiphenomenon.

WHAT IS THE ROLE OF T-CELLS IN ENL?

T-lymphocytes are part of the adaptive immune response which help to eliminate bacterial, viral, parasitic infections or malignant cells. The antigen specificity of the T-cell is based on recognition through the T-cell receptor (TCR) of unique antigenic peptides presented by major histocompatibility complex (MHC)-molecules on APCs: B cells, macrophages, and dendritic cells. There are two major T-cell lineages, defined by the presence of two surface co-receptor molecules, namely, CD4 and CD8. $CD4^+$ cells when they are activated produce cytokines as effector T helper cells, whereas $CD8^+$ lymphocytes form effector cytotoxic T lymphocytes (CTL). Furthermore, activated $CD4^+$ T helper cells can be subdivided into Th1, Th2, Th17, and T regulatory (Treg) subsets based on the production of signature cytokines (83).

Early studies investigating T-cell biology in the pathophysiology of ENL reported that ENL patients had higher T-cell numbers in peripheral blood than uncomplicated LL patients, although both LL and ENL patients had a significantly lower percentage and absolute number of T-cells compared to healthy controls (84). In addition, the high numbers of T-cells observed during ENL remained high post-ENL treatment compared to the LL controls (85).

Patients with ENL had increased $CD4^+$ T cell numbers and a simultaneous decrease in $CD8^+$ T cell numbers and an increased $CD4^+/CD8^+$ ratio in the blood compared to LL controls (86, 87), while ENL patients had decreased $CD4^+/CD8^+$ ratio after successful treatment. This ratio increased in those patients who had an ENL recurrence (87). An increased $CD4^+/CD8^+$ ratio in ENL patients was reported by several subsequent studies (87–92). In acute SLE, it has been suggested that the failure of $CD8^+$ T-cell activity could lead to increased IgG production and to the subsequent formation of ICs (93). However, there are studies in ENL reporting a decreased $CD4^+/CD8^+$ ratio compared to non-reactional LL controls (94) or a similar ratio (95, 96).

The first immunohistological studies of T-cell subsets in skin lesions included small numbers of ENL patients and assessed the percentage and ratio of $CD4^+$ and $CD8^+$ T cells by comparing them to non-ENL lepromatous specimens (89, 91, 97–103). ENL skin lesions, like peripheral blood, were characterized by an increased $CD4^+/CD8^+$ ratio in all but one of these studies (89, 91, 97–100, 102, 103).

$CD4^+$ T cells differentiate according to the microenvironment into Th1, Th2 cells, or subsets of Th17 and Treg (104). Recent studies have reported the frequency of the newly described Th17 and Treg subsets in leprosy (105, 106). Using flow cytometry in ENL, the absolute numbers and proportion of Tregs were shown to be significantly lower during ENL although FoxP3 expression, a marker they used to define Tregs, was higher (107). Tregs suppress or downregulate induction and proliferation of effector T cells (108). Therefore, the observation of lower numbers of Tregs in ENL could account for the relatively higher proportion of T cells previously described in multiple ENL studies. Two more publications from the same group addressed the frequency of Tregs in ENL, defined as $CD4^+CD25^{high}$ $FoxP3^+$ cells and reported the ratio of Treg/Teffector cells to be low in ENL (109, 110). These results should be interpreted with caution since dichotomizing cells into $CD25^{high}$ and $CD25^{low}$ to identify Tregs is highly subjective. There is no consensus on the thresholds of CD25 expression to delineate Tregs within the $CD25^{high}$ population (111). Variations in FoxP3 expression within the $CD25^{high}$ population have been observed even in healthy individuals (112).

A recent study that used flow cytometry described a significant reduction in percentage of $CD4^+CD25^+FoxP3^+$ Treg cells and mean fluorescence intensity of FoxP3 in PBMC in patients with ENL compared to LL controls (113). The observed reduction of Tregs in ENL patients could lower the inhibitory effects on effector T cells and therefore lead to enhanced Th17 activity, tipping the balance toward inflammation, as previously described in other conditions such as tuberculous pleural effusion (114). Interestingly, an increase of FoxP3 mRNA expression by PBMC in ENL patients compared to LL controls has also been reported (113). The conflicting results for FoxP3 could be due to variation in the flow cytometry gating or the fact that FoxP3 mRNA may not be translated to functional FoxP3. A previous study measured the expression of Foxp3 by qPCR in skin biopsies and PBMC of five patients with ENL and detected Foxp3 in all skin and PBMC samples. An upward trend of Foxp3 in PBMC was described during the first 21 days of thalidomide treatment (115). The authors suggested that thalidomide may boost Tregs by T-cell costimulation *via* CD28 and therefore augment the IL-2-dependent number and/or function of Tregs (115). However, the changes in Foxp3 expression did not reach statistical significance, while no IL-2 mRNA was detected in any samples (115). Another study addressed FoxP3 expression by immunohistochemistry in skin but there was no difference in patients with ENL compared to non-reactional leprosy controls across the spectrum (116). Recent research suggests that Tregs constitute a stable cell lineage whose committed state in a changing environment is ensured by DNA demethylation of the Foxp3 locus irrespective of ongoing Foxp3 expression (117). Further investigation is needed to better define the role of Tregs in the pathogenesis of ENL.

Patients with ENL do not exhibit the phenomenon of "anergy" of cell-mediated immune response observed in untreated LL patients (118). Patients with ENL had elevated mean proliferative responses to several mitogens compared to uncomplicated LL patients (86, 87), while an enhancement in T-cell-related functions during the acute phase of an ENL reaction has also been described (94).

The interpretation of the role of T cell subsets in ENL is hampered by small sample sizes and methodological issues. 63.6% of the 44 studies investigating the role of T cells in ENL (**Table 3**) are cross-sectional and lack serial sampling before and after treatment for ENL. However, it appears that T cell subsets do play an important role in ENL because multiple studies report an increased $CD4^+/CD8^+$ ratio in ENL patients in both skin and peripheral blood.

TABLE 3 | Human studies on ENL investigating T-cell biology.

Reference; study site(s)	Study population	Timing of sampling	MDT status	ENL treatment status	Type of samples	Measures	Findings
Lim et al. (84); USA and Korea (mixed ethnic background)	7 LL ENL, 20 active LL, 9 inactive LL, 4 BB, 3 indeterminate leprosy	ND	All patients treated with Dapsone or Clofazimine or Rifampicin for varying durations	5 had received various doses of steroids and 3 were treated with steroids at the time of the study	Blood	T lymphocyte numbers by the rosette assay	ENL showed T-lymphocyte numbers significantly higher than LL LL had lower T-lymphocyte numbers than HC
Anders et al. (119); Papua New Guinea	31 leprosy: 13 BL/LL with amyloidosis (11/13 frequent ENL), 9 BL/LL ENL without amyloidosis, 9 BL/LL with few or no ENL episodes	ND	Approximately half patients on clofazimine and other half on dapsone	2 ENL at testing: 1 steroids and 1 stibophen	Blood	Lympohocyte transformation tests	Patients with a history of frequent ENL had greater cell-mediated responses to PHA than patients without ENL
Izumi et al. (70); Japan β	12 ENL, 49 active lepromatous, 24 inactive lepromatous, 7 borderline, 6 tuberculoid, 9 HC	ND	ND	ND	PBMC	Percentage and number of Tμ (T cells with Fc receptor for IgG) and Tγ (T cells with Fc receptor for IgM)	No significant differences between different clinical groups
Bach et al. (86); France (multiple ethnic groups)	9 BL/LL with no recent history of ENL, 9 BL/LL suffered from ENL less than 2 months prior to the investigation, 13 BT/TT, HC	ND	Some untreated and others on MDT	Certain ENL on antireactional treatment	Blood	T cell subsets; Proliferative responses to mitogens Con A-induced suppressive activity	Increased %age of helper T cells in ENL Decreased %age of suppressor T cells in ENL Elevated proliferative responses to mitogens in ENL Most ENL decrease of suppressive index, whereas none of the LL or TT patients had a diminished suppressive activity
Dubey et al. (120); India	41 untreated cases of leprosy, 64 TT and LL taking antileprosy treatment, reactional (8 ENL and 10T1Rs), 11/41 follow-up from untreated leprosy patients	ND	64 cases on antileprosy treatment	Untreated cases of ENL?	Blood	Lymphocytic culture: percentage of Blast transformation	Blast percentage in ENL slightly higher than T1R
Mshana et al. (90); Ethiopia	21 BL/LL, 10 BT, 5 ENL	ND	All patients received MDT but unclear whether sampled prior to MDT	No patient on thalidomide	Blood	Lymphoproliferative responses to PPD or PHA T-cell subsets	Higher responses to PPD or PHA in ENL Decreased number of suppressor cells prior to ENL, which increased with clinical recovery from ENL
Mshana et al. (88); Ethiopia	69 leprosy patients: 26 ENL, 13 HC	Untreated samples	Untreated samples	Untreated samples	Blood	T lymphocyte subpopulations; lymphoproliferation using *M. leprae*, PHA and PPD	ENL patients had decrease in suppressor cells and an increase of $CD4^+/CD8^+$ ratio compared to LL ENL had higher responses to both PHA and PPD BL/LL patients with or without ENL lower proliferative responses to *M. leprae* than BT patients and HC

(Continued)

TABLE 3 | Continued

Reference; study site(s)	Study population	Timing of sampling	MDT status	ENL treatment status	Type of samples	Measures	Findings
Wallach et al. (87); France (samples from multiple ethnic groups)	9 recent ENL, 6 bacteriologically positive patients of which 1 ENL more than 5 years ago, 9 treated leprosy patients of which 3 had ENL	Described in detail each patient duration of disease	All treatment described in detail	Some on antireactional treatment	Blood	T cell subsets; Lymphocyte transformation tests: proliferative responses to mitogens	ENL patients have elevated Helper/Suppressor ratio Mean proliferative responses elevated in ENL
Bach et al. (121); France	8 treated lepromatous without recent ENL with BI < 1 +, 6 lepromatous with BI > 2+ (untreated or suffering a relapse, without recent ENL reaction), 12 lepromatous who underwent at least one ENL episode, 13 tuberculoid, 41 HC	ND	ND	ND	PBMC	T-cell subsets; Proliferative response to *M. leprae* and PPD of isolated T-cell subsets	ENL decreased CD8⁺ T cell percentages and increased CD4⁺/CD8⁺ ratios T-cell subset percentages returned to normal either when the bacterial load was reduced by treatment or when the ENL reaction resolved ENL episodes associated with improvement of T-cell unresponsiveness to various antigens or mitogens
Modlin et al. (97); USA	15 non-reactional leprosy BT/BB/BL/LL, 17 reactional (6 T1R, 9 ENL, 2 Lucio's reaction)	ND	Results did not differ between treated and untreated subjects	3 ENL had no therapy	Skin	T lymphocyte subsets	The helper/suppressor ratio in ENL was significantly higher than in non-reactional lepromatous disease
Modlin et al. (98); USA	14 leprosy patients (4 tuberculoid, 2 borderline in T1R, 1 BL, 7 lepromatous of which 5 ENL), 8 HC	ND	6 treated patients	ND	Skin	T lymphocyte subsets	ENL lesions showed 2:1 predominance of helper cells whereas in the lesions without ENL the helper: suppressor ratio was 1:1 smaller
Sasiain et al. (122): Argentina	16 ENL, 12 HC	First blood sample ND; 9 ENL 20-30 days after stopping thalidomide	All patients on MDT	Thalidomide in patients with ENL	PBMC	ConA-induced suppressor response	Suppressor T-cell function was reduced during ENL and after ENL than HC
Narayanan et al. (89); India	7 LL ENL, 6 BT T1R, 5BL T1R, 18 BT-LL	ND	ND	ND	Skin	T cell phenotypes	Lesions of ENL showed increase in T cells with a predominance of the helper/inducer subset; CD4⁺/CD8⁺ ratio was higher in ENL and T1R than non-reactional lesions

(Continued)

TABLE 3 | **Continued**

Reference; study site(s)	Study population	Timing of sampling	MDT status	ENL treatment status	Type of samples	Measures	Findings
Rea et al. (96); USA ε	19 ENL, 24 LL non-reactional with treatment, 12 LL non-reactional no treatment, 18 LL with long-term treatment, 4 LL with Lucio's, 13 BL, 13 T1R, 18 Tuberculoid, 13 Tuberculoid with long-term treatment	ND	Some patients on MDT	ENL before receiving thalidomide	PBMC	T cell subsets	Active LL patients have lymphopenia, a proportionate reduction in the numbers of each of the three T cell subsets Insignificant changes in T cell subsets expressed as percentages and in the helper: suppressor ratio
Laal et al. (94); India ε	15 ENL, 13 LL	During active ENL and 1 week to 4 months after stopping treatment	On MDT	First sample before initiation of antireactional treatment Second sample 1 week to 4 months after stopping treatment	Blood	Leukocyte migration inhibition test Lymphoproliferation Suppressor cell activity; T cell subsets	ENL significant inhibition of antigen-induced leukocyte migration Lymphoproliferation enhanced during the acute phase of ENL Enhanced antigen-stimulated suppression of mitogen responses in ENL Leukocyte migration inhibition, lymphoproliferation, and suppressor cell activity were reduced in post-ENL to the unresponsive state seen in stable LL Lower $CD4^+/CD8^+$ ratio in ENL compared to LL
Modlin et al. (99); USA	12 ENL and 10 non-reactional leprosy; 19 ENL blood samples	ND	ENL biopsies: 8/12 treated with dapsone; ENL blood: 15/19 treated	Some ENL were treated	Blood Skin	T lymphocyte subsets	ENL tissue more cells of the helper-inducer phenotype and fewer of the suppressor-cytotoxic phenotype, as compared with non-reactional LL No correlation between tissue and blood helper-suppressor ratios
Wallach et al. (91); France	ND	ND	ND	ND	Blood Skin	T cell helper-suppressor (HS) ratio	HS ratio higher in ENL lesions and blood than non-ENL leprosy controls
Modlin et al. (100); USA	Biopsies: 25 ENL, 23 tuberculoid, 23 non-reactional lepromatous; Blood: 18 ENL	ND	Some patients received treatment	Some patients on treatment?	Blood Skin	Skin: number of T cells, T cell subsets; Blood: lepromin-induced suppression of the Con A stimulation	Increases in both $CD4^+/CD8^+$ ratio and the number of IL2-positive cells in ENL Suppressor activity decreased significantly in ENL Suppressor activity returned to normal after ENL subsided
Rao and Rao (123); India ε	44 ENL, 39 BL/LL, 22 post-ENL	ENL patients before starting ENL treatment, post-ENL after patient had not taken anti-inflammatory/ steroids for at least 3 and 7 days	From 39 non-reactional cases: 20 untreated and 19 with dapsone for less than a year	Before starting treatment for ENL with steroids or anti-inflammatory drugs, post-ENL: ensuring that the patient had not taken anti-inflammatory drugs or steroids for at least 3 and 7 days, respectively	Blood	Sub-population of T cells with receptors for Fc portion of IgG (Tr) and Fc portion of IgM (Tμ)	Tμ/Tr ratio higher in ENL than lepromatous and post-ENL patients

(Continued)

TABLE 3 | Continued

Reference; study site(s)	Study population	Timing of sampling	MDT status	ENL treatment status	Type of samples	Measures	Findings
Rao and Rao (85); India	77 leprosy: 44 ENL	ENL: before starting anti-ENL treatment, post-ENL: After patient had noe taken anti-inflammatory drugs or steroids for at least 3 and 7 days	19 patients treated with dapsone for less than 1 year	Before starting treatment for ENL with anti-inflammatory drugs or steroids	Blood	Leykocyte migration inhibition test (LMIT) Enumeration of early and total T lymphocytes	No significant difference in mean migratory index to PHA, PPD, sonicate *M. leprae* Whole *M. leprae* increased response in ENL compared to LL Lower migratory indices to whole *M. leprae* in post-ENL than LL %age of early T lymphocytes increased in ENL compared to LL %age of early T lymphocytes remained high in post-ENL compared to LL Cell-mediated immune responses enhanced during ENL and return to LL levels once the episode is over
Shen et al. (101); USA	10 ENL, 8TT/BT, 10 BL/LL, 10 T1R	ND	ND	ND	Skin PBMC	CD3, CD4, CD8 and Ta1 (memory) positive cells	CD3, CD4 and CD8 showed percentages of positive cells in lesions similar between patient groups No significant difference in%age of memory T-cells in ENL compared to LL
Bottasso et al. (124); Argentina	8 LL/ENL, 17 LL, 9 TT, 11 HC	ND	Patients on MDT	Patients with ENL were not on thalidomide treatment but unknown whether they were on steroids	Blood	T-Lymphocytes count absolute and relative; Lymphocyte functional assay: capacity of rosetta formation	Active LL showed a decrease in T-lymhocytes ENL showed a restoration of the levels of T-lymphocytes
Rasheed et al. (125); Zambia and Pakistan	167 leprosy of which 21 LL/ENL, 12 BL/T1R, 24 BT/T1R, 46 endemic HC	ND	ND	ND	Serum Lymphocytes	Lymphocytotoxic activity	Lymphocytotoxic activity scores were significantly raised in patients with reactions
Sasiain et al. (126); Argentina	53 leprosy patients TT/BT/BB/BL/LL and 9 LL/ENL, 23 HC	ND	Received MDT	Thalidomide for ENL	PBMC	Proportion of $CD8^+$ cells *M. leprae*-induced suppression of T-cell proliferation; Induction of IL-2R by culture with *M. leprae* PHA- and ConA-induced proliferation	Proportion of $CD8^+$ cells was low in LL patients and tended to normalize during ENL episodes
Bhoopat et al. (127); Thailand ε	57 ENL (19 acute/38 chronic), 61 active LL, 33 cured leprosy	26 BL/35 LL newly diagnosed and untreated	ND	If corticosteroid and/or thalidomide was initiated before or during the study, precise timing of medication was recorded with respect to the time of collection of laboratory specimens	Blisters induced over a representative skin lesion	T cell subsets *in situ*	The lesions of chronic ENL showed a decreased number of $CD8^+$ cells and increased helper/suppressor ratio compared to those in acute ENL and non-reactional leprosy; Systemic administration of corticosteroids caused a reduction in the $CD4^+$ cell population but did not change $CD8^+$ cell population

(Continued)

TABLE 3 | **Continued**

Reference; study site(s)	Study population	Timing of sampling	MDT status	ENL treatment status	Type of samples	Measures	Findings
Rea and Modlin (102); USA δ	ND	ND	ND	ND	Skin	T-cell phenotypes: $CD4^+$ versus $CD8^+$ cells, γ/δ and α/β receptor-bearing lymphocytes, T-memory and T-naïve cells	ENL lesions predominance of $CD4^+$ cells similar to those in tuberculoid (TT/BT?) and T1R LL patients showed an excess of $CD8^+$cells
Tyagi et al. (53); India	4 TT/BT, 5 BL/LL, 4 ENL	ND	ND	ND	Blood	Effect of isolated circulating ICs from BL/LL or ENL patients to lymphocyte transformation test on T cells of HC	PEG precipitates isolated from BL/LL or ENL subjects had a significant suppressive effect on lymphocyte proliferation in HC
Foss et al. (128); Brazil δ	28 lepromatous: 11 ENL, 23 tuberculoid, 19 HC	ND	lepromatous patients 86% treated with dapsone	11 ENL at time of blood collection no immunosuppressive drug	Blood	T lymphocyte response to concanavalin A	Marked reduction on concanavalin A-induced lymphoproliferation in patients with ENL
Santos et al. (129); Brazil ε	59 LL/BL, 10 ENL, 4 T1R, 4 post-reactional	ND	On MDT	No specific treatment for reactions before blood collection	PBMC	Lymphocyte proliferation after ConA and *M. leprae*	T1R showed greater lymphocyte proliferation compared to all other groups
de la Barrera et al. (130); Argentina	7 TT/BT, 20 BL/LL of which 3 ENL	ND	All patients on MDT	ND	PBMC	T-cell cytotoxic activity induced by *M. leprae* and *M.tb* heat shock protein (HSP)	*M. leprae* hsp65 induced cytotoxic responses only in those MB patients undergoing ENL
Vieira et al. (131); Brazil δ, ε	95 MB leprosy (30LL/65BL) of which 51 ENL	At leprosy diagnosis and at onset of reactional episode	Time of MDT for each ENL	Sample before thalidomide and steroids?	PBMC	Lymphocyte transformation test (LTT)	Some patients showed lymphoproliferative response during ENL
Mahaisavariya et al. (103); Thailand	17 non-reactional, 8 T1R, 12 ENL	Biopsy at the time of diagnosis and not the time of reaction	ND	ND	Skin	T-lymphocyte subsets	%age of CD8 infiltration reduced in ENL compared with non-reactional lepromatous The $CD4^+/CD8^+$ ratio of ENL statistically significant higher than from the non-reactional lepromatous group
Tadesse et al. (132); Ethiopia δ	33 leprosy: 14 BT, 11 BT T1R, 8 ENL, 11 HC	ND	Certain leprosy patients were treated on MDT	All ENL treated with steroids	PBMC	Lymphocyte blast transformation	Thalidomide treatment did not alter the lymphoproliferative response to the mycobacterial antigens during ENL
Mohanty et al. (133); India	21 BL/LL ENL, 38 TT/BT/BL/LL, 29 BT/BL T1R, 19 HC	ND	ND	ND	PBMC Serum	Immune responses against Stress proteins of *M. leprae* (lymphoproliferation)	ENL: no significant role of stress proteins except a heightened lymphoproliferative response to the 28 kDa antigen

(Continued)

TABLE 3 | Continued

Reference; study site(s)	Study population	Timing of sampling	MDT status	ENL treatment status	Type of samples	Measures	Findings
Villahermosa et al. (134); Philippines δ, ε	22 ENL	Before thalidomide and at study weeks 3 and 7 during thalidomide	MDT continued during the study	Samples untreated for antireactional drugs and during thalidomide treatment	Blood PMBC	Lymphocyte proliferation assays (LPA) to phytohemagglutinin and concanavalin A	Low LPA values pre-thalidomide in both PBMC and whole blood
Attia et al. (107); Egypt	38 leprosy: 6 ENL, 38 HC	Untreated samples	Untreated samples	Untreated samples; excluded patients on immunosuppressive drugs	Blood	Frequency of circulating Tregs; FoxP3 expression	Significantly lower frequency of Tregs but higher FoxP3 expression in ENL
Massone et al. (116); Brazil ε	20 leprosy: 3 ENL	Biopsies at the time of diagnosis	10, 12 and 13 months after beginning of MDT for LL	Untreated for antireactional treatment	Skin	Presence, frequency and distribution of Tregs	No statistical difference in FoxP3 expression between TT, BT, BL, and LL Significant increase in FoxP3 expression in T1R compared to ENL
Rada et al. (135); Venezuela ε	? ENL8[illegible] LL, 41 BL, 41 BB, 3[illegible] BT	ND	ND	ND	Blood	Cell-mediated immunological tests to mycobacterial proteins	T-lymphocyte proliferative response in reactional and non-reactional patients was negative
Saini et al. (136); India δ	21 MB: 16 ENL, 5 T1R	ENL blood during reaction and at 0.5 and 1 year after the onset of reaction	Duration of MDT described	ENL patients received steroids	PBMC	Lymphoproliferation of PBMC stimulated with *M. leprae*, recombinant Lsr2 and 6 synthetic peptides spanning the Lsr2	All patients with active ENL showed lymphoproliferation in response to peptides A and F
Abdallah et al. (109); Egypt δ	43 leprosy: 6 ENL, 40 HC	Untreated patients	Untreated samples	Untreated	Blood	Circulating Tregs	Tregs/Teffs lowest in ENL
Attia et al. (110); Egypt δ	43 leprosy: 6 ENL, 40 HC	Untreated patients	Untreated samples	Untreated	Blood	CD4(+) CD25(high)Foxp3 (+) regulatory cells	CD4(+)CD25(high)FoxP3(+) Treg levels lowest in ENL Treg/Teffs lowest in ENL
Hussain et al. (92); India	50 leprosy (28 without reactions, 11 T1R, 11 ENL), 50 HC, 50 pulmonary TB (25 HIV-TB co-infected and 25 without HIV infection), 50 HIV-positive	ND	Reactional episodes following antileprosy treatment	ND	Blood	$CD3^+$, $CD4^+$, $CD8^+$ and $CD4^+/CD8^+$ ratio with flow cytometry	$CD4^+$ counts raised during ENL compared to MB patients whereas $CD8^+$ counts lower The $CD4^+/CD8^+$ ratio doubled during reactional episodes of T1R and ENL
Parente et al. (137); Brazil	2 ENL, 103 leprosy TT/ET/BB/BL/LL 9 indeterminate, 8 T1R	2 ENL: 12 and 10 months after initiation of MDT	2 ENL after initiation of MDT	ND	Skin	Frequency and distribution of regulatory T cells	No significant differences in ENL

(Continued)

TABLE 3 | Continued

Reference; study site(s)	Study population	Timing of sampling	MDT status	ENL treatment status	Type of samples	Measures	Findings
Saini et al. (113); India δ	66 leprosy: 15 T1Rs, 15 ENL, 36 BT/LL	Newly diagnosed leprosy patients prior to institution of antireaction therapy	Freshly diagnosed patients: untreated subjects	Newly diagnosed leprosy patients prior to institution of antireaction therapy	PBMC	MLSA stimulated and unstimulated PBMC: gene expression with PCR array for 84 genes; T cell phenotypes	Increase in FOXP3 gene expression in ENL Th17 cells with intracellular IL-17A, F are increased in ENL and $CD4^{+}IL\text{-}21^{+}$ cells are higher in ENL Significant upregulation of $CD4^{+}CCR6^{+}$ cells in ENL Tregs decreased in ENL

α, also in ***Table 1****; β, also in* ***Table 2****; δ, also in* ***Table 4****; ε, also in* ***Table 5****.*
BB, mid-borderline leprosy; BL, borderline lepromatous leprosy; BT, borderline tuberculoid leprosy; ENL, erythema nodosum leprosum; HC, healthy controls; HS, helper-suppressor; HSP, heat shock protein; ICs, immune complexes; LL, lepromatous leprosy polar; LTT, lymphocyte transformation test; LPA, lymphocyte proliferation assay; MB, multibacillary; ND, not described; PEG, polyethylene glycol; PHA, purified phytohaemagglutinin; PPD, RT23 tuberculin-purified protein derivative; SLE, systemic lupus erythematosus; TB, tuberculosis; TT, tuberculoid leprosy polar.

WHAT IS THE ROLE OF TNF-α OR OTHER CYTOKINES IN ENL?

A role for TNF-α in ENL was first suggested by a Brazilian study that included 18 ENL patients at various stages of treatment with steroids or thalidomide (138). Serum TNF-α levels varied widely: from undetectable to extremely high levels (138). There was no obvious correlation between severity of ENL and cytokine levels, while patients who had received treatment had lower levels of TNF-α (138). High serum TNF-α levels were subsequently shown to decrease significantly during thalidomide treatment (139). These findings have been reproduced in other populations measuring serum TNF-α levels (128, 131, 140–147), whereas two studies failed to show increased levels of serum TNF-α during ENL (148, 149). The high variability in serum TNF-α between studies might be due to patient differences. Although genetic differences between different ethnic groups cannot be ruled out, it still remains unclear why there is such a high variability in the TNF-α levels between individuals presenting ENL.

A study of the plasma levels of TNF-α reported increased levels during ENL (150) while other studies contradicted this finding (115, 134, 151). In fact, Haslett et al., which included 20 male ENL patients excluding patients with moderate or severe ENL-associated neuritis, reported circulating plasma TNF-α levels to be lower at time of ENL diagnosis than LL controls (115). There was an upward trend in plasma TNF-α levels during thalidomide treatment which returned to baseline levels after discontinuation of thalidomide (115). This is an indication that thalidomide may in fact stimulate paradoxical overproduction of TNF-α (115). The inhibition of TNF-α by thalidomide may be prominent when macrophage production of this cytokine is high but in mild disease plasma levels may not reflect lesional TNF-α production (115). Increased TNF-α levels after thalidomide treatment has been described in other conditions such as toxic epidermal necrolysis (152) and aphthous ulcers in patients with human immunodeficiency virus infection (153). It has been suggested that the mechanism of the paradoxical overproduction of TNF-α by thalidomide could be due to the propensity of thalidomide to costimulate T-cells to produce cytokines including TNF-α (154). All the patients in the study of Haslett et al. showed improvement in ENL after receiving thalidomide during the first 21 days of treatment (115).

Interestingly, the studies that measured the *ex vivo* PBMC production of TNF-α in response to lipopolysaccharide, BCG, or *M. leprae* in patients with ENL as compared to BL/LL patients showed consistently greater amounts of TNF-α secretion in patients with ENL (150, 155–157).

The successful use of the anti-TNF therapy with infliximab and etanercept in three patients with ENL, resulting reduction of inflammation and treatment of ENL, is additional evidence of the inflammatory role of TNF-α in ENL (158–160).

The results of studies on IFN-γ are more consistent than those on TNF-α suggesting an important role for IFN-γ in the pathophysiology and occurrence of ENL. A clinical trial administered recombinant IFN-γ to BL/LL patients as a replacement therapy because LL is characterized by anergy to antigens of *M. leprae* and inability to produce IFN-γ (150). Repeated intradermal injection of recombinant IFN-γ induced ENL in 6 out of 10 BL/LL patients within 7 months compared to an incidence of 15% per year in patients who received MDT alone (150). Elevated serum IFN-γ was found in patients with ENL who also had high TNF-α levels (139). Other studies have demonstrated an increase of serum IFN-γ (143, 144, 148) and an increase of IFN-γ mRNA in PBMC (161–163) and in skin biopsies (161, 164) during ENL. There is a study reporting serum IFN-γ to be significantly lower in patients at the onset of ENL, which increased after thalidomide treatment (142). However, IFN-γ has been identified by Ingenuity Pathway Analysis networks as the second most significant upstream regulator (after CCL5) of the expression changes in microarrays performed in PBMC derived from patients with ENL (58).

There are contradictory findings about the role of serum IL-1β levels. Most studies have reported that serum IL-1β levels may have a prognostic value for developing ENL (144, 148, 165, 166) and that there is a statistically significant correlation between TNF-α and IL-1β (140). However, studies failed to show any association of serum IL-1β or plasma IL-1β with ENL (138, 151). IL-1β mRNA in PBMC was upregulated at the onset of ENL (161) but not in skin lesions (167).

IL-2 has a key role in the immune system primarily by its direct effects on T-cells such as promoting differentiation of different T-cell subsets and contributing to the development of T-cell immunological memory. IL-2 signals through the IL-2 receptor (IL2R), which is essential for the signaling in T-cells. There were no differences in the serum IL-2 or IL2 mRNA in skin biopsies between ENL and patients with LL (115, 148, 151). However, four studies reported an increase in soluble IL-2 receptor (sIL2R) levels (115, 131, 165, 168) or IL2Rp55 mRNA in PBMC (161) in patients with ENL.

Serum IL-6 (147, 151, 169, 170) and IL-6 mRNA in PBMC and skin (161) have been reported to be elevated during ENL. IL-6 tag single-nucleotide polymorphisms have been reported to be a risk factor for ENL (170) and IL-6 plasma levels were correlated with the IL-6 genotypes (170). A study reported increased serum IL-6 receptor (sIL6R) levels in ENL, which declined significantly after the completion of a corticosteroid treatment (143). However, other studies did not show associations of IL-6 serum levels with ENL (134, 139, 143).

An *ex vivo* study in PBMC isolated from ENL patients and LL controls showed a correlation of raised levels of cytokines IL-17A and its isomers as well as other Th17-associated cytokines IL-21, IL-22, and IL-23 with ENL (113). However, other studies failed to detect an association of ENL with serum IL-17 (110, 151, 171).

There are 49 studies measuring cytokines in ENL (**Table 4**), and the majority of these studies show a significant increase of the pro-inflammatory cytokines during ENL. TNF-α appears to be a regulator of the condition while there is substantial evidence supporting a role for IFN-γ as well. There is also evidence that other cytokines such as IL-1β and IL-6 or cytokine receptors such as sIL2R and sIL6R are also involved. Therefore, inhibitors of these molecules may be useful in a clinical setting. It is possible that genetic differences could account for differences observed between studies but methodological differences are also likely factors.

WHAT OTHER IMMUNE MECHANISMS ARE IMPLICATED IN ENL?

Sixty-four studies on other immunological factors in ENL have been performed (**Table 5**).

Innate Immunity

Genetic studies have shown associations between several single-nucleotide polymorphisms (SNP) of innate immunity genes such as *NOD2* (210), the natural resistance-associated macrophage protein (*NRAMP1*) (211), and *TLR1* (212, 213) with ENL.

A recent study from Brazil, which investigated whether DNA sensing *via* TLR9, constitutes a major inflammatory pathway during ENL (80) showed that both the skin lesions and peripheral leukocytes (B-cells, monocytes, and plasmacytoid dendritic cells) of ENL patients express higher TLR9 levels than BL/LL controls (80). In addition, the levels of endogenous human and pathogen-derived TLR9 ligands (human and mycobacterial DNA-histone complexes) were also higher in the circulation of ENL patients than BL/LL controls (80). Furthermore, stimulation of PBMC isolated from ENL patients with TLR9 agonist led to higher levels of TNF-α, IL-6, and IL-1β, than those of non-reactional leprosy and healthy controls. Usage of a TLR9 synthetic antagonist was able to significantly inhibit the secretion of pro-inflammatory cytokines after stimulation with *M. leprae* lysate (80). This is the first study to support the potential of TLR signaling inhibitors as a therapeutic strategy for ENL (80).

B-Lymphocytes and Immunoglobulins

Early studies enumerated B lymphocytes in skin lesions (89) and in peripheral blood (76, 78, 94, 123) of patients with ENL, while most of these studies did not find any association between B-cells and the development of ENL. Other studies looked at the IgM PGL-I in sera as a marker for ENL (40, 184, 192, 206), but most of these studies did not show an association (50, 82, 149, 187, 194, 205). Significantly lower serum levels of IgG1 and IgG3 subclasses of *M. leprae*-specific antibodies have been demonstrated in ENL patients compared to the BL/LL controls (190). This decrease of *M. leprae*-specific IgG1 and IgG3 antibodies in sera has not been related to downregulation of B cell responses since ENL episodes were characterized by an increase of polyclonal IgG1 antibody synthesis by the B cells, declining after subsidence of the reaction (191). The authors suggested that activation of B-cells is restricted to IgG1-secreting B cells in the blood of patients with lepromatous disease (191), while the lower serum concentrations of *M. leprae*-specific IgG1 and IgG3 (190) could be due to antibody deposition in the tissues (191). Interestingly, surface CD64 (FcγRI), the high-affinity receptor for monomeric IgG1 and IgG3 is expressed at higher levels on circulating neutrophils derived from ENL patients compared to non-reactional leprosy controls (25). The higher CD64 neutrophil expression could explain the presence of lower serum IgG1 and IgG3 levels in ENL patients compared to BL/LL controls.

TABLE 4 | Human studies on ENL investigating cytokines.

Reference; study site(s)	Study population	Timing of sampling	MDT status	ENL treatment status	Type of samples	Measures	Findings
Filley et al. (168); India ε	7 ENL	Before, during and after the episode	All patients on MDT	ENL treated with steroids and/or thalidomide	Serum	IL2R	IL2R increase during ENL
Rea and Modlin (102); USA γ	ND	ND	ND	ND	Skin	IL-2 positive and IFN-γ positive mRNA-bearing lymphocytes	IL2- positive lymphocytes prevalent in ENL and in tuberculoid lesions Cells expressing IFN-γ mRNA in ENL lesions slightly increased compared to lepromatous
Sarno et al. (138); Brazil	18 ENL, 39 BT/BL/BB/LL, 4 T1R	ND	16/18 patients on various stages of MDT/2 untreated	3 ENL on thalidomide and 7 ENL on prednisone; others untreated for reaction	Serum	Tumor necrosis factor (TNF)-α and IL-1	TNF varied from undetectable to extremely high levels in ENL No correlation between severity of ENL and cytokine level Neither TNF nor IL-1 correlate with number or duration of ENL episodes Treated patients with steroids or thalidomide lower TNF
Sehgal et al. (172); India	11 ENL, 14 T1R, 20 leprosy non-reactional, 10 HC	Before starting antireactional treatment and when clinical signs of reaction had abated	On MDT	Samples before and after starting antireactional treatment	Serum	IL-2R	T1R upgrading group higher IL-2R than ENL
Sullivan et al. (173); USA ε	ND	ND	ND	ND	Skin	IFN-γ and TNF-α mRNA	IFN-γ mRNA in ENL similar to tuberculoid In LL and ENL lesions about 0.2% of cells expressed TNF-α
Barnes et al. (155); USA	12 active ENL, 14 inactive ENL, 6 T1R; 11 LL	ND	All patients had received less than 5 years chemotherapy	ND	PBMC	TNF-α	ENL: the levels of TNF-α release by PBMC were higher than any other leprosy Thalidomide reduced TNF-α by more than 90%
Parida et al. (140); India	12 ENL, 64 leprosy TT/BT/BB/BL/LL, 14 T1R	ND	Most patients before MDT treatment	ND	Serum	TNF and IL-1	Patients undergoing T1R or ENL showed high TNF levels Significant correlation between TNF and IL-1 in reaction
Sampaio et al. (150); Brazil and USA	13 LL ENL, 15 LL, 9 HC	ND	All patients were receiving MDT during the study.	7 ENL patient blood samples before starting treatment with thalidomide and 6 1-2 weeks after thalidomide	Plasma PBMC Monocytes	TNF-α	ENL patients greater release of TNF-α from monocytes High plasma TNF-α in ENL
Bhattacharaya et al. (146); India	11 ENL, 14 T1R, 20 leprosy without reactions, 20 HC	Before treatment and after clinical remission of reaction	on MDT	Before antireactional treatment with steroids	Serum	TNF	TNF levels in acute ENL were higher but not significant and rose to become significant following treatment and clinical remission than HC and MB controls
Foss et al. (128); Brazil γ	28 lepromatous: 11 ENL, 23 tuberculoid, 19 HC	ND	86% of lepromatous patients treated with dapsone	Time of blood collection no immunosuppressive drug	Serum	TNF-α	TNF was elevated in the serum of ENL patients

(Continued)

TABLE 4 | Continued

Reference; study site(s)	Study population	Timing of sampling	MDT status	ENL treatment status	Type of samples	Measures	Findings
Sampaio et al. (139); Brazil ε	49 BL/LL: 24 developed ENL	At the time of developing ENL, during thalidomide treatment, or after thalidomide treatment was discontinued; collected at 1-3, 6-7, and/or 13-21 days of thalidomide and 1-2 months after thalidomide	MDT was continued through the study	Thalidomide treatment for ENL	Sera	TNF-α, IL-6, IFN-γ	ENL highest TNF-α levels, which decreased significantly during thalidomide treatment Serum IFN-γ elevated in patients with high TNF-α levels
Santos et al. (156); Brazil	14 ENL (4 BL/10 LL), 12 BL/LL, 11 HC: 4 ENL post-reactions	ND	Half untreated and the other half treated with MDT	ENL patients were treated with thalidomide?	PBMC	TNF-α: spontaneous and *M. leprae* stimulated	ENL patients showed significantly greater release of TNF-α both spontaneously and induced by *M. leprae*-induced release in ENL patients
Vieira et al. (131); Brazil γ, ε	95 MB (30 LL/65 BL) of which 51 ENL	At leprosy diagnosis and at onset of reactional episode	Time of MDT for each ENL	Sample before thalidomide and steroids?	Serum	TNF-α, soluble IL-2R	TNF-α increased in 70.6% of ENL patients
Memon et al. (141); Pakistan	12 ENL, 27 leprosy (TT/BT/BL/LL), 14 household contacts and 22 endemic HC with no known leprosy contact	At the onset of ENL before initiation of treatment for reaction and after the reaction had subsided	10/12 ENL received previous MDT	Samples before antireactional treatment	Serum	TNF-α	TNF levels higher during acute phase of ENL and declined after clinical remission of the reaction
Moubasher et al. (148); Egypt	35 reactional (19 ENL/16 T1R), 55 leprosy, 20 HC	ND	Untreated ENL?	Untreated ENL?	Serum	IFN-γ,IL-2, IL-2R, IL-10, TNF-α, IL-1β	Both T1R and ENL showed significantly higher serum IFN-γ, IL-2R and IL-1β compared to non-reactional leprosy ENL showed increased levels of IL-10 compared to T1R
Moubasher et al. (165); Egypt	35 reactional (19 ENL, 36 non-reactional, 20 HC	PB patients assessed after 6 and 12 months of MDT/MB assessed after 12 months of MDT; Before and at the end of treatment with MDT	Before and after treatment with MDT	Corticosteroids were given to control the reactions	Serum	IL-2R, IL-10, IL-1β	IL-1β levels may have a prognostic marker for the development of reactions
Partida-Sanchez et al. (142); Mexico ε	9 ENL, 10 non-ENL, 10 HC	Beginning of reaction and after 1 and 2 months of thalidomide	All patients on MDT	Untreated samples and after 1 and 2 months of thalidomide	Serum	TNF-α, IFN-γ	TNF-α was significantly higher in ENL compared to non-ENL TNF levels decreased after ENL treatment IFN-γ significantly lower in patients at the onset of ENL and increased after thalidomide

(Continued)

TABLE 4 | Continued

Reference; study site(s)	Study population	Timing of sampling	MDT status	ENL treatment status	Type of samples	Measures	Findings
Sampaio et al. (147); Brazil	18 MB with ENL (5BL/13LL)	Biopsies at diagnosis, at onset of reaction, and after 3 and/or 7 days of pentoxifylline; Serum: day 0 (during ENL), 3-7, 10-14, 30 and 60 days after pentoxifylline	7 patients with ENL newly diagnosed; others on MDT	Pentoxyfylline, 2 ENL patients on thalidomide	PBMC	Serum TNF-α, IL-6, IL-10	Elevated TNF-α in the sera of ENL
					Serum	TNF-α, IL-6, IL-10 release by PBMC following *M. leprae* stimulation or LPS stimulation	Treatment with pentoxifylline reduced TNF-α Serum levels of IL-6 increased during ENL High TNF-α mRNA expression in lesions during ENL which decreased following treatment with pentoxifylline
					Skin	TNF-α, IL-6, IL-10 gene expression at skin	IL-6 mRNA reduced by up to 50-fold after treatment
Moraes et al. (161); Brazil	53 leprosy: 20 ENL, 11 T1R	At the time of leprosy diagnosis (unreactional) and at the onset of first reactional episode (reactional)	MDT was continued through the study	No anti-inflammatory drugs at the time of sample collection	PBMC	IL-1β, IL-6, IL-8, GM-CSF, IFN-γ,IL-2Rp55, perforin, TNFβ, TNF-α mRNA in PBMC;	In 7 ENL higher incidence of IFN-γ, perforin, GM-CSF, IL2R mRNA in blood Upregulation of IL-1β, IL-6, GM-CSF, IL-2R, IFN-γ mRNA in blood at onset of ENL at 3 ENL follow-up
				3 patients sequential sampling and after thalidomide	Skin	IL-4, IL-6, IL-8, IL10, IL-12, IFNγ, TNFα mRNA in skin	Skin lesions ENL: IFNγ and IL-4 differentially expressed
Oliveira et al. (33);Brazil α	10 BL/LL: 6 ENL, 10 HC	ND	On MDT	ND	Blood, P.B.Neutrophils	TNF-α, IL-8	Stimulated neutrophils secrete IL-8 and TNFα Increased TNF-α secretion from neutrophils after LPS stimulation Thalidomide inhibited TNF-α by neutrophils
Goulart et al. (174); Brazil	19 leprosy: 5 ENL/3 T1R, 9 HC	Untreated samples	Untreated samples	Untreated samples	PBMC	TGF-β1 in supernatants from adherent PBMC after stimulation with PGL-1, LPS or serum-free RPMI	Adherent PBMC from ENL secrete higher TGF-β1
Moraes et al. (164); Brazil	13 MB: 10 ENL, 3 T1R	Before and during pentoxyfilline or thalidomide	All patients on MDT	Before and during pentoxyfilline or thalidomide	Skin	mRNA expression: IFN-γ, IL-6, IL-10, IL-12 p40, TNF-α, IL-4	Expression of IFN-γ, IL-6, IL-10, IL-12 p40, TNF-α at the onset of reactional episodes (T1R and ENL) but IL-4 rarely detected Follow-up: TNF-α mRNA and IFN-γ, IL-6 and IL12p40 mRNA decreased after thalidomide or pentoxyfylline
Nath et al. (162); India	36 ENL, 105 TT/ BL/LL 7T1R, 9 HC	ND	All patients on MDT	ENL patients before antireactional treatment	PMBC	IFN-γ, IL-4, IL-10, IL-12	ENL: 58% demonstrated a polarized Th1 pattern with only 30% expressing both cytokines
Nath et al. (163); India	1 BL/7 LL ENL, 2 BL/6 LL8 stable	ND	Most patients on MDT	ENL patients prior to antireactional therapy	PBMC	Real-time PCR forIFN-γ, IL4, IL10, p40 IL12	IFN-γ detectable in all and IL12p40 in half of ENL IL12p40 mRNA higher in ENL compared to stable lepromatous
Sampaio et al. (157); Brazil	15 leprosy: 10 ENL	ND	On MDT	ND	PBMC, monocytes, monocytes/T-lymphocytes cocultures	TNF-α after stimulation with *M. leprae*	Isolated monocytes from ENL released significantly more TNF-α in response to *M. leprae* than monocytes from non-reactional

(*Continued*)

TABLE 4 | **Continued**

Reference; study site(s)	Study population	Timing of sampling	MDT status	ENL treatment status	Type of samples	Measures	Findings
Tadesse et al. (132); Ethiopia γ	14 BT, 11 BT T1R, 8 ENL 11 HC	ND	ND	All ENL treated with steroids	PBMC	TNF-α in culture supernatants	Thalidomide resulted in suppression of TNF-α production
Haslett et al. (115); Nepal	20 ENL, 20 LL with no history of ENL within the preceding 30 days	Blood samples: days 0, 3, 7, 14, and 28 of thalidomide; ELISPOT: days 0, 7, 21, and 28 Flow cytometry: days 0, 7, and 21; qRT-PCR: PBMCs days 0, 7, 21	All (except 1 patient) on MDT	Excluded patients who had received immunomodulating therapy within the preceding month	Plasma T-cells Skin	Plasma levels of IFN-γ, TNF-α, soluble IL2R, IL-12, IL-12 p40 and IL-12 p70 ELISPOT for IFN-γ; Flow cytometry for cytokine production by T cells qPCR: IL-2 genes	Circulating TNF-α levels lower at ENL diagnosis than controls Upward trend during thalidomide ENL baseline plasma levels of IL-12 lower than control Baseline levels of sIL2R higher in ENL than controls Thalidomide increased T cell subsets expressing both IL-2 and IFN-γ
Villahermosa et al. (134); Philippines γ, ε	22 ENL	Before thalidomide and at study weeks 3 and 7 during thalidomide	MDT was continued	Samples untreated for antireactional drugs and during thalidomide	Plasma	TNF-α, IL-6	TNF-α levels not detected IL-6 unchanged or reduced following thalidomide from week 0 to week 3 IL-6 undetectable at weeks 3 and 7
Belgaumkar et al. (169); India	71 BT/BB/BL, 11 pure neuritic, 6 T1R, 1 ENL, 30 HC	Untreated samples	Untreated samples	Patients on antileprosy treatment or steroids were excluded	Serum	IL-6, IFN-γ	The one patient with ENL had higher levels of IL-6 and IFN-γ in comparison to the BL/LL patients without reactions
Iyer et al. (143); Indonesia ε	131 TT/BT/BB/BL/LL, 44 ENL, 5 T1R 112 HC	ND	Patients on MDT	Prednisolone to treat reactions	Serum	IL-6, IFN-γ, TNF-α, IL-6R, IL-10, IL-4, sCD27	IFN-γ and IL-6R increased in ENL compared to non-ENL Completion of corticosteroid treatment: IFN-γ, TNF-α, sIL6R declined
Stefani et al. (151); Brazil	10 ENL, 10 T1R, 29 non-reactional controls	Newly detected untreated patients	Untreated samples	Untreated samples	Plasma	TNF-α, IFN-γ, IL12p70, IL-2, IL-17, IL-1β, IL-6, IL-15, IL-5, IL-8, MIP-α, MIP-β, RANTES, MCPI, CCL11/ eotaxin, CXCL10, IL-4, IL-10, IL13, IL-1Rα, IL-7, IL-9, G-CSF, PDGF BB, bFGF, VEGF	IL-6, IL-7 and PDGF BB elevated in ENL
Motta et al. (175); Brazil	44 leprosy of which 15 ENL, 10 HC	Baseline and 7 days after therapy for oral infection	ND	ND	Serum	IL-1, TNF-α, IL-6, IFN-γ,IL-10	No specific finding for ENL
Teles et al. (176); Brazil ε	32 leprosy: 10 ENL, 8 T1R	4 ENL patients before and during reaction	All patients on MDT	ND	Skin PBMC	TNF-α gene expression and levels in supernatants	PBMC stimulated with *M. leprae*: upregulation of gene expression of TNF-α and increase of TNF-α in supernatants after 1, 3, and 6 h

(Continued)

TABLE 4 | Continued

Reference; study site(s)	Study population	Timing of sampling	MDT status	ENL treatment status	Type of samples	Measures	Findings
Jadhav et al. (149); India ε	303 MB: 5 ENL	Serum samples at the time of recruitment	Newly registered: no MDT	Untreated	Serum	TNF-α	No significant outcome for ENL
Madan et al. (144); India	61 leprosy: 4 ENL and 2 ENL during study	Untreated samples, during reactional episodes and after completion of treatment	Untreated patients	Patients on steroids were excluded	Serum	TNF-α, IFN-γ, IL-1β, IL-10	All cytokines were raised in reactional (both T1R and ENL) compared to non-reactional IFN-γ, IL-1β and IL-10 were higher in ENL but only IL-10 was statistically significant compared to T1R Levels of all cytokines decreased after MDT
Rodrigues et al. (145); Brazil	18 LL with ENL during treatment; 13 non-reactional BT, 37 non-reactional BL/LL, 25 BL with T1R during treatment; 21 HC	Beginning of leprosy treatment, at diagnosis of reactional episode and at 3-5 years post-treatment	Samples before and during MDT	Untreated samples and after treatment with prednisolone	Serum	TNF-α	TNF-α higher during ENL than prior to the reaction
Chaitanya et al. (177); India	21 ENL, 80 T1R, 80 leprosy without reaction, 94 non-leprosy	Untreated samples	Untreated samples	Untreated samples	Serum	IL-17F	IL-17F elevated during T1R but no significant difference in ENL
Lockwood et al. (178); India ε	303 MB leprosy: 13 ENL	Skin biopsies at enrollment	Before MDT	Before antireactional treatment	Skin	TNF-α and TGF-β immunostaining	TNF-α: similar levels ENL and non-ENL TGF-β: no difference in ENL and non-ENL
Martiniuk et al. (179); Nepal and USA ε	7 ENL	Pre- and post-treatment with thalidomide	ND	Pre- and post-treatment with thalidomide	Skin biopsies	RT-PCR for hIL-17A, hIL-17B, hIL-17C, hIL-17D, hIL-17E, hIL17F	IL17A, was consistently seen before and after thalidomide Reduction in IL17B, IL17E and increase of IL17C following thalidomide
Sousa et al. (170); Brazil	33 ENL, 54 T1R, 16 reaction-free leprosy	ND	63.8% presented ENL during MDT	ND	Plasma	IL-6	Higher IL-6 in ENL and T1R compared to non-reactional
Abdallah et al. (171); Egypt	43 leprosy: 6 ENL, 43 HC	Untreated samples	Untreated samples	Untreated samples	Serum	IL-17, IL-4	Overproduction of IL-4 in LL patients
Saini et al. (136); India γ	21 MB: 16 ENL, 5 T1R	ENL blood during reaction and at 0.5 and 1 year after the onset of reaction	Duration of MDT described	ENL patients received steroids	PBMC	PBMC stimulated with *M. leprae*, recombinant Lsr2 and 6 synthetic peptides spanning the Lsr2 sequence: IFN-γ	During ENL stimulated PBMC showed IFN-γ release
Abdallah et al. (109); Egypt γ	43 leprosy: 6 ENL, 40 HC	Untreated patients	Untreated samples	Untreated samples	Serum	IL-1β, IL-4, IL12p70, IFN-γ	IL-4 highest among LL compared to ENL

(Continued)

TABLE 4 | Continued

Reference; study site(s)	Study population	Timing of sampling	MDT status	ENL treatment status	Type of samples	Measures	Findings
Attia et al. (110); Egypt γ	43 leprosy: 6 ENL, 40 HC	Untreated samples	Untreated samples	Untreated samples	Serum	IL-17, IL-22, IL-10, TGF-β	No statistically significant difference between groups
Berrington et al. (167); Nepal	85 leprosy: 9 ENL, 35 BL/LL non-reactional	ND	ND	ND	Skin	RT-PCR for CCL1, CCL2, CCL17, CCL18, IFNA1, IFNA8, IFNB1, IFNG, IL10, IL12a, IL12b, IL13, IL17a, IL18, IL1b, IL1ra, IL21, IL22, IL23, IL27, IL29, IL4, IL6, TNF	CCL18, IL12b and CD14 elevated in lesions of ENL but failed to reach significance when adjusted for multiple comparisons
Sallam et al. (166); Egypt	43 leprosy: 6 ENL, 43 HC	Untreated samples	Untreated samples	Untreated samples, excluded patients on corticosteroids	Serum	IL-1β, IL-12	Higher IL-1β in ENL compared to non- reactional No significant difference for IL-12
Dupnik et al. (58); Brazil β, ε	11 ENL, 11 T1R, 19 leprosy without reactions for microarray; 6 ENL, 11 T1R, 11 non-reactional for qPCR; 3 ENL for ICH	ND	3/11 ENL pre-treatment, 2/11 ENL on treatment and 6/11 post-treatment; leprosy controls matched for length of treatment	Excluded patients on steroids within 7 days and thalidomide within 28 days of enrollment	PBMC	Microarrays followed by qPCR	Cytokine-cytokine receptor interaction has been in the top 3 KEGG pathways in ENL CCL5 followed by IFN-γ was the most significant upstream regulator of the expression changes in the array
Saini et al. (113); India γ	66 leprosy: 15 T1R, 15 ENL, 36 stable leprosy without previous history or clinical evidence of reactions	Newly diagnosed leprosy patients prior to institution of antireaction therapy	Untreated samples	Untreated samples	PBMC	Antigen (MLSA) stimulated and unstimulated PBMC: gene expression with PCR array for 84 genes ELISA for cytokines IL-17A/F, IL-21, IL-22, IL-23A, IL-6, IL-1β, IFN-γ, TGF-β in supernatants	IL-23A mRNA expression increased in ENL IL-23R expression increased in ENL High expression of CCL20 and CCL22 in ENL ENL significant fold increase in IFN-γ Culture supernatants: Higher IL-17A/F in ENL patients compared to LL IL23A increased compared to LL IL-1β increased in ENL
Dias et al. (80); Brazil ε	30 ENL, 24 BL/LL, 31 HC	Upon diagnosis of reaction	BL/LL before MDT but most ENL patients on MDT	Before treatment with thalidomide or steroids	PBMC	TNF, IL-6 and IL-1β in response to TLR9 agonist	Higher production of TNF-α, IL-6, IL-1β in response to TLR9 agonist TLR9 antagonist inhibited the secretion of cytokines in response to *M. leprae* lysate

*α, also in **Table 1**; β, also in **Table 2**; γ, also in **Table 3**; ε, also in **Table 5**.*

BB, mid-borderline leprosy; BL, borderline lepromatous leprosy; BT, borderline tuberculoid leprosy; ENL, erythema nodosum leprosum; HC, healthy controls; ICs, immune complexes; LL, lepromatous leprosy polar; ND, not described; P.B.neutrophils, peripheral blood neutrophils; SLE, systemic lupus erythematosus; TB, tuberculosis; TT, tuberculoid leprosy polar.

TABLE 5 | Human studies on ENL investigating other immunological factors.

Reference; study site(s)	Study population	Timing of sampling	MDT status	ENL treatment status	Type of samples	Measures	Findings
Waters et al. (22); United Kingdom and Malaysia β	38 lepromatous ENL	ND	ND	ND	Serum	Immunoglobulins	No differences in immunoglobulin levels
Reichlin et al. (180); Malaysia	13 LL of which 7 ENL	ND	ND	ND	Blood	Euglobulin IgG Serum IgG	Levels of euglobulin IgG higher in the ENL-positive patients than in ENL-negative patients
Anthony et al. (60); India β	25 LL ENL, 10 LL without ENL	Active ENL lesions?	ND	ND	Serum	Immunoglobulins	High levels of immunoglobulins in both LL and ENL
Harikrishan et al. (71); India β	20 active LL; 15 ENL during active and subsided phase; 20 HC	ENL: during the active and subsided phase	ND	ND	Plasma	Immunoglobulins IgG, IgM, IgA	Serum levels of IgG and IgM during subsidence of ENL were significantly lower compared to that during the active phase of ENL
Humphres et al. (181); USA	14 LL ENL, 28 BL/LL, 21 HC	Multiple Serial sampling	10/19 LL patients untreated, 9/19 LL patients Dapsone	Corticosteroids day prior to initial assay for NK activity and was continued through treatment	PBMC	Natural killer cell activity	Natural killer cell activity significantly depressed in ENL
Rea and Yoshida (182); USA	108 leprosy (4 untreated ENL, 14 dapsone-treated active ENL, 10 dapsone-treated inactive ENL), 25 HC	ND	54 untreated patients and others dapsone-treated	Untreated	Blood	Macrophage migration inhibition activity	Positive serum inhibitory activity strongly associated with reactional states (ENL or T1R or Lucio's reaction) in both treated and untreated patients
Miller et al. (183); USA	9 leprosy: 3 T1Rs and 2 ENL	Serial sampling from date of initiation of therapy until the first year of treatment	On MDT	Reactional episodes were treated with corticosteroids and 1 ENL received thalidomide	Plasma	Antibodies to Mycobacterial Arabinomannan	High levels of antibody to Arabinomannan in 2 ENL patients
Narayanan et al. (89); India γ	35 leprosy patients: 7 LL with ENL, 6 BT, 6 BT with T1R, 4 BL, 5BL with T1R, 8 LL	ND	ND	ND	Skin	B cells	No increase of B cells in any of the lesions
Rea et al. (96); USA γ	19 ENL, 67 BL/LL 4 LL with Lucio's, 13 T1R, 18 Tuberculoid, 13 Tuberculoid long-term treatment	ND	Some patients on MDT	ENL before receiving thalidomide	PBMC	B-cells	B-cell percentage in the PBMC of ENL similar LL

(Continued)

TABLE 5 | Continued

Reference; study site(s)	Study population	Timing of sampling	MDT status	ENL treatment status	Type of samples	Measures	Findings
Schwerer et al. (184); USA	121 leprosy (including ENL), 28 contacts, 15 HC	ND	ND	ND	Serum	Anti-PGL I IgM levels	Serum anti-PGL I IgM levels lower in ENL compared to patients with comparable BI
Andreoli et al. (40); India	12 ENL	ND	All patients on MDT with specified duration	Treated with prednisone and/or thalidomide	Serum	Circulatory IgM antibody levels to the PGL I; IgM, IgG, IgA antibody levels to *M. leprae* antigenic preparation	During ENL: decrease of circulatory IgM antibody levels to PGL I but no significant change to IgG, IgM or IgA antibody levels to the soluble antigens from *M. leprae*
Laal et al. (94); India γ	15 ENL 13 LL	During active ENL and 1 week to 4 months after stopping treatment	Treatment with combination antileprosy drugs was continued throughout	First sample before initiation of antireactional treatment; second sample 1 week to 4 months after treatment	PBMC	B cells	B-cell percentages in PBMC of ENL patients were similar to those of uncomplicated LL
Blavy et al. (185); Senegal	34 ENL and 50 leprosy patients	ND	ND	ND	Lymphocytes	HLA phenotyping	Not significant findings of any HLA phenotype regarding ENL
Levis et al. (186); USA	ND	ND	ND	ND	Serum	IgM and IgG antibodies to PGL-I	ENL lower anti-PGL-I IgM than non-ENL of comparable BI
Rao and Rao (123); India γ	44 ENL, 39 LL, 22 post-ENL	ENL cases before starting treatment for ENL, post-ENL after the patient had not taken anti-inflammatory drugs or steroids for at least 3 and 7 days	From 39 non-reactional: 20 untreated and 19 with dapsone for less than a year	ENL before starting ENL treatment, post-ENL after the patient had not taken anti-inflammatory drugs or steroids for at least 3 and 7 days	Blood	B lymphocytes in peripheral blood	B cells: no difference between groups
Sehgal et al. (76); India β	21 patients with leprosy reactions either T1R or ENL	ND	ND	ND	B-cells Serum	Percentage and absolute count of B-cells; Immunoglobulins IgG, IgA, IgM	During ENL a significant increase in the percentage and absolute count of B-lymphocytes Significantly elevated serum immunoglobulin values after subsidence of ENL
Levis et al. (187); USA	40 ENL, 63 leprosy without ENL, HC	ND	ND	ND	Serum	IgM antibody to PGL-I; IgM and IgG Abs to *M.tb* and *M. leprae* LAM	No correlation between IgM or IgG Ab to LAM and bacillary index

(*Continued*)

TABLE 5 | Continued

Reference; study site(s)	Study population	Timing of sampling	MDT status	ENL treatment status	Type of samples	Measures	Findings
Rao and Rao (85); India	44 ENL, 39 lepromatous, 22 post-ENL	ENL blood before starting ENL treatment, post-ENL after patient does not take any anti-inflammatory drugs or steroids for the last 3 or 7 days	20 patients no previous MDT and 19 treated with dapsone	Before starting treatment with anti-inflammatory drugs or steroids	Blood	Leukocyte migration inhibition test	Lower migratory indices to whole *M. leprae* during ENL
Rao and Rao (78); India β	44 ENL, 39 BL/LL, 22 post-ENL	ENL before starting anti-ENL treatment, post-ENL ensuring that the patient had not taken anti-inflammatory drugs or steroids for at least 3 or 7 days	20 BL/LL untreated and 19 BL/LL treated with dapsone	Before starting treatment with steroids or anti-inflammatory drugs	Serum	IgG, IgA, IgM	IgG and IgM decreased in ENL than lepromatous and post-ENL Serum IgA elevated in ENL than lepromatous group and further increase post-ENL
Filley et al. (168); India δ	7 ENL	Before, during and after the episode	All patients on MDT	ENL was treated with steroids and/or thalidomide	Serum	%GO	During ENL%GO transiently raised, and this rise parallels an increase in circulating IL2R
Bhoopat et al. (127); Thailand γ	57 ENL (19 acute/38 chronic), 61 active LL, 33 control patients whose leprosy had been cured	26 BL and 35 LL newly diagnosed and untreated	ND	When treatment with corticosteroids and/or thalidomide was initiated precise timing was recorded with respect to the time of collection of specimens	Blisters induced over a representative skin lesion	IgM antibody to PGL-I and Tac peptide	IgM antibody to PGL-I and Tac peptide levels were elevated in chronic ENL lesions Corticosteroids reduced IgM antibody to PGL-I but did not change the levels of Tac peptide
Ramanathan et al. (49); India β	26 BL/LL of which 11 ENL, 24 HC	Blood was taken before initiation of treatment and then to 2-month intervals up to 20 months	Untreated and then on MDT samples every 2 months	Treated but after blood sampling	Serum	IgG, IgA and IgM	ENL no significant relation with immunoglobulin levels
Sullivan et al. (173); USA δ	ND	ND	ND	ND	Skin	ICAM-1, ICAM-1 ligand LFA-1	Prominent keratinocyte ICAM-1 expression

(Continued)

TABLE 5 | Continued

Reference; study site(s)	Study population	Timing of sampling	MDT status	ENL treatment status	Type of samples	Measures	Findings
Scollard et al. (82); Thailand β	4 cured leprosy, 10 leprosy (5BT, 3BL, 2LL), 8 ENL patients (5LL and 3BL), 3 T1R, 4 HC	ND	ND	ND	Blisters induced over representative skin lesion Sera	Immunoglobulins (IgG, IgA, IgM) to whole *M. leprae and to* PGL-I	No statistically significant difference regarding immunoglobulins
Sehgal et al. (188); India	25 leprosy with reactions (of which 11 ENL), 20 leprosy without reactions, 10 HC	ND	On MDT	Reactional patients on prednisolone	Lymphocytes	Lymphocyte adenosine deaminase activity (L-ADA)	The patients with leprosy reactions (both ENL and T1Rs) had higher enzyme L-ADA than controls (the enzyme has a role in activation, differentiation and proliferation of lymphocytes)
Sampaio et al. (139); Brazil δ	49 BL/LL of which 24 ENL	ENL at the time of developing ENL, during thalidomide treatment, or after thalidomide treatment	MDT was continued through the study	Certain ENL during thalidomide treatment	Skin biopsies	MCH II and ICAM-1 in histology	MHC II and ICAM-1 on epidermal keratinocytes in ENL downregulated with thalidomide
Santos et al. (129); Brazil γ	10 ENL, 59 LL/BL, 4 T1R, 4 post-reactional	ND	On MDT	No antireactional treatment before blood collection	PBMC, Monocytes	Monocyte activation by procoagulant activity, HLA-DR	No significant difference in monocyte activation between the different groups No significant differences in HLA-DR between groups
Singh et al. (189); India	44 active ENL, 48 prior history of ENL, 125 stable lepromatous, 40 HC not endemic	ND	ND	Untreated samples	Serum	Antibodies against B cell epitopes of *M. leprae* recombinant protein LSR	Antibodies against a specific distinct peptide region only in patients undergoing ENL
Kifayet and Hussain (190); Pakistan	67 BL/LL acute ENL, 83 non-reactional BL/LL, 77 endemic HC	ND	Most on MDT but 83 non-reactional less than 2 weeks of MDT	ND	Plasma	*M. leprae*-specific IgG subclasses	Lower concentrations of all IgG subclasses during ENL but lower IgG1 and IgG3 during ENL before treatment
Kifayet et al. (191); Pakistan	13 ENL acute and post-remission of reaction, 16 non-reactional stable LL, 32 endemic HC	During acute ENL (n = 13) and after the reaction has subsided	ND	ND	Plasma B-cells	IgG subclasses *M. leprae*-specific antibodies; Detection and enumeration of antibody-secreting B cells by ELISPOT	Polyclonal IgG1 elevated in acute ENL compared LL controls and decreased when ENL subsided IgG2 antibodies lower during acute ENL and increased after reaction has subsided Discrepancy in serum concentrations and B cell frequency
Vieira et al. (131); Brazil γ, δ	95 MB leprosy (30 LL and 65 BL) of which 51 ENL	At leprosy diagnosis and at onset of reactional episode	Time of MDT for each ENL patients in study	Sample before thalidomide and steroids?	Serum	Circulating anti-neural and antimycobacterial antibodies	Detection of anti-neural (anti-ceramide and anti-galactocerebroside) antibodies in ENL sera No difference between reactional and non-reactional lepromatous patients regarding IgM antibodies Higher levels of anti-ceramide IgM and diminished levels of anti-galactocerebroside antibodies in reactional compared to non-reactional patients

(Continued)

TABLE 5 | Continued

Reference; study site(s)	Study population	Timing of sampling	MDT status	ENL treatment status	Type of samples	Measures	Findings
Rojas et al. (50); Brazil β	19 ENL, 10 BL/LL non-ENL patients, 13 family contacts; 15 healthy non-contacts	ND	Both untreated patients and patients on MDT for 1-72 months	ND	Serum	Anti-PGL-I IgM, IgG responses to recombinant 10-kDa heat shock protein	IgM anti-PGL-I and IgG anti-10-kDa heat shock protein antibodies were constituents of the immune complexes in patients with ENL while free antibody levels did not differentiate between ENL and non-ENL patients
Beuria et al. (192); India	18 ENL, 44 BL/LL, 62 BT/TT, 17 HC	ND	Most patients on MDT	ND	Serum	IgG subclass levels to *M. leprae* sonicated antigens (MLSA) and PGL-I	ENL patients showed a significant fall in IgG3 antibody to MLSA and PGL-I compared to BL/LL leprosy controls
Freire et al. (193); Brazil	59 leprosy (including 12 ENL), 60 HC	ND	11/12 ENL were on MDT and 1/12 with dapsone	ND	Serum	Anti-neutrophil cytoplasmic antibodies (ANCA)	ANCA are present in 28.8% of leprosy patients but are not related to vasculitis in the ENL reaction and are not a marker of a specific clinical from
Partida-Sanchez et al. (142); Mexico δ	9 ENL, 10 non-ENL leprosy, 10 HC	Beginning of reaction and after 1 and 2 months of thalidomide	All patients on MDT	Before thalidomide, second sample after 1 month of thalidomide and third after 2 months	Plasma	IgM and IgG antibody subclasses to *M. leprae* sonicated extract	ENL at the onset of reaction had slightly higher anti-*M. leprae* IgG1 and IgG2 antibodies compared to non-ENL but not statistically significant
Stefani et al. (194); Brazil	600 leprosy: 31 ENL, 45 T1R, HC	Untreated	Before MDT treatment	Untreated	Serum	IgM and IgG anti-PGL-I	Patients presenting with T1R or ENL at leprosy diagnosis have same level of IgM anti-PGL-I antibody response as leprosy patients without reactions at diagnosis
Beuria et al. (195); India	44 BL/LL, 62 TT/BT, 18 ENL, 15 T1R, 17 HC	ND	BL/LL: 90% on MDT TT/BT: mostly untreated	Steroids after collection of samples	Serum	IgG1, IgG2, IgG3 and IgG4 to LAM	Reduction in IgG3 in ENL compared to active BL/LL Higher IgG1 in ENL than T1R
Hamerlinck et al. (196); Philippines, Netherlands	13 ENL, 22 T1R, 26 leprosy unreactional, 10 HC	Serial samples during MDT: 2 ENL follow-up and received corticosteroids, 14 leprosy free of reactions, 4 T1R 6, 12, 18, 30 months during follow-up	13 ENL before MDT, 2 ENL during MDT	During ENL before treatment	Serum	Neopterin	T1R and ENL higher neopterin levels compared to non-reactional individuals Corticosteroid treatment reduces levels of neopterin
Mahaisavariya et al. (197); Thailand	95 leprosy patients: 63 non-reactional, 19 T1R, 13 ENL	A biopsy at time of diagnosis and an additional biopsy later, in some cases at the time of reaction	ND	Before antireactional treatment?	Skin	Mast cells	Reduction of mast cell counts in both T1R and ENL compared to non-reactional patients

(Continued)

TABLE 5 | Continued

Reference; study site(s)	Study population	Timing of sampling	MDT status	ENL treatment status	Type of samples	Measures	Findings
Schon et al. (198); Ethiopia	4 ENL, 5 T1R	ND	ENL cases: 2 MDT untreated and 2 MDT-treated	Steroids	Urine	Urinary levels of metabolites of NO	Urinary nitric oxide metabolites decreased significantly after steroid treatment
Antunes et al. (199); Brazil	3 ENL, 3 T1R	First biopsy during reactional episode and second during remission	On MDT	Thalidomide for ENL	Skin	Neuropeptides; quantification of mast cells and their subsets	Increase mast cells in the inflammatory infiltrate of the reactional (both T1R and ENL) biopsies compared to non-reactional
Rada et al. (200); Venezuela	29 ENL, 19 MB not reactional, 11 PB, 28 HC	Before treatment	Untreated samples	Untreated samples	Serum PBMC	Nitrite/Nitrate levels	Supernatants of PBMC from ENL patients significantly elevated levels of nitrite/nitrates compared to LL or tuberculoid leprosy
Sunderkotter et al. (201); Brazil	Skin Biopsies: 41 non-reactional leprosy, 8 ENL, 10 T1R; Serum samples: 16ENL, 5RR, 7TT, 13BT/BB/LL, 19 HC	Untreated samples	Skin biopsies: 42 untreated non-reactional leprosy 8 ENL of which 5 MDT	Before treatment with steroids or thalidomide	Serum Skin	MRP8, MRP14	Increase of serum levels of MRP8 and MRP14 in ENL Higher percentage of MRP8⁺ and MRP14⁺ cells in ENL skin lesions than non-reactional
Nigam et al. (202); India	80 leprosy: 10 ENL and 10 T1R; 20 HC	ND	ND	ND	Serum	Deaminase	Deaminase levels were higher in patients with reaction
Villahermosa et al. (134); Philippines γ, δ	22 ENL	Before thalidomide administration and at study weeks 3 and 7 during thalidomide treatment	MDT was continued during the study	Samples untreated for antireactional drugs and during thalidomide treatment	Urine	Neopterin	ENL higher neopterin values in urine than HC
Iyer et al. (143); Indonesia δ	131 leprosy patients (44 ENL), 112 HC	ND	Patients were classified irrespective of MDT status	Prednisolone to treat reactions	Plasma	Neopterin	Neopterin no significant difference between ENL and non-ENL
Mohanty et al. (203); India	14 ENL before and after resolution of ENL, 5 LL	Before commencing antireactional therapy and after resolution of ENL	All patients on MDT	Before commencing antireactional treatment	Urine	Urinary nitric oxide metabolites	Urinary nitric oxide metabolite higher in ENL compared to non-reactional LL These levels were reduced with resolution of reaction following antireactional therapy
Santos et al. (204); Brazil	8 leprosy: 3 ENL	ND	MDT during the study: length of MDT described	Thalidomide during the study	PBMC Skin	B7-1 expression (flow cytometry and IHC)	Higher B7 expression in ENL and T1R patients than non-reactional in both PBMC and cutaneous lesions

(Continued)

TABLE 5 | Continued

Reference; study site(s)	Study population	Timing of sampling	MDT status	ENL treatment status	Type of samples	Measures	Findings
Silva et al. (205); Brazil	25 leprosy: 5 ENL and 8 T1R	0, 2, 4, 6, and 12 months of MDT	All patients on MDT	Untreated	Plasma	PGL-I levels Neopterin	Serum PGL-I levels did not differ significantly between ENL and non-ENL No significant correlation of neopterin between ENL and non-ENL
Brito Mde et al. (206); Brazil	104 reactions after completion of MDT (44 ENL), 104 with no post-treatment reactions (8 ENL)	ND	All patients were treated with MDT; half had finished MDT	ND	Plasma	ML flow (IgM anti-PGL-I positive serology)	The patients with positive serology after MDT presented a 10.4 fold greater chance of developing post-treatment reactions (ENL or T1R)
Iyer et al. (207); Indonesia	78 leprosy (36 ENL and 3 T1R), 36 HC	ND	30 untreated and 48 treated patients	Reactions were treated using prednisolone	Serum Skin	Chitotriocidase	Serum chitotriosidase activity elevated in ENL compared with HC but not with non-ENL leprosy Significant decline of serum chitotriosidase following corticosteroid treatment in ENL
Lee et al. (24); USA α	6 ENL, 7 LL	ND	ND	ND	Skin	Microarrays and gene expression; IHC for E-selectin	Upregulation of gene expression: in ENL lesions of the selectin family of adhesion molecules IHC: higher levels of E-selectin in ENL lesions
Massone et al. (116); Brazil γ	20 leprosy biopsies (3 ENL)	Biopsies at the time of diagnosis	10, 12 and 13 months after beginning of MDT for LL	Untreated	Skin	Presence, frequency and distribution of plasmacytoid dendritic cells	CD123 expression was observed in 2/3 ENL biopsies
Rada et al. (135); Venezuela γ	81 LL, 41 BL, 41 BB, 3% BT	ND	ND	ND	Blood	Serological immunological tests to various mycobacterial proteins	Mean antibody values against complete mycobacterial proteins higher in non-reactional individuals
Teles et al. (176); Brazil δ	32 leprosy: 10 ENL, 8 T1R	4 ENL patients before and during reaction biopsy samples	All patients were receiving MDT	ND	Skin Serum	MMP-2, MMP-9, TIMP-1	RT-PCR for MMP-2 and MMP-9 versus TIMP-1 in ENL sequential samples in 4 ENL patients: TNF-α, MMP-2 and MMP-9 mRNA enhanced IHC and confocal microscopy: absence of MMP positivity in ENL epidermis ELISA in sera of ENL: elevated MMP-9 but not TIMP-1 compared to non-reactional patients
Jadhav et al. (149); India δ	303 MB followed up for 2 years: 5 ENL	Serum samples at the time of recruitment	Newly registered MB patients: no MDT	Untreated	Serum	Antibodies to PGL-I, LAM, ceramide, S100	No statistically significant outcome for ENL
Lockwood et al. (178); India δ	303 new MB leprosy (13 ENL)	Skin biopsies at enrollment	Before MDT treatment started	Before antireactional treatment	Skin Nerve	Immunostaining for CD68 and iNOS	Reactional biopsies had significantly fewer $CD68^+$ cells than non-reactional Nearly all biopsies in the LL group had $CD68^+$ cells present and these were not altered in ENL ENL showed some iNOS staining but not significant differences with non-ENL

(Continued)

TABLE 5 | Continued

Reference; study site(s)	Study population	Timing of sampling	MDT status	ENL treatment status	Type of samples	Measures	Findings
Martiniuk et al. (179); Nepal and USA δ	7 ENL	Pre- and post-treatment with thalidomide	ND	Pre- and post-treatment with thalidomide	Skin	RT-PCR for hRORγT, hCD70, hCD27, hPLZF-1, hCTLA4, hAHR, hiNOS2, hARNT, hIDO, hGARP, hCD46	Reduction in CD70, GARP, IDO and increase of RORγT, ARNT following thalidomide treatment
Singh et al. (208); India	240 leprosy: 19 ENL, 69 BL/LL	ND	ND	ND	Serum	IgG antibodies against keratin	No significant difference in ENL
Dupnik et al. (58); Brazil β, δ	11 ENL, 11 T1R, 19 leprosy controls without reactions for microarray; additional 28 leprosy (6 ENL, 11 T1R, 11 non-reactional) for qPCR validation; 3 ENL for IHC	ND	3/11 ENL pre-treatment, 2/11 ENL on treatment and 6/11 post-treatment; leprosy controls matched for stage of treatment	Excluded patients on steroids within 7 days and thalidomide within 28 days of enrollment	PBMC	Microarray and qPCR for transcriptional profile of PBMC; Flow cytometry for monocyte populations	Top 3 KEGG pathways in ENL were *S.aureus* infection, SLE, cytokine-cytokine receptor interaction No significant difference in the proportion of circulating monocytes between reactional and non-reactional PBMC
Mandal et al. (209); India	15 reactional (both ENL and T1R), 15 HC	ND	ND	ND	PBMC	Vitamin D receptor (VDR) mRNA	All the individuals with low VDR expression manifested ENL
Dias et al. (80); Brazil δ	30 ENL, 24 BL/LL, 31 HC	Upon diagnosis of reaction	BL/LL before MDT but most ENL on MDT	Before treatment with thalidomide or steroids	PBMC (monocytes, B-cells, pDCs) Skin	Expression of TLR9	Skin lesions and PBMC of ENL express higher levels of TLR-9
Schmitz et al. (25); Brazil α	62 leprosy: 22 ENL, 16 HC	ENL: before and 7 days after thalidomide	Patients before and after MDT	Before and after thalidomide	Skin	CD64 expression by qPCR and IHC	CD64 mRNA and protein expressed in ENL lesions Thalidomide reduced CD64 expression

*α, also in **Table 1**; β, also in **Table 2**; γ, also in **Table 3**; δ, also in **Table 4**.*

BB, mid-borderline leprosy; BL, borderline lepromatous leprosy; BT, borderline tuberculoid leprosy; ENL, erythema nodosum leprosum; HC, healthy controls; ICAM-1, keratinocyte intracellular adhesion molecule 1; ICs, immune complexes; LAM, lipoarabinomannan; LL, lepromatous leprosy polar; MLSA, M. leprae sonicated antigens; ND, not described; PGL I, phenolic glycolipid I; SLE, systemic lupus erythematosus; TB, tuberculosis; TT, tuberculoid leprosy; polar,%GO, proportion of oligosaccharide chains on the Fc fragment of IgG which terminate with N-acetylglucosamine and not galactose.

New Suggested Pathogenetic Mechanisms

Two recent studies of gene expression provide evidence of activation of novel molecular pathways in ENL.

Lee et al performed bioinformatic pathways analysis of gene expression profiles in leprosy skin lesions and found "cell movement" as the top biological pathway characterizing ENL (24). The study further described a neutrophil recruitment pathway including genes of key molecules that mediate neutrophil binding to endothelial cells (24). This neutrophil recruitment pathway characterizing ENL was inhibited by thalidomide (24). Consistent with these findings is a study of transcriptional profiles in PBMC of leprosy patients by Dupnik et al which identified "granulocyte adhesion and diapedesis" as one of the top canonical pathways characterizing ENL (58). Dupnik et al. identified 517 differentially expressed genes in patients with ENL (58). The pathway analysis revealed that the top three Kyoto Encyclopedia of Genes and Genomes (KEGG) pathways that changed in ENL were *Staphylococcus aureus* infection, systemic lupus erythematosus (SLE), and cytokine-cytokine receptor interaction, while the complement and coagulation pathway was also associated with ENL (58). CCL5 was the most significant upstream regulator in the array followed by IFN-γ (58). Transcripts uniquely increased in ENL included the complement receptors C3AR1 and C5AR1 while uniquely decreased transcripts in ENL included IL-10 and cytotoxic T-lymphocyte associates protein 4 (CTLA-4), modulators of T-cell responses (58). Hepcidin, catholicidin, antimicrobial peptides, C1q, and defencins had also an increased expression in ENL, while CCL2, CCL3, and SOD2 could be potential biomarkers for ENL (58). Transcripts increased in PBMC from ENL patients also included FcγR1 (CD64), FPR1, and FPR2, which recognize formylated peptides produced by bacteria triggering receptor on myeloid cells 1 (TREM1) and the related molecule triggering receptor expressed on myeloid cells-like 1 (TREML-1) (58).

The microarray studies performed in skin lesions and PBMC have generated a long list of candidate genes that regulate immune function to be associated with ENL. These merit further research.

LIMITATIONS OF THE SYSTEMATIC REVIEW

PubMed was the only database used to identify eligible studies. This will have resulted in studies published in journals not listed in PubMed being omitted from our review. A search of gray literature may also have contributed data which may have influenced our conclusions. The high heterogeneity of included studies in terms of study questions and outcomes and the different methodologies used meant that a meta-analysis was not possible.

METHODOLOGICAL CONSIDERATIONS OF THE STUDIES INCLUDED IN THE SYSTEMATIC REVIEW

Many of the studies of immunological features of ENL contain significant limitations in both design and reporting. Most seriously 66% of the studies did not have a case definition of ENL.

More than 70% of studies sampled individuals at a single time point. Sampling at two time points was seen in 21.2% of studies, 3 time points in 2.7% of studies, whereas 4 or more time points was described in only 5.5% of studies. Some studies did not have appropriate controls- patients with uncomplicated BL and LL. Although 93.2% of studies used BL/LL patients as controls, the remaining 6.8% of studies used other control groups such as healthy volunteers or leprosy contacts or tuberculoid leprosy patients or patients with Type 1 reaction. Often the controls were not matched for age, sex or treatment status. Controls should be matched for age and sex since these factors may influence T cell and neutrophil numbers and functions (214–216) as well as TNF-α and other cytokine levels (217).

ENL is a condition that can be acute, recurrent, or chronic, and therefore, the timing of sample collection is crucial. No information on the timing of the sampling is described in 54.8% of all studies. The importance of timing for sample collection during ENL could explain the discrepancies observed in multiple studies as has been suggested in the studies addressing the role of neutrophils in ENL. Studies using serial sampling yield more meaningful data compared to cross-sectional studies. The interval between time points is important and needs to be kept as consistent as possible for all study subjects.

Only one study matched BL/LL controls and ENL cases for length of MDT. Patients may develop ENL prior to the diagnosis of leprosy, during MDT or after successful completion of MDT. MDT may affect the immune status of leprosy patients and thus the matching of cases and controls for this variable is important. Two of the components of MDT, dapsone and clofazimine, have been associated with alterations in neutrophil and lymhocyte function (218–220). Dapsone stimulates neutrophil migration (218) and inhibits production of Prostaglandin E_2 by neutrophils (220). In addition, dapsone inhibits lymphocyte transformation (218). On the other hand, clofazimine enhances production of Prostaglandin E_2 by neutrophils (220). Dapsone and anti-dapsone antibodies have been identified in circulating ICs of leprosy patients (221). Circulating cytokine and chemokine levels also change with MDT (165, 222, 223). In addition, gene expression studies could be affected by MDT since the MDT component rifampicin may modify the expression of certain housekeeping genes (224). A total of 30.8% of studies did not report the MDT status of their cases or controls, 12.3% collected untreated patient samples, whereas 56.2% collected patient samples at various stages of MDT.

The effect of immunosuppressive drugs used to treat ENL on the findings of studies is an important factor which should be considered. In 37.7% of studies, there was no reporting of whether participants were on ENL treatment when samples were collected. Treatment with corticosteroids affects T-cells and neutrophil function (225, 226) and also gene expression studies by influencing housekeeping genes (224). Treatment with thalidomide may increase the neutrophil numbers, at least partially through differentially modulating the surface expression of markers CD18 and CD44 by the neutrophils in the bone marrow and the spleen (227). Thalidomide treatment may also affect T-cell functions by suppressing $CD4^+$ T-cell proliferation while increasing their conversion to $CD4^+FoxP3^+$ Tregs (228).

Moreover, thalidomide treatment may reduce cytokine levels (229). Less than half (34.2%) of studies indicate that samples were obtained prior to the start of ENL treatment.

Only 17.8% of all studies collected samples from more than one system, while samples from both blood and skin were described in only 12.3% of all studies.

FUTURE STUDY DESIGN

Studies of ENL may be difficult to design and conduct. In addition, no animal model of ENL is available. Obtaining sufficient numbers of patients so that studies are adequately powered is difficult unless multicenter studies are performed which increase the logistical complexity and cost of the research. Patients are often on treatment (both MDT and immunosuppression) which may influence the study outcomes.

A large cohort study of newly diagnosed patients with BL and LL would be optimal in allowing matching of cases and controls. Some BL/LL patients who have not developed ENL at enrollment in the study should be recruited and followed until they develop the disorder. Detailed clinical information which includes demographic data, ENL severity using a robust measure, treatment status, in conjunction with well-timed and documented specimen collection (preferably of blood and skin), effective specimen storage, and transportation. ENL is a systemic disease and ideally samples from more than one system, i.e., both blood and skin should be obtained where appropriate. Well-designed laboratory experiments using a wide range of techniques should be used to interrogate such important specimens.

CONCLUSION

Figure 2 gives an overview of the immunology of ENL.

Our understanding of the causes of ENL is limited. The factors that initiate and/or sustain it might help to identify strategies to prevent or control the associated inflammation.

There is some evidence to support a role for neutrophils and ICs/complement in the inflammation associated with ENL; however, their role in the initiation of ENL remains unclear. The increase of TNF-α and other pro-inflammatory cytokines during ENL has been shown in multiple investigations, while suppression of TNF-α leads to clinical improvement. T-cell subsets appear to be important in ENL since multiple reports describe an increased $CD4^+/CD8^+$ ratio in ENL patients compared to BL/LL controls.

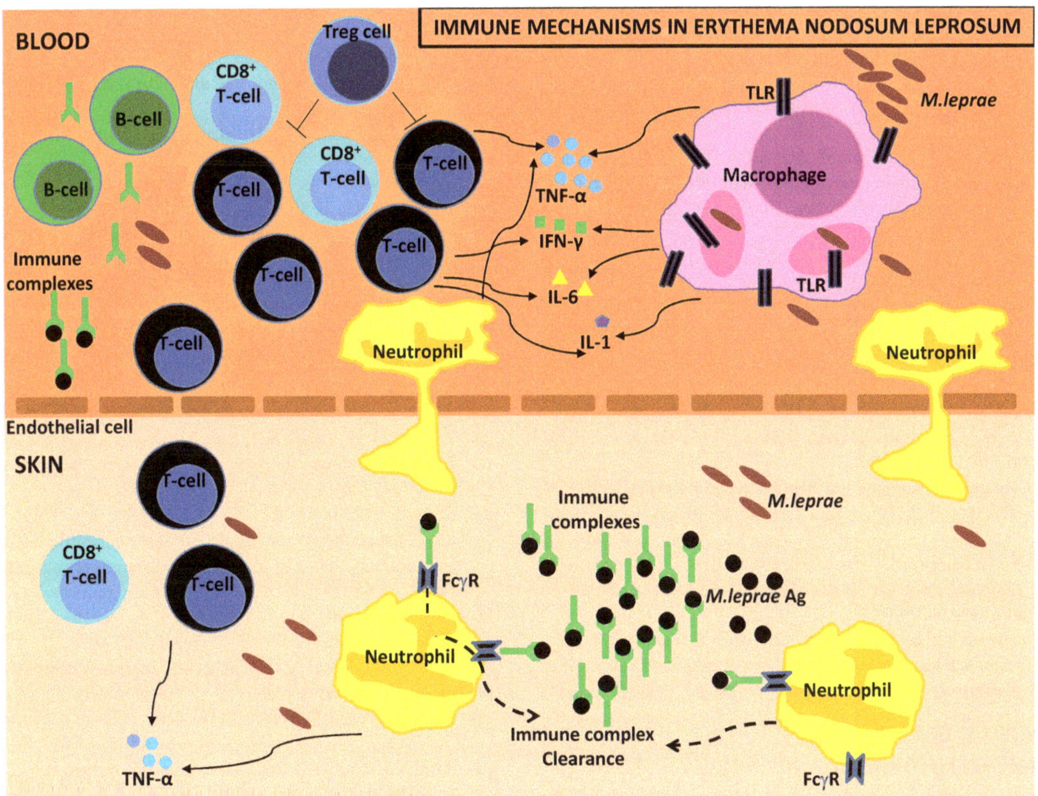

FIGURE 2 | Immune mechanisms in erythema nodosum leprosum (ENL). The diagram illustrates the different immune mechanisms which have been described in the literature of ENL. High volume of immune complexes (ICs) are formulated due to the increased antibody formation by the B cells and the increased mycobacterial antigens by fragmentation of the *M. leprae* bacilli. ICs are deposited in the skin. Neutrophils are drawn to the skin where they help in the IC clearance using their surface Fcγ receptors. An increase of $CD4^+/CD8^+$ T cell subset ratio in both peripheral blood and skin characterizes the disorder. Macrophages form the *M. leprae* intracellular niche and in concert with neutrophils and T-cells secret high levels of tumor necrosis factor (TNF)-α and other pro-inflammatory cytokines to further complicate the phenotype of ENL.

New technologies such as microarray studies pave the way and may lead to novel immunological pathways associated with ENL. Further research of the association of ENL with pathophysiological pathways such as the SLE pathway or the *S. aureus* infection pathway may improve our understanding of the disorder and potentially lead to novel therapeutic strategies. There are still large gaps in our understanding of this severe complication of leprosy despite the large number of studies examining the immunology of ENL. A systems biology approach may provide new insights.

This systematic review has highlighted the complex interactions at play in ENL and the difficulty in elucidating the various inflammatory pathways. We should rise to the challenge of understanding how these mechanisms operate and interact so that we can improve the treatment of patients with ENL.

AUTHOR CONTRIBUTIONS

AP and SW were responsible for the study concept and design; made critical revision of the manuscript for important intellectual content. AP was responsible for acquisition, analysis, and interpretation of data and for drafting the manuscript. AP, SW, and DL edited the manuscript.

ACKNOWLEDGMENTS

We would like to thank Anna and Andreas Goulielmos for proof reading and grammatical error corrections.

FUNDING

This work was supported by Hospital and Homes of St. Giles.

REFERENCES

1. Britton WJ, Lockwood DN. Leprosy. *Lancet* (2004) 363(9416):1209–19. doi:10.1016/S0140-6736(04)15952-7
2. Ridley DS, Jopling WH. Classification of leprosy according to immunity. A five-group system. *Int J Lepr Other Mycobact Dis* (1966) 34(3):255–73.
3. Turk JL, Waters MF. Cell-mediated immunity in patients with leprosy. *Lancet* (1969) 2(7614):243–6. doi:10.1016/S0140-6736(69)90009-9
4. Moran CJ, Ryder G, Turk JL, Waters MF. Evidence for circulating immune complexes in lepromatous leprosy. *Lancet* (1972) 2(7777):572–3. doi:10.1016/S0140-6736(72)91962-9
5. Scollard DM, Smith T, Bhoopat L, Theetranont C, Rangdaeng S, Morens DM. Epidemiologic characteristics of leprosy reactions. *Int J Lepr Other Mycobact Dis* (1994) 62(4):559–67.
6. Walker SL, Balagon M, Darlong J, Doni SN, Hagge DA, Halwai V, et al. ENLIST 1: an international multi-centre cross-sectional study of the clinical features of erythema nodosum leprosum. *PLoS Negl Trop Dis* (2015) 9(9):e0004065. doi:10.1371/journal.pntd.0004065
7. Mabalay MC, Helwig EB, Tolentino JG, Binford CH. The histopathology and histochemistry of erythema nodosum leprosum. *Int J Lepr* (1965) 33:28–49.
8. Anthony J, Vaidya MC, Dasgupta A. Ultrastructure of skin in erythema nodosum leprosum. *Cytobios* (1983) 36(141):17–23.
9. Murphy GF, Sanchez NP, Flynn TC, Sanchez JL, Mihm MC Jr, Soter NA. Erythema nodosum leprosum: nature and extent of the cutaneous microvascular alterations. *J Am Acad Dermatol* (1986) 14(1):59–69. doi:10.1016/S0190-9622(86)70008-X
10. Sehgal VN, Gautam RK, Koranne RV, Beohar PC. The histopathology of type I (lepra) and type II (ENL) reactions in leprosy. *Indian J Lepr* (1986) 58(2):240–3.
11. Adhe V, Dongre A, Khopkar U. A retrospective analysis of histopathology of 64 cases of lepra reactions. *Indian J Dermatol* (2012) 57(2):114–7. doi:10.4103/0019-5154.94278
12. Sarita S, Muhammed K, Najeeba R, Rajan GN, Anza K, Binitha MP, et al. A study on histological features of lepra reactions in patients attending the Dermatology Department of the Government Medical College, Calicut, Kerala, India. *Lepr Rev* (2013) 84(1):51–64.
13. Job CK, Gude S, Macaden VP. Erythema nodosum leprosum. A clinicopathologic study. *Int J Lepr* (1964) 32:177–84.
14. Pocaterra L, Jain S, Reddy R, Muzaffarullah S, Torres O, Suneetha S, et al. Clinical course of erythema nodosum leprosum: an 11-year cohort study in Hyderabad, India. *Am J Trop Med Hyg* (2006) 74(5):868–79.
15. Chandler DJ, Hansen KS, Mahato B, Darlong J, John A, Lockwood DN. Household costs of leprosy reactions (ENL) in rural India. *PLoS Negl Trop Dis* (2015) 9(1):e0003431. doi:10.1371/journal.pntd.0003431
16. Walker SL, Lebas E, Doni SN, Lockwood DN, Lambert SM. The mortality associated with erythema nodosum leprosum in ethiopia: a retrospective hospital-based study. *PLoS Negl Trop Dis* (2014) 8(3):e2690. doi:10.1371/journal.pntd.0002690
17. Shamseer L, Moher D, Clarke M, Ghersi D, Liberati A, Petticrew M, et al. Preferred reporting items for systematic review and meta-analysis protocols (PRISMA-P) 2015: elaboration and explanation. *BMJ* (2015) 349:g7647. doi:10.1136/bmj.g7647
18. Kruger P, Saffarzadeh M, Weber AN, Rieber N, Radsak M, von Bernuth H, et al. Neutrophils: between host defence, immune modulation, and tissue injury. *PLoS Pathog* (2015) 11(3):e1004651. doi:10.1371/journal.ppat.1004651
19. Leliefeld PH, Koenderman L, Pillay J. How neutrophils shape adaptive immune responses. *Front Immunol* (2015) 6:471. doi:10.3389/fimmu.2015.00471
20. Abalos RM, Tolentino JG, Bustillo CC. Histochemical study of erythema nodosum leprosum (ENL) lesions. *Int J Lepr Other Mycobact Dis* (1974) 42(4):385–91.
21. Hussain R, Lucas SB, Kifayet A, Jamil S, Raynes J, Uqaili Z, et al. Clinical and histological discrepancies in diagnosis of ENL reactions classified by assessment of acute phase proteins SAA and CRP. *Int J Lepr Other Mycobact Dis* (1995) 63(2):222–30.
22. Waters MF, Turk JL, Wemambu SN. Mechanisms of reactions in leprosy. *Int J Lepr Other Mycobact Dis* (1971) 39(2):417–28.
23. Pepler WJ, Kooij R, Marshall J. The histopathology of acute panniculitis nodosa leprosa (erythema nodosum leprosum). *Int J Lepr* (1955) 23(1):53–60.
24. Lee DJ, Li H, Ochoa MT, Tanaka M, Carbone RJ, Damoiseaux R, et al. Integrated pathways for neutrophil recruitment and inflammation in leprosy. *J Infect Dis* (2010) 201(4):558–69. doi:10.1086/650318
25. Schmitz V, Prata RB, Barbosa MG, Mendes MA, Brandao SS, Amadeu TP, et al. Expression of CD64 on circulating neutrophils favoring systemic inflammatory status in erythema nodosum leprosum. *PLoS Negl Trop Dis* (2016) 10(8):e0004955. doi:10.1371/journal.pntd.0004955
26. Hogg N. The structure and function of Fc receptors. *Immunol Today* (1988) 9(7–8):185–7. doi:10.1016/0167-5699(88)91206-6
27. Herra CM, Keane CT, Whelan A. Increased expression of Fc gamma receptors on neutrophils and monocytes may reflect ongoing bacterial infection. *J Med Microbiol* (1996) 44(2):135–40. doi:10.1099/00222615-44-2-135
28. Song SH, Kim HK, Park MH, Cho HI. Neutrophil CD64 expression is associated with severity and prognosis of disseminated intravascular coagulation. *Thromb Res* (2008) 121(4):499–507. doi:10.1016/j.thromres.2007.05.013
29. Buckle AM, Hogg N. The effect of IFN-gamma and colony-stimulating factors on the expression of neutrophil cell membrane receptors. *J Immunol* (1989) 143(7):2295–301.
30. Voorend CG, Post EB. A systematic review on the epidemiological data of erythema nodosum leprosum, a type 2 leprosy reaction. *PLoS Negl Trop Dis* (2013) 7(10):e2440. doi:10.1371/journal.pntd.0002440

31. Goihman-Yahr M, Rodriguez-Ochoa G, Aranzazu N, Convit J. Polymorphonuclear activation in leprosy. I. Spontaneous and endotoxin-stimulated reduction of nitroblue tetrazolium: effects of serum and plasma on endotoxin-induced activation. *Clin Exp Immunol* (1975) 20(2): 257–64.
32. Sher R, Anderson R, Glover A, Wadee AA. Polymorphonuclear cell function in the various polar types of leprosy and erythema nodosum leprosum. *Infect Immun* (1978) 21(3):959–65.
33. Oliveira RB, Moraes MO, Oliveira EB, Sarno EN, Nery JA, Sampaio EP. Neutrophils isolated from leprosy patients release TNF-alpha and exhibit accelerated apoptosis in vitro. *J Leukoc Biol* (1999) 65(3):364–71.
34. Wiggins RC, Cochrane CG. Immune-complex-mediated biologic effects. *N Engl J Med* (1981) 304(9):518–20. doi:10.1056/NEJM198102263040904
35. Hoiby N, Doring G, Schiotz PO. The role of immune complexes in the pathogenesis of bacterial infections. *Annu Rev Microbiol* (1986) 40:29–53. doi:10.1146/annurev.mi.40.100186.000333
36. Schifferli JA, Ng YC, Peters DK. The role of complement and its receptor in the elimination of immune complexes. *N Engl J Med* (1986) 315(8):488–95. doi:10.1056/NEJM198608213150805
37. Wemambu SN, Turk JL, Waters MF, Rees RJ. Erythema nodosum leprosum: a clinical manifestation of the arthus phenomenon. *Lancet* (1969) 2(7627):933–5. doi:10.1016/S0140-6736(69)90592-3
38. Anthony J, Vaidya MC, Dasgupta A. Immunological methods employed in an attempt to induce erythema nodosum leprosum (ENL) in mice. *Lepr India* (1978) 50(3):356–62.
39. Ridley MJ, Ridley DS. The immunopathology of erythema nodosum leprosum: the role of extravascular complexes. *Lepr Rev* (1983) 54(2): 95–107.
40. Andreoli A, Brett SJ, Draper P, Payne SN, Rook GA. Changes in circulating antibody levels to the major phenolic glycolipid during erythema nodosum leprosum in leprosy patients. *Int J Lepr Other Mycobact Dis* (1985) 53(2):211–7.
41. Chakrabarty AK, Maire M, Saha K, Lambert PH. Identification of components of IC purified from human sera. II. Demonstration of mycobacterial antigens in immune complexes isolated from sera of lepromatous patients. *Clin Exp Immunol* (1983) 51(2):225–31.
42. Furukawa F, Ozaki M, Imamura S, Yoshida H, Pinrat A, Hamashima Y. Associations of circulating immune complexes, clinical activity, and bacterial index in Japanese patients with leprosy. *Arch Dermatol Res* (1982) 274(1–2):185–8. doi:10.1007/BF00510372
43. Rojas-Espinosa O, Mendez-Navarrete I, Estrada-Parra S. Presence of C1q-reactive immune complexes in patients with leprosy. *Clin Exp Immunol* (1972) 12(2):215–23.
44. Wager O, Penttinen K, Almeida JD, Opromolla DV, Godal T, Kronvall G. Circulating complexes in leprosy studied by the platelet aggregation test. The platelet aggregation test and its relation to the Rubino test and other sero-immunological parameters in 135 patients with leprosy. *Clin Exp Immunol* (1978) 34(3):326–37.
45. Geniteau M, Adam C, Verroust P, Pasticier A, Saimot G, Coulaud JP, et al. [Immune complexes and complement in leprosy (author's transl)]. *Nouv Presse Med* (1981) 10(45):3697–700.
46. Bjorvatn B, Barnetson RS, Kronvall G, Zubler RH, Lambert PH. Immune complexes and complement hypercatabolism in patients with leprosy. *Clin Exp Immunol* (1976) 26(3):388–96.
47. Jayapal N, Shanmugasundaram N, Thomas PA, Valli PR, Thyagarajan SP, Subramanian S. A simple method to quantitate circulating immune complexes in different diseases. *Indian J Pathol Microbiol* (1989) 32(1):33–9.
48. Ramanathan VD, Sharma P, Ramu G, Sengupta U. Reduced complement-mediated immune complex solubilization in leprosy patients. *Clin Exp Immunol* (1985) 60(3):553–8.
49. Ramanathan VD, Tyagi P, Ramanathan U, Katoch K, Sengupta U, Ramu G. Persistent reduced solubilization of immune complexes in lepromatous leprosy patients with reactions. *Int J Lepr Other Mycobact Dis* (1991) 59(1):5–11.
50. Rojas RE, Demichelis SO, Sarno EN, Segal-Eiras A. IgM anti-phenolic glycolipid I and IgG anti-10-kDa heat shock protein antibodies in sera and immune complexes isolated from leprosy patients with or without erythema nodosum leprosum and contacts. *FEMS Immunol Med Microbiol* (1997) 19(1):65–74. doi:10.1111/j.1574-695X.1997.tb01073.x
51. Tung KS, Kim B, Bjorvatn B, Kronvall G, McLaren LC, Williams RC Jr. Discrepancy between Clq deviation and Raji cell tests in detection of circulating immune complexes in patients with leprosy. *J Infect Dis* (1977) 136(2):216–21. doi:10.1093/infdis/136.2.216
52. Tyagi P, Ramanathan VD, Girdhar BK, Katoch K, Bhatia AS, Sengupta U. Activation of complement by circulating immune complexes isolated from leprosy patients. *Int J Lepr Other Mycobact Dis* (1990) 58(1):31–8.
53. Tyagi P, Patil SA, Girdhar BK, Katoch K, Sengupta U. Suppressive effect of circulating immune complexes from leprosy patients on the lymphocyte proliferation induced by *M. leprae* antigens in healthy responders. *Int J Lepr Other Mycobact Dis* (1992) 60(4):562–9.
54. Valentijn RM, Faber WR, Lai AFRF, Chan Pin Jie JC, Daha MR, van Es LA. Immune complexes in leprosy patients from an endemic and a nonendemic area and a longitudinal study of the relationship between complement breakdown products and the clinical activity of erythema nodosum leprosum. *Clin Immunol Immunopathol* (1982) 22(2):194–202. doi:10.1016/0090-1229(82)90037-X
55. Penttinen K, Myllyla G, Vaheri A, Vesikari T, Kaariainen L. The platelet aggregation test (PA) as an immunological method in virology. *Prog Immunobiol Stand* (1970) 4:672–5.
56. Yanase K, Imamura S. Detection of circulating immune complexes in some skin diseases by platelet aggregation test. *Br J Dermatol* (1979) 100(2):227–8. doi:10.1111/j.1365-2133.1979.tb05567.x
57. Rojas RE, Segal-Eiras A. Characterization of circulating immune complexes in leprosy patients and their correlation with specific antibodies against *Mycobacterium leprae*. *Clin Exp Dermatol* (1997) 22(5):223–9. doi:10.1046/j.1365-2230.1997.2620675.x
58. Dupnik KM, Bair TB, Maia AO, Amorim FM, Costa MR, Keesen TS, et al. Transcriptional changes that characterize the immune reactions of leprosy. *J Infect Dis* (2015) 211(10):1658–76. doi:10.1093/infdis/jiu612
59. de Azevedo MP, de Melo PH. A comparative study of the complementary activity of serum in the polar forms of leprosy and in the leprosy reaction. *Int J Lepr Other Mycobact Dis* (1966) 34(1):34–8.
60. Anthony J, Vaidya MC, Dasgupta A. Immunoglobulin deposits in erythema nodosum leprosum (ENL). *Hansen Int* (1978) 3(1):12–7.
61. Lange K, Wasserman E, Slobody LB. The significance of serum complement levels for the diagnosis and prognosis of acute and subacute glomerulonephritis and lupus erythematosus disseminatus. *Ann Intern Med* (1960) 53:636–46. doi:10.7326/0003-4819-53-4-636
62. Lewis EJ, Carpenter CB, Schur PH. Serum complement component levels in human glomerulonephritis. *Ann Intern Med* (1971) 75(4):555–60. doi:10.7326/0003-4819-75-4-555
63. Ohi H, Tamano M. Decreased apolipoprotein levels are associated with decreased complement levels in acute glomerulonephritis. *Nephron* (2001) 88(4):389–90. doi:10.1159/000046028
64. Baatrup G, Petersen I, Kappelgaard E, Jepsen HH, Svehag SE. Complement-mediated solubilization of immune complexes. Solubilization inhibition and complement factor levels in SLE patients. *Clin Exp Immunol* (1984) 55(2):313–8.
65. Grevink ME, Horst G, Limburg PC, Kallenberg CG, Bijl M. Levels of complement in sera from inactive SLE patients, although decreased, do not influence in vitro uptake of apoptotic cells. *J Autoimmun* (2005) 24(4):329–36. doi:10.1016/j.jaut.2005.03.004
66. Schifferli JA, Morris SM, Dash A, Peters DK. Complement-mediated solubilization in patients with systemic lupus erythematosus, nephritis or vasculitis. *Clin Exp Immunol* (1981) 46(3):557–64.
67. de Messias IJ, Santamaria J, Brenden M, Reis A, Mauff G. Association of C4B deficiency (C4B*Q0) with erythema nodosum in leprosy. *Clin Exp Immunol* (1993) 92(2):284–7. doi:10.1111/j.1365-2249.1993.tb03393.x
68. Hauptmann G, Tappeiner G, Schifferli JA. Inherited deficiency of the fourth component of human complement. *Immunodefic Rev* (1988) 1(1):3–22.
69. Gelber RH, Drutz DJ, Epstein WV, Fasal P. Clinical correlates of C1Q-precipitating substances in the sera of patients with leprosy. *Am J Trop Med Hyg* (1974) 23(3):471–5.
70. Izumi S, Sugiyama K, Matsumoto Y, Nagai T. Numerical changes in T cell subsets (T gamma and T mu) in leprosy patients. *Microbiol Immunol* (1980) 24(8):733–40. doi:10.1111/j.1348-0421.1980.tb02874.x

71. Harikrishan S, Balakrishnan S, Bhatia VN. Serum immunoglobulin profile and C3 level in lepromatous leprosy patients. *Lepr India* (1982) 54(3):454–60.
72. Saha K, Chakraborty AK, Sharma V, Sehgal VN. An appraisal of third complement component (C3) and breakdown product (C3d) in erythema nodosum leprosum (ENL). *Lepr Rev* (1982) 53(4):253–60.
73. Mshana RN, Humber DP, Belehu A, Harboe M. Immunohistological studies of skin biopsies from patients with lepromatous leprosy. *J Clin Immunol* (1983) 3(1):22–9. doi:10.1007/BF00919135
74. Ramanathan VD, Parkash O, Ramu G, Parker D, Curtis J, Sengupta U, et al. Isolation and analysis of circulating immune complexes in leprosy. *Clin Immunol Immunopathol* (1984) 32(3):261–8. doi:10.1016/0090-1229(84)90270-8
75. Saha K, Chakrabarty AK, Sharma VK, Sehgal VN. Polyethylene glycol precipitates in serum during and after erythema nodosum leprosum – study of their composition and anticomplementary activity. *Int J Lepr Other Mycobact Dis* (1984) 52(1):44–8.
76. Sehgal VN, Gautam RK, Sharma VK. Immunoprofile of reactions in leprosy. *Int J Dermatol* (1986) 25(4):240–4. doi:10.1111/j.1365-4362.1986.tb02233.x
77. Chakrabarty AK, Kashyap A, Sehgal VN, Saha K. Solubilization of preformed immune complexes in sera of patients with type 1 and type 2 lepra reactions. *Int J Lepr Other Mycobact Dis* (1988) 56(4):559–65.
78. Rao TD, Rao PR. Serum immune complexes in erythema nodosum leprosum reactions of leprosy. *Indian J Lepr* (1988) 60(2):189–95.
79. Sehgal VN, Sharma V, Sharma VK. The effect of anti-reactional drugs on complement components in the type II, erythema nodosum leprosum, reaction. *Br J Dermatol* (1988) 119(2):255–8. doi:10.1111/j.1365-2133.1988.tb03209.x
80. Dias AA, Silva CO, Santos JP, Batista-Silva LR, Acosta CC, Fontes AN, et al. DNA sensing via TLR-9 constitutes a major innate immunity pathway activated during erythema nodosum leprosum. *J Immunol* (2016) 197(5):1905–13. doi:10.4049/jimmunol.1600042
81. Sehgal VN, Sharma V, Sharma VK. Comprehensive evaluation of complement components in the course of type I (Lepra) and type II (ENL) reactions. *Int J Dermatol* (1989) 28(1):32–5. doi:10.1111/j.1365-4362.1989.tb01306.x
82. Scollard DM, Bhoopat L, Kestens L, Vanham G, Douglas JT, Moad J. Immune complexes and antibody levels in blisters over human leprosy skin lesions with or without erythema nodosum leprosum. *Clin Immunol Immunopathol* (1992) 63(3):230–6. doi:10.1016/0090-1229(92)90227-F
83. Kim HJ, Cantor H. CD4 T-cell subsets and tumor immunity: the helpful and the not-so-helpful. *Cancer Immunol Res* (2014) 2(2):91–8. doi:10.1158/2326-6066.CIR-13-0216
84. Lim SD, Kiszkiss DF, Jacobson RR, Choi YS, Good RA. Thymus-dependent lymphocytes of peripheral blood in leprosy patients. *Infect Immun* (1974) 9(2):394–9.
85. Rao TD, Rao PR. Enhanced cell-mediated immune responses in erythema nodosum leprosum reactions of leprosy. *Int J Lepr Other Mycobact Dis* (1987) 55(1):36–41.
86. Bach MA, Chatenoud L, Wallach D, Phan Dinh Tuy F, Cottenot F. Studies on T cell subsets and functions in leprosy. *Clin Exp Immunol* (1981) 44(3):491–500.
87. Wallach D, Cottenot F, Bach MA. Imbalances in T cell subpopulations in lepromatous leprosy. *Int J Lepr Other Mycobact Dis* (1982) 50(3):282–90.
88. Mshana RN, Haregewoin A, Harboe M, Belehu A. Thymus dependent lymphocytes in leprosy. I. T lymphocyte subpopulations defined by monoclonal antibodies. *Int J Lepr Other Mycobact Dis* (1982) 50(3):291–6.
89. Narayanan RB, Laal S, Sharma AK, Bhutani LK, Nath I. Differences in predominant T cell phenotypes and distribution pattern in reactional lesions of tuberculoid and lepromatous leprosy. *Clin Exp Immunol* (1984) 55(3):623–8.
90. Mshana RN, Haregewoin A, Belehu A. Thymus-dependent lymphocytes in leprosy. II. Effect of chemotherapy on T-lymphocyte subpopulations. *J Clin Immunol* (1982) 2(2):69–74. doi:10.1007/BF00916889
91. Wallach D, Flageul B, Cottenot F, Bach MA. Patients with erythema nodosum leprosum lack T-suppressor cells. *Arch Dermatol* (1985) 121(11):1379. doi:10.1001/archderm.1985.01660110027002
92. Hussain T, Kulshreshtha KK, Yadav VS, Katoch K. CD4+, CD8+, CD3+ cell counts and CD4+/CD8+ ratio among patients with mycobacterial diseases (leprosy, tuberculosis), HIV infections, and normal healthy adults: a comparative analysis of studies in different regions of India. *J Immunoassay Immunochem* (2015) 36(4):420–43. doi:10.1080/15321819.2014.978082
93. Sakane T, Steinberg AD, Green I. Studies of immune functions of patients with systemic lupus erythematosus. I. Dysfunction of suppressor T-cell activity related to impaired generation of, rather than response to, suppressor cells. *Arthritis Rheum* (1978) 21(6):657–64. doi:10.1002/art.1780210608
94. Laal S, Bhutani LK, Nath I. Natural emergence of antigen-reactive T cells in lepromatous leprosy patients during erythema nodosum leprosum. *Infect Immun* (1985) 50(3):887–92.
95. Bullock WE. Leprosy: a model of immunological perturbation in chronic infection. *J Infect Dis* (1978) 137(3):341–54. doi:10.1093/infdis/137.3.341
96. Rea TH, Bakke AC, Parker JW, Modlin RL, Horwitz DA. Peripheral blood T lymphocyte subsets in leprosy. *Int J Lepr Other Mycobact Dis* (1984) 52(3):311–7.
97. Modlin RL, Gebhard JF, Taylor CR, Rea TH. In situ characterization of T lymphocyte subsets in the reactional states of leprosy. *Clin Exp Immunol* (1983) 53(1):17–24.
98. Modlin RL, Hofman FM, Taylor CR, Rea TH. T lymphocyte subsets in the skin lesions of patients with leprosy. *J Am Acad Dermatol* (1983) 8(2):182–9. doi:10.1016/S0190-9622(83)70021-6
99. Modlin RL, Bakke AC, Vaccaro SA, Horwitz DA, Taylor CR, Rea TH. Tissue and blood T-lymphocyte subpopulations in erythema nodosum leprosum. *Arch Dermatol* (1985) 121(2):216–9. doi:10.1001/archderm.121.2.216
100. Modlin RL, Mehra V, Jordan R, Bloom BR, Rea TH. In situ and in vitro characterization of the cellular immune response in erythema nodosum leprosum. *J Immunol* (1986) 136(3):883–6.
101. Shen JY, Hofman FM, Gunter JR, Modlin RL, Rea TH. In situ identification of activated Ta1+ T lymphocytes in human leprosy skin lesions. *Int J Lepr Other Mycobact Dis* (1987) 55(3):494–8.
102. Rea TH, Modlin RL. Immunopathology of leprosy skin lesions. *Semin Dermatol* (1991) 10(3):188–93.
103. Mahaisavariya P, Kulthanan K, Khemngern S, Pinkaew S. Lesional T-cell subset in leprosy and leprosy reaction. *Int J Dermatol* (1999) 38(5):345–7. doi:10.1046/j.1365-4362.1999.00621.x
104. Fehervari Z, Sakaguchi S. CD4+ Tregs and immune control. *J Clin Invest* (2004) 114(9):1209–17. doi:10.1172/JCI23395
105. Saini C, Prasad HK, Rani R, Murtaza A, Misra N, Narayan NP, et al. Lsr 2 of *Mycobacterium leprae* and its synthetic peptides elicit restitution of in vitro T cell responses in erythema nodosum leprosum and reversal reactions in lepromatous leprosy patients. *Clin Vaccine Immunol* (2013) 20(5):673–82. doi:10.1128/CVI.00762-12
106. Saini C, Ramesh V, Nath I. CD4+ Th17 cells discriminate clinical types and constitute a third subset of non Th1, Non Th2 T cells in human leprosy. *PLoS Negl Trop Dis* (2013) 7(7):e2338. doi:10.1371/journal.pntd.0002338
107. Attia EA, Abdallah M, Saad AA, Afifi A, El Tabbakh A, El-Shennawy D, et al. Circulating CD4+ CD25 high FoxP3+ T cells vary in different clinical forms of leprosy. *Int J Dermatol* (2010) 49(10):1152–8. doi:10.1111/j.1365-4632.2010.04535.x
108. Boer MC, Joosten SA, Ottenhoff TH. Regulatory T-cells at the interface between human host and pathogens in infectious diseases and vaccination. *Front Immunol* (2015) 6:217. doi:10.3389/fimmu.2015.00217
109. Abdallah M, Attia EA, Saad AA, El-Khateeb EA, Lotfi RA, Abdallah M, et al. Serum Th1/Th2 and macrophage lineage cytokines in leprosy; correlation with circulating CD4(+) CD25(high) FoxP3(+) T-regs cells. *Exp Dermatol* (2014) 23(10):742–7. doi:10.1111/exd.12529
110. Attia EA, Abdallah M, El-Khateeb E, Saad AA, Lotfi RA, Abdallah M, et al. Serum Th17 cytokines in leprosy: correlation with circulating CD4(+) CD25 (high)FoxP3 (+) T-regs cells, as well as down regulatory cytokines. *Arch Dermatol Res* (2014) 306(9):793–801. doi:10.1007/s00403-014-1486-2
111. Santegoets SJ, Dijkgraaf EM, Battaglia A, Beckhove P, Britten CM, Gallimore A, et al. Monitoring regulatory T cells in clinical samples: consensus on an essential marker set and gating strategy for regulatory T cell analysis by flow cytometry. *Cancer Immunol Immunother* (2015) 64(10):1271–86. doi:10.1007/s00262-015-1729-x

112. Brusko T, Wasserfall C, McGrail K, Schatz R, Viener HL, Schatz D, et al. No alterations in the frequency of FOXP3+ regulatory T-cells in type 1 diabetes. *Diabetes* (2007) 56(3):604–12. doi:10.2337/db06-1248
113. Saini C, Siddiqui A, Ramesh V, Nath I. Leprosy reactions show increased Th17 cell activity and reduced FOXP3+ Tregs with concomitant decrease in TGF-beta and increase in IL-6. *PLoS Negl Trop Dis* (2016) 10(4):e0004592. doi:10.1371/journal.pntd.0004592
114. Ye ZJ, Zhou Q, Du RH, Li X, Huang B, Shi HZ. Imbalance of Th17 cells and regulatory T cells in tuberculous pleural effusion. *Clin Vaccine Immunol* (2011) 18(10):1608–15. doi:10.1128/CVI.05214-11
115. Haslett PA, Roche P, Butlin CR, Macdonald M, Shrestha N, Manandhar R, et al. Effective treatment of erythema nodosum leprosum with thalidomide is associated with immune stimulation. *J Infect Dis* (2005) 192(12):2045–53. doi:10.1086/498216
116. Massone C, Nunzi E, Ribeiro-Rodrigues R, Talhari C, Talhari S, Schettini AP, et al. T regulatory cells and plasmocytoid dentritic cells in hansen disease: a new insight into pathogenesis? *Am J Dermatopathol* (2010) 32(3):251–6. doi:10.1097/DAD.0b013e3181b7fc56
117. Miyao T, Floess S, Setoguchi R, Luche H, Fehling HJ, Waldmann H, et al. Plasticity of Foxp3(+) T cells reflects promiscuous Foxp3 expression in conventional T cells but not reprogramming of regulatory T cells. *Immunity* (2012) 36(2):262–75. doi:10.1016/j.immuni.2011.12.012
118. Rea TH, Levan NE. Variations in dinitrochlorobenzene responsivity in untreated leprosy: evidence of a beneficial role for anergy. *Int J Lepr Other Mycobact Dis* (1980) 48(2):120–5.
119. Anders EM, McAdam KP, Anders RF. Cell-mediated immunity in amyloidosis secondary to lepromatous leprosy. *Clin Exp Immunol* (1977) 27(1):111–7.
120. Dubey GK, Joglekar VK, Hardas UD, Chaubey BS. A study of cell mediated immunity in leprosy. *Lepr India* (1981) 53(2):197–203.
121. Bach MA, Hoffenbach A, Lagrange PH, Wallach D, Cottenot F. Mechanisms of T-cell unresponsiveness in leprosy. *Ann Immunol* (1983) 134D(1):75–84.
122. Sasiain MC, Ruibal Ares B, Balina LM, Valdez R, Bachmann AE. ConA-induced suppressor cells in lepromatous leprosy patients during and after erythema nodosum leprosum. *Int J Lepr Other Mycobact Dis* (1983) 51(3):321–7.
123. Rao TD, Rao PR. Tr, T mu and B lymphocytes in erythema nodosum leprosum reactions of leprosy. *Indian J Lepr* (1986) 58(4):601–8.
124. Bottasso O, Puig N, Amerio N, Morini JC. [Study of T lymphocyte subpopulations in patients with leprosy, using incubation with theophylline]. *Med Cutan Ibero Lat Am* (1988) 16(5):397–401.
125. Rasheed FN, Locniskar M, McCloskey DJ, Hasan RS, Chiang TJ, Rose P, et al. Serum lymphocytotoxic activity in leprosy. *Clin Exp Immunol* (1989) 76(3):391–7.
126. Sasiain MD, de la Barrera S, Valdez R, Balina LM. Reduced suppressor cell response to *Mycobacterium leprae* in lepromatous leprosy. *Infect Immun* (1989) 57(3):951–6.
127. Bhoopat L, Scollard DM, Theetranont C, Chiewchanvit S, Nelson DL, Utaipat U. Studies of human leprosy lesions in situ using suction-induced blisters: cell changes with IgM antibody to PGL-1 and interleukin-2 receptor in clinical subgroups of erythema nodosum leprosum. *Asian Pac J Allergy Immunol* (1991) 9(2):107–19.
128. Foss NT, de Oliveira EB, Silva CL. Correlation between TNF production, increase of plasma C-reactive protein level and suppression of T lymphocyte response to concanavalin A during erythema nodosum leprosum. *Int J Lepr Other Mycobact Dis* (1993) 61(2):218–26.
129. Santos DO, Suffys PN, Moreira AL, Bonifacio K, Salgado JL, Esquenazi D, et al. Evaluation of chemiluminescence, procoagulant activity and antigen presentation by monocytes from lepromatous leprosy patients with or without reactional episodes. *Lepr Rev* (1994) 65(2):88–99.
130. de la Barrera S, Fink S, Finiasz M, Minnucci F, Valdez R, Balina LM, et al. Lack of cytotoxic activity against *Mycobacterium leprae* 65-kD heat shock protein (hsp) in multibacillary leprosy patients. *Clin Exp Immunol* (1995) 99(1):90–7. doi:10.1111/j.1365-2249.1995.tb03477.x
131. Vieira LM, Sampaio EP, Nery JA, Duppre NC, Albuquerque EC, Scheinberg MA, et al. Immunological status of ENL (erythema nodosum leprosum) patients: its relationship to bacterial load and levels of circulating IL-2R. *Rev Inst Med Trop Sao Paulo* (1996) 38(2):103–11. doi:10.1590/S0036-46651996000200004
132. Tadesse A, Taye E, Sandoval F, Shannon EJ. Thalidomide does not modify the ability of cells in leprosy patients to incorporate [3H]-thymidine when incubated with *M. leprae* antigens. *Lepr Rev* (2003) 74(3):206–14.
133. Mohanty KK, Joshi B, Katoch K, Sengupta U. Leprosy reactions: humoral and cellular immune responses to *M. leprae*, 65kDa, 28kDa, and 18 kDa antigens. *Int J Lepr Other Mycobact Dis* (2004) 72(2):149–58. doi:10.1489/1544-581X(2004)072<0149:LRHACI>2.0.CO;2
134. Villahermosa LG, Fajardo TT Jr, Abalos RM, Balagon MV, Tan EV, Cellona RV, et al. A randomized, double-blind, double-dummy, controlled dose comparison of thalidomide for treatment of erythema nodosum leprosum. *Am J Trop Med Hyg* (2005) 72(5):518–26.
135. Rada E, Aranzazu N, Rodriguez V, Borges R, Convit J. [Serological and cellular reactivity to mycobacterial proteins in Hansen's disease]. *Invest Clin* (2010) 51(3):325–40.
136. Saini C, Prasad HK, Rani R, Murtaza A, Misra N, Shanker Narayan NP, et al. Lsr2 of *Mycobacterium leprae* and its synthetic peptides elicit restitution of T cell responses in erythema nodosum leprosum and reversal reactions in patients with lepromatous leprosy. *Clin Vaccine Immunol* (2013) 20(5):673–82. doi:10.1128/CVI.00762-12
137. Parente JN, Talhari C, Schettini AP, Massone C. T regulatory cells (TREG) (TCD4+CD25+FOXP3+) distribution in the different clinical forms of leprosy and reactional states. *An Bras Dermatol* (2015) 90(1):41–7. doi:10.1590/abd1806-4841.20153311
138. Sarno EN, Grau GE, Vieira LM, Nery JA. Serum levels of tumour necrosis factor-alpha and interleukin-1 beta during leprosy reactional states. *Clin Exp Immunol* (1991) 84(1):103–8.
139. Sampaio EP, Kaplan G, Miranda A, Nery JA, Miguel CP, Viana SM, et al. The influence of thalidomide on the clinical and immunologic manifestation of erythema nodosum leprosum. *J Infect Dis* (1993) 168(2):408–14. doi:10.1093/infdis/168.2.408
140. Parida SK, Grau GE, Zaheer SA, Mukherjee R. Serum tumor necrosis factor and interleukin 1 in leprosy and during lepra reactions. *Clin Immunol Immunopathol* (1992) 63(1):23–7. doi:10.1016/0090-1229(92)90088-6
141. Memon RA, Kifayet A, Shahid F, Lateef A, Chiang J, Hussain R. Low serum HDL-cholesterol is associated with raised tumor necrosis factor-alpha during ENL reactions. *Int J Lepr Other Mycobact Dis* (1997) 65(1):1–11.
142. Partida-Sanchez S, Favila-Castillo L, Pedraza-Sanchez S, Gomez-Melgar M, Saul A, Estrada-Parra S, et al. IgG antibody subclasses, tumor necrosis factor and IFN-gamma levels in patients with type II lepra reaction on thalidomide treatment. *Int Arch Allergy Immunol* (1998) 116(1):60–6. doi:10.1159/000023926
143. Iyer A, Hatta M, Usman R, Luiten S, Oskam L, Faber W, et al. Serum levels of interferon-gamma, tumour necrosis factor-alpha, soluble interleukin-6R and soluble cell activation markers for monitoring response to treatment of leprosy reactions. *Clin Exp Immunol* (2007) 150(2):210–6. doi:10.1111/j.1365-2249.2007.03485.x
144. Madan NK, Agarwal K, Chander R. Serum cytokine profile in leprosy and its correlation with clinico-histopathological profile. *Lepr Rev* (2011) 82(4):371–82.
145. Rodrigues LS, Hacker MA, Illarramendi X, Pinheiro MF, Nery JA, Sarno EN, et al. Circulating levels of insulin-like growth factor-I (IGF-I) correlate with disease status in leprosy. *BMC Infect Dis* (2011) 11:339. doi:10.1186/1471-2334-11-339
146. Bhattacharya SN, Chattopadhaya D, Saha K. Tumor necrosis factor: status in reactions in leprosy before and after treatment. *Int J Dermatol* (1993) 32(6):436–9. doi:10.1111/j.1365-4362.1993.tb02816.x
147. Sampaio EP, Moraes MO, Nery JA, Santos AR, Matos HC, Sarno EN. Pentoxifylline decreases in vivo and in vitro tumour necrosis factor-alpha (TNF-alpha) production in lepromatous leprosy patients with erythema nodosum leprosum (ENL). *Clin Exp Immunol* (1998) 111(2):300–8. doi:10.1046/j.1365-2249.1998.00510.x
148. Moubasher AD, Kamel NA, Zedan H, Raheem DD. Cytokines in leprosy, I. Serum cytokine profile in leprosy. *Int J Dermatol* (1998) 37(10):733–40. doi:10.1046/j.1365-4362.1998.00381.x
149. Jadhav R, Suneetha L, Kamble R, Shinde V, Devi K, Chaduvula MV, et al. Analysis of antibody and cytokine markers for leprosy nerve damage and reactions in the INFIR cohort in India. *PLoS Negl Trop Dis* (2011) 5(3):e977. doi:10.1371/journal.pntd.0000977

150. Sampaio EP, Moreira AL, Sarno EN, Malta AM, Kaplan G. Prolonged treatment with recombinant interferon gamma induces erythema nodosum leprosum in lepromatous leprosy patients. *J Exp Med* (1992) 175(6):1729–37. doi:10.1084/jem.175.6.1729
151. Stefani MM, Guerra JG, Sousa AL, Costa MB, Oliveira ML, Martelli CT, et al. Potential plasma markers of Type 1 and Type 2 leprosy reactions: a preliminary report. *BMC Infect Dis* (2009) 9:75. doi:10.1186/1471-2334-9-75
152. Wolkenstein P, Latarjet J, Roujeau JC, Duguet C, Boudeau S, Vaillant L, et al. Randomised comparison of thalidomide versus placebo in toxic epidermal necrolysis. *Lancet* (1998) 352(9140):1586–9. doi:10.1016/S0140-6736(98)02197-7
153. Jacobson JM, Greenspan JS, Spritzler J, Ketter N, Fahey JL, Jackson JB, et al. Thalidomide for the treatment of oral aphthous ulcers in patients with human immunodeficiency virus infection. National Institute of Allergy and Infectious Diseases AIDS Clinical Trials Group. *N Engl J Med* (1997) 336(21):1487–93. doi:10.1056/NEJM199705223362103
154. Haslett PA, Hanekom WA, Muller G, Kaplan G. Thalidomide and a thalidomide analogue drug costimulate virus-specific CD8+ T cells in vitro. *J Infect Dis* (2003) 187(6):946–55. doi:10.1086/368126
155. Barnes PF, Chatterjee D, Brennan PJ, Rea TH, Modlin RL. Tumor necrosis factor production in patients with leprosy. *Infect Immun* (1992) 60(4):1441–6.
156. Santos DO, Suffys PN, Bonifacio K, Marques MA, Sarno EN. In vitro tumor necrosis factor production by mononuclear cells from lepromatous leprosy patients and from patients with erythema nodosum leprosum. *Clin Immunol Immunopathol* (1993) 67(3 Pt 1):199–203. doi:10.1006/clin.1993.1065
157. Sampaio EP, Oliveira RB, Warwick-Davies J, Neto RB, Griffin GE, Shattock RJ. T cell-monocyte contact enhances tumor necrosis factor-alpha production in response to *Mycobacterium leprae*. *J Infect Dis* (2000) 182(5):1463–72. doi:10.1086/315902
158. Faber WR, Jensema AJ, Goldschmidt WF. Treatment of recurrent erythema nodosum leprosum with infliximab. *N Engl J Med* (2006) 355(7):739. doi:10.1056/NEJMc052955
159. Ramien ML, Wong A, Keystone JS. Severe refractory erythema nodosum leprosum successfully treated with the tumor necrosis factor inhibitor etanercept. *Clin Infect Dis* (2011) 52(5):e133–5. doi:10.1093/cid/ciq213
160. Chowdhry S, Shukla A, D'Souza P, Dhali T, Jaiswal P. Treatment of severe refractory erythema nodosum leprosum with tumor necrosis factor inhibitor etanercept. *Int J Mycobacteriol* (2016) 5(2):223–5. doi:10.1016/j.ijmyco.2016.02.002
161. Moraes MO, Sarno EN, Almeida AS, Saraiva BC, Nery JA, Martins RC, et al. Cytokine mRNA expression in leprosy: a possible role for interferon-gamma and interleukin-12 in reactions (RR and ENL). *Scand J Immunol* (1999) 50(5):541–9. doi:10.1046/j.1365-3083.1999.00622.x
162. Nath I, Vemuri N, Reddi AL, Bharadwaj M, Brooks P, Colston MJ, et al. Dysregulation of IL-4 expression in lepromatous leprosy patients with and without erythema nodosum leprosum. *Lepr Rev* (2000) 71(Suppl):S130–7.
163. Nath I, Vemuri N, Reddi AL, Jain S, Brooks P, Colston MJ, et al. The effect of antigen presenting cells on the cytokine profiles of stable and reactional lepromatous leprosy patients. *Immunol Lett* (2000) 75(1):69–76. doi:10.1016/S0165-2478(00)00271-6
164. Moraes MO, Sarno EN, Teles RM, Almeida AS, Saraiva BC, Nery JA, et al. Anti-inflammatory drugs block cytokine mRNA accumulation in the skin and improve the clinical condition of reactional leprosy patients. *J Invest Dermatol* (2000) 115(6):935–41. doi:10.1046/j.1523-1747.2000.00158.x
165. Moubasher AD, Kamel NA, Zedan H, Raheem DD. Cytokines in leprosy, II. Effect of treatment on serum cytokines in leprosy. *Int J Dermatol* (1998) 37(10):741–6. doi:10.1046/j.1365-4362.1998.00381.x
166. Sallam MA, Attia EA, Soliman MS. Assessment of serum level of interleukin-1b and interleukin-12 in leprosy: impact of previous bacillus calmitte guerin vaccination. *Arch Dermatol Res* (2014) 306(2):189–95. doi:10.1007/s00403-013-1411-0
167. Berrington WR, Kunwar CB, Neupane K, van den Eeden SJ, Vary JC Jr, Peterson GJ, et al. Differential dermal expression of CCL17 and CCL18 in tuberculoid and lepromatous leprosy. *PLoS Negl Trop Dis* (2014) 8(11):e3263. doi:10.1371/journal.pntd.0003263
168. Filley E, Andreoli A, Steele J, Waters M, Wagner D, Nelson D, et al. A transient rise in agalactosyl IgG correlating with free interleukin 2 receptors, during episodes of erythema nodosum leprosum. *Clin Exp Immunol* (1989) 76(3):343–7.
169. Belgaumkar VA, Gokhale NR, Mahajan PM, Bharadwaj R, Pandit DP, Deshpande S. Circulating cytokine profiles in leprosy patients. *Lepr Rev* (2007) 78(3):223–30.
170. Sousa AL, Fava VM, Sampaio LH, Martelli CM, Costa MB, Mira MT, et al. Genetic and immunological evidence implicates interleukin 6 as a susceptibility gene for leprosy type 2 reaction. *J Infect Dis* (2012) 205(9):1417–24. doi:10.1093/infdis/jis208
171. Abdallah M, Emam H, Attia E, Hussein J, Mohamed N. Estimation of serum level of interleukin-17 and interleukin-4 in leprosy, towards more understanding of leprosy immunopathogenesis. *Indian J Dermatol Venereol Leprol* (2013) 79(6):772–6. doi:10.4103/0378-6323.120723
172. Sehgal VN, Bhattacharya SN, Shah Y, Sharma VK, Gupta CK. Soluble interleukin-2 receptors: levels in leprosy, and during and after type 1 (lepra) and type 2 (ENL) reactions. *Lepr Rev* (1991) 62(3):262–8.
173. Sullivan L, Sano S, Pirmez C, Salgame P, Mueller C, Hofman F, et al. Expression of adhesion molecules in leprosy lesions. *Infect Immun* (1991) 59(11):4154–60.
174. Goulart IM, Mineo JR, Foss NT. Production of transforming growth factor-beta 1 (TGF-beta1) by blood monocytes from patients with different clinical forms of leprosy. *Clin Exp Immunol* (2000) 122(3):330–4. doi:10.1046/j.1365-2249.2000.01376.x
175. Motta AC, Furini RB, Simao JC, Ferreira MA, Komesu MC, Foss NT. The recurrence of leprosy reactional episodes could be associated with oral chronic infections and expression of serum IL-1, TNF-alpha, IL-6, IFN-gamma and IL-10. *Braz Dent J* (2010) 21(2):158–64. doi:10.1590/S0103-64402010000200012
176. Teles RM, Teles RB, Amadeu TP, Moura DF, Mendonca-Lima L, Ferreira H, et al. High matrix metalloproteinase production correlates with immune activation and leukocyte migration in leprosy reactional lesions. *Infect Immun* (2010) 78(3):1012–21. doi:10.1128/IAI.00896-09
177. Chaitanya S, Lavania M, Turankar RP, Karri SR, Sengupta U. Increased serum circulatory levels of interleukin 17F in type 1 reactions of leprosy. *J Clin Immunol* (2012) 32(6):1415–20. doi:10.1007/s10875-012-9747-3
178. Lockwood DN, Nicholls P, Smith WC, Das L, Barkataki P, van Brakel W, et al. Comparing the clinical and histological diagnosis of leprosy and leprosy reactions in the INFIR cohort of Indian patients with multibacillary leprosy. *PLoS Negl Trop Dis* (2012) 6(6):e1702. doi:10.1371/journal.pntd.0001702
179. Martiniuk F, Giovinazzo J, Tan AU, Shahidullah R, Haslett P, Kaplan G, et al. Lessons of leprosy: the emergence of TH17 cytokines during type II reactions (ENL) is teaching us about T-cell plasticity. *J Drugs Dermatol* (2012) 11(5):626–30.
180. Reichlin M, Pranis RA, Gelber RH, Rees RJ, Taverne J, Turk JL. Correlation of euglobulin immunoglobulin G levels with erythema nodosum leprosum in lepromatous leprosy. *Clin Immunol Immunopathol* (1977) 8(2):335–44. doi:10.1016/0090-1229(77)90123-4
181. Humphres RC, Gelber RH, Krahenbuhl JL. Suppressed natural killer cell activity during episodes of erythema nodosum leprosum in lepromatous leprosy. *Clin Exp Immunol* (1982) 49(3):500–8.
182. Rea TH, Yoshida T. Serum macrophage migration inhibition activity in patients with leprosy. *J Invest Dermatol* (1982) 79(5):336–9. doi:10.1111/1523-1747.ep12500088
183. Miller RA, Harnisch JP, Buchanan TM. Antibodies to mycobacterial arabinomannan in leprosy: correlation with reactional states and variation during treatment. *Int J Lepr Other Mycobact Dis* (1984) 52(2):133–9.
184. Schwerer B, Meeker HC, Sersen G, Levis WR. IgM antibodies against phenolic glycolipid I from *Mycobacterium leprae* in leprosy sera: relationship to bacterial index and erythema nodosum leprosum. *Acta Leprol* (1984) 2(2–4):394–402.
185. Blavy G, Thiam D, Ndoye B, Diakhate L, Millan J. [HLA and leprosy in Dakar: distribution of histocompatibility antigens in leprous patients and their relationship to ENL reactions]. *Acta Leprol* (1986) 4(1):93–9.
186. Levis WR, Meeker HC, Schuller-Levis G, Sersen E, Schwerer B. IgM and IgG antibodies to phenolic glycolipid I from *Mycobacterium leprae* in leprosy: insight into patient monitoring, erythema nodosum leprosum, and bacillary

persistence. *J Invest Dermatol* (1986) 86(5):529–34. doi:10.1111/1523-1747.ep12354963
187. Levis WR, Meeker HC, Schuller-Levis G, Sersen E, Brennan PJ, Fried P. Mycobacterial carbohydrate antigens for serological testing of patients with leprosy. *J Infect Dis* (1987) 156(5):763–9. doi:10.1093/infdis/156.5.763
188. Sehgal VN, Bhattacharya SN, Shah Y, Rao YN, Gupta CK. Lymphocyte adenosine deaminase activity (L-ADA) in leprosy, during and after treatment of reactions. *Clin Exp Dermatol* (1992) 17(1):20–3. doi:10.1111/j.1365-2230.1992.tb02526.x
189. Singh S, Jenner PJ, Narayan NP, Ramu G, Colston MJ, Prasad HK, et al. Critical residues of the *Mycobacterium leprae* LSR recombinant protein discriminate clinical activity in erythema nodosum leprosum reactions. *Infect Immun* (1994) 62(12):5702–5.
190. Kifayet A, Hussain R. Selective decrease of *M. leprae*-specific IgG1 and IgG3 antibodies in leprosy patients associated with ENL. *Int J Lepr Other Mycobact Dis* (1996) 64(2):105–14.
191. Kifayet A, Shahid F, Lucas S, Hussain R. Erythema nodosum leprosum is associated with up-regulation of polyclonal IgG1 antibody synthesis. *Clin Exp Immunol* (1996) 106(3):447–53. doi:10.1046/j.1365-2249.1996.d01-860.x
192. Beuria MK, Parkash O, Joshi B, Mohanty KK, Katoch K, Sengupta U. Levels of IgG subclasses in active and inactive cases in the disease spectrum of leprosy. *Int Arch Allergy Immunol* (1998) 115(1):61–6. doi:10.1159/000023831
193. Freire BF, Ferraz AA, Nakayama E, Ura S, Queluz TT. Anti-neutrophil cytoplasmic antibodies (ANCA) in the clinical forms of leprosy. *Int J Lepr Other Mycobact Dis* (1998) 66(4):475–82.
194. Stefani MM, Martelli CM, Morais-Neto OL, Martelli P, Costa MB, de Andrade AL. Assessment of anti-PGL-I as a prognostic marker of leprosy reaction. *Int J Lepr Other Mycobact Dis* (1998) 66(3):356–64.
195. Beuria MK, Mohanty KK, Katoch K, Sengupta U. Determination of circulating IgG subclasses against lipoarabinomannan in the leprosy spectrum and reactions. *Int J Lepr Other Mycobact Dis* (1999) 67(4):422–8.
196. Hamerlinck FF, Klatser PR, Walsh DS, Bos JD, Walsh GP, Faber WR. Serum neopterin as a marker for reactional states in leprosy. *FEMS Immunol Med Microbiol* (1999) 24(4):405–9. doi:10.1111/j.1574-695X.1999.tb01312.x
197. Mahaisavariya P, Jiamton S, Manonukul J, Khemngern S. Mast cells in leprosy and leprosy reaction. *Int J Dermatol* (2000) 39(4):274–7. doi:10.1046/j.1365-4362.2000.00908.x
198. Schon T, Leekassa R, Gebre N, Sundqvist T, Bizuneh E, Britton S. High dose prednisolone treatment of leprosy patients undergoing reactions is associated with a rapid decrease in urinary nitric oxide metabolites and clinical improvement. *Lepr Rev* (2000) 71(3):355–62.
199. Antunes SL, Liang Y, Neri JA, Sarno EN, Haak-Frendscho M, Johansson O. Mast cell subsets and neuropeptides in leprosy reactions. *Arq Neuropsiquiatr* (2003) 61(2A):208–19. doi:10.1590/S0004-282X2003000200010
200. Rada E, Marzal M, Aranzazu N, Convit J. [Increase in nitric oxide concentrations in serum and mononuclear cell cultures from patients with Type II reaction state of Hansen's disease]. *Invest Clin* (2003) 44(2):129–36.
201. Sunderkotter CH, Tomimori-Yamashita J, Nix V, Maeda SM, Sindrilaru A, Mariano M, et al. High expression of myeloid-related proteins 8 and 14 characterizes an inflammatorily active but ineffective response of macrophages during leprosy. *Immunology* (2004) 111(4):472–80. doi:10.1111/j.0019-2805.2004.01836.x
202. Nigam PK, Srivastava P, Patra PK. Serum adenosine deaminase levels in reactional and non-reactional leprosy. *Indian J Dermatol Venereol Leprol* (2005) 71(1):20–2. doi:10.4103/0378-6323.13780
203. Mohanty KK, Gupta M, Girdhar BK, Girdhar A, Chakma JK, Sengupta U. Increased level of urinary nitric oxide metabolites in leprosy patients during type 2 reactions and decreased after antireactional therapy. *Lepr Rev* (2007) 78(4):386–90.
204. Santos DO, Castro HC, Bourguignon SC, Bastos OM, Rodrigues CR, Van Heuverswyn H, et al. Expression of B7-1 costimulatory molecules in patients with multibacillary leprosy and reactional states. *Clin Exp Dermatol* (2007) 32(1):75–80. doi:10.1111/j.1365-2230.2006.02291.x
205. Silva EA, Iyer A, Ura S, Lauris JR, Naafs B, Das PK, et al. Utility of measuring serum levels of anti-PGL-I antibody, neopterin and C-reactive protein in monitoring leprosy patients during multi-drug treatment and reactions. *Trop Med Int Health* (2007) 12(12):1450–8. doi:10.1111/j.1365-3156.2007.01951.x
206. Brito Mde F, Ximenes RA, Gallo ME, Buhrer-Sekula S. Association between leprosy reactions after treatment and bacterial load evaluated using anti PGL-I serology and bacilloscopy. *Rev Soc Bras Med Trop* (2008) 41 (Suppl 2):67–72.
207. Iyer A, van Eijk M, Silva E, Hatta M, Faber W, Aerts JM, et al. Increased chitotriosidase activity in serum of leprosy patients: association with bacillary leprosy. *Clin Immunol* (2009) 131(3):501–9. doi:10.1016/j.clim.2009.02.003
208. Singh I, Yadav AR, Mohanty KK, Katoch K, Bisht D, Sharma P, et al. Molecular mimicry between HSP 65 of *Mycobacterium leprae* and cytokeratin 10 of the host keratin; role in pathogenesis of leprosy. *Cell Immunol* (2012) 278(1–2):63–75. doi:10.1016/j.cellimm.2012.06.011
209. Mandal D, Reja AH, Biswas N, Bhattacharyya P, Patra PK, Bhattacharya B. Vitamin D receptor expression levels determine the severity and complexity of disease progression among leprosy reaction patients. *New Microbes New Infect* (2015) 6:35–9. doi:10.1016/j.nmni.2015.04.001
210. Berrington WR, Macdonald M, Khadge S, Sapkota BR, Janer M, Hagge DA, et al. Common polymorphisms in the NOD2 gene region are associated with leprosy and its reactive states. *J Infect Dis* (2010) 201(9):1422–35. doi:10.1086/651559
211. Teixeira MA, Silva NL, Ramos Ade L, Hatagima A, Magalhaes V. [NRAMP1 gene polymorphisms in individuals with leprosy reactions attended at two reference centers in Recife, northeastern Brazil]. *Rev Soc Bras Med Trop* (2010) 43(3):281–6. doi:10.1590/S0037-86822010000300014
212. de Sales Marques C, Brito-de-Souza VN, Albuquerque Guerreiro LT, Martins JH, Amaral EP, Cardoso CC, et al. Toll-like receptor 1 (TLR1) N248S single nucleotide polymorphism is associated with leprosy risk and regulates immune activation during mycobacterial infection. *J Infect Dis* (2013) 208(1):120–9. doi:10.1093/infdis/jit133
213. Schuring RP, Hamann L, Faber WR, Pahan D, Richardus JH, Schumann RR, et al. Polymorphism N248S in the human toll-like receptor 1 gene is related to leprosy and leprosy reactions. *J Infect Dis* (2009) 199(12):1816–9. doi:10.1086/599121
214. Scotland RS, Stables MJ, Madalli S, Watson P, Gilroy DW. Sex differences in resident immune cell phenotype underlie more efficient acute inflammatory responses in female mice. *Blood* (2011) 118(22):5918–27. doi:10.1182/blood-2011-03-340281
215. Tan Trao V, Huong PL, Thuan AT, Long HT, Trach DD, Wright EP. Responses to *Mycobacterium leprae* by lymphocytes from new and old leprosy patients: role of exogenous lymphokines. *Ann Inst Pasteur Immunol* (1988) 139(2):121–33. doi:10.1016/0769-2625(88)90034-7
216. Beyrau M, Bodkin JV, Nourshargh S. Neutrophil heterogeneity in health and disease: a revitalized avenue in inflammation and immunity. *Open Biol* (2012) 2(11):120134. doi:10.1098/rsob.120134
217. Fatima N, Faisal SM, Zubair S, Ajmal M, Siddiqui SS, Moin S, et al. Role of pro-inflammatory cytokines and biochemical markers in the pathogenesis of type 1 diabetes: correlation with age and glycemic condition in diabetic human subjects. *PLoS One* (2016) 11(8):e0161548. doi:10.1371/journal.pone.0161548
218. Anderson R, Gatner EM, van Rensburg CE, Grabow G, Imkamp FM, Kok SK, et al. In vitro and in vivo effects of dapsone on neutrophil and lymphocyte functions in normal individuals and patients with lepromatous leprosy. *Antimicrob Agents Chemother* (1981) 19(4):495–503. doi:10.1128/AAC.19.4.495
219. Anderson R, Gatner EM. Changes in neutrophil motility accompanying dapsone and rifampicin therapy. *Lepr Rev* (1981) 52(1):19–22.
220. Anderson R. Enhancement by clofazimine and inhibition by dapsone of production of prostaglandin E2 by human polymorphonuclear leukocytes in vitro. *Antimicrob Agents Chemother* (1985) 27(2):257–62. doi:10.1128/AAC.27.2.257
221. Das PK, Klatser PR, Pondman KW, Huikeshoven H, Landheer JE, Leiker DL, et al. Dapsone and anti-dapsone antibody in circulating immune complexes in leprosy patients. *Lancet* (1980) 1(8181):1309–11. doi:10.1016/S0140-6736(80)91772-9
222. Trao VT, Huong PL, Thuan AT, Anh DD, Trach DD, Rook GA, et al. Changes in cellular response to mycobacterial antigens and cytokine production

patterns in leprosy patients during multiple drug therapy. *Immunology* (1998) 94(2):197–206. doi:10.1046/j.1365-2567.1998.00485.x
223. Mendonca VA, Costa RD, Lyon S, Penido RA, Borges VO, Bretas TL, et al. Plasma levels of chemokines during leprosy specific treatment. *Acta Trop* (2010) 113(2):151–4. doi:10.1016/j.actatropica.2009.10.010
224. Nishimura M, Koeda A, Suzuki E, Shimizu T, Kawano Y, Nakayama M, et al. Effects of prototypical drug-metabolizing enzyme inducers on mRNA expression of housekeeping genes in primary cultures of human and rat hepatocytes. *Biochem Biophys Res Commun* (2006) 346(3):1033–9. doi:10.1016/j.bbrc.2006.06.012
225. Libert C, Dejager L. How steroids steer T cells. *Cell Rep* (2014) 7(4):938–9. doi:10.1016/j.celrep.2014.04.041
226. Langereis JD, Oudijk EJ, Schweizer RC, Lammers JW, Koenderman L, Ulfman LH. Steroids induce a disequilibrium of secreted interleukin-1 receptor antagonist and interleukin-1beta synthesis by human neutrophils. *Eur Respir J* (2011) 37(2):406–15. doi:10.1183/09031936.00170409
227. Auttachoat W, Zheng JF, Chi RP, Meng A, Guo TL. Differential surface expression of CD18 and CD44 by neutrophils in bone marrow and spleen contributed to the neutrophilia in thalidomide-treated female B6C3F1 mice. *Toxicol Appl Pharmacol* (2007) 218(3):227–37. doi:10.1016/j.taap.2006.11.019
228. Kim BS, Kim JY, Lee JG, Cho Y, Huh KH, Kim MS, et al. Immune modulatory effect of thalidomide on T cells. *Transplant Proc* (2015) 47(3):787–90. doi:10.1016/j.transproceed.2014.12.038
229. Hernandez Mde O, Fulco Tde O, Pinheiro RO, Pereira Rde M, Redner P, Sarno EN, et al. Thalidomide modulates *Mycobacterium leprae*-induced NF-kappaB pathway and lower cytokine response. *Eur J Pharmacol* (2011) 670(1):272–9. doi:10.1016/j.ejphar.2011.08.046

10

Host Intracellular Signaling Events and Pro-inflammatory Cytokine Production in African Trypanosomiasis

Shiby M. Kuriakose[†]*, Rani Singh*[†‡] *and Jude E. Uzonna**

Department of Immunology, Faculty of Health Sciences, University of Manitoba, Winnipeg, MB, Canada

Edited by:
Abhay Satoskar, The Ohio State University, USA

Reviewed by:
Sreenivas Gannavaram, Food and Drug Administration, USA
Nurittin Ardic, Gulhane Military Medical Academy, Turkey

***Correspondence:**
Jude E. Uzonna
jude.uzonna@umanitoba.ca

[†]*Shiby M. Kuriakose and Rani Singh contributed equally to the manuscript.*

[‡]Present address:
Rani Singh, Department of Immunology and Infectious Diseases, The Forsyth Institute, Harvard School of Dental Medicine, Cambridge, MA, USA

Pathogens, such as bacteria, viruses, and parasites, possess specific molecules or proteins that are recognized by several host innate immune receptors, leading to the activation of several intracellular signaling molecules and pathways. The magnitude and quality of these events significantly affect the outcome of infection. African trypanosomes, including *Trypanosoma congolense*, are capable of manipulating the host immune response, including the activity of macrophages, which are the key immune cells that contribute to the immunopathogenesis of African trypanosomiasis. Although it is known that immune hyperactivation and excessive pro-inflammatory cytokine production are the hallmarks of African trypanosomiasis, the mechanisms through which these events are triggered are poorly defined. However, it is known that macrophages may play a significant role in these processes, because phagocytosis of trypanosomes by macrophages initiates intracellular signal transduction cascades that lead to the release of pro-inflammatory cytokines and alteration in cell function. This review highlights recent progress in our understanding of the innate immune receptors, signaling pathways, and transcription factors involved in *T. congolense*-induced pro-inflammatory cytokine production in macrophages. It will reveal the existence of complex signaling events through which the parasite modulates the host immune response, thus identifying novel targets that could aid in designing strategies to effectively control the disease.

Keywords: African trypanosomiasis, intracellular signaling peptides and proteins, pro-inflammatory cytokines, innate immune response, host–parasite interactions

INTRODUCTION

African trypanosomes are flagellated protozoan parasites that cause disease in both humans and animals. The disease in humans is called human African trypanosomiasis (HAT) or sleeping sickness and is considered as a neglected disease along with other parasitic diseases, such as leishmaniasis and schistosomiasis (1). Other than malaria and schistosomiasis, HAT is the third significant contributor to the global burden of parasitic diseases (1). The disease in humans is mainly caused by *Trypanosoma (T.) brucei rhodesiense* and *T. brucei gambiense*. Animal African trypanosomiasis (AAT), otherwise called Nagana, is caused by *T. congolense, T. brucei brucei*, and *T. vivax*. *T. congolense* is the major species that primarily cause animal trypanosomiasis, particularly in domestic livestock. The disease is transmitted to the mammalian host by the bite of several species of infected tsetse fly vector belonging to the genus *Glossina*. The incidence of animal trypanosomiasis coincides with the distribution of their tsetse fly vectors and according to the 2014 WHO report, tsetse fly transmitted trypanosomiasis occurs in 36 sub-Saharan countries and covers almost 10 million square kilometers of the African continent (2).

African trypanosomiasis is of considerable economic and social importance and is one of the most important factors restricting economic development in Africa (3). The disease threatens the health of about 70 million people and 50 million cattle (4). According to the 2014 WHO fact sheet, the estimated number of actual cases is 20,000, with an estimated 70 million people at risk of developing the disease (2). Due to the continued control efforts, the number of cases has dropped for the first time in last 50 years (2). However, sporadic cases and small outbreaks are reported increasingly in patients in non-endemic countries despite the decline in the number of new cases among Africans (5). It is estimated that about 50 cases of HAT occur annually outside Africa, and 94 cases of HAT were reported in 19 non-disease endemic African countries during the period of 2000–2010 (5, 6).

Although animal trypanosomiasis can occur in all domesticated animals, cattle are the main species affected due to the feeding preferences of tsetse flies. The disease adversely affects livestock production and farming and has a major impact on human and economic development. It is estimated that three million cattle die annually from the disease, and the estimated economic loss in cattle production is about 4 billion US dollars per year (7). Despite the fact that the disease kills or disables hundreds of thousands of people and animals in endemic areas, the treatment for the disease is not satisfactory. Currently, there are no vaccines available to prevent the disease in both humans and animals, and the current treatment methods have several limitations, and the development of drug resistance is a growing problem.

IMMUNITY TO AFRICAN TRYPANOSOMIASIS

Innate Immunity

Studies have shown that various animal species respond differently to infections with various species of African trypanosomes. For example, the West African cattle breeds, such as the N'Dama, are naturally resistant (trypanotolerant) to African trypanosomiasis, whereas the zebu-type cattle breeds, such as Boran, are comparatively susceptible (trypanosusceptible). Similarly, different mouse strains show varying resistance to infections with African trypanosomes. The relatively resistant C57BL/6 represents the trypanotolerant phenotype, whereas highly susceptible Balb/c mice represent trypanosusceptible phenotype. The differential resistance and susceptibility in animals have been associated with difference in genetic factors and immune responses, including innate immunity. Both humoral (serum-associated) and cellular factors contribute to the innate immunity to African trypanosomes.

It has long been known that humans and some primates are naturally resistant to infections with certain species of African trypanosomes like *T. brucei brucei*, *T. congolense*, and *T. vivax* (8). This resistance is due to the presence of trypanosome lytic factors (TLF) that are present in the serum (9, 10). The key components of TLFs are apolipoprotein L-1 (APOL-1) and hepatoglobin-related protein (Hpr) (9). Evidences suggest that APOL-1 kills trypanosomes through the binding of its N-terminal domain into the parasite lysosomal membrane, leading to the generation of ionic pores (11). APOL-1 exists in two forms, the high-density lipoprotein-bound TLF, termed as TLF1, and a large lipid poor IgM/APOL-1 complex, termed as TLF2 (12, 13). These two complexes enter the parasites through different uptake modes, but their mechanism of cytotoxicity is same (14).

Apart from soluble serum factors, the innate immune cells also play a crucial role in the immunopathogenesis of African trypanosomiasis. Macrophages are perhaps one of the most important innate cells involved in this process. Studies show that the activation of macrophages in infected animals may have a dual role (protection and disease exacerbation) in African trypanosomiasis. Activated macrophages release reactive nitrogen and oxygen species that have toxic and cytostatic effects on the parasites (15, 16). In addition, macrophages also actively participate in the phagocytosis and clearance of opsonized parasites (17). However, excessive macrophage activation leads to the release of large amount of pro-inflammatory cytokines and induces immunopathology.

The activation of macrophages in African trypanosomiasis is initiated by the receptor-mediated interaction with parasite molecules that could result in phagocytosis. In particular, the interaction between the GPI membrane that anchors the parasite VSG molecule and the pattern recognition receptors (PRRs) results in macrophage activation, triggering an inflammatory response. In *T. brucei* infection, the activation of the innate immune system is mediated through the adaptor protein MyD88 (18). Although the parasite molecule that binds to macrophages in *T. congolense* infection has not been identified yet, studies from our lab have shown that immune activation is initiated by the binding of the parasite to the toll-like receptor 2 (TLR2) expressed on macrophages. This binding leads to the activation of various intracellular signaling molecules and pathways associated with inflammation. Thus, we showed that *T. congolense*-induced activation of mitogen-activated protein kinases (MAPKs) and signal transducer and activator of transcription (STAT) proteins in macrophages results in the production of various pro-inflammatory cytokines, including IL-6, IL-12, and TNF, and collectively activates the adaptive immune response. *T. congolense*-induced innate immune receptor activation and signaling events are explained later in this review.

Adaptive Immunity

Effective immunity to any microbe requires efficient recognition of antigens derived from the microbe by the host immune system, which is followed by the generation of effector T and B cells, leading to the production of cytokines and antibodies, respectively. Due to their extracellular life style, effective immunity against African trypanosomes depends on optimal antibody production by B cells. However, this is severely hindered by the inherent ability of the parasites to undergo antigenic variation.

The variant surface glycoprotein (VSG) is a dense, protective surface coat that covers the plasma membrane of African trypanosomes. Although VSG induces strong B cell (antibody) response and anti-VSG variant-specific antibodies are effective in clearing clones expressing that specific VSG, the ability of the parasite to switch from expressing one VSG to another on its surface helps it to effectively evade the host immune system by being one step ahead of the host's immune response (19). The constant cycling of parasites (expression of new VSGs) and antibody responses throughout the infection result in undulating parasitemia and chronic infections in African trypanosomiasis.

Since African trypanosomes are extracellular parasites, antibodies are required for protection. In *T. congolense*-infected animals, the production of antibodies against the parasite VSG occurs early after infection. Previous studies have shown that the clearance of parasites in trypanosome-infected mice is correlated with the serum levels of parasite-specific antibodies, particularly IgG2a and IgG3, but not with IgM antibodies (20). It is also clear from several studies that resistant mouse strains mount earlier and higher anti-VSG antibody response than the relatively susceptible mouse strains (20). Although anti-VSG antibodies are necessary, they are not sufficient for controlling *T. congolense* infection in the highly susceptible Balb/c mice. Interestingly, enhanced resistance to *T. congolense*-infected mice is correlated with serum levels of antibodies against common trypanosomal antigens (20).

Dysregulation of cytokine network is a hallmark of African trypanosomiasis and excessive production of inflammatory cytokines, and the release of inflammatory mediators has been proposed as a major cause of death in infected animals (20–22). The effect of cytokines in the course of trypanosomiasis is dependent on timing, microenvironment, quality, and quantity being produced. For example, type I cytokines have been shown to confer resistance to African trypanosomes by limiting parasite growth during the early stage of infection (20, 21). However, sustained type I cytokine response may be harmful and promote disease progression (23, 24). In line with this, it has been shown that anti-IFN-γ treatment resulted in very low parasitemia, control of several waves of parasitemia, and over fourfold increase in survival period in the highly susceptible BALB/c mice (25). However, infection of IFN-γ-knockout mice showed that IFN-γ is required for the survival of the relatively resistant mice to *T. congolense* or *T. brucei* infections. Thus, infected, relatively resistant C57BL/6 mice failed to control their first wave of parasitemia and succumbed acutely to the infection (21, 26). Also, the induction of IFN-γ production in $CD8^+$ T cells by trypanosomal-activating molecule has been shown to lead to susceptibility (27). IL-10 is another cytokine that has been shown to be critical for trypanotolerance. Deficiency of IL-10 resulted in a striking reduction in the survival time of *T. congolense*- and *T. brucei*-infected mice, and this was associated with increased levels of serum pro-inflammatory cytokines like IFN-γ, TNF, and NO (28, 29). This has led to the conclusion that IL-10 is critically important for dampening excessive inflammatory state caused by overproduction of pro-inflammatory cytokines.

BALB/c mice are highly susceptible to experimental *T. congolense* and *T. brucei* infections, whereas C57BL/6 mice are relatively resistant. Death in the highly susceptible BALB/c mice is associated with systemic inflammatory response syndrome and shock-like state due, in part, to excessive production of pro-inflammatory cytokines and immune hyperactivation (22, 30). Thus, death of the susceptible animals is due to disease-associated complications and not primarily related to excessive parasite load itself. A recent report shows that while *T. congolense*-infected wild-type (WT) and $CD4^{-/-}$ ($CD4^+$ T cell-deficient) BALB/c mice have similar parasitemia and survival time, partial depletion of $CD4^+$ T cells in WT mice leads to lower parasitemia and longer survival than infected normal WT mice (28). This partial depletion of $CD4^+$ T cells also resulted in reduced IFN-γ production without affecting IL-10 and parasite-specific IgG antibody production (28). This observation further supports the notion that the early mortality of the infected BALB/c mice is due to the excessive IFN-γ production, which in turn exerts pathology by causing macrophage hyperactivation and excessive pro-inflammatory cytokine production. Thus, an optimal/balanced cytokine response is needed at the initial stage of the infection for parasite clearance, and IFN-γ-producing $CD4^+$ T cells play an important role in the immunopathogenesis of African trypanosomiasis.

It has been known for long that many indigenous African mammals can harbor natural trypanosome infections without developing severe disease symptoms (31). In particular, while the indigenous West African N'Dama cattle are relatively resistant, the exotic zebu breeds are highly susceptible to African trypanosomiasis (32). The extreme susceptibility of exotic cattle breeds imported to sub-Saharan Africa may be related to excessive (rather than impaired) immune response and/or activation. Likewise, in HAT, death is usually due to the nature of the host immune reaction, as inflammatory cytokines play an important role in the development of trypanosomiasis-associated disease complications (33, 34).

Regulatory T cells (Tregs) have been shown to play an important role in the pathogenesis of several infectious diseases (35–37). However, their role in the pathogenesis of African trypanosomiasis is controversial. For example, one report suggests that Tregs play a crucial role in the enhanced resistance of mice to *T. congolense* infections (38). In this study, it was shown that $CD4^+CD25^+Foxp3^+$ (Tregs) limit excessive IFN-γ production by $CD4^+$ and $CD8^+$ T cells and thereby downregulate macrophage activation and subsequent production of pro-inflammatory cytokines (38). However, overwhelming evidence from several studies suggest that Tregs may contribute to susceptibility to infection with African trypanosomes (39, 40). For example, Wei and Tabel showed that depletion of $CD4^+CD25^+Foxp3^+$ T cells by anti-CD25 mAb leads to complete control of infection in highly susceptible BALB/c mice (39). In line with this, studies from our lab have shown that Tregs contribute to enhanced susceptibility to experimental *T. congolense* infection in mice (40). Following *T. congolense* infection, Tregs contributed to enhanced disease in both relatively resistant C57BL/6 and highly susceptible Balb/c mice. Depletion of Tregs by using anti-CD25 mAb showed that Tregs negatively affect efficient parasite control, whereas adoptive transfer of highly enriched $CD4^+$ $CD25^+$ $Foxp3^+$ T cells resulted in increased peak parasitemia and production of disease exacerbating inflammatory cytokines like IFN-γ and IL-6 during early infection in the resistant C57BL/6 mice (40, 41).

ROLE OF MACROPHAGES IN AFRICAN TRYPANOSOMIASIS

Macrophages play a crucial role in the control of many protozoan parasitic infections, including African trypanosomiasis. Intact monocytic cell system is important for the initiation and maintenance of anti-trypanosome responses, because the mononuclear phagocytic system is crucial in the phagocytosis of opsonized trypanosomes (13), which is the major mechanism for the removal of trypanosomes from the blood stream. During the course of trypanosome infection, the numbers of macrophages are greatly increased in many organs, including the liver, spleen, and lymph nodes, and these cells display morphological and functional features of activation (42, 43). Because phagocytosis of trypanosomes by macrophages (mainly Kupffer cells) leads to the production of pro-inflammatory cytokines (22), macrophages also play a critical role in mediating immunopathology in infected mice. The activation of macrophages during trypanosome infection is due, in part, to their exposure to parasite components and host-derived IFN-γ produced in response to parasite antigens by T cells (44). The parasite antigens include invariant membrane, cytoplasmic, nuclear antigens, and the VSG. Although the exact mechanism of attachment and phagocytosis of trypanosomes by macrophages is not fully understood, optimal phagocytosis has been shown to occur in the presence of variant-specific antibodies (45, 46).

Several studies have investigated the activation of macrophages following infection with different species of African trypanosomes. In experimental *T. brucei* infection, a large percentage of splenic macrophages exhibit membrane and functional characteristics associated with immune activation, release large amounts of IL-12, TNF, and nitric oxide, and contribute to the modulation of host immunity and resistance (47). In *T. congolense* infection, the complement and antibody-mediated phagocytosis by splenic and liver (Kupffer cells) macrophages is the major mechanism by which trypanosomes are cleared from an infected host (17, 48). The uptake of antibody-coated (including IgM and IgG) parasites by macrophages results in their activation and production of nitric oxide and pro-inflammatory cytokines (16). In addition, activated macrophages present trypanosomal antigens to $CD4^+$ T cells in an MHC class II-dependent manner resulting in the production of IFN-γ (28), further enhancing macrophage activation and pro-inflammatory cytokines production. Furthermore, the massive phagocytosis of parasites by splenic and hepatic macrophages at the peak parasitemia further leads to their hyperactivation and increased production of nitric oxide, monokines, and pro-inflammatory cytokines. This systemic cytokine overproduction leads to systemic inflammatory response like syndrome, which contributes to the death of trypanosome-infected mice (28). Indeed, our group previously showed that following *T. congolense* infection, both splenic and hepatic macrophages produce copious amounts of pro-inflammatory cytokines that in turn contribute to disease and mortality (49). Treatment of infected mice with berenil (a trypanocide) significantly downregulated pro-inflammatory cytokine production and led to survival from an otherwise lethal infection in the highly susceptible BALB/c mice (49).

Direct evidence for the involvement of macrophages in the production of pro-inflammatory cytokines, following interaction with *T. congolense*, has been demonstrated *in vitro*. Kaushik et al. found that bone marrow-derived macrophages (BMDMs) produce an array of cytokines, including IL-6, TNF, and IL-12, upon stimulation with *T. congolense* and *T. brucei* whole cell lysate (50). Interestingly, they also showed that BMDMs from the highly susceptible BALB/c mice produced significantly more of these cytokines than those from the relatively resistant C57BL/6 mice (50). This is consistent with the notion that acute death observed in the infected BALB/c mice might be due to unregulated immune activation and excessive pro-inflammatory cytokine production. We have confirmed that the stimulation of BMDMs with *T. congolense* induces IL-6, IL-12, and TNF production, and this is dependent on intracellular signals resulting from activation of MAPKs and STAT proteins (STAT1 and STAT3). We further delineated the innate receptor involved in the recognition of the parasite constituent(s) and showed the critical involvement of the adaptor molecule, MyD88, in these processes.

M1 versus M2 Macrophages

Macrophage activation is a term used to describe macrophages that have been primed or stimulated to show enhanced effector activities. Depending on the nature of activating stimuli and the eventual phenotype displayed, activated macrophages are classified into two major groups: classically activated macrophages or M1 type and alternatively activated macrophages or M2 type. The classical activation denotes effector macrophages that are produced during cell-mediated immune responses. Classically activated macrophages secreted high levels of pro-inflammatory cytokines and mediators and exhibit enhanced microbicidal and tumoricidal capacity. IFN-γ, produced by innate or adaptive immune cells, is the major molecule involved in the classical activation of macrophages (51). In contrast, Th2 cytokines, such as IL-4 and IL-13, are the key drivers of alternative macrophage activation (52). Alternatively activated macrophages are poor at antigen presentation to T cells, are less efficient at producing toxic oxygen and nitrogen radicals, and do not effectively kill intracellular pathogens (53). However, they are important in the clearance of helminth and nematode infections (54, 55).

Several studies have shown that the cytokine environment during African trypanosomiasis is key at influencing the activation of macrophages, resulting in the development of M1 and M2 subsets that are regulated antagonistically (56). Classically activated macrophages (M1) develop during the early stage of infection in the presence of type I cytokine environment (IFN-γ and TNF). In contrast, alternatively activated macrophages (M2) develop during the late stage of infection when a type II cytokine environment (IL-4, IL-10, TGF-β, etc.) predominates (56). As in other infections, M1 macrophages from mice infected with African trypanosomes have been shown to possess pro-inflammatory properties and produce nitric oxide, reactive oxygen species, and TNF, whereas M2 macrophages have anti-inflammatory properties due to increased arginase activity and IL-10 production (56). The development of M1 macrophages during the early stage of the disease appears to be responsible for the early release of pro-inflammatory mediators associated with type I immune response, and this could contribute to the control of initial parasitemic waves. Also, the activation of M1 macrophages could result in collateral damage to tissues as seen in the liver and spleen (57). A switch from M1 to M2 macrophages occurs after the clearance of initial waves of parasitemia in the relatively resistant mice (13). The resultant secretion of anti-inflammatory mediators by these M2 macrophages is thought to actively participate in tissue healing (56, 57).

INNATE IMMUNE RECEPTORS AND ADAPTOR PROTEINS IN AFRICAN TRYPANOSOMIASIS

The germline-encoded innate receptor families, also known as PRRs, are critical for initiating innate immune responses in the host. They include nucleotide-binding oligomerization domain (NOD)-like receptors (NLRs), RIG-1-like receptors (RLRs), membrane-bound C-type lectin receptors (CLRs), and transmembrane toll-like receptors (TLRs). TLRs are the most widely studied PRRs and play an important role in the recognition of molecular signatures on microbes resulting in the initiation of inflammatory response (58). They are the first group of PRRs identified and have been shown to recognize a wide range of pathogen-associated molecular patterns (PAMPs) (59, 60). By inducing distinct gene expression patterns, stimulation of different TLRs leads to the activation of innate immunity, which in turn instructs the development of antigen-specific acquired immunity (61). Ten human and twelve mice functional TLRs have been identified till date, and each TLR detects distinct PAMPs derived from bacteria, viruses, fungi, and parasites (62, 63).

Toll-like receptor-mediated intracellular signaling events are initiated by the binding of different TLRs with their corresponding ligands, resulting in the recruitment of a number of adaptor proteins. Myeloid differentiation primary response protein 88 (MyD88) is the major adaptor protein that is involved in almost all TLR signaling pathways. The other adaptor proteins are TIR-domain-containing adaptor protein (TIRAP), TIR-domain-containing adaptor-inducing IFN-β (TRIF), and TRIF-related adaptor molecule (TRAM) (58, 64). The activation of these adaptor molecules, following ligation of TLRs with their specific ligands, leads to the activation of various intracellular signaling molecules (including MAPKs and STATs) and transcription factors, such as nuclear factor kappa B (NFκB) and activating protein-1 (AP-1), and a resultant initiation of specific immune responses. Mice with genetic ablation of MyD88 are highly susceptible to *T. cruzi* and *T. brucei* infections due, in part, to impaired IL-12 and IFN-γ production (18, 65). In these experimental models, it was proposed that signaling through MyD88 in innate immune cells has a protective role by activating Th1 response (18, 65). In contrast, in some experimental models, decreased pro-inflammatory responses resulting from the lack of MyD88 signaling is beneficial to the host (66). Similarly, the GPI anchor from *T. cruzi* trypomastogotes activates phosphorylation of ERK, p38, and IκB in mouse peritoneal macrophages (67), and mice deficient in the adaptor protein MyD88 signaling are highly susceptible to *T. cruzi* infection due, in part, to impaired pro-inflammatory cytokine production, suggesting that MyD88 is critical for the activation of MAPKs and NFκB by *T. cruzi* (65). Recently, we showed that deficiency of MyD88 results in inhibition of *T. congolense*-induced p38 and STAT1 phosphorylation and a concomitant downregulation of IL-6 and IL-12 by macrophages. These findings indicate that this adaptor molecule plays an important role in *T. congolense*-induced pro-inflammatory cytokine production in macrophages. They further suggest the involvement of TLR-mediated recognition and signaling in *T. congolense*-induced pro-inflammatory cytokine production in macrophages.

Although TLRs have been shown to play an important role in the innate immune responses to several parasitic infections (18, 68–71), their role in immunity to African trypanosomiasis is largely unknown. However, the findings that synergistic recognition of *T. cruzi* DNA by TLR9 and glycosylphosphatidylinositol (GPI) by TLR2 cooperatively lead to the induction of pro-inflammatory cytokine production in macrophages (72) and suggest that TLRs may play an important role in the recognition of African trypanosomes, which are related to *T. cruzi*. Indeed, a recent report showed that, in *T. brucei* infections, TLR2-mediated signaling contributes to intra-cerebral control of parasite load in the brain (73).

No report has investigated the innate receptor(s) through which the innate immune system detects and responds to *T. congolense* infection. Given that MAPKs, STATs, and NFκB pathways are activated, following stimulation of macrophages and is dependent on MyD88 signaling, we speculated that TLR(s) is/are involved in *T. congolense* recognition by macrophages. We found that the deficiency of either TLR4 or TLR9 does not affect intracellular signaling events and pro-inflammatory cytokine production in macrophages in response to *T. congolense* stimulation. In contrast, *T. congolense*-induced MAPK and STAT phosphorylation was significantly downregulated in TLR2-deficient macrophages compared to the WT cells, and this was associated with a concomitant reduction in IL-6 and IL-12 production. This indicates that *T. congolense*-induced activation of important signaling pathways leading to optimal immune response is mediated through TLR2 and may have an important role in regulating the outcome of the disease. Indeed, *T. congolense*-infected, TLR2-deficient mice on the usually relatively resistant C57BL/6 background developed uncontrolled parasitemia and died within 10 days of infection in contrast to their WT counterpart mice that controlled several waves of parasitemia and survived for more than 100 days. Altogether, these observations show that TLR2 plays a critical role in *T. congolense* recognition by macrophages and resistance to the parasite. However, it remains to be determined what parasite moiety is recognized by TLR, although the GPI, which anchors VSG to the membrane, is a prime candidate.

INTRACELLULAR SIGNALING PATHWAYS IN AFRICAN TRYPANOSOMIASIS

The major signaling pathways that are known to induce cytokine production in immune cells include MAPK pathway, JAK–STAT pathway, and NFκB pathway. MAPKs are a group of highly conserved serine/threonine protein kinases that mediate intracellular signaling events necessary for carrying out a variety of fundamental cellular processes, such as proliferation, differentiation, motility, stress response, apoptosis, and survival. Three major families of MAPKs are recognized including, extracellular signal-regulated kinases (ERK), p38 MAPKs, and c-Jun N-terminal kinase/stress-activated protein kinases (JNK-SAPKs) (74, 75). STATs were first described by Darnell et al. (76) as transcription factors induced by ligands in IFN-treated cells. Later, several other groups showed the critical role of STATs in signal transduction pathways by cytokines and growth factors (77). All STATs have seven well-defined domains, including an N-terminal conserved domain, dimerization domain, SH2 domain, and a C-terminal transactivation domain. The amino-terminal region prevents dimerization of STATs in their inactive state (78), and the SH2 domain is critical for the recruitment of STATs to activate receptor complexes and also for the interaction with Janus (JAK) and Src kinases. This region is the most conserved domain among STATs and plays an important role in STAT signaling by facilitating homodimerization and heterodimerization, which are crucial for nuclear localization and DNA-binding activities (79). The activation of MAPKs and STATs is important in regulating pro-inflammatory cytokine production in immune cells (80). Their activation initiates a cascade of intracellular signaling events culminating in the expression of various pro-inflammatory genes. Thus, MAPK and STAT family members coordinate and propagate multiple inflammatory immune responses (81, 82).

Nuclear factor kappa B is a family of transcription factors that play important roles in inflammation, immunity, cell proliferation, differentiation, and survival. In the inactive state, NFκB is complexed with the inhibitory protein IκBα. IκB kinase complex is activated by inducing stimuli, leading to the phosphorylation, ubiquitination, and degradation of IκB proteins. The degradation of IκB proteins releases NFκB dimers resulting in their translocation to nucleus, where they bind to specific DNA sequences and promote transcription of target genes. NFκB family is composed of five members in mammals, such as RelA/p65, RelB, c-Rel, p50, and p52. All the members are characterized by the presence of Rel homology domain (RHD), which is essential for dimerization and binding to cognate DNA elements (83).

Although infection with African trypanosomes leads to profound production of pro-inflammatory cytokines, the intracellular signaling pathways leading to the production of these cytokines are poorly studied. There are sporadic and inconsistent reports on the role of MAPK and STAT family proteins in trypanosome-induced pro-inflammatory cytokine production. Soluble VSG of *T. brucei rhodesiense* has been shown to initiate a cascade of ERK, p38, JNK MAPK, and NFκB pathways, eventually leading to the expression of various pro-inflammatory genes, such as TNF, IL-12, IL-6, and iNOS (84). Consistent with this, several protozoan infections, other than African trypanosomes, have also been shown to induce the phosphorylation of various MAPKs, leading to enhanced cytokine production. One study shows that ERK and p38 phosphorylation was triggered by *T. cruzi* (a close relative of African trypanosomes) GPI anchor, leading to the activation of NFκB and culminating in the activation of pro-inflammatory genes (67). Another report shows that the *T. cruzi* induces STAT1 phosphorylation, both at mRNA and protein levels, and this is associated with increased binding of STAT1 homodimers to gamma-activated site (GAS) elements, leading to NO production (85). In addition, another report showed that infection with *Toxoplasma gondii* leads to MAPK and STAT3 phosphorylation and subsequent production of inflammatory cytokines (86, 87). These studies suggest that MAPK and STAT family proteins may be involved in the induction of pro-inflammatory cytokine production by African trypanosomes.

Recent studies from our lab have provided strong evidence showing that MAPK and STAT proteins play a critical role in *T. congolense*-induced pro-inflammatory cytokine production in macrophages. Stimulation of immortalized macrophages and primary BMDMs with whole cell lysate of *T. congolense* leads to time-dependent phosphorylation of ERK, p38, JNK, STAT1, and STAT3 proteins, which was associated with a concomitant production of pro-inflammatory cytokines *in vivo* and *in vitro*. We confirmed the involvement of these pathways in *T. congolense*-induced pro-inflammatory cytokine production by showing that pretreatment of the cells with specific inhibitors of ERK (U0126), p38 (SB203580), JNK (SP600125), STAT1 (Fludarabine), and STAT3 (S31-201) cause a significant downregulation of IL-6 and IL-12 production in response to trypanosome stimulation[1].

Given that the heterodimer consisting of p65 and p50 NFκB is critically important in the transcription of pro-inflammatory genes, we also assessed whether this key transcription factor is involved in *T. congolense*-induced pro-inflammatory cytokine production in macrophages. We found a time-dependent increase in the phosphorylation of NFκB p65 upon stimulation with *T. congolense* in macrophages, which corresponded with the production of IL-6 and IL-12. Collectively, our results show that the stimulation of macrophages with *T. congolense* extracts initiates intracellular signaling cascades, involving phosphorylation of MAPKs and STATs, leading to the eventual activation of the master transcription factor, of NFκB. The overall result is the transcription of pro-inflammatory genes and subsequent release of pro-inflammatory cytokines. However, it remains to be determined which innate receptor and adaptor molecules are involved in initiating these intracellular signaling events.

TRYPANOSOME-INDUCED SIGNALING IN RELATIVELY RESISTANT VERSUS HIGHLY SUSCEPTIBLE MURINE MACROPHAGES

Although macrophages are critical for clearing parasites in the blood following infection with African trypanosomes, their classical activation and production of pro-inflammatory cytokines (IL-6, IL-12, and TNF), nitric oxide, and increased expression of MHC class II and co-stimulatory molecules (15, 44, 88, 89) have also been shown to contribute to disease pathogenesis and severity. The degree of this activation and effector function of macrophages has some relationship to the degree of susceptibility of

[1]Kuriakose et al. *Trypanosoma congolense* induced pro-inflammatory cytokine production is by the activation of TLR signaling system through MAPK and STAT proteins (submitted).

different inbred mouse strains to experimental trypanosomiasis. Indeed, we previously reported that several cytokine genes are differentially regulated in trypanosome-infected macrophages from the highly susceptible BALB/c and relatively resistant C57BL/6 mice (50). IFN-γ-primed primary BMDMs from BALB/c mice, when infected with *T. brucei* or *T. congolense* whole cell lysate, produced significantly higher IL-6 than cells from C57BL/6 mice. Similarly, although PBMCs from both highly susceptible Boran and relatively resistant N'Dama cattle infected with *T. congolense* showed increased levels of IL-6 mRNA, this rise occurred earlier and lasted longer in the Boran than in the N'Dama cattle (90). This suggested the possibility of using IL-6 as a predictive marker of disease severity in bovine trypanosomiasis. Recent findings from our lab suggest the involvement of ERK, JAK–STAT3 pathways in *T. congolense*-induced release of IL-6 by macrophages from both BALB/c and C57BL/6 mice.

In contrast to IL-6, macrophages from BALB/c mice produced relatively lower TNF and IL-12 than those from C57BL/6 mice following trypanosome stimulation. Similar observations have been made in monocytes from the relatively resistant and highly susceptible cattle breeds (91). TNF is one of the most crucial pro-inflammatory cytokines that confer immunity against African trypanosomiasis. Magez and colleagues (92) found that the inhibition of trypanosomiasis-associated pathology in C57BL/6 and BALB/c mice is correlated with the shedding of soluble p75 TNF-receptor during peak parasitemia stages. Indeed, infection-associated pathologies were strongly reduced in both *T. brucei brucei*-infected $TNF^{-/-}$ and $TNF\text{-}R2^{-/-}$ mice (92). Because TNF signals through both TNF-R1 (55 kDa) and TNF-R2 (75 kDa), it has been suggested that TNF-R2 signaling in trypanosomiasis mediates infection-associated pathology, whereas TNF-R1 signaling has little or no impact on the infection outcome (92).

In addition to the production of pro-inflammatory cytokines, macrophages also produce nitric oxide, which plays a critical role in resistance against African trypanosomes *via* its cytostatic and cytotoxic effects (93–95). This is not surprising, given that the intracellular signaling pathways that result in the production of pro-inflammatory cytokines also lead to the activation of the iNOS gene and subsequent production of nitric oxide (96). Recently, we showed that in an IFN-γ-enriched environment, *T. congolense*-treated immortalized macrophages and BMDMs from the relatively resistant and highly susceptible mice show differential phosphorylation of MAPKs and STATs, suggesting that African trypanosomes induce differential signaling events in macrophages from different strains of mice *in vitro* (96). In addition to differential MAPK and STAT phosphorylation, *T. congolense* induces significantly higher levels of NO production in IFN-γ-primed immortalized macrophages and BMDMs the relatively resistant than from highly susceptible mice (96). Interestingly, we found that nitric oxide production in macrophages occurred *via* the activation of MAPK (including p38, Erk1/2, and JNK), and this was significantly inhibited by specific MAPK inhibitors in BALB/c but not in C57BL/6 cells (96). In addition, *T. congolense* and IFN-γ-induced NO production was dependent on STAT1 phosphorylation and was totally suppressed by fludarabine (a specific STAT1 inhibitor). We further showed that *T. congolense*-induced differential iNOS transcriptional promoter activation in IFN-γ-primed cells is dependent on the activation of both GAS1 and GAS2 transcription factors in BALB/c but only on GAS1 in C57BL/6 cells (96). Collectively, these observations suggest coordinated, but differentially regulated, signaling pathways that lead to NO production in macrophages from the susceptible and relatively resistant mice following their interaction with *T. congolense*.

CONCLUDING REMARKS

It is clear from the experimental animal studies that susceptibility to African trypanosomiasis is related to immune hyperactivation, leading to the production of pathogenic pro-inflammatory cytokines. However, little is known about the intracellular signaling events and pathways involved in the production of these pro-inflammatory cytokines. Understanding these events could reveal novel and unexpected aspects of the parasite pathobiology that could provide new concepts for understanding host–parasite interactions, disease pathogenesis, and the major cause of susceptibility to the disease. Our recent studies clearly show the involvement of TLR2–MyD88-dependent pathway, leading to activation of MAPKs, STATs, and NFκB in *T. congolense*-induced pro-inflammatory cytokine production and resistance to the infection (**Figure 1**). The involvement of MAPKs and STATs in

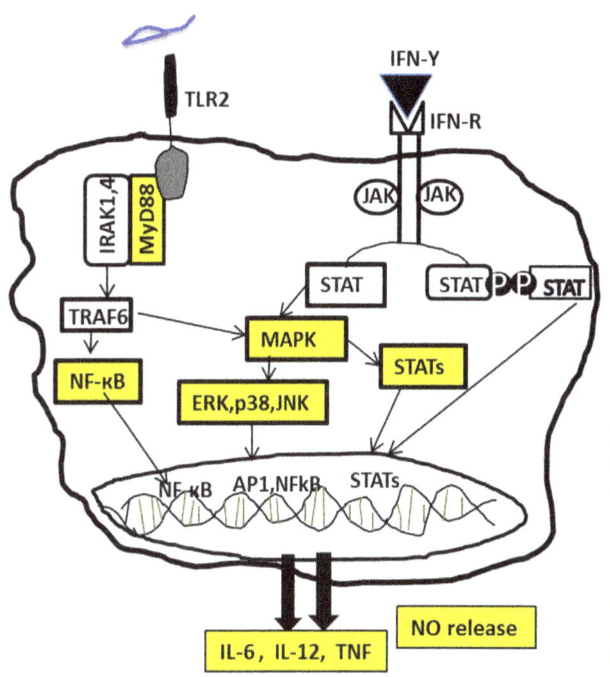

FIGURE 1 | Signaling pathways involved in *Trypanosoma congolense*-induced pro-inflammatory cytokine production in macrophages. The recognition of *T. congolense* by macrophages through toll-like receptor 2 (TLR2) results in the recruitment of the adaptor protein MyD88 and subsequent activation of several intracellular signaling molecules, including MAPKs and STATs. Activation of these molecules in turn leads to the activation of the transcription factor NFkB, which results in the production of IL-6, IL-12, TNF, and nitric oxide (NO). Stimulation of macrophages with *T. congolense* in the presence of IFN-γ produced by $CD4^+$ T cells and other immune cells further enhances the activation of these pathways, leading to increased proinflammatory cytokine and nitric oxide production.

T. congolense-induced cytokine production suggests that targeting these pathways offer novel strategies for managing the disease in animals infected with African trypanosomes. Indeed, we have found that the beneficial effects of berenil (a trypanocide), may in addition to its trypanocidal effects, involve its ability to downregulate pro-inflammatory cytokine production *via* inhibiting intracellular signaling events, including p38, JNK MAPK, STAT1, and STAT3 phosphorylation (49, 97). Thus, targeting these pathways may be beneficial for treating disease process in animals infected with African trypanosomes. Although we have made an attempt to delineate some signaling mechanisms in macrophages, following their interaction with trypanosomal molecules, there are still important questions that remain unanswered and require extensive investigation. For example, are the host immune signaling mechanisms different in response to diverse *Trypanosoma* species and its components? How are complex signaling molecules and their interactions involved in either resistance or susceptibility of the host postinfection? What parasite antigens are recognized by TLR2, leading to activation of the innate immune system and immunity to *T. congolense* infection? Providing answers to these questions may pave the way for identification of novel targets that regulate intracellular events and pathways, leading to resistance and/or susceptibility to African trypanosomiasis.

AUTHOR CONTRIBUTIONS

SK and Dr. RS contributed equally to the manuscript and are co-first authors.

FUNDING

Funding for this work was provided, in part, by the Natural Sciences and Engineering Research Council of Canada and Research Manitoba (JU). SK is supported by Research Manitoba Doctoral Fellowship.

REFERENCES

1. Kennedy PGE. The fatal sleep: Africa's killer disease that went undiscovered for centuries. Edition: Illustrated, reprint. Luath Press Limited (2010). 192 p.
2. WHO. *Trypanosomiasis, Human African (Sleeping Sickness)*. (2014). Available from: http://www.who.int/mediacentre/factsheets/fs259/en/
3. Wilson SG, Morris KR, Lewis IJ, Krog E. The effects of trypanosomiasis on rural economy with special reference to the Sudan, Bechuanaland and West Africa. *Bull World Health Organ* (1963) **28**(5–6):595–613.
4. (FAO) FaAootUN. Programme against African trypanosomiasis. (2007). Available from: ftp://ftp.fao.org/docrep/fao/010/a1289e/a1289e01.pdf
5. Simarro PP, Franco JR, Cecchi G, Paone M, Diarra A, Ruiz Postigo JA, et al. Human African trypanosomiasis in non-endemic countries (2000-2010). *J Travel Med* (2012) **19**(1):44–53. doi:10.1111/j.1708-8305.2011.00576.x
6. Lejon V, Boelaert M, Jannin J, Moore A, Buscher P. The challenge of *Trypanosoma brucei gambiense* sleeping sickness diagnosis outside Africa. *Lancet Infect Dis* (2003) **3**(12):804–8. doi:10.1016/S1473-3099(03)00834-X
7. Hursey BS. The programme against African trypanosomiasis: aims, objectives and achievements. *Trends Parasitol* (2001) **17**(1):2–3. doi:10.1016/S1471-4922(00)01851-1
8. Xong HV, Vanhamme L, Chamekh M, Chimfwembe CE, Van Den Abbeele J, Pays A, et al. A VSG expression site-associated gene confers resistance to human serum in *Trypanosoma rhodesiense*. *Cell* (1998) **95**(6):839–46. doi:10.1016/S0092-8674(00)81706-7
9. Lugli EB, Pouliot M, Portela Mdel P, Loomis MR, Raper J. Characterization of primate trypanosome lytic factors. *Mol Biochem Parasitol* (2004) **138**(1):9–20. doi:10.1016/j.molbiopara.2004.07.004
10. Vanhollebeke B, Truc P, Poelvoorde P, Pays A, Joshi PP, Katti R, et al. Human *Trypanosoma evansi* infection linked to a lack of apolipoprotein L-I. *N Engl J Med* (2006) **355**(26):2752–6. doi:10.1056/NEJMoa063265
11. Perez-Morga D, Vanhollebeke B, Paturiaux-Hanocq F, Nolan DP, Lins L, Homble F, et al. Apolipoprotein L-I promotes trypanosome lysis by forming pores in lysosomal membranes. *Science* (2005) **309**(5733):469–72. doi:10.1126/science.1114566
12. Vanhamme L, Paturiaux-Hanocq F, Poelvoorde P, Nolan DP, Lins L, Van Den Abbeele J, et al. Apolipoprotein L-I is the trypanosome lytic factor of human serum. *Nature* (2003) **422**(6927):83–7. doi:10.1038/nature01461
13. Namangala B. Contribution of innate immune responses towards resistance to African trypanosome infections. *Scand J Immunol* (2012) **75**(1):5–15. doi:10.1111/j.1365-3083.2011.02619.x
14. Wheeler RJ. The trypanolytic factor-mechanism, impacts and applications. *Trends Parasitol* (2010) **26**(9):457–64. doi:10.1016/j.pt.2010.05.005
15. Leppert BJ, Mansfield JM, Paulnock DM. The soluble variant surface glycoprotein of African trypanosomes is recognized by a macrophage scavenger receptor and induces I kappa B alpha degradation independently of TRAF6-mediated TLR signaling. *J Immunol* (2007) **179**(1):548–56. doi:10.4049/jimmunol.179.1.548
16. Pan W, Ogunremi O, Wei G, Shi M, Tabel H. CR3 (CD11b/CD18) is the major macrophage receptor for IgM antibody-mediated phagocytosis of African trypanosomes: diverse effect on subsequent synthesis of tumor necrosis factor alpha and nitric oxide. *Microbes Infect* (2006) **8**(5):1209–18. doi:10.1016/j.micinf.2005.11.009
17. Shi M, Wei G, Pan W, Tabel H. *Trypanosoma congolense* infections: antibody-mediated phagocytosis by Kupffer cells. *J Leukoc Biol* (2004) **76**(2):399–405. doi:10.1189/jlb.1003500
18. Drennan MB, Stijlemans B, Van den Abbeele J, Quesniaux VJ, Barkhuizen M, Brombacher F, et al. The induction of a type 1 immune response following a *Trypanosoma brucei* infection is MyD88 dependent. *J Immunol* (2005) **175**(4):2501–9. doi:10.4049/jimmunol.175.4.2501
19. Donelson JE. Antigenic variation and the African trypanosome genome. *Acta Trop* (2003) **85**(3):391–404. doi:10.1016/S0001-706X(02)00237-1
20. Uzonna JE, Kaushik RS, Gordon JR, Tabel H. Cytokines and antibody responses during *Trypanosoma congolense* infections in two inbred mouse strains that differ in resistance. *Parasite Immunol* (1999) **21**(2):57–71. doi:10.1046/j.1365-3024.1999.00202.x
21. Hertz CJ, Filutowicz H, Mansfield JM. Resistance to the African trypanosomes is IFN-gamma dependent. *J Immunol* (1998) **161**(12):6775–83.
22. Shi M, Pan W, Tabel H. Experimental African trypanosomiasis: IFN-gamma mediates early mortality. *Eur J Immunol* (2003) **33**(1):108–18. doi:10.1002/immu.200390013
23. Magez S, Radwanska M, Beschin A, Sekikawa K, De Baetselier P. Tumor necrosis factor alpha is a key mediator in the regulation of experimental *Trypanosoma brucei* infections. *Infect Immun* (1999) **67**(6):3128–32.
24. Okomo-Assoumou MC, Daulouede S, Lemesre JL, N'Zila-Mouanda A, Vincendeau P. Correlation of high serum levels of tumor necrosis factor-alpha with disease severity in human African trypanosomiasis. *Am J Trop Med Hyg* (1995) **53**(5):539–43.
25. Uzonna JE, Kaushik RS, Gordon JR, Tabel H. Experimental murine *Trypanosoma congolense* infections. I. Administration of anti-IFN-gamma antibodies alters trypanosome-susceptible mice to a resistant-like phenotype. *J Immunol* (1998) **161**(10):5507–15.
26. Magez S, Radwanska M, Drennan M, Fick L, Baral TN, Brombacher F, et al. Interferon-gamma and nitric oxide in combination with antibodies are key protective host immune factors during *Trypanosoma congolense* Tc13 Infections. *J Infect Dis* (2006) **193**(11):1575–83. doi:10.1086/503808

27. Olsson T, Bakhiet M, Edlund C, Hojeberg B, Van der Meide PH, Kristensson K. Bidirectional activating signals between *Trypanosoma brucei* and CD8+ T cells: a trypanosome-released factor triggers interferon-gamma production that stimulates parasite growth. *Eur J Immunol* (1991) **21**(10):2447–54. doi:10.1002/eji.1830211022
28. Shi M, Wei G, Pan W, Tabel H. Experimental African trypanosomiasis: a subset of pathogenic, IFN-gamma-producing, MHC class II-restricted CD4+ T cells mediates early mortality in highly susceptible mice. *J Immunol* (2006) **176**(3):1724–32. doi:10.4049/jimmunol.176.3.1724
29. Namangala B, Noel W, De Baetselier P, Brys L, Beschin A. Relative contribution of interferon-gamma and interleukin-10 to resistance to murine African trypanosomosis. *J Infect Dis* (2001) **183**(12):1794–800. doi:10.1086/320731
30. Uzonna JE, Kaushik RS, Gordon JR, Tabel H. Immunoregulation in experimental murine *Trypanosoma congolense* infection: anti-IL-10 antibodies reverse trypanosome-mediated suppression of lymphocyte proliferation in vitro and moderately prolong the lifespan of genetically susceptible BALB/c mice. *Parasite Immunol* (1998) **20**(6):293–302. doi:10.1046/j.1365-3024.1998.00156.x
31. Wilson AJ, Dar FK, Paris J. A study on the transmission of salivarian trypanosomes isolated from wild tsetse flies. *Trop Anim Health Prod* (1972) **4**(1):14–22. doi:10.1007/BF02357090
32. Dwinger RH, Clifford DJ, Agyemang K, Gettinby G, Grieve AS, Kora S, et al. Comparative studies on N'Dama and zebu cattle following repeated infections with *Trypanosoma congolense*. *Res Vet Sci* (1992) **52**(3):292–8. doi:10.1016/0034-5288(92)90027-Y
33. Ziegelbauer K, Multhaup G, Overath P. Molecular characterization of two invariant surface glycoproteins specific for the bloodstream stage of *Trypanosoma brucei*. *J Biol Chem* (1992) **267**(15):10797–803.
34. Magez S, Radwanska M. African trypanosomiasis and antibodies: implications for vaccination, therapy and diagnosis. *Future Microbiol* (2009) **4**(8):1075–87. doi:10.2217/fmb.09.65
35. Periasamy S, Dhiman R, Barnes PF, Paidipally P, Tvinnereim A, Bandaru A, et al. Programmed death 1 and cytokine inducible SH2-containing protein dependent expansion of regulatory T cells upon stimulation With *Mycobacterium tuberculosis*. *J Infect Dis* (2011) **203**(9):1256–63. doi:10.1093/infdis/jir011
36. Sarfo BY, Wilson NO, Bond VC, Stiles JK. *Plasmodium berghei* ANKA infection increases Foxp3, IL-10 and IL-2 in CXCL-10 deficient C57BL/6 mice. *Malar J* (2011) **10**:69. doi:10.1186/1475-2875-10-69
37. Okwor I, Liu D, Beverley SM, Uzonna JE. Inoculation of killed *Leishmania major* into immune mice rapidly disrupts immunity to a secondary challenge via IL-10-mediated process. *Proc Natl Acad Sci U S A* (2009) **106**(33):13951–6. doi:10.1073/pnas.0905184106
38. Guilliams M, Oldenhove G, Noel W, Herin M, Brys L, Loi P, et al. African trypanosomiasis: naturally occurring regulatory T cells favor trypanotolerance by limiting pathology associated with sustained type 1 inflammation. *J Immunol* (2007) **179**(5):2748–57. doi:10.4049/jimmunol.179.5.2748
39. Wei G, Tabel H. Regulatory T cells prevent control of experimental African trypanosomiasis. *J Immunol* (2008) **180**(4):2514–21. doi:10.4049/jimmunol.180.4.2514
40. Okwor I, Onyilagha C, Kuriakose S, Mou Z, Jia P, Uzonna JE. Regulatory T cells enhance susceptibility to experimental *Trypanosoma congolense* infection independent of mouse genetic background. *PLoS Negl Trop Dis* (2012) **6**(7):e1761. doi:10.1371/journal.pntd.0001761
41. Onyilagha C, Okwor I, Kuriakose S, Singh R, Uzonna J. Low-dose intradermal infection with *Trypanosoma congolense* leads to expansion of regulatory T cells and enhanced susceptibility to reinfection. *Infect Immun* (2014) **82**(3):1074–83. doi:10.1128/IAI.01028-13
42. Clayton CE, Selkirk ME, Corsini CA, Ogilvie BM, Askonas BA. Murine trypanosomiasis: cellular proliferation and functional depletion in the blood, peritoneum, and spleen related to changes in bone marrow stem cells. *Infect Immun* (1980) **28**(3):824–31.
43. Grosskinsky CM, Ezekowitz RA, Berton G, Gordon S, Askonas BA. Macrophage activation in murine African trypanosomiasis. *Infect Immun* (1983) **39**(3):1080–6.
44. Paulnock DM, Coller SP. Analysis of macrophage activation in African trypanosomiasis. *J Leukoc Biol* (2001) **69**(5):685–90.
45. Shakibaei M, Frevert U. Cell surface interactions between *Trypanosoma congolense* and macrophages during phagocytosis in vitro. *J Protozool* (1992) **39**(1):224–35. doi:10.1111/j.1550-7408.1992.tb01306.x
46. Mosser DM, Roberts JF. *Trypanosoma brucei*: recognition in vitro of two developmental forms by murine macrophages. *Exp Parasitol* (1982) **54**(3):310–6. doi:10.1016/0014-4894(82)90040-6
47. Schleifer KW, Mansfield JM. Suppressor macrophages in African trypanosomiasis inhibit T cell proliferative responses by nitric oxide and prostaglandins. *J Immunol* (1993) **151**(10):5492–503.
48. Dempsey WL, Mansfield JM. Lymphocyte function in experimental African trypanosomiasis. V. Role of antibody and the mononuclear phagocyte system in variant-specific immunity. *J Immunol* (1983) **130**(1):405–11.
49. Kuriakose S, Muleme HM, Onyilagha C, Singh R, Jia P, Uzonna JE. Diminazene aceturate (Berenil) modulates the host cellular and inflammatory responses to *Trypanosoma congolense* infection. *PLoS One* (2012) **7**(11):e48696. doi:10.1371/journal.pone.0048696
50. Kaushik RS, Uzonna JE, Zhang Y, Gordon JR, Tabel H. Innate resistance to experimental African trypanosomiasis: differences in cytokine (TNF-alpha, IL-6, IL-10 and IL-12) production by bone marrow-derived macrophages from resistant and susceptible mice. *Cytokine* (2000) **12**(7):1024–34. doi:10.1006/cyto.2000.0685
51. Mackaness GB. Cellular immunity and the parasite. *Adv Exp Med Biol* (1977) **93**:65–73. doi:10.1007/978-1-4615-8855-9_5
52. Martinez FO, Gordon S. The M1 and M2 paradigm of macrophage activation: time for reassessment. *F1000Prime Rep* (2014) **6**:13. doi:10.12703/P6-13
53. Edwards JP, Zhang X, Frauwirth KA, Mosser DM. Biochemical and functional characterization of three activated macrophage populations. *J Leukoc Biol* (2006) **80**(6):1298–307. doi:10.1189/jlb.0406249
54. Anthony RM, Urban JF Jr, Alem F, Hamed HA, Rozo CT, Boucher JL, et al. Memory T(H)2 cells induce alternatively activated macrophages to mediate protection against nematode parasites. *Nat Med* (2006) **12**(8):955–60. doi:10.1038/nm1451
55. Reyes JL, Terrazas LI. The divergent roles of alternatively activated macrophages in helminthic infections. *Parasite Immunol* (2007) **29**(12):609–19. doi:10.1111/j.1365-3024.2007.00973.x
56. Namangala B, De Baetselier P, Noel W, Brys L, Beschin A. Alternative versus classical macrophage activation during experimental African trypanosomosis. *J Leukoc Biol* (2001) **69**(3):387–96.
57. Gomez-Rodriguez J, Stijlemans B, De Muylder G, Korf H, Brys L, Berberof M, et al. Identification of a parasitic immunomodulatory protein triggering the development of suppressive M1 macrophages during African trypanosomiasis. *J Infect Dis* (2009) **200**(12):1849–60. doi:10.1086/648374
58. Medzhitov R. Toll-like receptors and innate immunity. *Nat Rev Immunol* (2001) **1**(2):135–45. doi:10.1038/35100529
59. Akira S, Uematsu S, Takeuchi O. Pathogen recognition and innate immunity. *Cell* (2006) **124**(4):783–801. doi:10.1016/j.cell.2006.02.015
60. Beutler BA. TLRs and innate immunity. *Blood* (2009) **113**(7):1399–407. doi:10.1182/blood-2008-07-019307
61. Akira S, Takeda K. Toll-like receptor signalling. *Nat Rev Immunol* (2004) **4**(7):499–511. doi:10.1038/nri1391
62. Kawai T, Akira S. The roles of TLRs, RLRs and NLRs in pathogen recognition. *Int Immunol* (2009) **21**(4):317–37. doi:10.1093/intimm/dxp017
63. Kawai T, Akira S. Toll-like receptors and their crosstalk with other innate receptors in infection and immunity. *Immunity* (2011) **34**(5):637–50. doi:10.1016/j.immuni.2011.05.006
64. Takeda K, Akira S. Toll-like receptors in innate immunity. *Int Immunol* (2005) **17**(1):1–14. doi:10.1093/intimm/dxh186
65. Campos MA, Closel M, Valente EP, Cardoso JE, Akira S, Alvarez-Leite JI, et al. Impaired production of proinflammatory cytokines and host resistance to acute infection with *Trypanosoma cruzi* in mice lacking functional myeloid differentiation factor 88. *J Immunol* (2004) **172**(3):1711–8. doi:10.4049/jimmunol.172.3.1711
66. Adachi K, Tsutsui H, Kashiwamura S, Seki E, Nakano H, Takeuchi O, et al. *Plasmodium berghei* infection in mice induces liver injury by an IL-12- and toll-like receptor/myeloid differentiation factor 88-dependent mechanism. *J Immunol* (2001) **167**(10):5928–34. doi:10.4049/jimmunol.167.10.5928

67. Ropert C, Almeida IC, Closel M, Travassos LR, Ferguson MA, Cohen P, et al. Requirement of mitogen-activated protein kinases and I kappa B phosphorylation for induction of proinflammatory cytokines synthesis by macrophages indicates functional similarity of receptors triggered by glycosylphosphatidylinositol anchors from parasitic protozoa and bacterial lipopolysaccharide. *J Immunol* (2001) **166**(5):3423–31.
68. Muraille E, De Trez C, Brait M, De Baetselier P, Leo O, Carlier Y. Genetically resistant mice lacking MyD88-adapter protein display a high susceptibility to *Leishmania major* infection associated with a polarized Th2 response. *J Immunol* (2003) **170**(8):4237–41. doi:10.4049/jimmunol.170.8.4237
69. Del Rio L, Butcher BA, Bennouna S, Hieny S, Sher A, Denkers EY. *Toxoplasma gondii* triggers myeloid differentiation factor 88-dependent IL-12 and chemokine ligand 2 (monocyte chemoattractant protein 1) responses using distinct parasite molecules and host receptors. *J Immunol* (2004) **172**(11):6954–60. doi:10.4049/jimmunol.172.11.6954
70. Krishnegowda G, Hajjar AM, Zhu J, Douglass EJ, Uematsu S, Akira S, et al. Induction of proinflammatory responses in macrophages by the glycosylphosphatidylinositols of *Plasmodium falciparum*: cell signaling receptors, glycosylphosphatidylinositol (GPI) structural requirement, and regulation of GPI activity. *J Biol Chem* (2005) **280**(9):8606–16. doi:10.1074/jbc.M413541200
71. Naik RS, Branch OH, Woods AS, Vijaykumar M, Perkins DJ, Nahlen BL, et al. Glycosylphosphatidylinositol anchors of *Plasmodium falciparum*: molecular characterization and naturally elicited antibody response that may provide immunity to malaria pathogenesis. *J Exp Med* (2000) **192**(11):1563–76. doi:10.1084/jem.192.11.1563
72. Bafica A, Santiago HC, Goldszmid R, Ropert C, Gazzinelli RT, Sher A. Cutting edge: TLR9 and TLR2 signaling together account for MyD88-dependent control of parasitemia in *Trypanosoma cruzi* infection. *J Immunol* (2006) **177**(6):3515–9. doi:10.4049/jimmunol.177.6.3515
73. Amin DN, Vodnala SK, Masocha W, Sun B, Kristensson K, Rottenberg ME. Distinct toll-like receptor signals regulate cerebral parasite load and interferon alpha/beta and tumor necrosis factor alpha-dependent T-cell infiltration in the brains of *Trypanosoma brucei*-infected mice. *J Infect Dis* (2012) **205**(2):320–32. doi:10.1093/infdis/jir734
74. Cobb MH. MAP kinase pathways. *Prog Biophys Mol Biol* (1999) **71**(3–4):479–500. doi:10.1016/S0079-6107(98)00056-X
75. Force T, Bonventre JV. Growth factors and mitogen-activated protein kinases. *Hypertension* (1998) **31**(1 Pt 2):152–61. doi:10.1161/01.HYP.31.1.152
76. Darnell JE Jr, Kerr IM, Stark GR. Jak-STAT pathways and transcriptional activation in response to IFNs and other extracellular signaling proteins. *Science* (1994) **264**(5164):1415–21. doi:10.1126/science.8197455
77. Murray PJ. The JAK-STAT signaling pathway: input and output integration. *J Immunol* (2007) **178**(5):2623–9. doi:10.4049/jimmunol.178.5.2623
78. Mertens C, Zhong M, Krishnaraj R, Zou W, Chen X, Darnell JE Jr. Dephosphorylation of phosphotyrosine on STAT1 dimers requires extensive spatial reorientation of the monomers facilitated by the N-terminal domain. *Genes Dev* (2006) **20**(24):3372–81. doi:10.1101/gad.1485406
79. Rawlings JS, Rosler KM, Harrison DA. The JAK/STAT signaling pathway. *J Cell Sci* (2004) **117**(Pt 8):1281–3. doi:10.1242/jcs.00963
80. Makela SM, Strengell M, Pietila TE, Osterlund P, Julkunen I. Multiple signaling pathways contribute to synergistic TLR ligand-dependent cytokine gene expression in human monocyte-derived macrophages and dendritic cells. *J Leukoc Biol* (2009) **85**(4):664–72. doi:10.1189/jlb.0808503
81. Levy DE, Darnell JE Jr. Stats: transcriptional control and biological impact. *Nat Rev Mol Cell Biol* (2002) **3**(9):651–62. doi:10.1038/nrm909
82. Rao KM. MAP kinase activation in macrophages. *J Leukoc Biol* (2001) **69**(1):3–10.
83. May MJ, Ghosh S. Rel/NF-kappa B and I kappa B proteins: an overview. *Semin Cancer Biol* (1997) **8**(2):63–73. doi:10.1006/scbi.1997.0057
84. Paulnock DM, Freeman BE, Mansfield JM. Modulation of innate immunity by African trypanosomes. *Parasitology* (2010) **137**(14):2051–63. doi:10.1017/S0031182010001460
85. Stahl P, Ruppert V, Schwarz RT, Meyer T. *Trypanosoma cruzi* evades the protective role of interferon-gamma-signaling in parasite-infected cells. *PLoS One* (2014) **9**(10):e110512. doi:10.1371/journal.pone.0110512
86. Valere A, Garnotel R, Villena I, Guenounou M, Pinon JM, Aubert D. Activation of the cellular mitogen-activated protein kinase pathways ERK, P38 and JNK during *Toxoplasma gondii* invasion. *Parasite* (2003) **10**(1):59–64. doi:10.1051/parasite/2003101p59
87. Denkers EY, Butcher BA, Del Rio L, Kim L. Manipulation of mitogen-activated protein kinase/nuclear factor-kappaB-signaling cascades during intracellular *Toxoplasma gondii* infection. *Immunol Rev* (2004) **201**:191–205. doi:10.1111/j.0105-2896.2004.00180.x
88. Coller SP, Mansfield JM, Paulnock DM. Glycosylinositolphosphate soluble variant surface glycoprotein inhibits IFN-gamma-induced nitric oxide production via reduction in STAT1 phosphorylation in African trypanosomiasis. *J Immunol* (2003) **171**(3):1466–72. doi:10.4049/jimmunol.171.3.1466
89. Dagenais TR, Freeman BE, Demick KP, Paulnock DM, Mansfield JM. Processing and presentation of variant surface glycoprotein molecules to T cells in African trypanosomiasis. *J Immunol* (2009) **183**(5):3344–55. doi:10.4049/jimmunol.0802005
90. Mertens B, Taylor K, Muriuki C, Rocchi M. Cytokine mRNA profiles in trypanotolerant and trypanosusceptible cattle infected with the protozoan parasite *Trypanosoma congolense*: protective role for interleukin-4? *J Interferon Cytokine Res* (1999) **19**(1):59–65. doi:10.1089/107999099314423
91. O'Gorman GM, Park SD, Hill EW, Meade KG, Mitchell LC, Agaba M, et al. Cytokine mRNA profiling of peripheral blood mononuclear cells from trypanotolerant and trypanosusceptible cattle infected with *Trypanosoma congolense*. *Physiol Genomics* (2006) **28**(1):53–61. doi:10.1152/physiolgenomics.00100.2006
92. Magez S, Truyens C, Merimi M, Radwanska M, Stijlemans B, Brouckaert P, et al. P75 tumor necrosis factor-receptor shedding occurs as a protective host response during African trypanosomiasis. *J Infect Dis* (2004) **189**(3):527–39. doi:10.1086/381151
93. Kaushik RS, Uzonna JE, Gordon JR, Tabel H. Innate resistance to *Trypanosoma congolense* infections: differential production of nitric oxide by macrophages from susceptible BALB/c and resistant C57Bl/6 mice. *Exp Parasitol* (1999) **92**(2):131–43. doi:10.1006/expr.1999.4408
94. Sternberg J, Mabbott N, Sutherland I, Liew FY. Inhibition of nitric oxide synthesis leads to reduced parasitemia in murine *Trypanosoma brucei* infection. *Infect Immun* (1994) **62**(5):2135–7.
95. Vincendeau P, Daulouede S, Veyret B, Darde ML, Bouteille B, Lemesre JL. Nitric oxide-mediated cytostatic activity on *Trypanosoma brucei gambiense* and *Trypanosoma brucei brucei*. *Exp Parasitol* (1992) **75**(3):353–60. doi:10.1016/0014-4894(92)90220-5
96. Singh R, Kone BC, Gounni AS, Uzonna JE. Molecular regulation of *Trypanosoma congolense*-induced nitric oxide production in macrophages. *PLoS One* (2013) **8**(3):e59631. doi:10.1371/journal.pone.0059631
97. Kuriakose S, Muleme H, Onyilagha C, Okeke E, Uzonna JE. Diminazene aceturate (Berenil) modulates LPS induced pro-inflammatory cytokine production by inhibiting phosphorylation of MAPKs and STAT proteins. *Innate Immun* (2013) **20**:760–73.

11

Modulation of Innate Immune Mechanisms to Enhance *Leishmania* Vaccine-Induced Immunity: Role of Coinhibitory Molecules

Sreenivas Gannavaram[1]*, Parna Bhattacharya[1], Nevien Ismail[1], Amit Kaul[1], Rakesh Singh[2] and Hira L. Nakhasi[1]*

[1]Laboratory of Emerging Pathogens, Division of Emerging and Transfusion Transmitted Diseases, Food and Drug Administration, Silver Spring, MD, USA, [2]Department of Biochemistry, Banaras Hindu University, Varanasi, India

Edited by:
Diana Bahia,
Universidade Federal de Minas Gerais, Brazil

Reviewed by:
Simona Stäger,
University of Quebec, Canada
Patricia Talamás-Rohana,
CINVESTAV, Mexico

***Correspondence:**
Sreenivas Gannavaram
sreenivas.gannavaram@fda.hhs.gov;
Hira Nakhasi
hira.nakhasi@fda.hhs.gov

No licensed human vaccines are currently available against any parasitic disease including leishmaniasis. Several antileishmanial vaccine formulations have been tested in various animal models, including genetically modified live-attenuated parasite vaccines. Experimental infection studies have shown that *Leishmania* parasites utilize a broad range of strategies to undermine effector properties of host phagocytic cells, i.e., dendritic cells (DCs) and macrophages (MΦ). Furthermore, *Leishmania* parasites have evolved strategies to actively inhibit T_H1 polarizing functions of DCs and to condition the infected MΦ toward anti-inflammatory/alternative/M2 phenotype. The altered phenotype of phagocytic cells is characterized by decreased production of antimicrobial reactive oxygen, nitrogen molecules, and pro-inflammatory cytokines, such as IFN-γ, IL-12, and TNF-α. These early events limit the activation of T_H1-effector cells and set the stage for pathogenesis. Furthermore, this early control of innate immunity by the virulent parasites results in substantial alteration in the adaptive immunity characterized by reduced proliferation of $CD4^+$ and $CD8^+$ T cells and T_H2-biased immunity that results in production of anti-inflammatory cytokines, such as TGF-β, and IL-10. More recent studies have also documented the induction of coinhibitory ligands, such as CTLA-4, PD-L1, CD200, and Tim-3, that induce exhaustion and/or non-proliferation in antigen-experienced T cells. Most of these studies focus on viral infections in chronic phase, thus limiting the direct application of these results to parasitic infections and much less to parasitic vaccines. However, these studies suggest that vaccine-induced protective immunity can be modulated using strategies that enhance the costimulation that might reduce the threshold necessary for T cell activation and conversely by strategies that reduce or block inhibitory molecules, such as PD-L1 and CD200. In this review, we will focus on the polarization of antigen-presenting cells and subsequent role of costimulatory and coinhibitory molecules in mediating vaccine-induced immunity using live-attenuated *Leishmania* parasites as specific examples.

Keywords: leishmaniasis, vaccine, coinhibitory molecules, CD200R, PD-L1, CTLA-4, TIM-3, CD200

INTRODUCTION

Visceral leishmaniasis (VL) has an annual incidence of 0.2–0.4 million cases globally and results in about 60,000 deaths (1). It has been reported that 0.7–1.2 million cases of cutaneous leishmaniasis (CL) occur around the globe (2). The disease currently affects 12 million people with 350 million people at risk of infection. The majority of patients with CL or VL develop a long-term protective immunity after cure from infection, which suggests that development of an effective vaccine against leishmaniasis is possible (3, 4). High frequency of asymptomatic infections, toxicity of currently available drugs, and increasing parasite resistance underline the need for an effective prophylactic vaccine against leishmaniasis (5).

Extensive studies in murine models that acquire resistance to *Leishmania major* infection have resulted in a broader understating of the mediators of protection, primarily a T_H1-biased response (6). However, the principal determinants for inducing a strong T_H1-type response following infection with virulent *Leishmania* parasites including expression of MHC-I/II, CD40, CD80/CD86, and cytokines, such as IL-12, by the antigen-presenting cells (APCs) [dendritic cells (DCs) and macrophages (MΦ)] are targets for alteration by the parasites (7, 8). *Leishmania* parasites can survive in a wide range of cell types. The parasites are phagocytosed by neutrophils first and are taken up by the MΦ and DCs without causing an overt immunological reaction (6). The multifarious interactions between *Leishmania* and the host APCs have profound effects on the final outcome of the interaction, either host resistance or susceptibility. MΦ are not only the primary host cell for *Leishmania* that permit parasite proliferation but also the major effector cells in eliminating the infection. The effective clearance of parasites by MΦ depends on the activation of an appropriate immune response usually initiated by the DCs (8).

REPROGRAMING OF MACROPHAGE/DC DIFFERENTIATION

Leishmania donovani resides predominantly in host MΦ where it enters by phagocytosis and establishes itself within parasitophorous vacuoles (9). Macrophage responses to parasites lead to discrete, stereotyped phenotypes, which are usually a combination of inflammatory and anti-inflammatory functions (8). This plasticity in macrophage function has been defined either as classically activated (M1 phenotype) representing leishmanicidal activity or an alternatively activated state (M2 phenotype) that confers susceptibility to infection (10).

Classically activated macrophages (CAM) are primed by T_H1 (or pro-inflammatory) cytokines and triggered by microbial products to produce antimicrobial molecules, such as reactive oxygen species (ROS) and nitric oxide (NO), through the action of inducible nitric oxide synthase (iNOS) and subsequently acquire a heightened effector function (11–13). CAM activation is also characterized by the induction of an array of pro-inflammatory cytokines, such as tumor necrosis factor [TNFα, IL-1β, and interleukin (IL)-12], which amplify T_H1 immune responses (14–16). Specifically, IL-12 is a pivotal cytokine required for $CD4^+$ T_H1 development and production of IFN-γ (17). Since CAMs acquire properties necessary for the destruction of invading pathogens and priming the innate immune response, *Leishmania* parasites have evolved mechanisms to subvert microbicidal functions of the CAMs through depletion of antimicrobial molecules, such as ROS and NO (18–20), thereby reprograming the infected MΦ to an alternative activation state (21). Alternative macrophage activation is mainly induced by T_H2 cytokines (22, 23) that antagonize the microbicidal properties of CAMs (8).

The enhanced T_H2 response during virulent *Leishmania* infection leads to enhanced arginase activities in the MΦ, a prototypic alternative activation marker in mouse MΦ that allows parasite survival (22–24). Of the T_H2 cytokines, IL-10 has emerged as the principal cytokine responsible for disease pathogenesis (25). IL-10 induced during *Leishmania* infection inhibits microbicidal activity of MΦ by attenuating the generation of NO and pro-inflammatory cytokines (26). Therefore, reprograming of the MΦ enables *Leishmania* parasites to evade the antimicrobial innate immune response and to proliferate within the phagolysosome of the macrophage.

The critical balance between host-protective T_H1 (or pro-inflammatory) versus disease-promoting T_H2 (or anti-inflammatory) effector responses determines the outcome of infection in leishmaniasis. This outcome is dictated by the relative levels of IL-12 and IL-10 produced by the classical/M1 and alternative/M2 MΦ, respectively (8, 16, 21). Previous studies have also shown that the MΦ are able to regulate immunologic outcomes on their own *via* directing the T cell response (27, 28). For example, infection of MΦ with virulent *Leishmania* induces a parasite-favoring T_H2 response instead of a host-protective T_H1 response (29). Interestingly, studies have shown that classical/pro-inflammatory/M1 MΦ have a direct role in the induction of T_H1-polarized response (16, 27, 28, 30). Such M1 macrophage-induced T cell polarization was further linked with an induction of protection in several vaccine studies including recombinant BCG, attenuated *West Nile virus* (WNV), and live-attenuated measles virus (31–33). These recent studies demonstrated that the vaccine antigens, indeed, reprogram the MΦ to induce a pro-inflammatory response and result in improved protective immunity. Likewise, our studies with genetically modified live-attenuated *L. donovani* parasites revealed that these attenuated parasites induce classical activation of MΦ that direct host-protective T_H1 response in mice (21). A recent study has also reported that M2 polarization of monocytes–macrophages is a hallmark of post kala-azar dermal leishmaniasis that sustains chronic lesions (10). Additionally, repolarization of the monocytes to M1 type by antileishmanial drugs as stated above suggests that switching from M2 to M1 phenotype might also be important in a therapeutic setting similar to what our studies indicated with prophylactic vaccination (21).

Although an effective clearance of parasites involves antimicrobial activity of the MΦ, the main host cell for *Leishmania* parasites, it also requires the activation of an appropriate immune response that is initiated by DCs (7, 8). Since MΦ are not the major producers of IL-12, potentiation of adaptive immune response depends primarily on the DC-derived IL-12 (34). In addition, parasite-infected MΦ are incompetent in priming naive $CD4^+$ T cells as well as stimulating antigen (Ag)-specific $CD4^+$ T cells (35). Thus, DCs play a crucial role in coordinating immune responses in leishmaniasis by providing the IL-12 necessary for the induction of protective T_H1 response (36, 37). In both *L. major* and *L. donovani* mouse studies, the central role of DCs in determining the T_H1/T_H2 balance as well as the outcome of disease is demonstrated (36, 38). Specifically, *L. donovani* infection is shown to cause DCs to produce IL-12 that leads to NK cell activation (36). In the spleen, interaction between DCs and T cells occurs in the periarteriolar lymphoid sheath (PALS) into which DC and T cells migrate from the marginal zone (MZ) along chemokine gradients, especially CCL19/21, and this interaction between DCs and the T cells is necessary to produce protective immunity against *L. donovani* (39). Leon et al. also reported that monocyte–DCs formed at the infection site control the induction of protective T_H1 responses against *Leishmania* (37). Nevertheless, infection with virulent *Leishmania* parasites is shown to prevent development of protective T_H1 immunity by dysregulating DC function (7, 40). Additionally, Moll et al. reported that *L. major* parasites inhibit T_H1 polarizing functions of epidermal-derived DCs (Langerhans cells) by upregulating IL-4 receptor expression along with concomitant downregulation of IL-12p40 production (41). Of interest, virulent *Leishmania* interferes with intracellular signaling in DCs, which affects their antigen-presenting functions, and hence, their ability to induce optimal cell-mediated immunity against the parasite (7, 42). DCs infected with intracellular pathogens, such as *Leishmania*, have been shown to elicit an MHC class-I-dependent $CD8^+$ T-cell response, a process referred to as cross-presentation (43). Further studies have demonstrated the role of *Leishmania*-induced cleavage of proteins, such as SNAREs, that mediate the fusion of phagosomes with lysosomes, a key process in the antigen processing and presentation to the T cells (44, 45).

The role of DCs in priming a protective T_H1 response is also illustrated by DC vaccination studies where exogenous administration of antigen-loaded DCs showed promising results in the treatment of different forms of leishmaniasis (46, 47). Curiously, adoptive transfer of DCs pulsed *ex vivo* with soluble *L. donovani* Ags (SLDA) to naive mice induced the Ag-specific production of IFN-γ and increased the percentage of activation markers on spleen lymphocytes. Moreover, SLDA-pulsed DCs engineered by retroviral gene transfer techniques to secrete high levels of biologically active murine-IL-12 augmented this immune response, further indicating the central role of DC secreted IL-12 in potentiating a protective response (48). Additionally, Schnitzer et al. have reported that DCs pulsed *ex vivo* with *L. major* antigen-induced protection in otherwise susceptible mice against subsequent challenges with the parasites (49). While it is well documented that antigen-loaded DC-based vaccines can induce protective immunity against *Leishmania* pathogenesis (50), live-attenuated parasites might be more practical in inducing protective immunity by modulating the DC function as illustrated by studies performed using DC-based vaccines. Taken together, these studies indicate that for an antileishmanial vaccine to induce protective immunity, T_H1 favoring conditions in the DCs is a necessary requirement and live-attenuated parasite vaccines might be able to set in those conditions.

DETERMINANTS OF T CELL IMMUNITY

An effective long-term protection requires activation of adaptive immunity mediated by T lymphocytes. APCs, primarily DCs, present processed antigens in combination with major histocompatibility complex to naive T cells displaying the corresponding T cell receptor. However, this presentation also requires additional signals arising from positive and negative coreceptors, i.e., costimulatory and coinhibitory molecules. The response of the T cell thus activated requires an amalgam of signals from the immunological synapse, and T cell activation can only occur when the stimulatory/inhibitory signals are able to overcome a certain threshold (51). Of the stimulatory signals, CD28 is the best-studied costimulatory molecule. Involvement of CD28 in the immunological synapse decreases the amount of antigen necessary to elicit T cell activation. Importantly, inflammatory signals regulate expression of CD28-binding partners, such as B7-1 (CD80) and B7-2 (CD86). Previous studies have identified suppression of costimulatory signals in DCs infected with virulent *Leishmania* parasites that results in poor IL-12 production and $CD4^+$ T cell priming (36, 52–54). A T_H2-polarized response was observed in $CD40^{-/-}$ mice as indicated by excessive IL-4 and low IFN-γ. A direct role for CD40:CD40L ligation was shown not only in production of IL-12p70 but also in activation of T cells by the DCs (55, 56). Deficient expression of costimulatory CD80 was observed in *L. donovani*-infected MΦ (57). Similarly, CD86 has been shown to orchestrate either a T_H1 or T_H2 type response depending on the relative contribution of CD80 (58). More recently, several immune inhibitory mechanisms have been explored, which control exacerbated immune response of the host to prevent self-damage from unchecked inflammation (59). These mechanisms not only control the effector function of immune cells but also reprogram them for their alternate functions, such as humoral immunity and tissue remodeling, to maintain homeostasis between an immune response and immune tolerance. The interplay between signals arising from costimulatory molecules and coinhibitory molecules has been identified as a critical determinant in T cell activation. It has been hypothesized that costimulatory signals may act like a "rheostat" to modulate T cell activation in that costimulatory molecules reduce the TCR signaling threshold necessary for T cell activation, whereas inhibitory molecules restrict T cell activation (60).

Although there is considerable debate over the requirements for maintaining protection against reinfection in *Leishmania*, studies have shown that antigen-specific memory T cells are a principal component of protective immunity to intracellular pathogens, such as *L. major* (61–65). The memory T cells are distinguished by their ability (i) to survive long term in secondary lymphoid tissues and (ii) to undergo rapid and robust proliferation upon reinfection and acquisition of effector function. Yet, our understanding of T cell differentiation and memory formation is mainly derived from models of acute viral and bacterial infections, such as *Lymphocytic Choriomeningitis Virus*, *Vaccinia virus*, and *Listeria monocytogenes*. Persistent infections, such as *Leishmania*, may differ in significant ways from these models in the T cell response dynamics. Several dysfunctions including severe limitation in T cell expansion, delay in peak T cell expansion, anergy, and expression of exhaustion markers in chronic infections have been reported (66–69).

In an acute infection, the T cell response typically follows three phases: expansion, contraction, and memory. During the first phase, upon presentation of antigens by DCs, naive T cells divide and differentiate into effector cells that acquire the ability to produce the pro-inflammatory cytokines IFN-γ and TNF, as well as cytotoxic proteins such as granzymes and perforin. This cascade of events by which T lymphocytes undergo differentiation and clonal expansion is regulated by signaling *via* antigens, costimulation, and cytokine receptors that induce the expression of transcription factors that dictate the fate of the T cells to acquire either an effector function or memory precursors (70). In case of CD8+ T cells, the cell fates are controlled by a coordinated set of changes in the expression of the transcription factors Id2, T-bet, and Blimp-1, which promote terminally differentiated effector cells, and Foxo1, TCF-1, Eomes and Bcl-6, which promote development of memory precursors (71–73). Earlier studies in *L. donovani* demonstrated the role of CD8+ T cells in the resolution of infection (74). In virulent *L. donovani* infection of mice, induction of exhausted CD8+ T cells has been demonstrated (66). More recent studies have shown a role for the transcription factor IRF-5 in regulating the antigen-specific CD8+ T cells responses during murine *L. donovani* infection (75). Inflammatory response generated by IRF-5 is shown to induce the expression of HIF-α in DCs and to limit CD8+ T cell expansion (75). Development of CD4+ T cells responses following either a chronic infection in comparison are less well studied (76). However, the dynamics of T cell effector/memory responses in a prophylactic vaccine setting are less well understood. It may be argued that in contrast to virulent parasites, prophylactic vaccines are composed of attenuated parasites due to their inherent immunomodulatory attributes may be able to induce conditions optimal for the immune system to generate memory cells that confer protection against subsequent infection (**Figure 1**). On the other hand, recombinant antigens may, and often, require adjuvants to enhance T cell responses. More systematic studies with prophylactic candidate vaccines are necessary to reveal the necessary conditions that must precede a strong protective response.

COINHIBITORY MOLECULES IN PATHOGENESIS AND VACCINE IMMUNITY

In a chronic infection, it is essential that the host immune response be appropriately controlled to respond to and remove pathogens while avoiding excessive production of inflammatory cytokines and chemical mediators, such as ROS. The immune inhibitory signals, therefore, can dampen the effects of excessive immune reactions, which can lead to increased tissue damage and morbidity and mortality. In the following sections, we describe the role of various coinhibitory molecules and discuss their potential roles in inducing protective immunity with implication to live-attenuated vaccines.

CTLA-4

B7 molecules expressed by DCs and MΦ are upregulated following activation. These B7 molecules B7-1 (CD80) and B7-2 (CD86) have dual specificity for receptors CD28 and CTLA-4 on T cells (77). CTLA-4 is an inhibitory coreceptor that is induced rapidly by activated T cells (78). Because of its higher affinity to CD80/86 than CD28, CTLA-4 often outcompetes and excludes CD28 from the immunological synapse, thereby causing suppression of T cell activation. In symptomatic VL, both the splenic aspirates and PBMCs from *L. donovani*-infected humans showed higher level of CTLA-4 expression (69). Higher CTLA-4 expression was also observed in PKDL lesions suggesting an association with persistence of parasites (79). Higher frequencies of CTLA-4+ CD4+ T cells have been identified in HIV-*Leishmania* coinfections that are associated with a poor immunological profile that might explain the persistence and relapse of the *Leishmania* infection (80). However, treatment with anti-CTLA-4 has shown mixed results. Blockade of CTLA-4 resulted in enhanced granuloma maturation indicating parasite-killing activity in the liver; however, in splenic aspirate cultures of human VL samples, CTLA-4 blockade did not show an increase in IFN-γ production (69, 81). The role of CTLA-4 in T cell activation has not been studied in a *Leishmania* vaccine setting. Reports have shown that preventing CTLA-4 interactions can improve T cell activation in other vaccine models. For example, *in vivo* blockade of CTLA-4 enhances antigen-specific CD4+ T cell responses after peptide immunization in complete Freund's adjuvant for cryptococcal infection (82). Transient CTLA-4 blockade increased the number of memory CD8+ T cells during low-dose *Listeria* infection in mice, and CTLA-4 blockade enhanced the response of memory CD8+ T cells (83). The mechanism by which blockade of CTLA-4 might exert its functions is not completely understood. It has been proposed that blocking CTLA-4 might dampen the IL-10-mediated suppressive effects since CTLA-4 is very highly expressed on Treg cells (84). CTLA-4 plays an important role in the homeostasis and function of a population of suppressive activities of Treg cells (85). A direct role for CTLA-4 in T cell activation is suggested by studies that showed removal of B7 molecules from the surface of APCs by blocking CTLA-4 and, thus, increase the signal threshold necessary for T cell activation (86). Taken together, these studies suggest that CTLA-4 might be a potential target for modulation by prophylactic vaccines because of its effects first on T cell activation by its effects on B7 molecules and second by its ability to affect Treg cells.

PROGRAMED DEATH-1

Programed death-1 (PD-1) is an inhibitory receptor that is inducibly expressed on stimulated CD4+ T cells, CD8+ T cells, B cells, and monocytes (66, 68, 87). PD-1 modulates T cell receptors and CD28 signaling through recruitment of the phosphatase SHP2. PD-1 binds to B7 family ligands, PD-L1 (B7-H1/CD274) and PD-L2 (B7-DC/CD273) (88). PD-L1 expression occurs on a wide variety of cells, whereas PD-L2 expression is restricted to APCs (DCs, monocytes, and some B cell subsets). Expression of both PD-1 ligands is modulated by cytokines, such as IL-2, IL-7, IL-15, IL-21, and IFN-γ (87, 89). Expression of PD-1 is induced by antigen receptor ligation and is dampened after resolution of infection in the absence of TCR signaling (90). However, sustained PD-1 expression is reported in several chronic disease

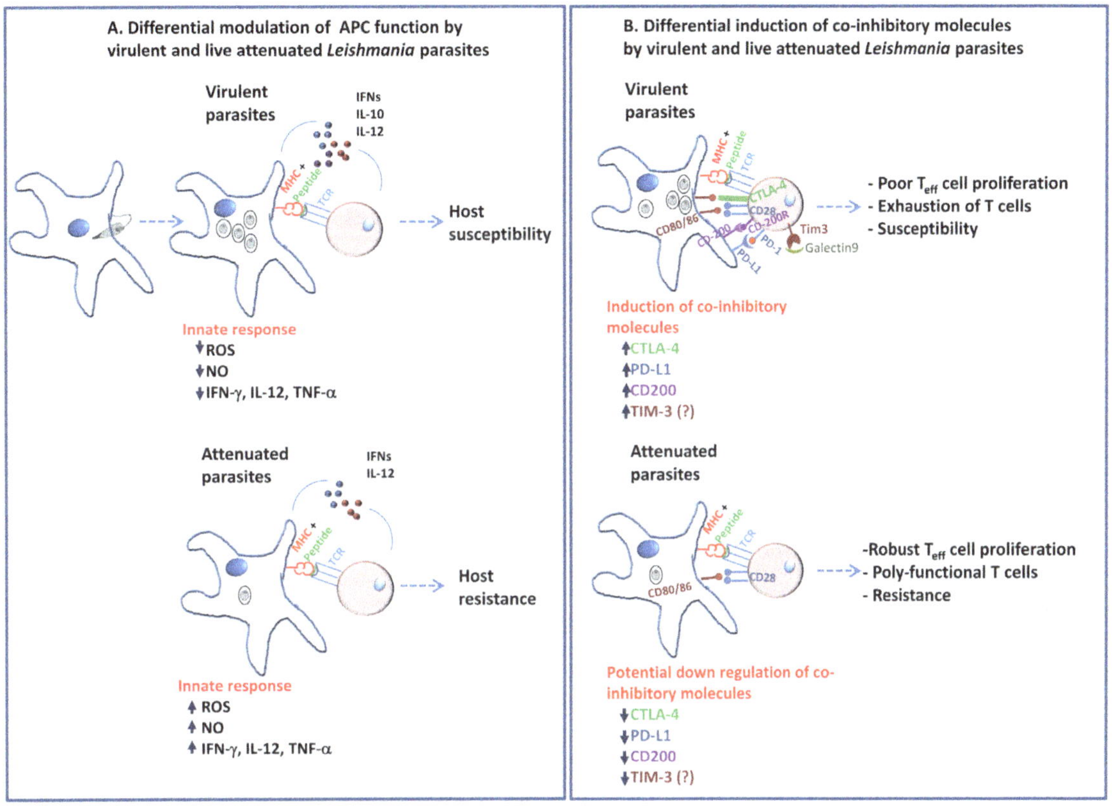

FIGURE 1 | A schematic diagram describing the differential modulation of antigen-presenting cells in the early innate response and role of coinhibitory molecules in adaptive immunity in *Leishmania* pathogenesis is shown. (A) Virulent *Leishmania donovani* parasites are known to suppress antigen-presenting activity of DCs and macrophages. This is shown to be accompanied by reduction in the inflammatory cytokines as compared to infection with live-attenuated *L. donovani* parasites. These differences in the early interaction with the host APCs by the parasites determine the outcome, i.e., host susceptibility in case of virulent parasites or host resistance in case of attenuated parasites. **(B)** Previous studies have shown the role of coinhibitory molecules, such as CTLA-4, PD-L1, and CD200, in facilitating the parasite survival by inducing a restrained pro-inflammatory (T_H1) environment. The role of Tim-3 in *Leishmania* pathogenesis remains to be explored. The role of coinhibitory molecules in shaping the T cell immunity suggests that attenuated parasites might downregulate these pathways compared to virulent parasites in order to induce protective immunity. Blockade of the coinhibitory signals during priming by use of adjuvants might show potential improvement in the vaccine-induced immunity.

models, including leishmaniasis, malaria, and Chagas (91–95). Specific parasite molecules (LPG from *Leishmania mexicana*) have also been shown to induce PD-1 expression in $CD8^+$ T cells and PD-L2 in MΦ (96). Sustained PD-1 expression is maintained primarily due to continuous TCR ligation, and PD-1 ligation dramatically shifts the dose-response curve, making T cells much less sensitive to T-cell receptor-generated signals (97). TCR signals promote demethylation of regulatory regions of the PD-l locus. Additionally, transcription factor T-bet binds upstream of the PD-1 gene and represses its transcription. Since T-bet expression is downregulated in persistent TCR stimulation, low T-bet expression enables PD-1 transcription (98). Elevated expression of T-bet and PD-1 in $CD4^+$ T cells in the patients of tuberculosis was also reported (99). In addition, recent studies have also identified transcriptional factors, hypoxia-inducible factor (HIF)-1α and signal transducer and activation of transcription 3 (STAT-3), that act on the promoter of PD-L1 to regulate its expression. In addition, microRNAs, including miR-570, miR-513, miR-197, miR-34a, and miR-200, negatively regulate PD-L1 (100).

Sustained PD-1 expression is a hallmark of dysfunctional T cells and commonly found in chronic infections. Importantly, several studies have shown that interfering with the PD-1 pathway rescues function in exhausted T cells including *Leishmania* and *Plasmodium* infections (66, 68, 87, 93, 101). Similar observations were also described in hepatitis C virus-infected patients that interfering with the PD-1/PD-L1 pathway during the early stage of immune responses can result in improved T cell responses (102). PD-L1 blockade during acute herpes simplex virus (HSV)-1 infection is shown to increase the magnitude and polyfunctionality of the HSV-specific $CD8^+$ effector responses (103).

An important role as a clinical intervention for the PD-1 pathway in T cell exhaustion was first shown in LCMV infection in mice (104) and later demonstrated to occur in a diverse range of clinically relevant chronic human infections, such as HIV infection, hepatitis B and hepatitis C virus infections, and cancer (105–107). Several studies also tested combination of PD-1 and other ligands in blockade therapies (108, 109). Compared to clinical advances in treating cancer, blockade of inhibitory receptors as a strategy to treat chronically infected patients has lagged behind.

In addition to the role of PD-L1/PD-1 in regulating T cell function, it has been proposed that PD-L1/PD-1 interaction through the regulation by IFN-γ might allow for recognition of minor epitopes that otherwise would not be selected. This is consistent with the observation that PD-1 engagement preferentially inhibits low-avidity antigen receptors and thus allows expansion of immunodominant clones (89). In addition, in the absence of PD-L1/PD-1 interactions, APCs provide stronger stimulation to T cells. This is supported by experiments showing a significant enhancement of HIV-specific T cell responses by blocking PD-L1 (110). As a result, blockade of the PD-1 pathway has more significant effects in promoting T cell activation during conditions of suboptimal antigen presentation, such as with low antigen dose or with weak or low numbers of APCs (89).

Taken together, these studies suggest that PD-1 pathway blockade is an attractive strategy to improve prophylactic vaccination by allowing for better antigen selection and better T cell functional responses. Even though PD-1 pathway blockade is an attractive strategy to improve prophylactic vaccination, few studies have focused on the PD-1 pathway during early stages of T cell responses. Most importantly, PD-1 blockade has been studied in assessing $CD8^+$ T cell responses, whereas the role of the PD-1 pathway on $CD4^+$ T cell differentiation, an important mediator of antileishmanial immunity, has been relatively neglected.

CD200R

CD200R is an inhibitory receptor expressed in the cells of lymphoid lineage, such as NK cells, $CD4^+$, and $CD8^+$ T cells, especially upon stimulation (59). Its ligand, CD200 (OX2), is a glycoprotein expressed on a broad number of cell types, including solid tumors and hematologic malignancies. Studies have also shown differential expression of CD200R on T cell subsets in mice and humans including effector and central memory T cells (111).

CD200R1 is an Ig superfamily transmembrane glycoprotein expressed on the surface of myeloid cells; it can also be induced in certain T-cell subsets (112, 113). CD200R1 interacts with CD200, which is also an Ig superfamily transmembrane glycoprotein, to downregulate myeloid cell functions. CD200 is expressed on the surface of a variety of cells including neurons, epithelial cells, endothelial cells, fibroblasts, lymphoid cells, and astrocytes (112, 114–116). The regulation of CD200R1 signaling can occur by posttranslational modifications mainly by phosphorylation of tyrosine residues in the CD200R1 cytoplasmic tail or by the inducible expression or the downregulation of either CD200R1 or CD200. Each of these mechanisms can be exploited by pathogens.

Unlike most immune inhibitory receptors, CD200R1 does not contain an immunoreceptor tyrosine-based inhibitory motif (ITIM). Stimulation by CD200 leads to the phosphorylation of these tyrosines by Src kinases, which recruit the adapter protein downstream of tyrosine kinase (Dok2) through its PTB domain (117). Dok2 serves as the major initiator of signaling through CD200R1, beginning with binding to Ras-GTPase-activating protein (RasGAP) and is required for CD200R1 function (117). This is in contrast to ITIM containing inhibitory receptors, which utilize SHPs and SHIP-1 as the major initiator proteins and Dok proteins as secondary modulators of downstream signaling (117, 118).

Compared to other inhibitory molecules, CD200 is less well studied in parasitic infections. *Leishmania amazonensis*, which causes severe disease in both humans and mice, induces CD200 expression in bone marrow MΦ from wild-type mice (119). Induction of CD200 upon infection was essential for the parasite virulence and development of systemic Leishmaniasis. Addition of CD200-Fc restored the virulence of *L. amazonensis* in mice lacking CD200. Distinct differences in CD200 signaling have been identified with other *Leishmania* species. *L. major*, which causes only localized cutaneous lesions, does not induce CD200 in MΦ. Interestingly, however, treatment with CD200-Fc of *L. major*-infected mice caused the parasite to disseminate into a systemic infection similar to that of *L. amazonensis* (119). *L. amazonensis* has evolved to utilize CD200 expression as a mechanism for inhibiting both NO production and induction of iNOS during infection. Interestingly, *L. amazonensis* increased CD200 expression on MΦ. MΦ have generally been found to express CD200R1, which can then interact with non-myeloid cells expressing CD200. Interestingly, mice lacking CD200 induced robust antiviral immunity including virus-specific $CD4^+$ T cell responses (120). In two chronic infection models including *Salmonella* and *Schistosoma*, upregulated CD200R expression has been shown to result in poor multifunctional $CD4^+$ T cell responses (112). Upregulated CD200R expression was associated with the development of T_H2-type response. These results indicate that CD200 could be a candidate target for investigation as a regulator of T cell immunity in prophylactic antileishmanial vaccines since multifunctional $CD4^+$ T cells have been shown to be correlated with protection in several *Leishmania* vaccine studies (**Figure 1**) (121, 122).

TIM-3

The Tim-3 protein is a member of the T cell immunoglobulin and mucin domain (Tim) family, which encompasses a group of type-I transmembrane proteins expressed by both innate and adaptive cell types within the immune system. All Tim proteins are expressed on the cell surface and have been shown to function as receptors for soluble or cell-associated ligands. Additionally, certain Tim proteins can be shed from the cell surface and be found in soluble forms, suggesting a role as cell-free ligands. To date, the IgV domain of Tim-3 has been shown to interact with phosphatidylserine (PS) displayed on the surface of apoptotic cells, the alarmin protein high mobility group box 1 (HMGB1), and Galectin-9, a widely expressed soluble protein with specificity for carbohydrate chains containing β-galactoside sugars. Of the several ligands, the interaction between Tim-3 and Galectin-9 has been shown to impact $CD4^+$ T cell functions. Addition of Galectin-9 to cultured Tim-3 T_H1-type $CD4^+$ T (T_H1) cells induced apoptosis and necrosis, while injection of Galectin-9 into mice blunted immune responses driven by antigen-specific T_H1 cells. These studies indicated that binding of Galectin-9 to Tim-3 results in the suppression of T cell responses, which supports the notion that Tim-3 functions as an inhibitory receptor for T cells.

A role for Tim-3 in regulating responses by CD4$^+$ T cells has been suggested based on studies where addition of Galectin-9 induced the death of Tim-3$^+$ T_H1 cells *in vitro*. Other studies using autoimmune disease models suggest that ligation of Tim-3 by Galectin-9 leads to suppression of T_H1-dependent immune responses. Studies of microbial infections also sought to investigate how Tim-3 and Galectin-9 influence CD4$^+$ T cell responses. Tim-3 overexpression was observed on T cells that were senescent and dysfunctional in HCV infection, and blockade of Tim-3 rescued dysfunctional CD4$^+$ and CD8$^+$ T cells (123). In contrast, active tuberculosis patients exhibited increased numbers of Tim-3-expressing CD4$^+$ and CD8$^+$ T cells, which preferentially displayed polarized effector memory phenotypes. Tim-3$^+$ CD4$^+$ and CD8$^+$ T cell subsets showed greater effector functions for producing Th1/Th22 cytokines and CTL effector molecules than those lacking Tim-3 expression, and Tim-3$^+$ T cells controlled intracellular Mtb replication in MΦ (124). Overall, these findings suggest that Tim-3 can promote CD4$^+$ T cell responses mounted against Mtb infection. This conclusion contrasts with other studies that showed Tim-3 expression correlated with T cell exhaustion. Taken together, these findings raise the possibility that Tim-3 function is influenced by context and that Tim-3 may inhibit or promote CD4$^+$ T cell responses depending upon the microbe involved and the characteristics of the immune response elicited by the infection. Elevated expression of Tim-3 expression on T cells from HIV-1-infected individuals correlated positively with HIV-1 viral load and CD38 expression and inversely with CD4$^+$ T cell count. In progressive HIV-1 infection, Tim-3 expression was upregulated on HIV-1-specific CD8$^+$ T cells (125). Blocking the Tim-3 signaling pathway restored proliferation and enhanced cytokine production in HIV-1-specific T cells (125). In addition, blocking of Tim-3 rescued macrophage and T cell function in HIV positive tuberculosis patients (126). Persistence of HCV was associated with lower frequencies of IL-21-producing CD4$^+$ T cells, reduced proliferation, and increased expression of the inhibitory receptors Tim-3, PD-1, and CTLA-4 on HCV-specific CD8$^+$ T cells. Progression to persistent infection was accompanied by increased plasma levels of the Tim-3 ligand Galectin-9 and expansion of Gal-9 expressing regulatory Treg cells (127). Thus, studies in a diverse pathogen models suggests a net negative impact of Tim-3 expression on T cell-dependent antiviral immune responses. However, studies in Mtb point out that the Tim-3 expression and T cell function may be pathogen-specific. Consistent with this, in murine malaria, expression of both Tim-3 and Galectin-9 were associated with liver damage and acute lung injury (128, 129), suggesting a role in pathogenesis. It would be of interest to study if Tim-3 signaling has any role in prophylactic parasite vaccines.

CONCLUSION

Chronic infections with intracellular parasites have been shown to induce several inhibitory molecules that subvert the development of protective immunity in the host and favor the survival of the parasite by mainly preventing the development of functional T cell immunity. While the role of these coinhibitory signals in virulence is being explored, important questions regarding their roles in shaping the protective immunity in a prophylactic vaccine setting are being recognized. The role of inhibitor signals in not only regulating T cell functions but modifying the adaptive immunity suggests that these molecules could be potential targets for modulation by candidate vaccines. Importantly, studies using blockade of ligands, such as PD-L1 and CTLA-4, have revealed important insights as to the clinical importance of such interventions.

AUTHOR CONTRIBUTIONS

All authors listed have made substantial, direct, and intellectual contribution to the work and approved it for publication.

REFERENCES

1. Alvar J, Velez ID, Bern C, Herrero M, Desjeux P, Cano J, et al. Leishmaniasis worldwide and global estimates of its incidence. *PLoS One* (2012) **7**:e35671. doi:10.1371/journal.pone.0035671
2. Okwor I, Uzonna J. Social and economic burden of human leishmaniasis. *Am J Trop Med Hyg* (2016) **94**:489–93. doi:10.4269/ajtmh.15-0408
3. Kumar R, Goto Y, Gidwani K, Cowgill KD, Sundar S, Reed SG. Evaluation of *ex vivo* human immune response against candidate antigens for a visceral leishmaniasis vaccine. *Am J Trop Med Hyg* (2010) **82**:808–13. doi:10.4269/ajtmh.2010.09-0341
4. Selvapandiyan A, Dey R, Gannavaram S, Lakhal-Naouar I, Duncan R, Salotra P, et al. Immunity to visceral leishmaniasis using genetically defined live-attenuated parasites. *J Trop Med* (2012) **2012**:631460. doi:10.1155/2012/631460
5. Sundar S, Chakravarty J. Leishmaniasis: an update of current pharmacotherapy. *Expert Opin Pharmacother* (2013) **14**:53–63. doi:10.1517/14656566.2013.755515
6. Kaye P, Scott P. Leishmaniasis: complexity at the host-pathogen interface. *Nat Rev Microbiol* (2011) **9**:604–15. doi:10.1038/nrmicro2608
7. Soong L. Modulation of dendritic cell function by *Leishmania* parasites. *J Immunol* (2008) **180**:4355–60. doi:10.4049/jimmunol.180.7.4355
8. Liu D, Uzonna JE. The early interaction of *Leishmania* with macrophages and dendritic cells and its influence on the host immune response. *Front Cell Infect Microbiol* (2012) **2**:83. doi:10.3389/fcimb.2012.00083
9. Liévin-Le Moal V, Loiseau PM. *Leishmania* hijacking of the macrophage intracellular compartments. *FEBS J* (2016) **283**:598–607. doi:10.1111/febs.13601
10. Mukhopadhyay D, Mukherjee S, Roy S, Dalton JE, Kundu S, Sarkar A, et al. M2 polarization of monocytes-macrophages is a hallmark of Indian post kala-azar dermal leishmaniasis. *PLoS Negl Trop Dis* (2015) **9**:e0004145. doi:10.1371/journal.pntd.0004145
11. Gordon S. Alternative activation of macrophages. *Nat Rev Immunol* (2003) **3**:23–35. doi:10.1038/nri978
12. Muraille E, Leo O, Moser M. TH1/TH2 paradigm extended: macrophage polarization as an unappreciated pathogen-driven escape mechanism? *Front Immunol* (2014) **5**:603. doi:10.3389/fimmu.2014.00603
13. Osorio EY, Zhao W, Espitia C, Saldarriaga O, Hawel L, Byus CV, et al. Progressive visceral leishmaniasis is driven by dominant parasite-induced STAT6 activation and STAT6-dependent host arginase 1 expression. *PLoS Pathog* (2012) **8**:e1002417. doi:10.1371/journal.ppat.1002417
14. Arango Duque G, Descoteaux A. Macrophage cytokines: involvement in immunity and infectious diseases. *Front Immunol* (2014) **5**:491. doi:10.3389/fimmu.2014.00491
15. Lopes MF, Costa-Da-Silva AC, Dosreis GA. Innate immunity to *Leishmania* infection: within phagocytes. *Mediators Inflamm* (2014) **2014**:754965. doi:10.1155/2014/754965
16. Mills CD, Ley K. M1 and M2 macrophages: the chicken and the egg of immunity. *J Innate Immun* (2014) **6**:716–26. doi:10.1159/000364945

17. Hsieh CS, Macatonia SE, Tripp CS, Wolf SF, O'garra A, Murphy KM. Development of TH1 CD4+ T cells through IL-12 produced by *Listeria*-induced macrophages. *Science* (1993) **260**:547–9. doi:10.1126/science.8097338
18. Liew FY, Millott S, Parkinson C, Palmer RM, Moncada S. Macrophage killing of *Leishmania* parasite *in vivo* is mediated by nitric oxide from L-arginine. *J Immunol* (1990) **144**:4794–7.
19. Murray HW. Cell-mediated immune response in experimental visceral leishmaniasis. II. Oxygen-dependent killing of intracellular *Leishmania donovani* amastigotes. *J Immunol* (1982) **129**:351–7.
20. Olivier M, Gregory DJ, Forget G. Subversion mechanisms by which *Leishmania* parasites can escape the host immune response: a signaling point of view. *Clin Microbiol Rev* (2005) **18**:293–305. doi:10.1128/CMR.18.2.293-305.2005
21. Bhattacharya P, Dey R, Dagur PK, Kruhlak M, Ismail N, Debrabant A, et al. Genetically modified live attenuated *Leishmania donovani* parasites induce innate immunity through classical activation of macrophages that direct the Th1 response in mice. *Infect Immun* (2015) **83**:3800–15. doi:10.1128/IAI.00184-15
22. Iniesta V, Gomez-Nieto LC, Corraliza I. The inhibition of arginase by N(omega)-hydroxy-L-arginine controls the growth of *Leishmania* inside macrophages. *J Exp Med* (2001) **193**:777–84. doi:10.1084/jem.193.6.777
23. Kropf P, Fuentes JM, Fahnrich E, Arpa L, Herath S, Weber V, et al. Arginase and polyamine synthesis are key factors in the regulation of experimental leishmaniasis *in vivo*. *FASEB J* (2005) **19**:1000–2. doi:10.1096/fj.04-3416fje
24. Iniesta V, Gomez-Nieto LC, Molano I, Mohedano A, Carcelen J, Miron C, et al. Arginase I induction in macrophages, triggered by Th2-type cytokines, supports the growth of intracellular *Leishmania* parasites. *Parasite Immunol* (2002) **24**:113–8. doi:10.1046/j.1365-3024.2002.00444.x
25. Bhattacharya P, Ali N. Involvement and interactions of different immune cells and their cytokines in human visceral leishmaniasis. *Rev Soc Bras Med Trop* (2013) **46**:128–34. doi:10.1590/0037-8682-0022-2012
26. Noel W, Raes G, Hassanzadeh Ghassabeh G, De Baetselier P, Beschin A. Alternatively activated macrophages during parasite infections. *Trends Parasitol* (2004) **20**:126–33. doi:10.1016/j.pt.2004.01.004
27. Mills CD, Kincaid K, Alt JM, Heilman MJ, Hill AM. M-1/M-2 macrophages and the Th1/Th2 paradigm. *J Immunol* (2000) **164**:6166–73. doi:10.4049/jimmunol.164.12.6166
28. Rath M, Muller I, Kropf P, Closs EI, Munder M. Metabolism *via* arginase or nitric oxide synthase: two competing arginine pathways in macrophages. *Front Immunol* (2014) **5**:532. doi:10.3389/fimmu.2014.00532
29. Chakkalath HR, Titus RG. *Leishmania major*-parasitized macrophages augment Th2-type T cell activation. *J Immunol* (1994) **153**:4378–87.
30. Cassol E, Cassetta L, Alfano M, Poli G. Macrophage polarization and HIV-1 infection. *J Leukoc Biol* (2010) **87**:599–608. doi:10.1189/jlb.1009673
31. Rennick LJ, De Vries RD, Carsillo TJ, Lemon K, Van Amerongen G, Ludlow M, et al. Live-attenuated measles virus vaccine targets dendritic cells and macrophages in muscle of nonhuman primates. *J Virol* (2015) **89**:2192–200. doi:10.1128/JVI.02924-14
32. Saiga H, Nieuwenhuizen N, Gengenbacher M, Koehler AB, Schuerer S, Moura-Alves P, et al. The recombinant BCG deltaureC::hly vaccine targets the AIM2 inflammasome to induce autophagy and inflammation. *J Infect Dis* (2015) **211**:1831–41. doi:10.1093/infdis/jiu675
33. Xie G, Luo H, Tian B, Mann B, Bao X, Mcbride J, et al. A *West Nile virus* NS4B-P38G mutant strain induces cell intrinsic innate cytokine responses in human monocytic and macrophage cells. *Vaccine* (2015) **33**:869–78. doi:10.1016/j.vaccine.2014.12.056
34. Von Stebut E, Belkaid Y, Jakob T, Sacks DL, Udey MC. Uptake of *Leishmania major* amastigotes results in activation and interleukin 12 release from murine skin-derived dendritic cells: implications for the initiation of anti-*Leishmania* immunity. *J Exp Med* (1998) **188**:1547–52. doi:10.1084/jem.188.8.1547
35. Kima PE, Soong L, Chicharro C, Ruddle NH, Mcmahon-Pratt D. *Leishmania*-infected macrophages sequester endogenously synthesized parasite antigens from presentation to CD4+ T cells. *Eur J Immunol* (1996) **26**:3163–9. doi:10.1002/eji.1830261249
36. Gorak PM, Engwerda CR, Kaye PM. Dendritic cells, but not macrophages, produce IL-12 immediately following *Leishmania donovani* infection. *Eur J Immunol* (1998) **28**:687–95. doi:10.1002/(SICI)1521-4141(199802)28:02<687::AID-IMMU687>3.0.CO;2-N
37. Leon B, Lopez-Bravo M, Ardavin C. Monocyte-derived dendritic cells formed at the infection site control the induction of protective T helper 1 responses against *Leishmania*. *Immunity* (2007) **26**:519–31. doi:10.1016/j.immuni.2007.01.017
38. Suzue K, Kobayashi S, Takeuchi T, Suzuki M, Koyasu S. Critical role of dendritic cells in determining the Th1/Th2 balance upon *Leishmania major* infection. *Int Immunol* (2008) **20**:337–43. doi:10.1093/intimm/dxm147
39. Ato M, Maroof A, Zubairi S, Nakano H, Kakiuchi T, Kaye PM. Loss of dendritic cell migration and impaired resistance to *Leishmania donovani* infection in mice deficient in CCL19 and CCL21. *J Immunol* (2006) **176**:5486–93. doi:10.4049/jimmunol.176.9.5486
40. Jebbari H, Stagg AJ, Davidson RN, Knight SC. *Leishmania major* promastigotes inhibit dendritic cell motility *in vitro*. *Infect Immun* (2002) **70**:1023–6. doi:10.1128/IAI.70.2.1023-1026.2002
41. Moll H, Scharner A, Kampgen E. Increased interleukin 4 (IL-4) receptor expression and IL-4-induced decrease in IL-12 production by Langerhans cells infected with *Leishmania major*. *Infect Immun* (2002) **70**:1627–30. doi:10.1128/IAI.70.3.1627-1630.2002
42. Xin L, Li K, Soong L. Down-regulation of dendritic cell signaling pathways by *Leishmania amazonensis* amastigotes. *Mol Immunol* (2008) **45**:3371–82. doi:10.1016/j.molimm.2008.04.018
43. Bertholet S, Debrabant A, Afrin F, Caler E, Mendez S, Tabbara KS, et al. Antigen requirements for efficient priming of CD8+ T cells by *Leishmania major*-infected dendritic cells. *Infect Immun* (2005) **73**:6620–8. doi:10.1128/IAI.73.10.6620-6628.2005
44. Houde M, Bertholet S, Gagnon E, Brunet S, Goyette G, Laplante A, et al. Phagosomes are competent organelles for antigen cross-presentation. *Nature* (2003) **425**:402–6. doi:10.1038/nature01912
45. Matheoud D, Moradin N, Bellemare-Pelletier A, Shio MT, Hong WJ, Olivier M, et al. *Leishmania* evades host immunity by inhibiting antigen cross-presentation through direct cleavage of the SNARE VAMP8. *Cell Host Microbe* (2013) **14**:15–25. doi:10.1016/j.chom.2013.06.003
46. Flohe SB, Bauer C, Flohe S, Moll H. Antigen-pulsed epidermal Langerhans cells protect susceptible mice from infection with the intracellular parasite *Leishmania major*. *Eur J Immunol* (1998) **28**:3800–11. doi:10.1002/(SICI)1521-4141(199811)28:11<3800::AID-IMMU3800>3.3.CO;2-S
47. Moll H, Berberich C. Dendritic cell-based vaccination strategies: induction of protective immunity against leishmaniasis. *Immunobiology* (2001) **204**:659–66. doi:10.1078/0171-2985-00105
48. Ahuja SS, Reddick RL, Sato N, Montalbo E, Kostecki V, Zhao W, et al. Dendritic cell (DC)-based anti-infective strategies: DCs engineered to secrete IL-12 are a potent vaccine in a murine model of an intracellular infection. *J Immunol* (1999) **163**:3890–7.
49. Schnitzer JK, Berzel S, Fajardo-Moser M, Remer KA, Moll H. Fragments of antigen-loaded dendritic cells (DC) and DC-derived exosomes induce protective immunity against *Leishmania major*. *Vaccine* (2010) **28**:5785–93. doi:10.1016/j.vaccine.2010.06.077
50. Freitas-Silva R, Brelaz-De-Castro MC, Rezende AM, Pereira VR. Targeting dendritic cells as a good alternative to combat *Leishmania* spp. *Front Immunol* (2014) **5**:604. doi:10.3389/fimmu.2014.00604
51. Kamphorst AO, Araki K, Ahmed R. Beyond adjuvants: immunomodulation strategies to enhance T cell immunity. *Vaccine* (2015) **33**(Suppl 2):B21–8. doi:10.1016/j.vaccine.2014.12.082
52. Padigel UM, Perrin PJ, Farrell JP. The development of a Th1-type response and resistance to *Leishmania major* infection in the absence of CD40-CD40L costimulation. *J Immunol* (2001) **167**:5874–9. doi:10.4049/jimmunol.167.10.5874
53. Murray HW, Lu CM, Brooks EB, Fichtl RE, Devecchio JL, Heinzel FP. Modulation of T-cell costimulation as immunotherapy or immunochemotherapy in experimental visceral leishmaniasis. *Infect Immun* (2003) **71**:6453–62. doi:10.1128/IAI.71.11.6453-6462.2003
54. Tuladhar R, Natarajan G, Satoskar AR. Role of co-stimulation in leishmaniasis. *Int J Biol Sci* (2011) **7**:1382–90. doi:10.7150/ijbs.7.1382
55. Kamanaka M, Yu P, Yasui T, Yoshida K, Kawabe T, Horii T, et al. Protective role of CD40 in *Leishmania major* infection at two distinct phases of cell-mediated immunity. *Immunity* (1996) **4**:275–81. doi:10.1016/S1074-7613(00)80435-5
56. Caux C, Massacrier C, Vanbervliet B, Dubois B, Van Kooten C, Durand I, et al. Activation of human dendritic cells through CD40 cross-linking. *J Exp Med* (1994) **180**:1263–72. doi:10.1084/jem.180.4.1263

57. Kaye PM, Rogers NJ, Curry AJ, Scott JC. Deficient expression of co-stimulatory molecules on *Leishmania*-infected macrophages. *Eur J Immunol* (1994) **24**:2850–4. doi:10.1002/eji.1830241140
58. Elloso MM, Scott P. Expression and contribution of B7-1 (CD80) and B7-2 (CD86) in the early immune response to *Leishmania major* infection. *J Immunol* (1999) **162**:6708–15.
59. Norde WJ, Hobo W, Van Der Voort R, Dolstra H. Coinhibitory molecules in hematologic malignancies: targets for therapeutic intervention. *Blood* (2012) **120**:728–36. doi:10.1182/blood-2012-02-412510
60. Kamphorst AO, Ahmed R. Manipulating the PD-1 pathway to improve immunity. *Curr Opin Immunol* (2013) **25**:381–8. doi:10.1016/j.coi.2013.03.003
61. Sacks DL. Vaccines against tropical parasitic diseases: a persisting answer to a persisting problem. *Nat Immunol* (2014) **15**:403–5. doi:10.1038/ni.2853
62. Peters NC, Pagan AJ, Lawyer PG, Hand TW, Henrique Roma E, Stamper LW, et al. Chronic parasitic infection maintains high frequencies of short-lived Ly6C+CD4+ effector T cells that are required for protection against re-infection. *PLoS Pathog* (2014) **10**:e1004538. doi:10.1371/journal.ppat.1004538
63. Zaph C, Uzonna J, Beverley SM, Scott P. Central memory T cells mediate long-term immunity to *Leishmania major* in the absence of persistent parasites. *Nat Med* (2004) **10**:1104–10. doi:10.1038/nm1108
64. Colpitts SL, Scott P. The early generation of a heterogeneous CD4+ T cell response to *Leishmania major*. *J Immunol* (2010) **185**:2416–23. doi:10.4049/jimmunol.1000483
65. Glennie ND, Yeramilli VA, Beiting DP, Volk SW, Weaver CT, Scott P. Skin-resident memory CD4+ T cells enhance protection against *Leishmania major* infection. *J Exp Med* (2015) **212**:1405–14. doi:10.1084/jem.20142101
66. Joshi T, Rodriguez S, Perovic V, Cockburn IA, Stager S. B7-H1 blockade increases survival of dysfunctional CD8(+) T cells and confers protection against *Leishmania donovani* infections. *PLoS Pathog* (2009) **5**:e1000431. doi:10.1371/journal.ppat.1000431
67. Bunn PT, Stanley AC, De Labastida Rivera F, Mulherin A, Sheel M, Alexander CE, et al. Tissue requirements for establishing long-term CD4+ T cell-mediated immunity following *Leishmania donovani* infection. *J Immunol* (2014) **192**:3709–18. doi:10.4049/jimmunol.1300768
68. Esch KJ, Juelsgaard R, Martinez PA, Jones DE, Petersen CA. Programmed death 1-mediated T cell exhaustion during visceral leishmaniasis impairs phagocyte function. *J Immunol* (2013) **191**:5542–50. doi:10.4049/jimmunol.1301810
69. Gautam S, Kumar R, Singh N, Singh AK, Rai M, Sacks D, et al. CD8 T cell exhaustion in human visceral leishmaniasis. *J Infect Dis* (2014) **209**:290–9. doi:10.1093/infdis/jit401
70. Masopust D, Schenkel JM. The integration of T cell migration, differentiation and function. *Nat Rev Immunol* (2013) **13**:309–20. doi:10.1038/nri3442
71. Laidlaw BJ, Cui W, Amezquita RA, Gray SM, Guan T, Lu Y, et al. Production of IL-10 by CD4(+) regulatory T cells during the resolution of infection promotes the maturation of memory CD8(+) T cells. *Nat Immunol* (2015) **16**:871–9. doi:10.1038/ni.3224
72. Laidlaw BJ, Craft JE, Kaech SM. The multifaceted role of CD4(+) T cells in CD8(+) T cell memory. *Nat Rev Immunol* (2016) **16**:102–11. doi:10.1038/nri.2015.10
73. Kaech SM, Cui W. Transcriptional control of effector and memory CD8+ T cell differentiation. *Nat Rev Immunol* (2012) **12**:749–61. doi:10.1038/nri3307
74. Stern JJ, Oca MJ, Rubin BY, Anderson SL, Murray HW. Role of L3T4+ and LyT-2+ cells in experimental visceral leishmaniasis. *J Immunol* (1988) **140**:3971–7.
75. Hammami A, Charpentier T, Smans M, Stager S. IRF-5-mediated inflammation limits CD8+ T cell expansion by inducing HIF-1alpha and impairing dendritic cell functions during *Leishmania* infection. *PLoS Pathog* (2015) **11**:e1004938. doi:10.1371/journal.ppat.1004938
76. Jameson SC, Masopust D. Diversity in T cell memory: an embarrassment of riches. *Immunity* (2009) **31**:859–71. doi:10.1016/j.immuni.2009.11.007
77. Greenwald RJ, Freeman GJ, Sharpe AH. The B7 family revisited. *Annu Rev Immunol* (2005) **23**:515–48. doi:10.1146/annurev.immunol.23.021704.115611
78. Wang S, Chen L. T lymphocyte co-signaling pathways of the B7-CD28 family. *Cell Mol Immunol* (2004) **1**:37–42.
79. Katara GK, Ansari NA, Verma S, Ramesh V, Salotra P. Foxp3 and IL-10 expression correlates with parasite burden in lesional tissues of post kala azar dermal leishmaniasis (PKDL) patients. *PLoS Negl Trop Dis* (2011) **5**:e1171. doi:10.1371/journal.pntd.0001171
80. Vallejo A, Abad-Fernandez M, Moreno S, Moreno A, Perez-Elias MJ, Dronda F, et al. High levels of CD4(+) CTLA-4(+) Treg cells and CCR5 density in HIV-1-infected patients with visceral leishmaniasis. *Eur J Clin Microbiol Infect Dis* (2015) **34**:267–75. doi:10.1007/s10096-014-2229-1
81. Zubairi S, Sanos SL, Hill S, Kaye PM. Immunotherapy with OX40L-Fc or anti-CTLA-4 enhances local tissue responses and killing of *Leishmania donovani*. *Eur J Immunol* (2004) **34**:1433–40. doi:10.1002/eji.200324021
82. Mcgaha T, Murphy JW. CTLA-4 down-regulates the protective anticryptococcal cell-mediated immune response. *Infect Immun* (2000) **68**:4624–30. doi:10.1128/IAI.68.8.4624-4630.2000
83. Pedicord VA, Montalvo W, Leiner IM, Allison JP. Single dose of anti-CTLA-4 enhances CD8+ T-cell memory formation, function, and maintenance. *Proc Natl Acad Sci U S A* (2011) **108**:266–71. doi:10.1073/pnas.1016791108
84. Wing K, Onishi Y, Prieto-Martin P, Yamaguchi T, Miyara M, Fehervari Z, et al. CTLA-4 control over Foxp3+ regulatory T cell function. *Science* (2008) **322**:271–5. doi:10.1126/science.1160062
85. Sansom DM, Walker LS. The role of CD28 and cytotoxic T-lymphocyte antigen-4 (CTLA-4) in regulatory T-cell biology. *Immunol Rev* (2006) **212**:131–48. doi:10.1111/j.0105-2896.2006.00419.x
86. Qureshi OS, Zheng Y, Nakamura K, Attridge K, Manzotti C, Schmidt EM, et al. Trans-endocytosis of CD80 and CD86: a molecular basis for the cell-extrinsic function of CTLA-4. *Science* (2011) **332**:600–3. doi:10.1126/science.1202947
87. Kinter AL, Godbout EJ, Mcnally JP, Sereti I, Roby GA, O'shea MA, et al. The common gamma-chain cytokines IL-2, IL-7, IL-15, and IL-21 induce the expression of programmed death-1 and its ligands. *J Immunol* (2008) **181**:6738–46. doi:10.4049/jimmunol.181.10.6738
88. Jin HT, Ahmed R, Okazaki T. Role of PD-1 in regulating T-cell immunity. *Curr Top Microbiol Immunol* (2011) **350**:17–37. doi:10.1007/82_2010_116
89. Brown JA, Dorfman DM, Ma FR, Sullivan EL, Munoz O, Wood CR, et al. Blockade of programmed death-1 ligands on dendritic cells enhances T cell activation and cytokine production. *J Immunol* (2003) **170**:1257–66. doi:10.4049/jimmunol.170.3.1257
90. Blattman JN, Wherry EJ, Ha SJ, Van Der Most RG, Ahmed R. Impact of epitope escape on PD-1 expression and CD8 T-cell exhaustion during chronic infection. *J Virol* (2009) **83**:4386–94. doi:10.1128/JVI.02524-08
91. Mou Z, Muleme HM, Liu D, Jia P, Okwor IB, Kuriakose SM, et al. Parasite-derived arginase influences secondary anti-*Leishmania* immunity by regulating programmed cell death-1-mediated CD4+ T cell exhaustion. *J Immunol* (2013) **190**:3380–9. doi:10.4049/jimmunol.1202537
92. Gutierrez FR, Mariano FS, Oliveira CJ, Pavanelli WR, Guedes PM, Silva GK, et al. Regulation of *Trypanosoma cruzi*-induced myocarditis by programmed death cell receptor 1. *Infect Immun* (2011) **79**:1873–81. doi:10.1128/IAI.01047-10
93. Butler NS, Moebius J, Pewe LL, Traore B, Doumbo OK, Tygrett LT, et al. Therapeutic blockade of PD-L1 and LAG-3 rapidly clears established blood-stage *Plasmodium* infection. *Nat Immunol* (2012) **13**:188–95. doi:10.1038/ni.2180
94. Horne-Debets JM, Faleiro R, Karunarathne DS, Liu XQ, Lineburg KE, Poh CM, et al. PD-1 dependent exhaustion of CD8+ T cells drives chronic malaria. *Cell Rep* (2013) **5**:1204–13. doi:10.1016/j.celrep.2013.11.002
95. Wykes MN, Horne-Debets JM, Leow CY, Karunarathne DS. Malaria drives T cells to exhaustion. *Front Microbiol* (2014) **5**:249. doi:10.3389/fmicb.2014.00249
96. Martinez Salazar MB, Delgado Dominguez J, Silva Estrada J, Gonzalez Bonilla C, Becker I. Vaccination with *Leishmania mexicana* LPG induces PD-1 in CD8(+) and PD-L2 in macrophages thereby suppressing the immune response: a model to assess vaccine efficacy. *Vaccine* (2014) **32**:1259–65. doi:10.1016/j.vaccine.2014.01.016
97. Wei F, Zhong S, Ma Z, Kong H, Medvec A, Ahmed R, et al. Strength of PD-1 signaling differentially affects T-cell effector functions. *Proc Natl Acad Sci U S A* (2013) **110**:E2480–9. doi:10.1073/pnas.1305394110
98. Kao C, Oestreich KJ, Paley MA, Crawford A, Angelosanto JM, Ali MA, et al. Transcription factor T-bet represses expression of the inhibitory receptor PD-1 and sustains virus-specific CD8+ T cell responses during chronic infection. *Nat Immunol* (2011) **12**:663–71. doi:10.1038/ni.2046

99. Yang B, Zhai F, Jiang J, Wang X, Cao Z, Cheng X. Elevated expression of T-bet in mycobacterial antigen-specific CD4(+) T cells from patients with tuberculosis. *Cell Immunol* (2015) **298**:1–8. doi:10.1016/j.cellimm.2015.08.002
100. Chen J, Jiang CC, Jin L, Zhang XD. Regulation of PD-L1: a novel role of pro-survival signalling in cancer. *Ann Oncol* (2016) **27**:409–16. doi:10.1093/annonc/mdv615
101. Doe HT, Kimura D, Miyakoda M, Kimura K, Akbari M, Yui K. Expression of PD-1/LAG-3 and cytokine production by CD4(+) T cells during infection with *Plasmodium* parasites. *Microbiol Immunol* (2016) **60**:121–31. doi:10.1111/1348-0421.12354
102. Penna A, Pilli M, Zerbini A, Orlandini A, Mezzadri S, Sacchelli L, et al. Dysfunction and functional restoration of HCV-specific CD8 responses in chronic hepatitis C virus infection. *Hepatology* (2007) **45**:588–601. doi:10.1002/hep.21541
103. Jeon S, St Leger AJ, Cherpes TL, Sheridan BS, Hendricks RL. PD-L1/B7-H1 regulates the survival but not the function of CD8+ T cells in herpes simplex virus type 1 latently infected trigeminal ganglia. *J Immunol* (2013) **190**:6277–86. doi:10.4049/jimmunol.1300582
104. Barber DL, Wherry EJ, Masopust D, Zhu B, Allison JP, Sharpe AH, et al. Restoring function in exhausted CD8 T cells during chronic viral infection. *Nature* (2006) **439**:682–7. doi:10.1038/nature04444
105. Day CL, Kaufmann DE, Kiepiela P, Brown JA, Moodley ES, Reddy S, et al. PD-1 expression on HIV-specific T cells is associated with T-cell exhaustion and disease progression. *Nature* (2006) **443**:350–4. doi:10.1038/nature05115
106. Urbani S, Amadei B, Tola D, Pedrazzi G, Sacchelli L, Cavallo MC, et al. Restoration of HCV-specific T cell functions by PD-1/PD-L1 blockade in HCV infection: effect of viremia levels and antiviral treatment. *J Hepatol* (2008) **48**:548–58. doi:10.1016/j.jhep.2007.12.014
107. Ye B, Liu X, Li X, Kong H, Tian L, Chen Y. T-cell exhaustion in chronic hepatitis B infection: current knowledge and clinical significance. *Cell Death Dis* (2015) **6**:e1694. doi:10.1038/cddis.2015.42
108. Das R, Verma R, Sznol M, Boddupalli CS, Gettinger SN, Kluger H, et al. Combination therapy with anti-CTLA-4 and anti-PD-1 leads to distinct immunologic changes *in vivo*. *J Immunol* (2015) **194**:950–9. doi:10.4049/jimmunol.1401686
109. Kaufmann DE, Walker BD. PD-1 and CTLA-4 inhibitory cosignaling pathways in HIV infection and the potential for therapeutic intervention. *J Immunol* (2009) **182**:5891–7. doi:10.4049/jimmunol.0803771
110. Freeman GJ, Wherry EJ, Ahmed R, Sharpe AH. Reinvigorating exhausted HIV-specific T cells *via* PD-1-PD-1 ligand blockade. *J Exp Med* (2006) **203**:2223–7. doi:10.1084/jem.20061800
111. Rijkers ES, De Ruiter T, Baridi A, Veninga H, Hoek RM, Meyaard L. The inhibitory CD200R is differentially expressed on human and mouse T and B lymphocytes. *Mol Immunol* (2008) **45**:1126–35. doi:10.1016/j.molimm.2007.07.013
112. Caserta S, Nausch N, Sawtell A, Drummond R, Barr T, Macdonald AS, et al. Chronic infection drives expression of the inhibitory receptor CD200R, and its ligand CD200, by mouse and human CD4 T cells. *PLoS One* (2012) **7**:e35466. doi:10.1371/journal.pone.0035466
113. Wright GJ, Puklavec MJ, Willis AC, Hoek RM, Sedgwick JD, Brown MH, et al. Lymphoid/neuronal cell surface OX2 glycoprotein recognizes a novel receptor on macrophages implicated in the control of their function. *Immunity* (2000) **13**:233–42. doi:10.1016/S1074-7613(00)00023-6
114. Costello DA, Lyons A, Denieffe S, Browne TC, Cox FF, Lynch MA. Long term potentiation is impaired in membrane glycoprotein CD200-deficient mice: a role for toll-like receptor activation. *J Biol Chem* (2011) **286**:34722–32. doi:10.1074/jbc.M111.280826
115. Hoek RM, Ruuls SR, Murphy CA, Wright GJ, Goddard R, Zurawski SM, et al. Down-regulation of the macrophage lineage through interaction with OX2 (CD200). *Science* (2000) **290**:1768–71. doi:10.1126/science.290.5497.1768
116. Snelgrove RJ, Goulding J, Didierlaurent AM, Lyonga D, Vekaria S, Edwards L, et al. A critical function for CD200 in lung immune homeostasis and the severity of influenza infection. *Nat Immunol* (2008) **9**:1074–83. doi:10.1038/ni.1637
117. Mihrshahi R, Brown MH. Downstream of tyrosine kinase 1 and 2 play opposing roles in CD200 receptor signaling. *J Immunol* (2010) **185**:7216–22. doi:10.4049/jimmunol.1002858
118. Daeron M, Jaeger S, Du Pasquier L, Vivier E. Immunoreceptor tyrosine-based inhibition motifs: a quest in the past and future. *Immunol Rev* (2008) **224**:11–43. doi:10.1111/j.1600-065X.2008.00666.x
119. Cortez M, Huynh C, Fernandes MC, Kennedy KA, Aderem A, Andrews NW. *Leishmania* promotes its own virulence by inducing expression of the host immune inhibitory ligand CD200. *Cell Host Microbe* (2011) **9**:463–71. doi:10.1016/j.chom.2011.04.014
120. Stack G, Jones E, Marsden M, Stacey MA, Snelgrove RJ, Lacaze P, et al. CD200 receptor restriction of myeloid cell responses antagonizes antiviral immunity and facilitates *Cytomegalovirus* persistence within mucosal tissue. *PLoS Pathog* (2015) **11**:e1004641. doi:10.1371/journal.ppat.1004641
121. Darrah PA, Patel DT, De Luca PM, Lindsay RW, Davey DF, Flynn BJ, et al. Multifunctional TH1 cells define a correlate of vaccine-mediated protection against *Leishmania major*. *Nat Med* (2007) **13**:843–50. doi:10.1038/nm1592
122. Selvapandiyan A, Dey R, Nylen S, Duncan R, Sacks D, Nakhasi HL. Intracellular replication-deficient *Leishmania donovani* induces long lasting protective immunity against visceral leishmaniasis. *J Immunol* (2009) **183**:1813–20. doi:10.4049/jimmunol.0900276
123. Golden-Mason L, Palmer BE, Kassam N, Townshend-Bulson L, Livingston S, Mcmahon BJ, et al. Negative immune regulator Tim-3 is overexpressed on T cells in hepatitis C virus infection and its blockade rescues dysfunctional CD4+ and CD8+ T cells. *J Virol* (2009) **83**:9122–30. doi:10.1128/JVI.00639-09
124. Qiu Y, Chen J, Liao H, Zhang Y, Wang H, Li S, et al. Tim-3-expressing CD4+ and CD8+ T cells in human tuberculosis (TB) exhibit polarized effector memory phenotypes and stronger anti-TB effector functions. *PLoS Pathog* (2012) **8**:e1002984. doi:10.1371/journal.ppat.1002984
125. Jones RB, Ndhlovu LC, Barbour JD, Sheth PM, Jha AR, Long BR, et al. Tim-3 expression defines a novel population of dysfunctional T cells with highly elevated frequencies in progressive HIV-1 infection. *J Exp Med* (2008) **205**:2763–79. doi:10.1084/jem.20081398
126. Sada-Ovalle I, Ocana-Guzman R, Perez-Patrigeon S, Chavez-Galan L, Sierra-Madero J, Torre-Bouscoulet L, et al. Tim-3 blocking rescue macrophage and T cell function against *Mycobacterium tuberculosis* infection in HIV+ patients. *J Int AIDS Soc* (2015) **18**:20078. doi:10.7448/IAS.18.1.20078
127. Kared H, Fabre T, Bedard N, Bruneau J, Shoukry NH. Galectin-9 and IL-21 mediate cross-regulation between Th17 and Treg cells during acute hepatitis C. *PLoS Pathog* (2013) **9**:e1003422. doi:10.1371/journal.ppat.1003422
128. Xiao S, Liu J, Huang S, Lu F. Increased Gal-9 and Tim-3 expressions during liver damage in a murine malarial model. *Parasitol Res* (2016) **115**:663–72. doi:10.1007/s00436-015-4784-5
129. Liu J, Xiao S, Huang S, Pei F, Lu F. Upregulated Tim-3/galectin-9 expressions in acute lung injury in a murine malarial model. *Parasitol Res* (2016) **115**:587–95. doi:10.1007/s00436-015-4775-6

12

Staphylococcus epidermidis Biofilm-Released Cells Induce a Prompt and more Marked *In vivo* Inflammatory-Type Response than Planktonic or Biofilm Cells

Angela França[1,2], Begoña Pérez-Cabezas[3], Alexandra Correia[3,4], Gerald B. Pier[5], Nuno Cerca[1] and Manuel Vilanova[2,3,4]*

[1] Laboratory of Research in Biofilms Rosário Oliveira, Centre of Biological Engineering, University of Minho, Braga, Portugal, [2] Departamento de Imuno-Fisiologia e Farmacologia, Instituto de Ciências Biomédicas de Abel Salazar, Universidade do Porto, Porto, Portugal, [3] Instituto de Investigação e Inovação em Saúde, Universidade do Porto, Porto, Portugal, [4] Instituto de Biologia Molecular e Celular, Universidade de Porto, Porto, Portugal, [5] Division of Infectious Diseases, Department of Medicine, Brigham and Women's Hospital/Harvard Medical School, Boston, MA, USA

Edited by:
Olivier Dussurget,
University Paris Diderot, France

Reviewed by:
Glen C. Ulett,
Griffith University, Australia
Pietro Speziale,
Istituto Universitario Di Studi Superiori Di Pavia, Italy
Ying Zhang,
Johns Hopkins University, USA

***Correspondence:**
Nuno Cerca
nunocerca@ceb.uminho.pt

Staphylococcus epidermidis biofilm formation on indwelling medical devices is frequently associated with the development of chronic infections. Nevertheless, it has been suggested that cells released from these biofilms may induce severe acute infections with bacteraemia as one of its major associated clinical manifestations. However, how biofilm-released cells interact with the host remains unclear. Here, using a murine model of hematogenously disseminated infection, we characterized the interaction of cells released from *S. epidermidis* biofilms with the immune system. Gene expression analysis of mouse splenocytes suggested that biofilm-released cells might be particularly effective at activating inflammatory and antigen presenting cells and inducing cellular apoptosis. Furthermore, biofilm-released cells induced a higher production of pro-inflammatory cytokines, in contrast to mice infected with planktonic cells, even though these had a similar bacterial load in livers and spleens. Overall, these results not only provide insights into the understanding of the role of biofilm-released cells in *S. epidermidis* biofilm-related infections and pathogenesis, but may also help explain the relapsing character of these infections.

Keywords: *S. epidermidis,* biofilms, biofilm-released cells, splenocytes transcriptome, pro-inflammatory cytokines, tissue colonization

INTRODUCTION

Staphylococcus epidermidis is one of the most important etiological agents of device-associated infections due to its ability to adhere and form biofilms on the surface of indwelling medical devices (Vuong and Otto, 2002; Otto, 2009). When compared to planktonic cells, *S. epidermidis* cells within biofilms are known to be more tolerant to several classes of antibiotics (Cerca et al., 2005), as well as to the host immune effectors (Cerca et al., 2006; Kristian et al., 2008). Biofilms represent therefore a common cause of recurrent and relapsing infections (Costerton et al., 1999). Consequently, removal of the infected devices is often required to resolve these infections (von Eiff et al., 2002), which results in increased morbidity and, occasionally, mortality among infected patients (Otto, 2009). Due to the enormous impact of *S. epidermidis* biofilm-related infections on human health, the mechanisms underlying biofilm formation have been extensively studied in the last decades. It is currently accepted that biofilm formation involves three main stages: (1) initial adhesion, (2) maturation, and (3) disassembly (Otto, 2012). The later refers to the release of bacterial cells from the biofilm to the surrounding environment, and is the least understood stage of the biofilm lifecycle (Boles and Horswill, 2011). Importantly, biofilm disassembly has been associated with the emergence of severe acute infections such as bacteraemia (Wang et al., 2011) and the embolic events of endocarditis (Pitz et al., 2011). However, despite its clear importance in the clinical setting, little is known regarding the phenotype or interaction of these cells with the host immune system. In the first stages of biofilm formation, planktonic bacteria attached to medical devices undergo several physiological modifications that lead to the biofilm phenotype (Yao et al., 2005). Thus, it was thought that after disassembly biofilm-released cells would quickly revert to the planktonic phenotype (Kaplan, 2010; Chua et al., 2014). However, recent reports have shown that cells released from *Pseudomonas aeruginosa* (Rollet et al., 2009; Li et al., 2014), *Streptococcus mutans* (Liu et al., 2013), and *Streptococcus pneumoniae* (Marks et al., 2013) biofilms present features distinct from either the biofilm or planktonic phenotypes, showing higher virulence potential. Chua and collaborators showed that *P. aeruginosa* biofilm-released cells, when compared with their planktonic or biofilm counterparts, present higher expression level of genes associated with the bacterium virulence, namely Type 2 Secretion System (TSS) and T3SS *psc* gene and, more important, they showed that these genes are essential in eliciting full virulence against macrophages and in the rapid killing of *Caenorhabditis elegans* (Chua et al., 2014), respectively. In the case of *S. epidermidis,* it is only known that biofilm-released cells present higher tolerance than planktonic and biofilm cells to antibiotics (Franca et al., 2016). However, their full virulence potential remains unclear. A comprehensive analysis of the interaction between biofilm-released cells and the host would clarify their role in the pathogenesis of biofilm-related infections, and help to prevent the pathologic events associated with biofilm cells dissemination. Therefore, herein, a murine model of hematogenously disseminated infection was used to evaluate the capacity of *S. epidermidis* biofilm-released cells to (1) induce changes in the transcriptome of murine immune cells within the spleen, (2) stimulate the production of pro-inflammatory cytokines, and (3) colonize and persist in murine organs. Our results showed that *S. epidermidis* biofilm-released cells induce a prompt and more marked inflammatory-type response than do their planktonic or biofilm counterparts. In addition, these findings showed that particular properties of the biofilm-released cells need to be taken into account to efficiently target and treat acute infections originating from *S. epidermidis* biofilms.

MATERIALS AND METHODS

Ethics Statement

This study was performed in strict accordance with the recommendations of the European Convention for the Protection of Vertebrate Animals used for Experimental and Other Scientific Purposes (ETS 123), the 86/609/EEC directive and Portuguese rules (DL 129/92). All experimental protocols were approved by the competent national authority (Direcção-Geral de Veterinária), document 023517 (2010.11.25).

Mice

Female BALB/c mice, 8–12 weeks old, were purchased from Charles River (Barcelona, Spain) and kept under specific-pathogen-free conditions at the Animal Facility of the Instituto de Ciências Biomédicas Abel Salazar, Porto, Portugal. Mice were maintained in individually ventilated cages (5 animals per cage) with corncob bedding, and under controlled conditions of temperature (21 ± 1°C), relative humidity (between 45 and 65%) and light (12 h light/ dark cycle). Mice had *ad libitum* access to food and water. Hiding and nesting materials were provided for enrichment. All procedures such cage changing, water and food supply, as well as intravenous injections were always performed during the day cycle (between 7 and 19 h).

Bacteria and Growth Conditions

The biofilm forming strain *S. epidermidis* 9142, isolated from a blood culture (Mack et al., 1994), was used in this study. A single colony, from a Tryptic Soy Agar (TSA) plate, was inoculated into 2 mL of Tryptic Soy Broth (TSB, Liofilchem, Teramo, Italy) and incubated overnight at 37°C with shaking at 120 rpm. A suspension with $\sim 1 \times 10^8$ CFU/mL, prepared by adjusting the optical density (at 640 nm) of the overnight culture to 0.25 ± 0.05, was used to start both planktonic and biofilm cultures. For planktonic cultures 150 μL of 1×10^8 CFU/mL bacterial suspension was inoculated into 10 mL of TSB supplemented with 0.65% (v/v) glucose ($TSB_{0.65\%G}$) and incubated for 24 h at 37°C under agitation at 120 rpm. Biofilms were grown in 24-well plates made of polystyrene plastic (Orange Scientific, Braine-l' Alleud, Belgium) by inoculating 15 μL of the 1×10^8 CFU/mL bacterial suspension into 1 mL of $TSB_{0.65\%G,}$ then incubating at 37°C with agitation at 120 rpm. After 24 h of growth, biofilms were washed twice with apyrogenic Phosphate Buffered Saline (PBS, Gibco, MD, USA), 1 mL of fresh $TSB_{0.65\%G}$ was carefully added and biofilms allowed to grow, under the same temperature and agitation conditions, for additional 24 h. Biofilm-released cells, (i.e., the cells in the biofilm bulk-fluid), were collected as described before (Franca et al., 2016) from 12 originating biofilms and pooled together to decrease variability inherent to biofilm growth (Sousa et al., 2014). Four biofilms were washed twice with apyrogenic PBS, disrupted and also pooled together to reduce variability. Planktonic (4 mL of culture), biofilm and biofilm-released cells were then harvested by centrifugation, suspended in 4 mL of apyrogenic PBS (Gibco, MD, USA) and sonicated for 10 s at 18 W (Branson model W 185 D, Heat Systems Ultrasonics, CT, USA) in order to dissociate cell clusters. Cells viability was not reduced by this procedure as determined previously by

CFU counting and propidium iodide incorporation (Cerca et al., 2011).

Murine Model of Hematogenously Disseminated Infection

The inoculum of each of the bacterial populations was adjusted by flow cytometry to 5×10^8 total cells/mL, using SYBR Green (LifeTechnologies, MD, USA)/propidium iodide (Sigma, MO, USA) staining, as optimized before (Cerca et al., 2011). The number of cultivable cells was assessed by CFU counting. Adult mice, randomly allocated to each experimental group, were injected intravenously in the lateral tail vein, with the support of a restrainer, with 0.2 mL of 5×10^8 of planktonic, biofilm or biofilm-released cell suspensions. Control mice were injected intravenously with 0.2 mL of apyrogenic PBS. Sample size was determined based on the results of preliminary experiments. It was not possible to perform subsequent mouse studies in a blinded fashion. In order to address the alterations occurring during the acute phase of infection, the parameters evaluated in this study were assessed 2, 6, or 14 h after challenging the three *S. epidermidis* populations.

Serum Collection and Bacterial Load Determination in Organs

Two, 6, and 14 h post-infection, mice were anesthetized with isoflurane (Abbott laboratories, IL, USA) for terminal blood collection, and then euthanized by cervical dislocation. For serum collection, mouse blood was drawn through the retro-orbital route, incubated overnight at 4°C, and then centrifuged for 15 min at 4°C at 16,000 *g*. Serum was then transferred into a new tube and stored at −80°C until further use. Livers and spleens were aseptically removed and immediately transferred into tissue grinders with, respectively, 3 or 1 mL of PBS. Tissues were homogenized and quantitatively cultured on TSA plates. At all times during the procedure, samples were kept on ice. This experiment was performed 1 (for biofilms cells at all time points) to 3 (planktonic and biofilm-released cells, 6 h time point) independent times, with at least 5 animals per infected group.

Cytokines and Chemokines Quantification

Two, 6, and 14 h post-infection, the levels of the cytokines IL-6, TNF-α and the chemokines CXCL1 (KC), CCL2 (MCP-1), CCL3 (MIP-1α), CCL4 (MIP-1β) in mouse serum were quantified in a Bio-Plex® 200 using the kit Magnetic Custom Multiplex Bio-Plex Pro™ Mouse Cytokine Group I assay (Bio-Rad, CA, USA). The procedure was performed following the manufacturer's instructions. This experiment was performed 1 (all populations at 6 and 14 h time point) to 2 (all populations at 2 h time point) independent times, with at least 5 animals per infected group.

Microarray Analysis of Mouse Splenocytes

Spleen cells plays a major role in host immune response to blood-born pathogens, working in concert to activate mechanisms required for successful resolution of infection. Hence, in order to address the mechanisms activated during the first contact with the different *S. epidermidis* populations, the transcriptome of splenocytes was analyzed, by microarrays, 2 h after challenge. In brief, spleens were aseptically removed, transferred to 60 mm diameter sterile Petri dishes with 9 mL apyrogenic PBS and immediately placed on ice. Thereafter, using two sterile frosted glass slides, spleens were completely homogenized. The suspension was then passed through a sterile column of glass wool to remove fibrous tissue, the number of cells counted by flow cytometry, and 5×10^6 splenocytes harvested by 5 min centrifugation at 1200 rpm at 4°C. Cell pellets were immediately suspended in RLT buffer (QIAGEN, Heidelberg, Germany) and stored at −80°C until the next day. Total RNA was then isolated using the RNeasy Mini Kit (QIAGEN) following the manufacturer's instructions. Concentration and purity was determined using a NanoDrop™1000 and integrity was confirmed using an Agilent 2100 Bioanalyzer (Agilent Technologies, CA, USA). RNA integrity number values were above 8.5 for all samples. This experiment was performed once with 2 (control and planktonic cells) to 3 (biofilm and biofilm-released cells) animals per group.

Transcription levels in mouse splenocytes were determined using Affymetrix® Mouse Gene 2.1 ST Array Strip (Affymetrix, MA, USA). RNA was prepared for analysis using Ambion WT Expression Kit (ThermoFisher Scientific, MA, USA) and GeneChip® WT Terminal Labeling Kit (Affymetrix). Briefly, 100 ng of total RNA, containing spiked in Poly-A RNA controls, was used in a reverse transcription reaction to generate first-strand complementary DNA. After second-strand synthesis, double-stranded complementary DNA was used to generate cRNA. cRNA (15 μg) was then used for a second cycle of first-strand cDNA synthesis and the resultant single stranded cDNA (5.5 μg) was fragmented and end-labeled. Size distribution of the fragments was assessed using an Agilent 2100 Bioanalyzer (Agilent Technologies). End-labeled, fragmented cDNA (3.5 μg), was then used in a 150 μL hybridization cocktail containing hybridization controls (GeneAtlas® Hybridization, Wash, and Stain Kit for WT Array Strips, Affymetrix), of which 120 μL were hybridized on array strips for 20 h at 48°C. Standard post hybridization wash and double-stain protocols were used on an Affymetrix GeneAtlas system, followed by scanning of the array strips.

Microarray Data Analysis

The arrays were analyzed using Chipster 2 (Kallio et al., 2011) with a custom cdf file in mogene21stmmentrezg.db, as available from Brainarray database (version 17; Sandberg and Larsson, 2007). Following Robust Multi-array Average normalization and biomaRt annotation, differential expression was determined by empirical Bayes two-group test (Smyth, 2004) with Benjamini-Hochberg multiple testing correction and a *P*-value cut-off of 0.05. For further analyses, only genes with fold changes above 1.5 were included. The heatmap was constructed using matrix2png interface (Pavlidis and Noble, 2003). Venn diagram, created using VENNY 2.1 (Oliveros, 2007), was used to identify the genes that were uniquely and commonly expressed in mice infected with different *S. epidermidis* bacterial populations. Gene ontology (GO) terms enrichment was assessed using the Search Tool for the Retrieval of Interacting Genes/Proteins (STRING) (version 10; Franceschini et al., 2013). Only gene-sets passing

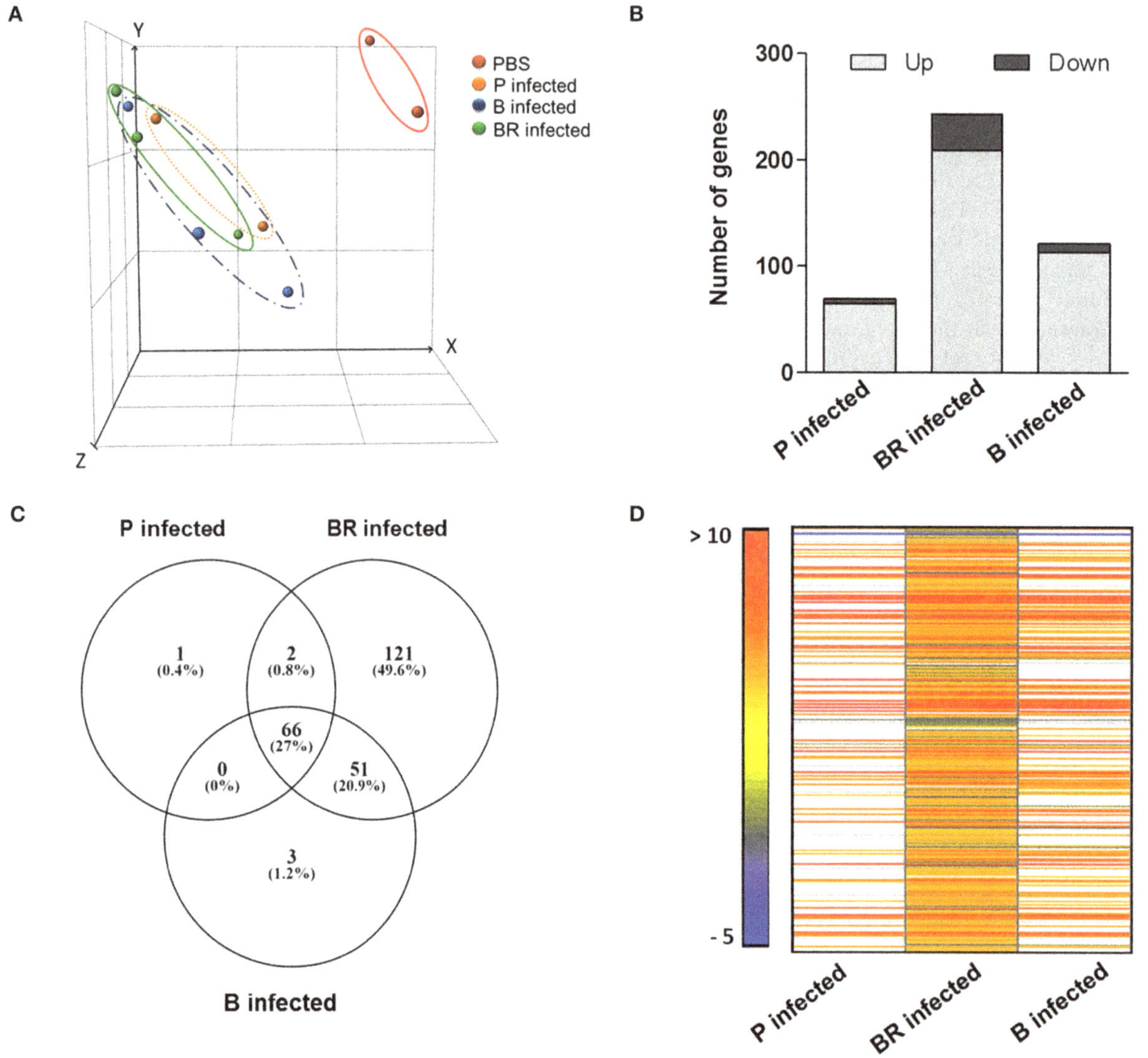

FIGURE 1 | Analysis of the transcriptome of the spleen of mice infected with different *S. epidermidis* populations. BALB/c mice were challenged intravenously with 1×10^8 planktonic (P) ($n = 2$), biofilm (B) ($n = 3$), biofilm-released (BR) cells ($n = 3$), or sham-infected treated with PBS alone (PBS) ($n = 2$). Two hours post-infection, spleens were collected and microarray analysis was performed. **(A)** Principal component analysis; **(B)** Number of genes with increased and decreased transcription in each condition ($P < 0.05$, Empirical Bayes two-group test with Benjamini-Hochberg multiple testing correction). **(C)** Venn diagram showing the number of genes that are commonly (overlapping circles) and uniquely expressed (non-overlapping circles) in each condition; **(D)** Heatmap of the differentially expressed genes. White lines indicate non-detected genes or genes with no significant alterations ($P > 0.05$, Empirical Bayes two-group test with Benjamini-Hochberg multiple testing correction).

significance thresholds ($P < 0.05$ with false discovery rate) were selected for further analysis. To reduce redundancy, GO terms found enriched in STRING were reanalyzed by REVIGO (Supek et al., 2011), allowing for small (0.5) similarity, using the species-specific *Mus musculus* database (in order to fine-tune the calculation of semantic distances which rely on information contents of GO terms for this particular species) and SimRel score. The complete list of the genes differentially and uniquely expressed in splenocytes of mice infected with planktonic, biofilm, or biofilm-released cells is available at GEO database repositorium, under the accession number GSE60992.

Statistical Analysis

Statistical analysis was carried out with GraphPad Prism (CA, USA). The normality of the data obtained was evaluated using Kolmogorov–Smirnov test. Accordingly, Kruskal–Wallis and Dunn's multiple comparison tests were applied and data depicted in median of all independent experiments. Differences among groups were considered significant when $P < 0.05$. Statistical differences found between planktonic and biofilm cells phenotype were not indicated as the aim of the study was not to compare the differences between them. Statistical analysis used for microarrays data evaluation

TABLE 1 | List of the top most transcribed genes in the spleen of mice infected with *S. epidermidis* planktonic, biofilm or biofilm-released cells.

Gene	Description	Fold change	*P*-value
PLANKTONIC CELLS INFECTED MICE			
Irg1	Immunoresponsive gene 1	39.08 ± 4.02	< 0.001
Clec4e	C-type lectin domain family 4, member e	17.46 ± 4.73	0.004
Cxcl2	Chemokine (C-X-C motif) ligand 2	16.63 ± 3.64	0.003
Cxcl10	Chemokine (C-X-C motif) ligand 10	8.73 ± 2.98	0.012
Clec5a	C-type lectin domain family 5, member a	7.80 ± 0.92	0.003
Slc7a11	Solute carrier family 7, member 11	7.88 ± 2.47	0.013
Ccrl2	Chemokine (C-C motif) receptor-like 2	7.20 ± 0.67	0.006
Mmrgpra2a	MAS-related GPR, member A2A	7.15 ± 2.17	0.012
Gpr84	G protein-coupled receptor 84	6.31 ± 1.23	0.009
Il1rn	Interleukin 1 receptor antagonist	5.93 ± 0.95	0.008
BIOFILM-RELEASED CELLS INFECTED MICE			
Irg1	Immunoresponsive gene 1	49.02 ± 5.41	< 0.001
Cxcl2	Chemokine (C-X-C motif) ligand 2	25.90 ± 3.85	< 0.001
Clec4e	C-type lectin domain family 4, member e	20.24 ± 0.24	< 0.001
Slc7a11	Solute carrier family 7, member 11	13.33 ± 1.99	< 0.001
Mrgpra2a	MAS-related GPR, member A2A	9.69 ± 0.14	< 0.001
Ccrl2	Chemokine (C-C motif) receptor-like 2	8.83 ± 1.03	< 0.001
Clec5a	C-type lectin domain family 5, member a	8.52 ± 1.10	< 0.001
Cxcl10	Chemokine (C-X-C motif) ligand 10	7.34 ± 0.46	< 0.001
Fpr1	Formyl peptide receptor 1	7.22 ± 1.18	< 0.001
Cd14	CD14 antigen	7.15 ± 0.47	< 0.001
BIOFILM CELLS INFECTED MICE			
Irg1	Immunoresponsive gene 1	42.97 ± 4.79	< 0.001
Clec4e	C-type lectin domain family 4, member e	18.49 ± 2.67	< 0.001
Cxcl2	Chemokine (C-X-C motif) ligand 2	18.22 ± 3.27	< 0.001
Slc7a11	Solute carrier family 7, member 11	10.51 ± 2.77	< 0.001
Cxcl10	Chemokine (C-X-C motif) ligand 10	10.24 ± 1.37	< 0.001
Ccrl2	Chemokine (C-C motif) receptor-like 2	8.54 ± 0.47	< 0.001
Clec5a	C-type lectin domain family 5, member a	8.61 ± 1.07	< 0.001
Mrgpra2a	MAS-related GPR, member A2A	7.68 ± 0.87	< 0.001
Fpr1	Formyl peptide receptor 1	6.59 ± 0.83	< 0.001
Gpr84	G protein-coupled receptor 84	6.18 ± 0.37	< 0.001

is particular and is specified in "microarrays data analysis" subsection.

RESULTS

Biofilm-Released Cells Induce a Particular Gene Expression Profile on Mouse Splenocytes

The transcriptomic profile of mice infected with *S. epidermidis* planktonic, biofilm and biofilm-released cells was compared with that of non-infected mice in order to identify the genes expressed during infection induced by each of the three populations. Principal components analysis revealed that infected mice displayed a markedly different gene expression profile from non-infected controls (**Figure 1A**). The differences among infected mouse groups, however, were not that evident. The genes with the highest or lowest levels of transcription were similar in the three groups of infected mice (**Tables 1, 2**). Nevertheless, despite these general similarities important differences were found in the number of genes with increased and decreased transcription (**Figure 1B**). Within the 243 genes found differentially expressed ($P < 0.05$) in mice infected with biofilm-released cells, 121 were exclusive to this infecting phenotype (**Figure 1C**), where 96 had increased transcription (above 1.5-fold change) and 25 had decreased transcription (above −1.5-fold change; **Figure 1D**). Among the genes with increased transcription in splenocytes of mice infected with biofilm-released cells, we found significant enrichment of several GO clusters (**Table 3**) including positive regulation of leukocyte cell cell adhesion, tumor necrosis factor-mediated signaling and T cell activation, and negative regulation of mitogen activated protein kinases (MAPK) cascade and interleukin-10 production. Interestingly, GO terms associated with programmed cell death such as regulation of intrinsic apoptotic signaling pathways, development of cell death, and cell killing were also enriched. Finally, we observed that the

TABLE 2 | List of the top less transcribed genes in the spleen of mice infected with *S. epidermidis* planktonic, biofilm, or biofilm-released cells.

Gene	Description	Fold change	*P*-value
PLANKTONIC CELLS INFECTED MICE			
Abcd2	ATP-binding cassette, sub-family D (ALD), member 2	−5.20 ± 1.41	0.013
Kctd12b	Potassium channel tetramerisation domain containing 12b	−2.14 ± 0.22	0.038
Mmp12	Matrix metallopeptidase 12	−2.03 ± 0.15	0.038
BIOFILM-RELEASED CELLS INFECTED MICE			
Abcd2	ATP-binding cassette, sub-family D (ALD), member 2	−5.31 ± 1.73	0.005
Ccr2	Chemokine (C-C motif) receptor 2	−2.43 ± 0.40	0.017
Sema6a	Sema domain, transmembrane domain (TM), and cytoplasmic domain, (semaphorin) 6A	−2.38 ± 0.37	0.014
Kctd12b	Potassium channel tetramerisation domain containing 12b	−2.35 ± 0.24	0.003
Pcdha7	Protocadherin alpha 7	−2.34 ± 0.30	0.005
Gm3376	Predicted gene 3376	−2.33 ± 0.23	0.004
Rgs2	Regulator of G-protein signaling 2	−2.33 ± 0.39	0.012
Tlr8	Toll-like receptor 8	−2.26 ± 0.36	0.028
Kcnj16	Potassium inwardly-rectifying channel, subfamily J, member 16	−2.17 ± 0.27	0.009
Mmp12	Matrix metallopeptidase 12	−2.18 ± 0.54	0.045
BIOFILM CELLS INFECTED MICE			
Abcd2	ATP-binding cassette, sub-family D (ALD), member 2	−4.83 ± 1.66	0.012
Kctd12b	Potassium channel tetramerisation domain containing 12b	−2.21 ± 0.27	0.013
Sema6a	Sema domain, transmembrane domain (TM), and cytoplasmic domain, (semaphorin) 6A	−2.20 ± 0.15	0.012
Rgs2	Regulator of G-protein signaling 2	−2.11 ± 0.21	0.012
Mmp12	Matrix metallopeptidase 12	−2.07 ± 0.03	0.005
Prr5l	Proline rich 5 like	−1.80 ± 0.15	0.040
Gm13710	Predicted gene 13710	−1.78 ± 0.06	0.016
Vstm4	V-set and transmembrane domain containing 4	−1.76 ± 0.13	0.032

great majority of the transcripts of the genes with increased expression in mice infected with biofilm-released cells were those encoding proteins mostly localized to the cytoplasm or in cells' organelles (**Table 3**). No enrichment was found among down-regulated genes, in any of the conditions tested. For further information regarding the GO terms found enriched in mice infected with planktonic or biofilm cells please see Supplementary Material.

A more comprehensive analysis revealed that within the greatest transcribed genes in mice infected with biofilm-released cells are genes with important functions in both innate and adaptive immune response such as those encoding the early activation marker CD69, and the co-stimulatory molecules CD80, CD86, and CD83, which are expressed on antigen-presenting cells and up-regulated upon exposure to pathogens. Furthermore, mRNA encoding the cytokine CCL17 or TARC (thymus and activation-regulated chemokine), a T cell attractant chemokine produced by dendritic cells, was found significantly up-regulated.

Biofilm-Released Cells Induce Higher Stimulation of Pro-Inflammatory Cytokines and Chemokines

As shown in **Figure 2**, mice infected with biofilm-released cells had significantly higher serum levels of the chemokines CCL3, CCL4, and CXCL1, as well as higher levels of TNF-α than mice infected with planktonic cells, 2 h after the bacterial challenge. At that time point, no differences were found in the levels of any assessed cytokines between biofilm and biofilm-released cells-infected mouse groups. In contrast, 6 h after infection, markedly higher serum levels of CXCL1, TNFα, and IL-6 were detected in mice infected with biofilm-released cells than in the biofilm cell-infected counterparts. By 14 h after infection, lower serum levels of CCL2 were detected in mice infected with biofilm-released cells, when compared with their planktonic infected counterparts. No significant differences were detected in the serum levels of any other assessed cytokine among the different infected groups.

Biofilm-Released Cells Present an Intermediate Ability to Colonize Murine Organs

Biofilm-released cells had an intermediate ability, between that of planktonic and biofilm cells, to colonize the liver and spleen (**Figure 3**). Interestingly, while in the first 6 h of infection, biofilm-released cell burden resembled that of planktonic cells, 14 h after infection the differences between planktonic and biofilm-released cells and the similarities between biofilm-released and biofilms cells become noticeable. It is important to note that although the inoculum was adjusted by flow cytometry the quantity of bacteria injected was also confirmed by CFU

TABLE 3 | GO term enrichment of the genes with increased transcription in the spleen of mice infected with *S. epidermidis* biofilm-released cells.

GO ID	Cluster representatives	N. of genes	*P*-value
BIOLOGICAL PROCESS			
GO:2001242	Regulation of intrinsic apoptotic signaling pathway	11	< 0.0001
GO:2000116	Regulation of cysteine-type endopeptidase activity	12	< 0.0001
GO:0043409	Negative regulation of MAPK cascade	9	< 0.0001
GO:1903039	Positive regulation of leukocyte cell-cell adhesion	9	< 0.001
GO:0050870	Positive regulation of T cell activation	8	0.001
GO:0044003	Modification by symbiont of host morphology or physiology	4	0.001
GO:0031329	Regulation of cellular catabolic process	15	0.003
GO:0010623	Developmental programmed cell death	4	0.004
GO:0002260	Lymphocyte homeostasis	5	0.005
GO:0031638	Zymogen activation	6	0.006
GO:0006986	Response to unfolded protein	5	0.007
GO:0031341	Regulation of cell killing	5	0.007
GO:0010243	Response to organonitrogen compound	13	0.007
GO:0032693	Negative regulation of interleukin-10 production	3	0.007
GO:1901698	Response to nitrogen compound	14	0.010
GO:0048646	Anatomical structure formation involved in morphogenesis	18	0.010
GO:0019724	B cell mediated immunity	5	0.011
GO:0035966	Response to topologically incorrect protein	5	0.011
GO:0010727	Negative regulation of hydrogen peroxide metabolic process	2	0.011
GO:0010743	Regulation of macrophage derived foam cell differentiation	3	0.014
GO:0043243	Positive regulation of protein complex disassembly	3	0.014
GO:0033043	Regulation of organelle organization	19	0.016
GO:0009628	Response to abiotic stimulus	16	0.017
GO:0009888	Tissue development	25	0.017
GO:0033209	Tumor necrosis factor-mediated signaling pathway	3	0.020
GO:0042940	D-amino acid transport	2	0.023
GO:0016192	Vesicle-mediated transport	16	0.030
GO:0018149	Peptide cross-linking	3	0.030
GO:0045787	Positive regulation of cell cycle	8	0.030
GO:0061028	Establishment of endothelial barrier	3	0.032
GO:0048147	Negative regulation of fibroblast proliferation	3	0.035
GO:0008637	Apoptotic mitochondrial changes	4	0.038
GO:0034976	Response to endoplasmic reticulum stress	5	0.046
GO:0009314	Response to radiation	9	0.046
GO:0051604	Protein maturation	7	0.047
GO:0051195	Negative regulation of cofactor metabolic process	2	0.048
GO:0048661	Positive regulation of smooth muscle cell proliferation	4	0.048
MOLECULAR PROCESS			
GO:0050786	RAGE receptor binding	4	< 0.001
GO:0016209	Antioxidant activity	6	0.005
GO:0048020	CCR chemokine receptor binding	4	0.005
GO:0046983	Protein dimerization activity	24	0.016
GO:0038024	Cargo receptor activity	5	0.038
GO:0042803	Protein homodimerization activity	17	0.046
CELLULAR COMPONENTS			
GO:0031988	Membrane-bounded vesicle	50	< 0.001
GO:0043226	Organelle	108	0.002
GO:0043227	Membrane-bounded organelle	116	0.016
GO:0044424	Intracellular part	115	0.017

(Continued)

TABLE 3 | Continued

GO ID	Cluster representatives	N. of genes	*P*-value
GO:0005737	Cytoplasm	99	0.017
GO:0005912	Adherens junction	12	0.018
GO:0072559	NLRP3 inflammasome complex	2	0.027
GO:0005622	Intracellular	114	0.032
GO:0005576	Extracellular region	46	0.042

Gene set enrichment was primary assessed with STRING (Franceschini et al., 2013) and then the GO terms only found in this condition were analyzed by REVIGO (Supek et al., 2011) to reduce redundancy.

counting, and the number of CFU was similar among the different populations.

DISCUSSION

Due to the significant role of biofilms in the emergence of nosocomial infections, previous studies have focused on comparisons between planktonic cultures and established biofilms in order to highlight particular features of biofilm-associated infections (Becker et al., 2001; Resch et al., 2005; Shemesh et al., 2007). The role of biofilm-released cells in the pathogenesis of biofilm infections is, however, poorly understood, with no prior studies addressing this issue in regard to *S. epidermidis* infection. We have recently shown that biofilm-released cells, obtained using the same experimental model used herein, are more tolerant than planktonic cells, or even biofilm cells, to antibiotics commonly used for staphylococcal infections treatment (Franca et al., 2016). Nevertheless, nothing is known about the interplay between these cells and the host immune system. Hence, we have evaluated the interaction between *S. epidermidis* biofilm-released cells and the host immune system, using planktonic and biofilm cells for comparative purposes. We first determined whether biofilm-released cells would induce a different transcriptional profile in splenocytes of mice infected through the hematogenous route. Transcriptomic data showed that mice challenged with biofilm-released cells responded distinctly from the ones infected with the other bacterial populations. Although, a striking difference was observed between control and infected mouse groups, less marked alterations were found within the mouse groups infected with the three *S. epidermidis* populations. Since we compared the response of the host to the same bacterium but in different stages of their lifecycle, fewer differences among infected groups were expected. However, a more exhaustive analysis revealed that the expression level of several genes encoding proteins involved, direct or indirectly, in the development of innate and adaptive immunity were significantly increased in biofilm-released cells-infected mice. An increased transcription of *S100a8* and *S100a9* genes, both encoding damage-associated proteins released mainly by degranulating neutrophils (Simard et al., 2011) and *Ly6g*, which encodes a neutrophil surface marker (Lee et al., 2013) were detected in splenocytes after 2 h of injection of biofilm-released cells. These mice also had the highest expression of *Cxcl2* and *Fpr1* encoding, respectively, neutrophil chemoattractant cytokine CXCL2 (Kobayashi, 2008) and chemotactic receptor formyl peptide receptor 1 that is also present on neutrophil cell membranes (Yang and Hwang, 2016). In accordance with the inflammatory-type response observed in microarrays analysis, these mice also obtained the highest serum levels of neutrophil chemo attractant cytokines CXCL1 and CCL3 (Kobayashi, 2008) 2 h after the challenge of biofilm-released cells. These results indicate that biofilm-released cells may be particularly effective in promoting neutrophil recruitment and activation. Neutrophils are very effective in eliminating extracellular bacteria (Nathan, 2006), and therefore the type and magnitude of response elicited by biofilm-released cells may explain their faster or more effective clearance from the liver and spleen of infected mice, as compared to planktonic cells. Moreover, biofilm-released cells were also more effective at inducing *Irg1* expression, a gene known to be highly expressed in macrophages in response to infections that limits bacterial survival (Cordes et al., 2015). In agreement with the pro-inflammatory response elicited, biofilm-released cell-infected mice showed down-regulated transcription of the anti-inflammatory cytokine interleukin-10. IL-10 is a key cytokine in decreasing inflammatory pathology (Saraiva and O'Garra, 2010), such as that resulting from infection (Duell et al., 2012) by negatively regulating inflammation (Couper et al., 2008). The impact of IL-10 repression in the context of *S. epidermidis* biofilm-released cells bloodstream infections would thus be worth to explore. Nevertheless, mice infected with biofilm cells, were the ones presenting the lowest bacterial burden although not eliciting the highest pro-inflammatory response as could be inferred from gene expression or cytokine levels. A possible explanation for the delayed clearance of biofilm-released cells as compared to biofilm cells may be an enhanced apoptosis of immune cells. This is supported by the significant enrichment of genes associated with this type of cell death observed in mice infected with biofilm-released cells such as the Caspase-4, Caspase-8, and FAS-associated death domain-like apoptosis regulator (Ulett and Adderson, 2006). Furthermore, enrichment of genes related to the assembly of the NLRP3 inflammasome complex, which has been associated in cell apoptosis and pyroptosis (Sagulenko et al., 2013), was also observed. Interestingly, it was recently shown that during early *Mycobacterium avium* biofilm infection, mononuclear cells phagocytic function was attenuated due to hyperstimulation of phagocytes and enhanced cell death by apoptosis induced by biofilm cells (Rose and Bermudez, 2014). Although, we have not specifically addressed this phenomenon in *S. epidermidis*

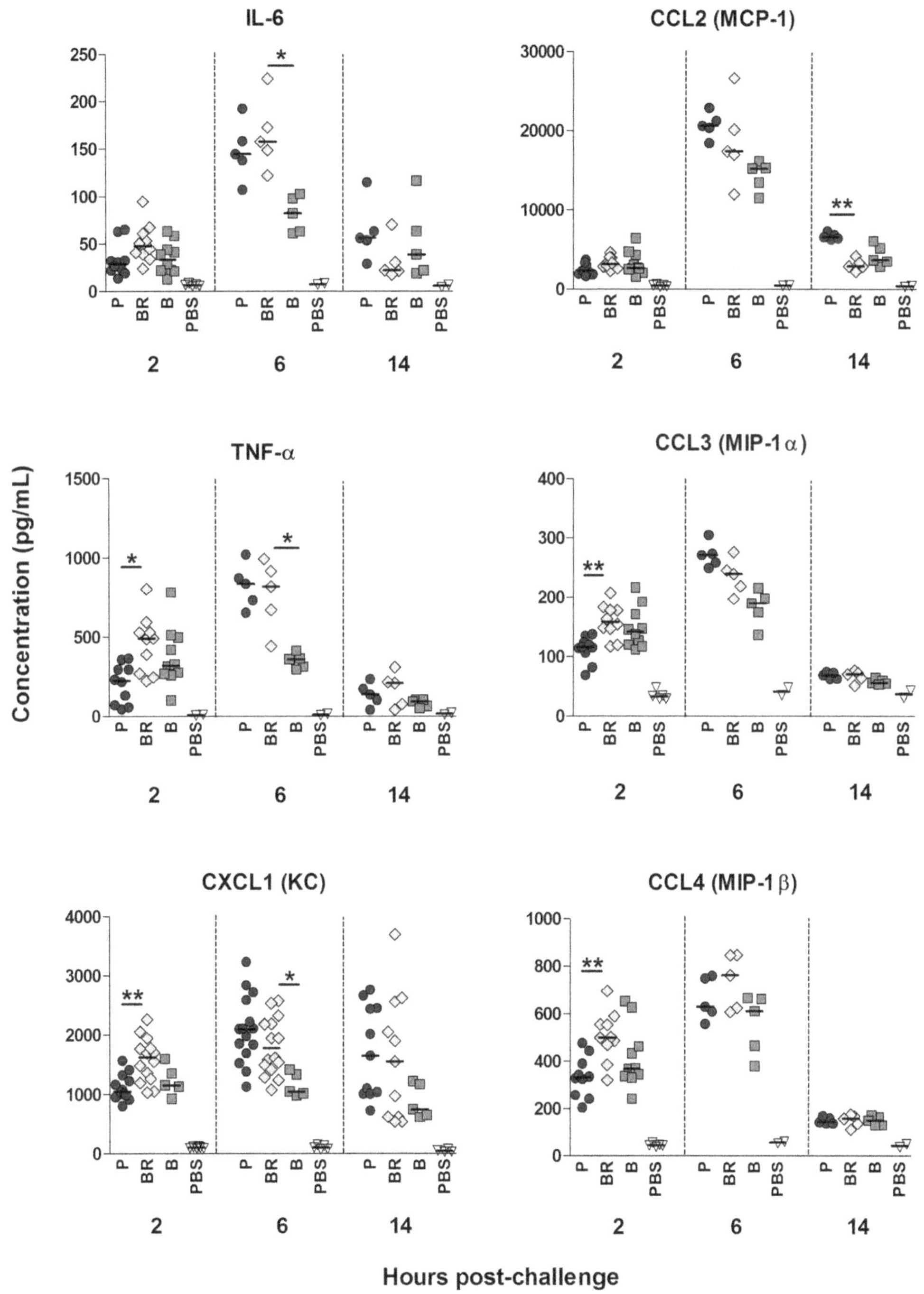

FIGURE 2 | Pro-inflammatory cytokines and chemokines induced by the different *S. epidermidis* populations. BALB/c mice were challenged intravenously with 1×10^8 planktonic (P), biofilm (B), biofilm-released (BR) cells, or sham-infected treated with PBS alone (PBS). The serum levels of the indicated cytokines were assessed 2, 6 and 14 h after infection. The obtained results are displayed as the concentration, in *p*g/mL, and the horizontal bars represent the median with range of 1 (6 and 14 h time points) to 2 independent (2 h time point) experiments that, per time point, presented the following number of animals: PBS $n = 2/2/2$; P $n = 10/5/5$; BR $n = 10/5/5$; B $n = 10/5/5$. Statistical differences among infected groups were evaluated using Kruskal–Wallis (Overall ANOVA $P < 0.05$) and *post hoc* Dunn's multiple comparison tests. $^*P < 0.05$, $^{**}P < 0.01$.

biofilm cells-infected mice, our results suggest that it would be worth investigate in future studies whether biofilm-released cells may employ a similar strategy to circumvent host inflammatory response.

Our results also suggest that biofilm-released cells might be particularly effective in activating antigen-presenting cells, specifically dendritic cells. This hypothesis is based on the significant increase in mRNA encoding the T cell co-stimulatory

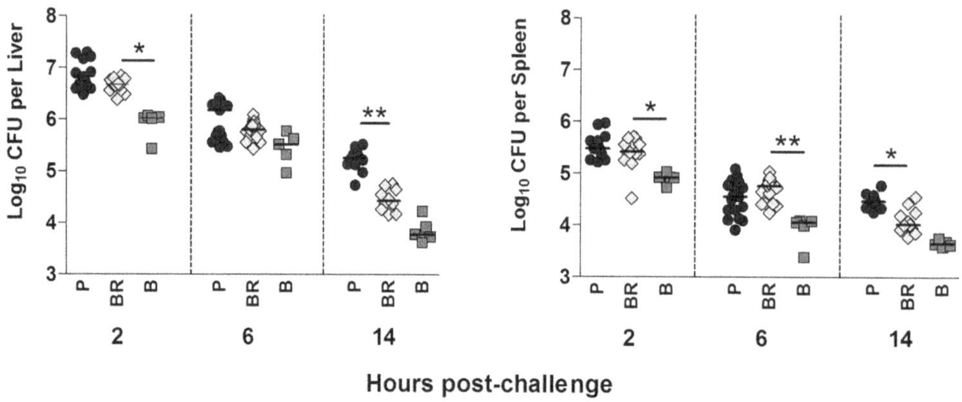

FIGURE 3 | Liver and spleen bacterial load after infection with the different *S. epidermidis* populations. BALB/c mice were challenged intravenously with 1 × 10^8 planktonic (P), biofilm (B), biofilm-released (BR) cells, or sham-infected treated with PBS alone (PBS). Liver and spleen bacterial burden was assessed 2, 6, and 14 h after intravenous infection. Each symbol represents an individual mouse and horizontal bars the median of 1 (biofilms) to 3 (P and BR, 6h time point) independent experiments that, per time point, presented the following number of animals: P $n = 14/16/11$; BR $n = 14/16/11$; B $n = 5/5/5$. Statistical differences among groups were evaluated with Kruskal–Wallis (Overall ANOVA $P < 0.05$) and *post hoc* Dunn's multiple comparison tests. *$P < 0.05$, **$P < 0.01$.

molecules CD80 and CD86 (Vasilevko et al., 2002; Sansom et al., 2003), the CD83 marker of mature dendritic cells (Lechmann et al., 2008), as well as CCL22 that encodes a chemokine secreted by both macrophages and dendritic cells (Yamashita and Kuroda, 2002) and CCL17 (TARC). Although, CCL17 has been associated with Th2-type responses (Xiao et al., 2003) how biofilm-released cells might affect T cell polarization should be determined in functional assays. In addition, since the spleen comprises cell types other than myeloid cells, including leukocytes and also non-hematopoietic cells, that may be able to produce pro-inflammatory mediators (Fritz and Gommerman, 2011; Bronte and Pittet, 2013), a better characterization of the cells particularly stimulated by biofilm-released cells is needed in order to identify the precise mechanism by which these cells interact with the host immune system.

It is important to emphasize that in this study only one *S. epidermidis* strain was used, and therefore, it was not possible to assess if these observations are transversal to the species or a strain-dependent phenomenon. Moreover, biofilm-released cells distinctive properties, in particular surface antigens, need to be fully characterized as these seem to have important consequences in the outcome of biofilm infections constituting interesting targets. Overall, our results indicate that *S. epidermidis* biofilm-released cells interact distinctly than planktonic or biofilm cells with the host immune system being particularly effective in inducing the production of pro-inflammatory cytokines and in stimulating neutrophils and monocytes. Biofilm-released cells might thus be of particular relevance in inducing deleterious inflammation frequently associated with *S. epidermidis* biofilm infections (Römling and Balsalbre, 2012) highlighting the urgent need to extend the study of *S. epidermidis* biofilm-originated infections by addressing the cells released by biofilms.

Finally, our findings also raise important concerns related to the current strategies proposed for the treatment of staphylococcal biofilm-related infections. The use of matrix-degrading enzymes, such as dispersin B, which is capable of dispersing cells from established biofilms (Kaplan, 2010), is one of the most frequently suggested strategies for the treatment for staphylococcal infections. However, as indicated by the data presented here, the use of matrix-degrading enzymes or other compounds leading to biofilm disassembly need to be carefully considered as biofilm-released cells can heighten the inflammatory response of the host consequently augmenting disease severity.

AUTHOR CONTRIBUTIONS

NC, GP, and MV designed the experiments. AF, BP, AC carried out the laboratory experiments. AF, GP, MV, and NC analyzed the data, interpreted the results, discussed analyses, interpretation and presentation. AF, AC, GP, NC, and MV wrote the paper. All authors have contributed to, seen and approved the manuscript.

FUNDING

This work was supported by European Union funds (FEDER/COMPETE) and by national funds (FCT) under the project with reference FCOMP-01-0124-FEDER-014309 (PTDC/BIA-MIC/113450/2009). The authors thank the FCT Strategic Project of UID/BIO/04469/2013 unit, and the project RECI/BBB-EBI/0179/2012 (FCOMP-01-0124-FEDER-027462). NC is an Investigator FCT. AF is supported by the FCT fellowship SFRH/BPD/99961/2014 and AC by the fellowship SFRH/BPD/91623/2012. The funders had no role in study design, data collection and interpretation, or the decision to submit the work for publication.

REFERENCES

Becker, P., Hufnagle, W., Peters, G., and Herrmann, M. (2001). Detection of differential gene expression in biofilm-forming versus planktonic populations of *Staphylococcus aureus* using micro-representational-difference analysis. *Appl. Environ. Microbiol.* 67, 2958–2965. doi: 10.1128/AEM.67.7.2958-2965.2001

Boles, B. R., and Horswill, A. R. (2011). Staphylococcal biofilm disassembly. *Trends Microbiol.* 19, 449–455. doi: 10.1016/j.tim.2011.06.004

Bronte, V., and Pittet, M. J. (2013). The spleen in local and systemic regulation of immunity. *Immunity* 39, 806–818. doi: 10.1016/j.immuni.2013.10.010

Cerca, F., Trigo, G., Correia, A., Cerca, N., Azeredo, J., and Vilanova, M. (2011). SYBR green as a fluorescent probe to evaluate the biofilm physiological state of *Staphylococcus epidermidis*, using flow cytometry. *Can. J. Microbiol.* 57, 850–856. doi: 10.1139/w11-078

Cerca, N., Jefferson, K. K., Oliveira, R., Pier, G. B., and Azeredo, J. (2006). Comparative antibody-mediated phagocytosis of *Staphylococcus epidermidis* cells grown in a biofilm or in the planktonic state. *Infect. Immun.* 74, 4849–4855. doi: 10.1128/IAI.00230-06

Cerca, N., Martins, S., Cerca, F., Jefferson, K. K., Pier, G. B., Oliveira, R., et al. (2005). Comparative assessment of antibiotic susceptibility of coagulase-negative staphylococci in biofilm versus planktonic culture as assessed by bacterial enumeration or rapid XTT colorimetry. *J. Antimicrob. Chemother.* 56, 331–336. doi: 10.1093/jac/dki217

Chua, S. L., Liu, Y., Yam, J. K., Chen, Y., Vejborg, R. M., Tan, B. G., et al. (2014). Dispersed cells represent a distinct stage in the transition from bacterial biofilm to planktonic lifestyles. *Nat. Commun.* 5:4462. doi: 10.1038/ncomms5462

Cordes, T., Michelucci, A., and Hiller, K. (2015). Itaconic acid: the surprising role of an industrial compound as a mammalian antimicrobial metabolite. *Annu. Rev. Nutr.* 35, 451–473. doi: 10.1146/annurev-nutr-071714-034243

Costerton, J. W., Stewart, P. S., and Greenberg, E. P. (1999). Bacterial biofilms: a common cause of persistent infections. *Science* 284, 1318–1322. doi: 10.1126/science.284.5418.1318

Couper, K. N., Blount, D. G., and Riley, E. M. (2008). IL-10: the master regulator of immunity to infection. *J. Immunol.* 180, 5771–5777. doi: 10.4049/jimmunol.180.9.5771

Duell, B. L., Tan, C. K., Carey, A. J., Cripps, A. W., and Ulett, G. C. (2012). Recent insights into microbial triggers of interleukin-10 production in the host and the impact on infectious disease pathogenesis. *FEMS Immunol. Med. Microbiol.* 64, 296–313. doi: 10.1111/j.1574-695x.2012.00931.x

Franca, A., Carvalhais, V., Vilanova, M., Pier, G. B., and Cerca, N. (2016). Characterization of an *in vitro* fed-batch model to obtain cells released from *S. epidermidis* biofilms. *AMB Express* 6, 23. doi: 10.1186/s13568-016-0197-9

Franceschini, A., Szklarczyk, D., Frankild, S., Kuhn, M., Simonovic, M., Roth, A., et al. (2013). STRING v9.1: protein-protein interaction networks, with increased coverage and integration. *Nucleic Acids Res.* 41, D808–D815. doi: 10.1093/nar/gks1094

Fritz, J. H., and Gommerman, J. L. (2011). Cytokine/stromal cell networks and lymphoid tissue environments. *J. Interferon. Cytokine Res.* 31, 277–289. doi: 10.1089/jir.2010.0121

Kallio, M. A., Tuimala, J. T., Hupponen, T., Klemela, P., Gentile, M., Scheinin, I., et al. (2011). Chipster: user-friendly analysis software for microarray and other high-throughput data. *BMC Genomics* 12:507. doi: 10.1186/1471-2164-12-507

Kaplan, J. B. (2010). Biofilm dispersal: mechanisms, clinical implications, and potential therapeutic uses. *J. Dent. Res.* 89, 205–218. doi: 10.1177/0022034509359403

Kobayashi, Y. (2008). The role of chemokines in neutrophil biology. *Front. Biosci.* 1, 2400–2407. doi: 10.2741/2853

Kristian, S. A., Birkenstock, T. A., Sauder, U., Mack, D., Gotz, F., and Landmann, R. (2008). Biofilm formation induces C3a release and protects *Staphylococcus epidermidis* from IgG and complement deposition and from neutrophil-dependent killing. *J. Infect. Dis.* 197, 1028–1035. doi: 10.1086/528992

Lechmann, M., Shuman, N., Wakeham, A., and Mak, T. W. (2008). The CD83 reporter mouse elucidates the activity of the CD83 promoter in B, T, and dendritic cell populations *in vivo*. *Proc. Natl. Acad. Sci. U.S.A.* 105, 11887–11892. doi: 10.1073/pnas.0806335105

Lee, P. Y., Wang, J. X., Parisini, E., Dascher, C. C., and Nigrovic, P. A. (2013). Ly6 family proteins in neutrophil biology. *J. Leukoc. Biol.* 94, 585–594. doi: 10.1189/jlb.0113014

Li, Y., Petrova, O. E., Su, S., Lau, G. W., Panmanee, W., Na, R., et al. (2014). BdlA, DipA and induced dispersion contribute to acute virulence and chronic persistence of *Pseudomonas aeruginosa*. *PLoS Pathog.* 10:e1004168. doi: 10.1371/journal.ppat.1004168

Liu, J., Ling, J. Q., Zhang, K., and Wu, C. D. (2013). Physiological properties of *Streptococcus mutans* UA159 biofilm-detached cells. *FEMS Microbiol. Lett.* 340, 11–18. doi: 10.1111/1574-6968.12066

Mack, D., Nedelmann, M., Krokotsch, A., Schwarzkopf, A., Heesemann, J., and Laufs, R. (1994). Characterization of transposon mutants of biofilm-producing *Staphylococcus epidermidis* impaired in the accumulative phase of biofilm production: genetic identification of a hexosamine-containing polysaccharide intercellular adhesin. *Infect. Immun.* 62, 3244–3253.

Marks, L. R., Davidson, B. A., Knight, P. R., and Hakansson, A. P. (2013). Interkingdom signaling induces *Streptococcus pneumoniae* biofilm dispersion and transition from asymptomatic colonization to disease. *MBio* 4, e00438–e00413. doi: 10.1128/mBio.00438-13

Nathan, C. (2006). Neutrophils and immunity: challenges and opportunities. *Nat. Rev. Immunol.* 6, 173–182. doi: 10.1038/nri1785

Oliveros, J. C. (2007). *VENNY. An Interactive Tool for Comparing Lists with Venn Diagrams.* Available online at: http://bioinfogp.cnb.csic.es/tools/venny/index.html

Otto, M. (2009). *Staphylococcus epidermidis*–the 'accidental' pathogen. *Nat. Rev. Microbiol.* 7, 555–567. doi: 10.1038/nrmicro2182

Otto, M. (2012). Staphylococcal infections: mechanisms of biofilm maturation and detachment as critical determinants of pathogenicity. *Annu. Rev. Med.* 64, 175–188. doi: 10.1146/annurev-med-042711-140023

Pavlidis, P., and Noble, W. S. (2003). Matrix2png: a utility for visualizing matrix data. *Bioinformatics* 19, 295–296. doi: 10.1093/bioinformatics/19.2.295

Pitz, A. M., Yu, F., Hermsen, E. D., Rupp, M. E., Fey, P. D., and Olsen, K. M. (2011). Vancomycin susceptibility trends and prevalence of heterogeneous vancomycin-intermediate *Staphylococcus aureus* in clinical methicillin-resistant *S. aureus* isolates. *J. Clin. Microbiol.* 49, 269–274. doi: 10.1128/JCM.00914-10

Resch, A., Rosenstein, R., Nerz, C., and Gotz, F. (2005). Differential gene expression profiling of *Staphylococcus aureus* cultivated under biofilm and planktonic conditions. *Appl. Environ. Microbiol.* 71, 2663–2676. doi: 10.1128/AEM.71.5.2663-2676.2005

Rollet, C., Gal, L., and Guzzo, J. (2009). Biofilm-detached cells, a transition from a sessile to a planktonic phenotype: a comparative study of adhesion and physiological characteristics in *Pseudomonas aeruginosa*. *FEMS Microbiol. Lett.* 290, 135–142. doi: 10.1111/j.1574-6968.2008.01415.x

Römling, U., and Balsalbre, C. (2012). Biofilm infections, their resilience to therapy and innovative treatment strategies. *J. Intern. Med.* 271, 541–561. doi: 10.1111/joim.12004

Rose, S. J., and Bermudez, L. E. (2014). Mycobacterium avium biofilm attenuates mononuclear phagocyte function by triggering hyperstimulation and apoptosis during early infection. *Infect. Immun.* 83, 405–412. doi: 10.1128/IAI.00820-13

Sagulenko, V., Thygesen, S. J., Sester, D. P., Idris, A., Cridland, J. A., Vajhala, P. R., et al. (2013). AIM2 and NLRP3 inflammasomes activate both apoptotic and pyroptotic death via ASC. *Cell Death Differ.* 20, 1149–1160. doi: 10.1038/cdd.2013.37

Sandberg, R., and Larsson, O. (2007). Improved precision and accuracy for microarrays using updated probe set definitions. *BMC Bioinformatics* 8:48. doi: 10.1186/1471-2105-8-48

Sansom, D. M., Manzotti, C. N., and Zheng, Y. (2003). What's the difference between CD80 and CD86? *Trends Immunol.* 24, 314–319. doi: 10.1016/S1471-4906(03)00111-X

Saraiva, M., and O'Garra, A. (2010). The regulation of IL-10 production by immune cells. *Nat. Rev. Immunol.* 10, 170–181. doi: 10.1038/nri2711

Shemesh, M., Tam, A., and Steinberg, D. (2007). Differential gene expression profiling of *Streptococcus mutans* cultured under biofilm and planktonic conditions. *Microbiology* 153, 1307–1317. doi: 10.1099/mic.0.2006/002030-0

Simard, J. C., Simon, M. M., Tessier, P. A., and Girard, D. (2011). Damage-associated molecular pattern S100A9 increases bactericidal activity of human neutrophils by enhancing phagocytosis. *J. Immunol.* 186, 3622–3631. doi: 10.4049/jimmunol.1002956

Smyth, G. K. (2004). Linear models and empirical bayes methods for assessing differential expression in microarray experiments. *Stat. Appl. Genet. Mol. Biol.* 3:3. doi: 10.2202/1544-6115.1027

Sousa, C., Franca, A., and Cerca, N. (2014). Assessing and reducing sources of genes expression variability in *Staphylococcus epidermidis* biofilms. *Biotechniques* 57, 295–301. doi: 10.2144/000114238

Supek, F., Bosnjak, M., Skunca, N., and Smuc, T. (2011). REVIGO summarizes and visualizes long lists of gene ontology terms. *PLoS ONE* 6:e21800. doi: 10.1371/journal.pone.0021800

Ulett, G. C., and Adderson, E. E. (2006). Regulation of apoptosis by Gram-positive bacteria: mechanistic diversity and consequences for immunity. *Curr. Immunol. Rev.* 2, 119–141. doi: 10.2174/157339506776843033

Vasilevko, V., Ghochikyan, A., Holterman, M. J., and Agadjanyan, M. G. (2002). CD80 (B7-1) and CD86 (B7-2) are functionally equivalent in the initiation and maintenance of CD4+ T-cell proliferation after activation with suboptimal doses of PHA. *DNA Cell Biol.* 21, 137–149. doi: 10.1089/10445490252925404

von Eiff, C., Peters, G., and Heilmann, C. (2002). Pathogenesis of infections due to coagulase-negative staphylococci. *Lancet Infect. Dis.* 2, 677–685. doi: 10.1016/S1473-3099(02)00438-3

Vuong, C., and Otto, M. (2002). *Staphylococcus epidermidis* infections. *Microbes Infect.* 4, 481–489. doi: 10.1016/S1286-4579(02)01563-0

Wang, R., Khan, B. A., Cheung, G. Y., Bach, T. H., Jameson-Lee, M., Kong, K. F., et al. (2011). *Staphylococcus epidermidis* surfactant peptides promote biofilm maturation and dissemination of biofilm-associated infection in mice. *J. Clin. Invest.* 121, 238–248. doi: 10.1172/JCI42520

Xiao, T., Fujita, H., Saeki, H., Mitsui, H., Sugaya, M., Tada, Y., et al. (2003). Thymus and activation-regulated chemokine (TARC/CCL17) produced by mouse epidermal Langerhans cells is upregulated by TNF-alpha and IL-4 and downregulated by IFN-gamma. *Cytokine* 23, 126–132. doi: 10.1016/S1043-4666(03)00221-7

Yamashita, U., and Kuroda, E. (2002). Regulation of macrophage-derived chemokine (MDC, CCL22) production. *Crit. Rev. Immunol.* 22, 105–114. doi: 10.1615/CritRevImmunol.v22.i2.10

Yang, S. C., and Hwang, T. L. (2016). The potential impacts of formyl peptide receptor 1 in inflammatory diseases. *Front. Biosci. (Elite. Ed).* 1, 436–449.

Yao, Y., Sturdevant, D. E., and Otto, M. (2005). Genomewide analysis of gene expression in *Staphylococcus epidermidis* biofilms: insights into the pathophysiology of *S. epidermidis* biofilms and the role of phenol-soluble modulins in formation of biofilms. *J. Infect. Dis.* 191, 289–298. doi: 10.1086/426945

13

Listeria ivanovii Infection in Mice: Restricted to the Liver and Lung with Limited Replication in the Spleen

*Mengying Zhou[1], Mingjuan Jiang[1], Chenyan Ren[1], Sijing Liu[1], Qikang Pu[1], Howard Goldfine[2], Hao Shen[2] and Chuan Wang[1]**

[1] Food Safety Monitoring and Risk Assessment Key Laboratory of Sichuan Province, Department of Public Health Laboratory Sciences, West China School of Public Health, Sichuan University, Chengdu, China, [2] Department of Microbiology, Perelman School of Medicine, University of Pennsylvania, Philadelphia, PA, USA

Edited by:
Olivier Dussurget,
Université Paris Diderot, France

Reviewed by:
Laurel L. Lenz,
University of Colorado School of Medicine, USA
Kenneth James Genovese,
United States Department of Agriculture, Agricultural Research Service, USA

***Correspondence:**
Chuan Wang
wangchuan@scu.edu.cn

Listeria monocytogenes (LM) vectors have shown much promise in delivery of viral and tumor antigens for the development of vaccines. *L. ivanovii* (LI) is a closely related bacterium with a similar intracellular life cycle that may offer advantages over LM because it is not a human pathogen, but can infect other animal species. Recent studies show that recombinant LI expressing *Mycobacterium tuberculosis* antigens is effective in inducing protective immunity in mouse models, demonstrating the potential of LI as a live vaccine vector. However, a key barrier in the development of LI into a live vaccine vector is that its pathogenic and immunogenic characteristics have yet to be fully understood. Therefore, in this research, C57BL/6J mice were inoculated with LM or LI intravenously or intranasally, and bacterial loads, histopathologic changes, and cytokine production were determined at indicated days post inoculation. Results showed that after intravenous infection with LM or LI, bacteria were found proliferating in the liver, spleen, and lung. However, LI could only reach a heavy burden in the liver and its ability to multiply and to resist host immunity seemed limited in the spleen and lung. After intranasal inoculation with LI, bacteria were mainly localized in the lung and failed to infect liver or spleen, while LM could. In organs with heavy LI burden, lesions were isolated, localized and densely packed, compared to lesions caused by LM, which were invasive. In the liver of intravenously inoculated mice and lung of intranasally inoculate mice, LI was able to elicit comparable cytokine production with LM and cause less severe histopathologic damages, and thus could be considered as a vector for treating or preventing hepatic or pulmonary diseases.

Keywords: *Listeria ivanovii*, *Listeria monocytogenes*, pathogenicity, immunogenicity, liver, lung

INTRODUCTION

Listeria monocytogenes (LM) is a Gram positive, food-borne pathogen responsible for human and animal listeriosis, characterized by severe gastroenteritis, central nervous system infections, and mother-to-child infections, with an overall mortality rate of 30% (Roberts and Wiedmann, 2003). LM is capable of multiplying in the cytoplasm of phagocytic and non-phagocytic cells and directly infecting adjacent cells, guaranteeing efficient antigen processing and presentation, which thereby enables it to elicit robust innate and antigen-specific cellular immunity (Stavru et al., 2011).

Consequently, LM is now being used as both a model to investigate host immune responses as well as a vaccine vector for T cell immunity (Singh and Wallecha, 2011; Stavru et al., 2011). Research into LM-based vaccines has made great progress in the last 20 years. Diverse antigen genes, such as genes of tumor related antigens and virus antigenic genes, have been genetically recombined into LM or attenuated strains, exhibiting desirable results *in vivo*, including tumor shrinkage, activated T cell immunity, and survival following higher challenging doses in mice (Mata et al., 2001; Brockstedt et al., 2004; Wood et al., 2011). So far, five LM-based vaccine candidates have been tested and are in various phases of clinical trials (Le et al., 2012).

In comparison, *L. ivanovii* (LI), the other pathogenic bacteria in the *Listeria* genus, is strictly limited to ruminant infection, except for extremely rare cases of infection in immunocompromised people (Dominguez-Bernal et al., 2006; Snapir et al., 2006; Guillet et al., 2010). Despite its different host tropism, LI is the bacterial species most similar to LM, sharing similar mechanisms for phagosome escape, intracellular mobility, reproduction, and cell-to-cell spreading (Vazquez-Boland et al., 2001b; Dominguez-Bernal et al., 2006). Therefore, LI could potentially be a promising vaccine vector even better than LM considering its lower pathogenicity in humans. Unfortunately, no research on LI-based vaccines has been reported until very recently, when two recombinant LI strains expressing *Mycobacterium tuberculosis* ESAT-6 or Ag85C protein were constructed and tested *in vivo* as tuberculosis vaccine candidates (Lin et al., 2015). Specific gamma interferon (IFN-γ) secretion of activated CD8 T cells was observed, which indicated the possibility of using LI as vaccine vector. However, the pathogenic and immunogenic features of LI and differences to LM remain poorly understood. Therefore, in order to correctly utilize LI as a novel live bacteria vaccine vector, it is important to conduct comprehensive studies such as the one detailed here.

In this study, C57BL/6J mice were inoculated intravenously (i.v.) or intranasally (i.n.), two effective routes of administration already established in C57BL/6J mice (Mayuko et al., 2002). Then, bacterial loads, histopathology and cytokine secretion kinetics along with infection course were systematically studied and compared between LM and LI infected mice.

MATERIALS AND METHODS

Animals

Female C57BL/6J mice were purchased from Beijing HFK Bioscience Company (Beijing, China). Mice aged 6–8 weeks were used in this study. All mice were maintained under specific-pathogen-free conditions throughout the experiments at the Animal Center of School of Public Health at Sichuan University. Mouse experiments were performed according to the guidelines of the Animal Care and Use Committee of Sichuan University.

Bacteria and Inoculum Preparation

LM 10403s strain and LI PAM55 strain were used in this study. All bacteria were stored in brain-heart infusion (BHI) broth containing 50% glycerol at −80°C. Bacteria were recovered in BHI broth from −80°C storage, and cultured in fresh BHI broth at 37°C and 200 rpm to mid-logarithmic phase (OD600≈0.4), then centrifuged at 4°C at 12000 *g* for 2 min, washed with normal saline, and then resuspended in normal saline. Titers of live bacteria were adjusted according to OD600 vs. bacteria concentration curves obtained in our earlier work (data not shown). Actual numbers of colony forming units (CFUs) of bacteria were verified by plating diluted inoculums on BHI agar plates.

Mouse Infection and Sacrifice

One hundred and sixty-two mice were randomly distributed into four infection groups (36 mice in each) and one naive group (18 mice). Two infection groups of mice were i.v. infected with 5×10^4 CFU LM or 5×10^5 CFU LI in a volume of 100 μl. The other two infection groups of mice were anesthetized by diethyl ether, and then i.n. inoculated with 5×10^5 CFU LM or 5×10^7 CFU LI in 25 μl of normal saline. All infection doses were 10-fold lower than the 50% lethal doses determined previously (data not shown). Naive group of mice were untreated. At 1, 2, 3, 5, and 8 each day(s) post inoculation (dpi), six mice in each i.v. infection group were sacrificed for analysis of bacterial loads and cytokine titers in organs. At 1, 3, 5, 8, and 11 each dpi, six mice in each i.n. infection group were sacrificed for the same analysis. At 4 and 6 each dpi, six mice in naive group were sacrificed for the same analysis. Additionally, six mice in each group were sacrificed for histopathologic analysis at 3 dpi (i.v. group), 5 dpi (i.n. group), or 4 dpi (naive group).

Determination of Bacteria Load in Organs from *Listeria*-Infected Mice

Liver, spleen, and lung were removed aseptically. Around 0.05 g tissue (the smallest lobe of liver, half spleen and one lobe of lung) was cut off and stored in liquid nitrogen for cytokine determinations. The rest of the tissue was weighed and homogenized in sterilized and pre-cooled phosphate buffer solution (PBS) containing 0.1% Triton-X100. Serial dilutions of homogenates were plated on BHI agar plates and cultured at 37°C. Colonies were counted after 24 h for LM or after 48 h for LI.

Histopathologic Analysis of Tissue from *Listeria*-Infected Mice

Liver, spleen, and lung were fixed in 10% buffered formalin, dehydrated and embedded in paraffin. Sections of 5 μm were cut, stained with hematoxylin-eosin and examined under 100 × magnification.

Preparation of Organ Extracts and Luminex Measurements of Cytokines

Frozen tissue was removed from liquid nitrogen, and ground in a cold mortar. Lysis buffer (50 mmol/L Tris-HCl pH7.4, 150 mmol/L NaCl, 1mmol/L EDTA, 1% Triton X-100, 1mmol/L phenylmethanesulfonyl fluoride, 5 μg/ml aprotinin, 5 μg/ml leupeptin) 100 μl, was added to lyse the tissue. Samples were incubated on ice for 10 min and vortexed for 2 min,

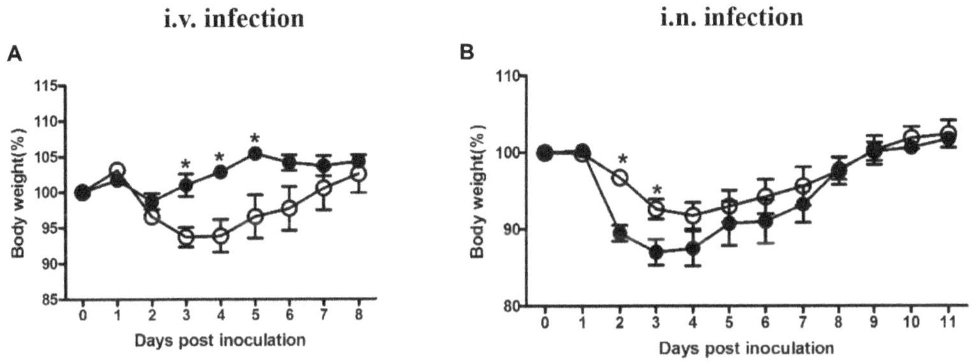

FIGURE 1 | Weight changing percentages after intravenous or intranasal inoculation with LM and LI. Mice were i.v. inoculated with 5 × 10^4 CFU LM or 5 × 10^5 CFU LI **(A)**, or i.n. inoculated with 5 × 10^5 CFU LM or 5 × 10^7 CFU LI **(B)**. Weight changing percentages were calculated and compared between LM (○) and LI (●) at the same time point by the method of *t* test (*$P < 0.05$). Each point represents mean ± standard error of the mean of six mice per group.

repeating twice. Samples were then centrifuged at 4°C and 16000 *g* for 10 min, and the supernatant was collected. Total protein concentration of each sample was measured using BCA Protein Assay kit (Beyotime, Shanghai, China) and was diluted to 10 mg/ml with PBS. All samples were stored at −80°C. Cytokine measurements were performed with a Luminex®100™ instrument using a ProcartaPlex Mouse Essential Th1/Th2 Cytokines Panel (eBioscience, San Diego, CA, USA). All procedures of cytokine determination were based on the manufacturer's instructions.

Statistical Analysis

Differences of daily weight between LM and LI infection mouse models were analyzed by *t*-test. Bacterial loads and cytokine titers in organs were compared by using the Mann–Whitney U non-parametric test. In all experiments, $P < 0.05$ was considered significant.

RESULTS

Kinetics of LM and LI Growth in Organs from Intravenous and Intranasal Infection Models

Four groups were either inoculated with 5 × 10^4 CFU LM (i.v.), 5 × 10^5 CFU LI (i.v.), 5 × 10^5 CFU LM (i.n.), or 5 × 10^7 CFU LI (i.n.). Each dose was 10-fold lower than the 50% lethal dose determined previously (data not shown). As an additional parameter of measurement during the course of infection, percentages of weight change after infection were taken daily (**Figure 1**). In intravenous infection model, mice infected either *Listeria* species started to lose weight at 1 dpi and then recovered gradually (**Figure 1A**). However, mice infected with LM reached maximum weight loss (6% of original weight) at 3 and 4 dpi and returned to original weight at 7 dpi, while mice infected with LI showed only slight weight loss (**Figure 1A**). After intranasal infection, mice infected with either LM or LI started to lose weight over 1 to 4 dpi and recovered to original weight at 9 dpi, but more weight loss was induced by LI infection (**Figure 1B**). The maximum weight loss for mice intranasally infected with LM was 8% of original weight while that for mice infected with LI was 13% of original weight (**Figure 1B**).

Bacterial loads in the liver, spleen, and lung were determined at indicated time points. In mice intravenously inoculated with 5 × 10^4 CFU LM, the bacterial load in the liver, though significantly less than that of LI at 1 dpi, kept increasing and reached maximum load of 4 × 10^4 CFU/g at 3 dpi (**Figure 2A**). By contrast, in mice intravenously inoculated with 5 × 10^5 CFU LI, about 88% of the bacteria invaded liver, causing the bacteria load in liver to reach 4 × 10^5 CFU per g then decreased gradually (**Figure 2A**). However, in the spleen after intravenous infection, LM showed notable and prolonged growth (**Figure 2C**). In mice intravenously inoculated with 5 × 10^5 CFU LI, the maximum load was 5 × 10^3 CFU per g spleen, while for mice inoculated with a smaller dose of LM, the maximum value was 3 × 10^6 CFU per g spleen (**Figure 2C**). Additionally, the load of LI in the spleen decreased below the detection limit after 3 dpi, but LM were still detectable at 8 dpi, showing that LI was cleared rapidly in host spleen (**Figure 2C**). In the lung after i.v. infection, though both LM and LI could reach a maximum load around 3 × 10^4 CFU per g lung, the former could resist host immunity and persist in the lung for over 5 days while the latter only persisted for 3 days, again showing its weak ability to maintain infection (**Figure 2E**).

In the liver from intranasal infection model, LM proliferated after 1 dpi, reached a peak value of 4 × 10^3 CFU per g liver at 5 dpi, and was undetectable at 8 dpi. In stark contrast, LI were hardly detectable for the duration of the experiment even though the dose was 100-fold higher than LM (**Figure 2B**). The same situation was also observed in the spleen, where levels of LM peaked at 3 dpi (10^5 CFU per g spleen), while LI remained at undetectable level for the duration of the entire experiment (**Figure 2D**). However, a difference was observed in the lung following intranasal inoculation. The bacterial load of LM increased after 1 dpi, peaked at 5 dpi, and gradually declined to undetectable levels at 11 dpi, presenting a distinct proliferative process (**Figure 2F**). In contrast, the load of LI

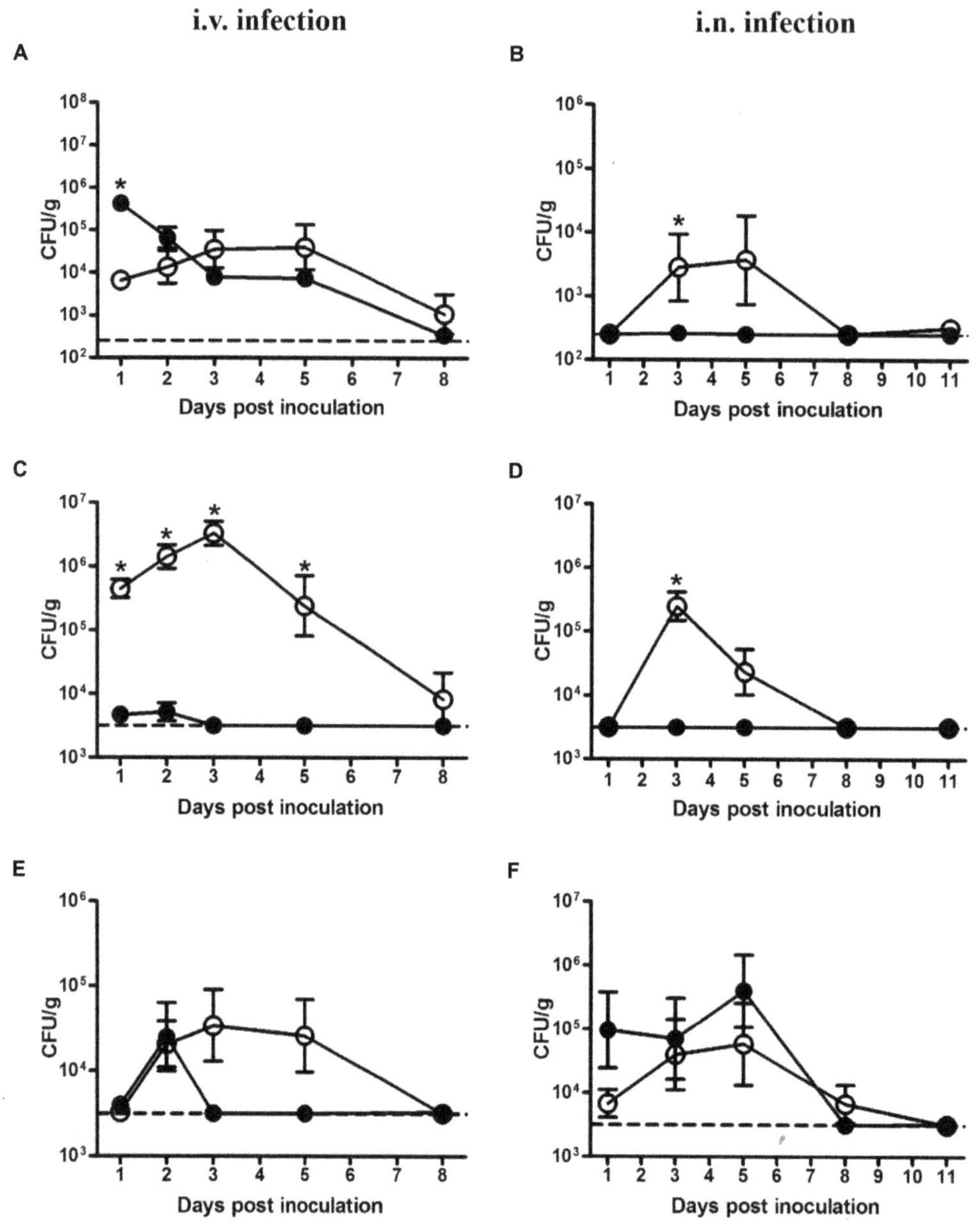

FIGURE 2 | Kinetics of LM and LI loads in liver, spleen, and lung from intravenous and intranasal infection models. Mice were i.v. inoculated with 5 × 10^4 CFU LM or 5 × 10^5 CFU LI, and bacterial loads in the liver **(A)**, spleen **(C),** and lung **(E)** were determined. Mice were i.n. inoculated with 5 × 10^5 CFU LM or 5 × 10^7 CFU LI, and bacterial loads in the liver **(B)**, spleen **(D),** and lung **(F)** were determined. The number of LM (○) and LI (●) in certain organ at indicated time points were compared by the method of the Mann–Whitney U non-parametric test (*$P < 0.05$) at the same time point. Each point represents the mean ± standard error of the mean for a group of six mice. The dotted lines represent the detection limits in each experiment.

maintained around 10^5 CFU per g lung for the first 5 days and sharply dropped below the detection limit at 8 dpi (**Figure 2F**). Though the maximum load of LI (4 × 10^5 CFU per g lung) in the lung were slightly higher than that of LM (6 × 10^4 CFU per g lung), the inoculation dose of LI was 100-fold higher, indicating that the LI could not resist host immunity and multiply as well as LM could in the lung. Taken together, after intranasal inoculation LI was mainly localized in the lung and could barely invade liver or spleen, while LM could multiply in these three organs, again demonstrating LI is less pathogenic.

Histopathologic Analysis of Organs Infected with LM or LI from Intravenous and Intranasal Infection Models

To study and compare tissue damage in host organs caused by LM or LI, mice that were intravenously infected were sacrificed at 3 dpi, mice that were intranasally infected were sacrificed at 5 dpi, and naive mice were sacrificed at 4 dpi. Histologic sections of liver, spleen, and lung samples were then made and analyzed (**Figure 3**). In the livers from intravenous infection model, lesions caused by LM were numerous and mainly located near vessels (**Figure 3D**), while lesions caused by LI were fewer but larger in size, featuring layers of necrotic hepatocytes and lymphocytes surrounding each other (**Figure 3G**). In the spleen from intravenous infection model, however, histopathology was consistent with bacteria load. LM infection led to severe necrosis which expanded to entirely efface the normal splenic architecture (**Figure 3E**), while no substantial changes were observed in LI infected spleens (**Figure 3H**). In the lung after i.v. infection, collapsed alveoli accompanied with lymphocytes were observed, along with more numerous and severe lesions for LM-infected mice (**Figure 3F**) than for LI-infected mice (**Figure 3I**). Compared to the intravenous infection model, both LM and LI inoculated intranasally presented smaller and fewer hepatic lesions and hardly observed splenic necrosis (**Figures 3J,K,M,N**). However, tissue damage of the lungs was severe and differed between mice infected with the two bacterial strains. Following intranasal infection with LM, pulmonary necrosis along with inflammatory infiltration expanded to a large area (**Figure 3L**). In contrast, lung lesions caused by LI were isolated and densely packed in a pattern similar to LI intravenously infected liver, where the bacterial burden was also heavy (**Figure 3O**). Histopathological results were consistent with bacterial loads, in that severe histopathologic changes were observed in tissues with high bacterial loads. However, lesions caused by LI were found isolated and densely packed which indicated the ability of LI to expand infection foci might be limited.

Kinetics of Cytokine Production in Organs Infected with LM or LI from Intravenous and Intranasal Infection Models

To investigate and compare inflammatory responses associated with LM and LI infection via intravenous (**Figure 4**) or intranasal (**Figure 5**) infection route, four groups of mice were inoculated accordingly. One group of mice was left uninfected as a naive control group. At indicated time points, Luminex beads assay was performed to determine cytokine titers in the liver, spleen, and lung, including tumor necrosis factor alpha (TNF-α), interleukin-12 (IL-12), interleukin-6 (IL-6), interleukin-4 (IL-4), and IFN-γ. The kinetics of cytokine production in mice infected with LM or LI differed by organs and inoculation routes.

Intravenous Infection Model

Liver

In the liver after intravenous injection, cytokines titers in LI inoculated mice were generally at comparable levels to titers in LM inoculated mice (**Figure 4**). TNF-α and IL-12 titers showed no substantial differences between mice infected with LM and LI at 1, 2, 3, and 5 dpi, except at 8 dpi (**Figures 4A,D**). IL-6 production was increased and reached a peak titer at 2 dpi after both *Listeria* infections and no difference of daily IL-6 production was found between LM and LI-infected mice (**Figure 4G**). IFN-γ titers in the liver were notably increased at 2 dpi in both LM and LI infected mice, but IFN-γ titer in LI infected mice decreased after 3 dpi while that in LM infected mice remain at a high level (**Figure 4J**). IL-4, an antagonistic factor in anti-listeriosis response, showed no differences between mice inoculated with the two *Listeria* strains within the 5 experimental days (**Figure 4M**).

Spleen

In the spleen following intravenous infection, however, cytokines production induced by LI were at significantly lower levels when compared with LM. For example, TNF-α titer for LM infected mice was significantly elevated (32.0 and 21.3 pg/mg total lysate protein) at 2 and 3 dpi, respectively, while the titer for LI for the same dpis were around 3.0 pg/mg total lysate protein, not significantly higher than it of naive mice (**Figure 4B**). The titers of IL-12 were increased after infection with either *Listeria* species, but the titer of LM infected mice was much higher (**Figure 4E**). Moreover, levels of both IL-6 and IFN-γ induced by LI were lower than those induced by LM at each experimental day, but they were still higher than those of naive mice and reached peak values of 4.5 pg/mg total lysate protein IL-6 and 17.3 pg/mg total lysate protein IFN-γ at 2 dpi (**Figures 4H,K**). IL-4 titer in the spleen kept declining during the course of LM infection, whereas titers for LI infection declined first and then returned to the level of naive control group since 5 dpi (**Figure 4N**).

Lung

No substantial differences in TNF-α, IL-12, and IL-4 secretion in the lung were found between LM and LI infected mice, but pulmonary IL-6 and IFN-γ induced by LM were significantly higher. No difference in titer of TNF-α was observed between the two bacteria in intravenously infected mice until 8 dpi, when a high titer of TNF-α (2.8 pg/mg total lysate protein) was found in the LM-infected mice (**Figure 4C**). Similarly, the difference of IL-12 titer between LM and LI occurred only at 8 dpi with IL-12 titer as low as 0.03 pg/mg total lysate protein in LI-infected mice (**Figure 4F**). IL-6 production in the lung after LM infection showed an obvious increase up to 9.5 pg/mg total lysate protein at 3 dpi, while the production after LI infection was barely increased (**Figure 4I**). Titers of pulmonary IFN-γ in LM infected mice was at least five times higher than that in LI infected mice (**Figure 4L**). No differences were found in IL-4 secretion in the two *Listeria* infected mice (**Figure 4O**).

Intranasal Infection Model

Liver

Cytokine titers in the liver and spleen were generally lower in the intranasal infection model than in the intravenous infection model, however, cytokine production in the lung were higher after i.n. infection than after i.v. infection, a finding consistent with bacterial burdens (**Figure 5**). In the liver from intranasal infection model, production of cytokines, except IFN-γ, were not notably promoted after infection with either *Listeria*. Kinetics of TNF-α, IL-12, and IL-4 induced by the two bacteria were the same (**Figures 5A,D,M**). IL-6 titer only differed at 1 dpi with 1.8 pg/mg

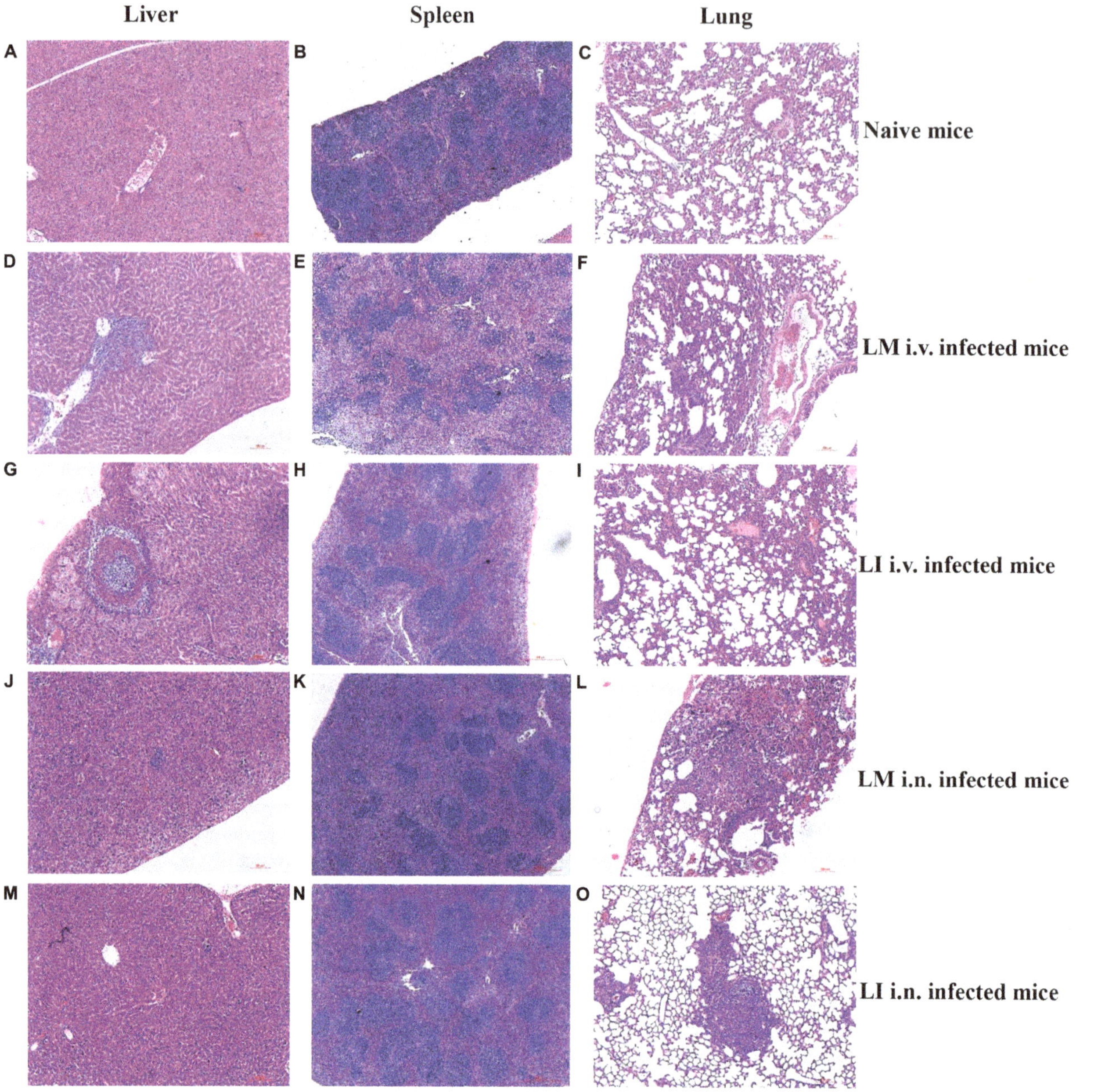

FIGURE 3 | Histopathologic analysis of liver, spleen, and lung from LM and LI intravenous and intranasal infection models. Mice were, respectively, inoculated with 5×10^4 CFU LM (i.v.), 5×10^5 CFU LI (i.v.), 5×10^5 CFU LM (i.n.) or 5×10^7 CFU LI, along with a group of naive mice. I.v. infected mice were sacrificed at 3 dpi, i.n. infected mice were sacrificed at 5 dpi, and naive mice were sacrificed at 4 dpi. Histologic sections of liver, spleen, and lung were made and stained with hematoxylin-eosin. Liver and lung sections were observed under a 40× microscope and spleen sections were observed under a 10× microscope. The images presented are representatives of changes seen in naïve mice **(A–C)**, LM i.v. **(D–F)**, LI i.v. **(G–I)**, LM i.n. **(J–L),** and LI i.n. **(M–O)**.

total lysate protein for the LM infected mice and 4.1 pg/mg total lysate protein for the LI infected mice (**Figure 5G**). IFN-γ production, however, were significantly different between LM and LI i.n. infected mice. Hepatic IFN-γ in LI inoculated mice slightly changed within the range of naïve mice, while LM induced IFN-γ highly secreted in the liver, 16.5 pg/mg total lysate protein at 3 dpi and 14.5 pg/mg total lysate protein at 11 dpi (**Figure 5J**).

Spleen

In the spleen from intranasal infection model, cytokine levels in LM infected mice were notably increased at 3 dpi while changes of cytokine levels in LI-infected mice were less significant. Titers of TNF-α and IL-12 were obviously increased to 10.1 pg/mg total lysate protein and 0.2 pg/mg total lysate protein at the third day after infection with LM while production of TNF-α and IL-12 barely changed after infection with LI (**Figures 5B,E**). The peak

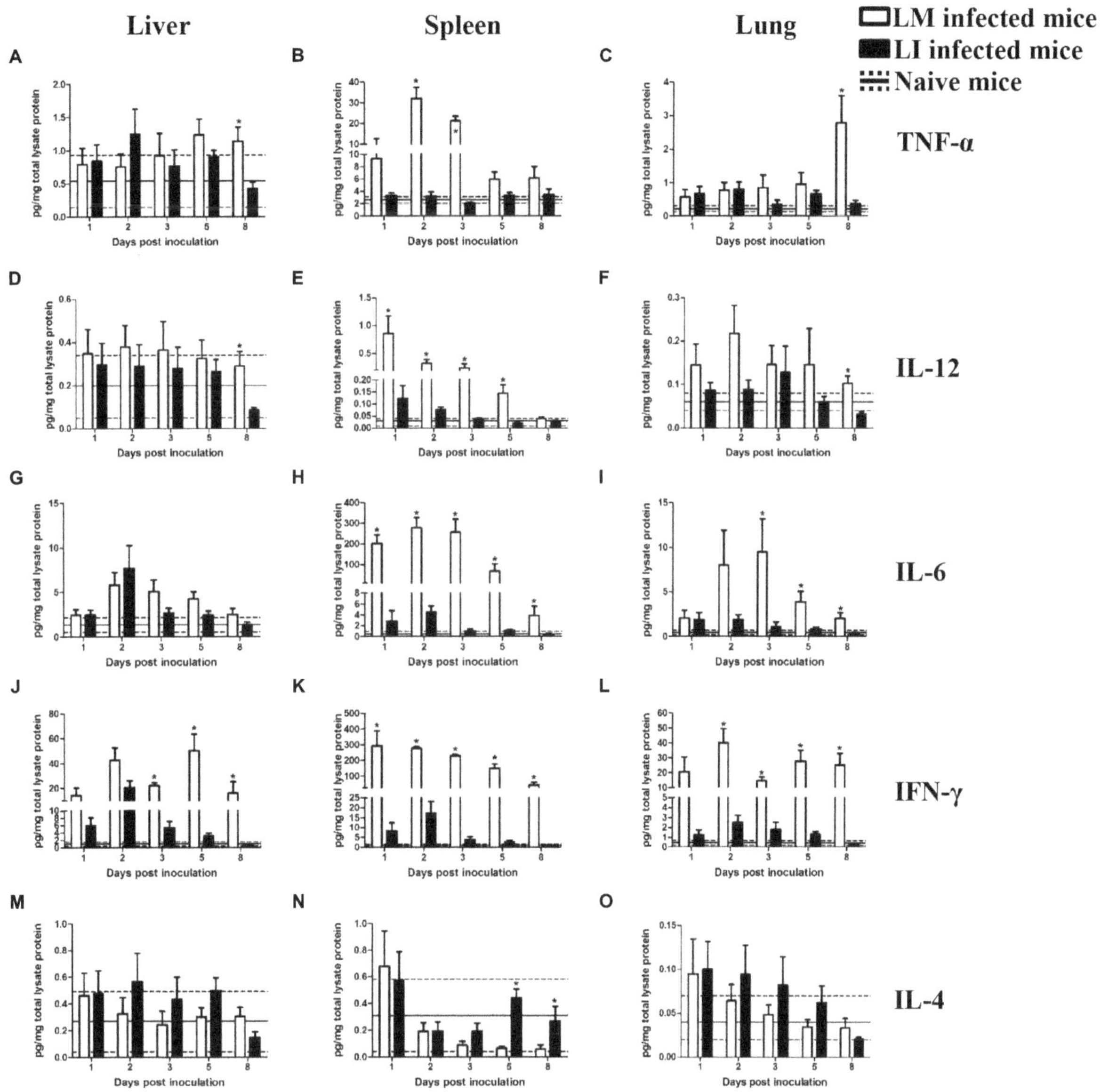

FIGURE 4 | Kinetics of cytokine production in organs from LM and LI intravenous infection models. Mice were, respectively, inoculated with 5 × 10^4 CFU LM (white bar) or 5 × 10^5 CFU LI (black bar), and one group of naive mice were set (solid line represents average value and the dotted line represents average value plus/minus standard deviation). Titers of TNF-α **(A–C)**, IL-12 **(D–F)**, IL-6 **(G–I)**, IFN-γ **(J–L)**, and IL-4 **(M–O)** in the liver, spleen, and lung were determined at 1, 2, 3, 5, and 8 dpi for *Listeria* infected mice or at 4 dpi for naive mice. The differences of titers at each test day were compared between LM and LI by the method of the Mann–Whitney U non-parametric test (*$P < 0.05$). Each bar represents the mean ± standard error of the mean for a group of six mice.

for IL-6 secretion occurred at 1 dpi with a titer of 22.5 pg/mg total lysate protein in LI infected mice, but in LM infected mice, the IL-6 peak occurred at 3 dpi with a titer of 29.3 pg/mg total lysate protein (**Figure 5H**). IFN-γ production was significantly higher following LM intranasal infection at 3, 5, 8, and 11 dpi (**Figure 5K**). Titers of IL-4 decreased after infection with both *Listeria* but it recovered earlier in LI infected mice (**Figure 5N**).

Lung

In the lung after intranasal infection, where the bacteria were mainly focused, cytokine titers were generally higher than those in the lung from the intravenous infection model, but fewer differences were found between mice infected with LM and LI. TNF-α, IL-6, and IL-4 production separately shared similar trends and levels in mice infected with LM or LI (**Figures 5C,I,O**). The peak of IL-12 was at 1 dpi in LI infected mice but occurred at 3 dpi in LM infected mice (**Figure 5F**). The titer of IFN-γ was higher in LM intranasally infected lung at 3, 5, 8, and 11 dpi, and the biggest difference presented at 5 dpi with 72.1 pg/mg total lysate protein in LM infected mice but only 9.4 pg/mg total lysate protein in LI infected mice (**Figure 5L**). However, the difference was not statistically significant.

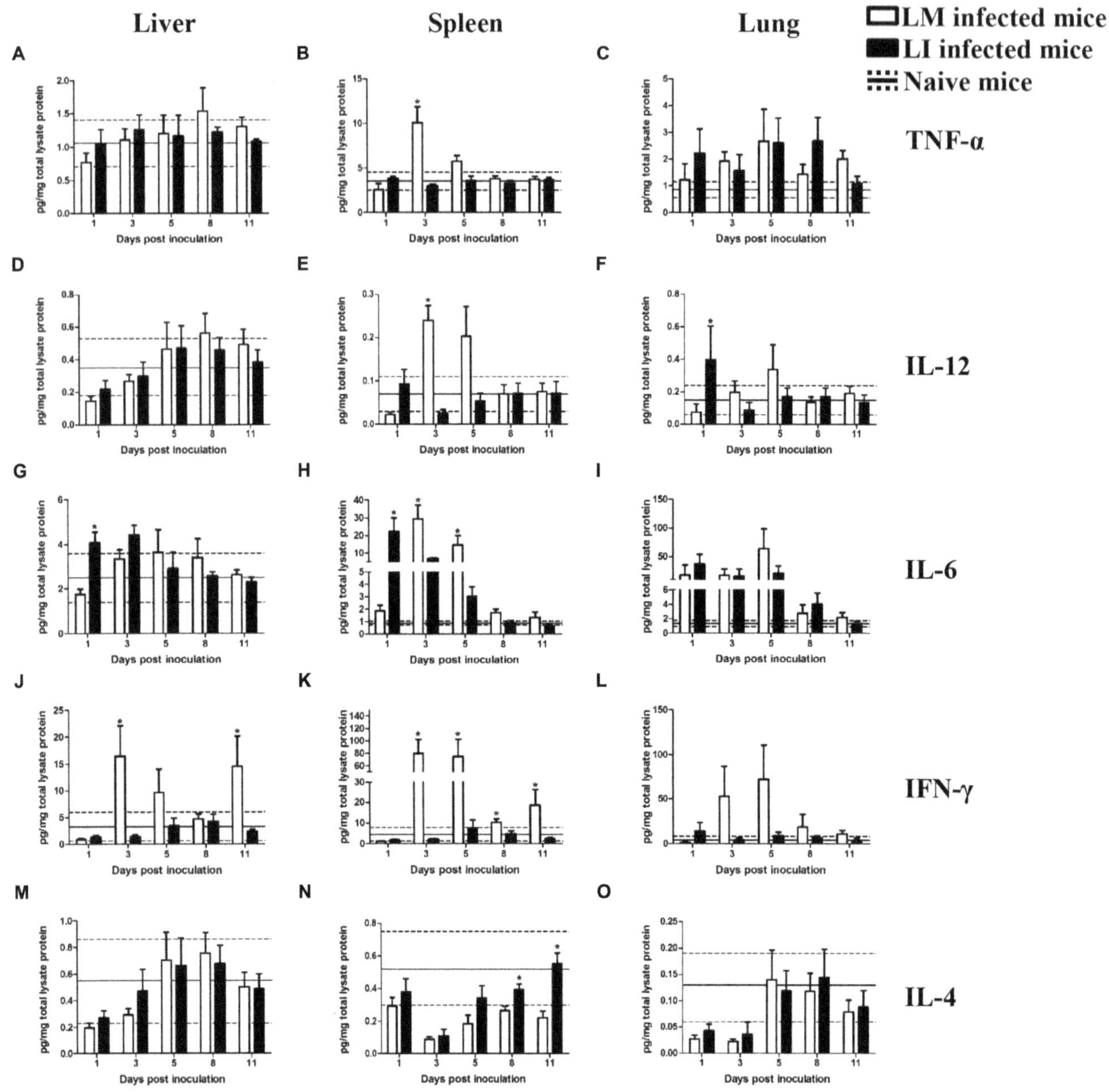

FIGURE 5 | Kinetics of cytokines production in organs from LM and LI intranasal infection models. Mice were, respectively, inoculated with 5×10^5 CFU LM (white bar) or 5×10^7 CFU LI (black bar), and one group of naive mice were set (solid line represents average value and the dotted line represents average value plus/minus standard deviation). Titers of TNF-α **(A–C)**, IL-12 **(D–F)**, IL-6 **(G–I)**, IFN-γ **(J–L)**, and IL-4 **(M–O)** in the liver, spleen, and lung were determined at 1, 3, 5, 8, and 11 dpi for *Listeria* infected mice or at 6 dpi for naive mice. The differences of titers at each test day were compared between LM and LI by the method of the Mann–Whitney U non-parametric test (*$P < 0.05$). Each bar represents the mean ± standard error of the mean for a group of six mice.

DISCUSSION

In this study pathogenic and immunologic outcomes of LI intravenous or intranasal infections in C57BL/6J mice were investigated in comparison with LM. In intravenous infection model, bacteria were delivered into the bloodstream and directly distributed to organs. In the spleen and lung, LI showed reduced pathogenicity compared with LM, as demonstrated by low and transient bacterial loads and minor histopathologic changes. In the liver, however, LI could rapidly reach a high burden, caused notable pathogenic damages and then cleared by host gradually. It indicated that LI in blood were mainly distributed to liver, which, to our knowledge, is related to a protein called Internalin B (InlB). In LM, InlB is found to trigger the entry of LM into hepatocytes by interacting with a hepatocyte growth factor receptor and two other cellular components (Bierne and

Cossart, 2002). LM mutant lacking the gene encoding InlB (*inlB*) displays a marked decrease in its capability to proliferate within mouse hepatocytes *in vivo* and *in vitro* (Gregory et al., 1997). The existence of InlB for LI has been proved by gene sequencing. Two genes named *i-inlB1* and *i-inlB2* were found to be genetically similar to *inlB* of LM, and were thought to code for an InlB ortholog in LI (Dominguez-Bernal et al., 2006). Comparing their amino acid sequences, both i-InlB1 and i-InlB2 share a similar architecture with InlB for the internalin domain and bacterial-surface anchoring domain, indicating that i-InlB1 and i-InlB2 may mediate the interaction between LI and hepatocytes. However, detailed differences in structure and function *in vivo* among these three internalins and how i-InlB1 and i-InlB2 interact hepatocytes are still wanted.

Another finding in intravenous infection model is that LI failed to keep multiplying as long as LM did and were cleared by host more rapidly, especially in the spleen and lung. After analyzing typical histopathologic changes of LI (**Figures 3G,O**), it was found that LI infection foci was isolated and densely packed while LM infection foci was invasive, indicating that, compared with LM, the abilities of LI to multiply intracellularly and/or to invade neighboring cells are limited. It has been proved that LI is not as capable of multiplying in splenic cells as LM *in vitro* (Kimoto et al., 2003). This flaw could be related to LI's poor ability to lyze phagosome, since LM mutant unable to effectively disrupt phagosome and escape into cytoplasma is soon cleared (Gaillard et al., 1986). Phagosome lysis is initiated by three LM virulence determinants: listeriolysin O (LLO) and two phospholipases (PlcA and PlcB; Vazquez-Boland et al., 2001b). These three molecules are also found in LI. Ivanolysin O (ILO) is the homolog of LLO in LI. A LM mutant expressing ILO instead of its own LLO failed to proliferate in the spleen, inducing a high-level of IFN-γ, as well as generating robust protective immunity (Frehel et al., 2003; Hideki et al., 2007), which indicates the weak ILO may be responsible for deficient intracellular multiplication of LI. PlcA and PlcB of LI are only verified genetically, and no details about their structure and function are available (Vazquez-Boland et al., 2001a). After escaping from the ruptured phagosome, LM replicates in the cytoplasm and expresses the surface molecule ActA which provides the bacteria with actin-based motility and propels it through membrane protrusions into neighboring cells to expend infection to a large area (Vazquez-Boland et al., 2001b). In LI, the ActA-like protein is called i-ActA and can restore mobility of LM *actA* deletion mutant strain *in vitro* (Gouin et al., 1995). However, detailed differences between i-ActA and ActA in the process of cell-to-cell infection remain unclear. So far, major virulence factors of LM has been genetically recognized in LI (Vazquez-Boland et al., 2001a). Similarities in genes between these homologs imply similar functions, but it is the differences that help to explain different behavior *in vivo* between LM and LI. More information is needed to make reliable conclusions.

In this research, immunologic features of LI in different organs after different inoculation were also conducted in comparison with LM. Five cytokines were adopted to reflect different aspects of host anti-*Listeria* immune response, due to sufficient data of host immunity to LM. TNF-α, IL-6, and IL-12 are secreted by LM infected macrophages to recruit and activate neutrophils and NK cells, which are both capable of killing infected cells (Harty et al., 1996; Mocci et al., 1997). IFN-γ is mainly secreted by activated NK cells to enhance bactericidal actions inside infected macrophages and promote expression of signal molecules on surface to further amplify immune stimulation (Pamer, 2004). Apart from innate immune response, T cell-mediated immune response also takes part in host immunity to first time infection of LM. IL-6 and IL-12 play a role in promoting Th0 cells to develop into Th1 cells, while IL-4 suppresses the development (Kaufman et al., 1997; Mocci et al., 1997). In this research, though the magnitude and duration of each cytokine differed between the two *Listeria*, the trends along with infection are same, which are increase of TNF-α, IL-6, IL-12, and IFN-γ and decrease of IL-4. The same secretion trend of cytokines implies host immune response to LI is similar with it to LM. However, in most cases, cytokine response to LM was stronger than that to LI, which was likely due to different behavior in tissues between LM and LI. In infected tissue, LM increased in number exponentially while LI was either cleared or not growing, and thus LI did not induce cytokine response as robust as LM did. This phenomenon, on the one hand, proves that LI is less virulent, and on the other hand indicates that LI, as a vaccine vector, needs optimization for stronger immunogenicity.

Nevertheless, cytokine secretion induced by LI infection was at comparable level of LM infectoin in these two situations: in the liver from i.v. inoculated mice and in the lung from i.n. inocualted mice. Titers of TNF-α, IL-6, IL-12, and IL-4 after infection changed at same level between LI and LM infections. IFN-γ in the liver after i.v. infection dropped earlier in LI infected mice than LM infected mice because LI could be cleared rapidly. IFN-γ in the lung after i.n. infection seemed higher in LM infected mice at 3 and 5 dpi, but the difference was not statistically significant due to individual differences. In general, LI could mainly infect mouse liver after i.v. inoculation and specifically infect mouse lung after i.n. inoculation. In these two situations, LI could induce immune response not significantly weaker than LM, cause less sever tissue damages and could be controlled by host more easily. Therefore, speculation is that LI, after being genetically improved for better immunogenicity, could be used as vaccine vector specific for hepatic or pulmonary disease, especially for immunocompromised people.

AUTHOR CONTRIBUTIONS

MZ, CW, and HS designed the study. MZ, MJ, CR, SL, and QP performed the experimental work. MZ analyzed the data and prepared the manuscript. CW, HG, and HS contributed to the final manuscript.

ACKNOWLEDGMENT

This work was supported by the National Natural Science Foundation of China (No. 31570924).

REFERENCES

Bierne, H., and Cossart, P. (2002). InlB, a surface protein of *Listeria monocytogenes* that behaves as an invasin and a growth factor. *J. Cell Sci.* 115, 3357–3368.

Brockstedt, D. G., Giedlin, M. A., Leong, M. L., Bahjat, K. S., Gao, Y., Luckett, W., et al. (2004). *Listeria*-based cancer vaccines that segregate immunogenicity from toxicity. *Proc. Natl. Acad. Sci. U.S.A.* 101, 13832–13837.

Dominguez-Bernal, G., Muller-Altrock, S., Gonzalez-Zorn, B., Scortti, M., Herrmann, P., Monzo, H. J., et al. (2006). A spontaneous genomic deletion in *Listeria ivanovii* identifies LIPI-2, a species-specific pathogenicity island encoding sphingomyelinase and numerous internalins. *Mol. Microbiol.* 59, 415–432. doi: 10.1111/j.1365-2958.2005.04955.x

Frehel, C., Lety, M., Autret, N., Beretti, J. L., Berche, P., and Charbit, A. (2003). Capacity of ivanolysin O to replace listeriolysin O in phagosomal escape and *in vivo* survival of *Listeria monocytogenes*. *Microbiology* 149, 611–620. doi: 10.1099/mic.0.25986-0

Gaillard, J. L., Berche, P., and Sansonetti, P. (1986). Transposon mutagenesis as a tool to study the role of hemolysin in the virulence of *Listeria monocytogenes*. *Infect. Immun.* 52, 50–55.

Gouin, E., Dehoux, P., Mengaud, J., Kocks, C., and Cossart, P. (1995). iactA of *Listeria ivanovii*, although distantly related to *Listeria monocytogenes* actA, restores actin tail formation in an L. *monocytogenes actA mutant*. *Infect. Immun.* 63, 2729–2737.

Gregory, S. H., Sagnimeni, A. J., and Wing, E. J. (1997). Internalin B promotes the replication of *Listeria monocytogenes* in mouse hepatocytes. *Infect. Immun.* 65, 5137–5141.

Guillet, C., Join-Lambert, O., Monnier, A., Leclercq, A., Mechaï, F., Mamzer-Bruneel, M., et al. (2010). Human listeriosis caused by *Listeria ivanovii*. *Emerg. Infect. Dis.* 16, 136–138. doi: 10.3201/eid1601.091155

Harty, J. T., Lenz, L. L., and Bevan, M. J. (1996). Primary and secondary immune responses to *Listeria monocytogenes*. *Curr. Opin. Immunol.* 8, 526–530. doi: 10.1016/S0952-7915(96)80041-0

Hideki, H., Ikuo, K., Takamasa, N., Takanari, T., Kohsuke, T., and Masao, M. (2007). Cytolysin-dependent escape of the bacterium from the phagosome is required but not sufficient for induction of the Th1 immune response against *Listeria monocytogenes* infection: distinct role of listeriolysin O determined by cytolysin gene replacement. *Infect. Immun.* 75, 3791–3801. doi: 10.1128/IAI.01779-06

Kaufman, S. H. E., Emoto, M., Szalay, G., Barsig, J., and Flesch, I. E. A. (1997). Interleukin-4 and listeriosis. *Immunol. Rev.* 158, 95–105. doi: 10.1111/j.1600-065X.1997.tb00995.x

Kimoto, T., Kawamura, I., Kohda, C., Nomura, T., Tsuchiya, K., Ito, Y., et al. (2003). Differences in gama interferon production induced by Listeriolysin O and Ivanolysin O result in different levels of protective immunity in mice infected with *Listeria monocytogenes* and *Listeria ivanovii*. *Infect. Immun.* 71, 2447–2454. doi: 10.1128/IAI.71.5.2447-2454.2003

Le, D. T., Dubenksy, T. W., and Brockstedt, D. G. (2012). Clinical development of *Listeria monocytogenes*-based immunotherapies. *Sem. Oncol.* 39, 311–322. doi: 10.1053/j.seminoncol.2012.02.008

Lin, Q., Zhou, M., Xu, Z., Khanniche, A., Hao, S., and Wang, C. (2015). Construction of two *Listeria ivanovii* attenuated strains expressing Mycobacterium tuberculosis antigens for TB vaccine purposes. *J. Biotechnol.* 19, 20–26. doi: 10.1016/j.jbiotec.2015.01.008

Mata, M., Yao, Z. J., Zubair, A., Syres, K., and Paterson, Y. (2001). Evaluation of a recombinant *Listeria monocytogenes* expressing an HIV protein that protects mice against viral challenge. *Vaccine* 19, 1435–1445. doi: 10.1016/S0264-410X(00)00379-0

Mayuko, M., Akio, N., Kenji, S., Yoh-Ich, T., and Yoichiro, I. (2002). Comparison of host resistance to primary and secondary Listeria monocytogenes infections in mice by intranasal and intravenous routes. *Infect. Immun.* 70, 4805–4811. doi: 10.1128/IAI.70.9.4805-4811.2002

Mocci, S., Dalrymple, S. A., Nishinakamura, R., and Murray, R. (1997). The cytokine stew and innate resistance to L. *monocytogenes*. *Immunol. Rev.* 158, 107–114. doi: 10.1111/j.1600-065X.1997.tb00996.x

Pamer, E. G. (2004). Immune responses to *Listeria monocytogenes*. *Nat. Rev. Immunol.* 4, 812–823. doi: 10.1038/nri1461

Roberts, A. J., and Wiedmann, M. (2003). Pathogen, host and environmental factors contributing to the pathogenesis of listeriosis. *Cell. Mol. Life Sci.* 60, 904–918. doi: 10.1007/s00018-003-2225-6

Singh, R., and Wallecha, A. (2011). Cancer immunotherapy using recombinant *Listeria monocytogenes*: transition from bench to clinic. *Hum. Vaccin.* 7, 497–505. doi: 10.4161/hv.7.5.15132

Snapir, Y. M., Vaisbein, E., and Nassar, F. (2006). Low virulence but potentially fatal outcome-*Listeria ivanovii*. *Eur. J. Intern. Med.* 17, 286–287. doi: 10.1016/j.ejim.2005.12.006

Stavru, F., Archambaud, C., and Cossart, P. (2011). Cell biology and immunology of *Listeria monocytogenes* infections: novel insights. *Immunol. Rev.* 240, 160–184. doi: 10.1111/j.1600-065X.2010.00993.x

Vazquez-Boland, J. A., Dominguez-Bernal, G., Gonzalez-Zorn, B., Kreft, J., and Goebel, W. (2001a). Pathogenicity islands and virulence evolution in *Listeria*. *Microbes Infect.* 3, 571–584. doi: 10.1016/S1286-4579(01)01413-7

Vazquez-Boland, J. A., Kuhn, M., Berche, P., Chakraborty, T., Domínguez-Bernal, G., Goebel, W., et al. (2001b). *Listeria* pathogenesis and molecular virulence determinants. *Clin. Microbiol. Rev.* 14, 584–640. doi: 10.1128/CMR.14.3.584-640.2001

Wood, L. M., Pan, Z. K., Guirnalda, P., Tsai, P., Seavey, M., and Paterson, Y. (2011). Targeting tumor vasculature with novel *Listeria*-based vaccines directed against cd105. *Cancer Immunol. Immunother.* 60, 931–942. doi: 10.1007/s00262-011-1002-x

14

Feeding a High Concentration Diet Induces Unhealthy Alterations in the Composition and Metabolism of Ruminal Microbiota and Host Response in a Goat Model

Canfeng Hua[1†], Jing Tian[1†], Ping Tian[1], Rihua Cong[2], Yanwen Luo[1], Yali Geng[1], Shiyu Tao[1], Yingdong Ni[1] and Ruqian Zhao[1]*

[1] Key Laboratory of Animal Physiology and Biochemistry, Ministry of Agriculture, Nanjing Agricultural University, Nanjing, China, [2] College of Veterinary Medicine, Northwest A&F University, Xianyang, China

Edited by:
Diana Bahia,
Universidade Federal de Minas Gerais, Brazil

Reviewed by:
Sukanya Narasimhan,
Yale School of Medicine, USA
Manuel Vilanova,
Instituto de Ciências Biomédicas Abel Salazar – Universidade do Porto, Portugal

***Correspondence:**
Yingdong Ni
niyingdong@njau.edu.cn

† These authors have contributed equally to this work.

There is limited knowledge about the impact of long-term feeding a high-concentrate (HC) diet on rumen microbiota, metabolome, and host cell functions. In this study, a combination of mass spectrometry-based metabolomics techniques, 454 pyrosequencing of 16S rDNA genes, and RT-PCR was applied to evaluate the changes of ruminal microbiota composition, ruminal metabolites, and related genes expression in rumen epithelial cells of lactating goats received either a 35% concentrate diet or a 65% concentrate diet for 4 or 19 weeks, respectively. Results show that feeding a HC diet reduced the microbiota diversity and led to the disorders of metabolism in the rumen. The concentrations of lactate, phosphorus, NH3-N and endotoxin Lipopolysaccharide in ruminal fluids, and plasma histamine, lactate and urine N (UN) were increased significantly in goats fed with a HC diet. A significant increase of genes expression related to volatile fatty acids transport, cell apoptosis, and inflammatory responses were also observed in goats fed with a HC diet. Correlation analysis revealed some potential relationships between bacteria abundance and metabolites concentrations. Our findings indicate that a HC diet can induce ruminal microbiota dysbiosis and metabolic disorders, thus increasing risks to host health and potential harm to the environment.

Keywords: bacteria, gene expression/regulation, metabolism, goat, rumen

INTRODUCTION

A vast ensemble of ruminal microbiota including bacteria, archaea, ciliate protozoa, and anaerobic fungi provide important metabolic capabilities to digest cellulose-rich feedstuffs and to convert them into a wide range of nutrient compounds in order to sustain body maintenance and performance. A high-fiber diet and a stable microbiota community are necessary for keeping ruminants health. However, in the current feeding systems, particularly in the intensive management systems, it is a common strategy to feed large amounts of grains diet to ruminants due to the lack of quality forage and pursuing high milk yield (Bal et al., 1997; Soriano et al., 2000; Boerman et al., 2015). It is well known that feeding excessive amounts of non-structural carbohydrates and highly fermentable forages to ruminants commonly result in subacute ruminal acidosis (SARA), a common metabolic disease prevalent in high-producing animals (Klevenhusen et al., 2013). Previous evidence showed that SARA can be experimentally induced by feeding 50~65% proportion of grain diet to ruminants (Tao et al., 2014a,b).

Animals suffering from SARA are usually accompanied with ruminal microbiota dysbiosis in bacteria, protozoa, anaerobic rumen fungi, archaea, and bacteriophages (Jami et al., 2013; Pitta et al., 2014). Currently, utilization of next generation sequencing technology (NGS) can describe the microbiome diversity and various factors that influence bacterial dynamics in greater resolution than ever before (Hristov et al., 2012). In previous studies (Mao et al., 2016), the alterations of ruminal microbiota were found in SARA ruminants fed with high-concentrate (HC) diets for a relatively short-term showing a decrease of Bacteroidetes, but an increase in Firmicutes. However, the effects of long-term feeding HC diets on the composition and structure of ruminal microbiota, metabolism and host responses are unknown. In this study, we used 454 pyrosequencing methods to investigate the changes of the structure and component of rumen microbiota in lactating dairy goats fed with a HC diet for short (4 weeks) or long (19 weeks) periods.

As one of the "core microbiome" (Mao et al., 2016), Firmicutes has a strong association with the biological fiber degradation (Koike et al., 2010). Methane is mainly produced by ruminal archaea, which is responsible for methane production in the rumen (Johnson and Johnson, 1995), and *Methanobrevibacter* is the most common genus in archaea (Poulsen et al., 2013). Methane is currently regarded as one of the most prevalent greenhouse gas, largely emitted from farm animal activities particularly from ruminants. Moreover, several toxic and inflammatory compounds were also found in the rumen (Saleem et al., 2012). Lipopolysaccharide (LPS) is typically released by the Gram-negative bacteria following the bacteria dying in the rumen (Mao et al., 2016). Histamine can alter rumen epithelial barrier and increases passive permeability (Mikelis et al., 2015).

Volatile fatty acids (VFAs) and microbial crude protein (MCP) are the principal products of bacteria fermentation (Russell et al., 1992). VFAs provide 70% of energy for ruminants, whereas excessive amount of VFAs will cause a considerable drop in rumen pH, push the activation of monocarboxylate transporters (MCTs), and other transport genes in the ruminal epithelium (Gabel et al., 2002). Mao et al. (2016) reported a significant decrease of saturated fatty acid and an increase of amine and phenylacetate concentrations in ruminal fluids of SARA animals. Rumen metabolic disorders associated with altering ruminal microbiota compositions are at high risks for developing diseases, particularly metabolic diseases including lameness, mastitis and laminitis (Zebeli and Metzler-Zebeli, 2012). Metabolomics can help us comprehensively understand the metabolism of microorganisms in the rumen.

Lipopolysaccharide may activate host cells via toll-like receptor 4 (TLR4) signaling pathway and induce the production and release of pro-inflammatory cytokines (Gruys et al., 2005). Feeding a HC diet for a short-term can induce the ruminal disturbance, a local inflammatory response in the ruminal epithelium, and even the systemic inflammation (Hook et al., 2011). However, information regarding the effects of long-term feeding HC diet on ruminal bacteria composition, metabolism, and the response of the epithelial cells has not been reported in ruminants. In this study, we used a combination of the 454 bar-coded pyrosequencing strategy and the gas chromatograph mass spectrometer (GC-MS) technique to investigate the effects of feeding a HC diet for short- and long-term on the alterations in ruminal microbiota and their metabolites, using goats as a model. Moreover, the relationships between microbiota abundance, ruminal metabolites, and genes expression related to host cells function were also analyzed in the present study.

MATERIALS AND METHODS

Ethics

The Institutional Animal Care and Use Committee (IACUC) of Nanjing Agricultural University approved all animal procedures. The "Guidelines on Ethical Treatment of Experimental Animals" (2006) No. 398 set by the Ministry of Science and Technology, China and the Regulation regarding the Management and Treatment of Experimental Animals" (2008) No. 45 set by the Jiangsu Provincial People's Government, was be strictly followed during the slaughter and sampling procedures.

Animals and Experimental Procedures

In brief, 15 healthy, mid-lactating goats (Guanzhong dairy goats) with an average initial body weight of 49.7 ± 5.5 kg (mean ± SD) were housed in individual pens in a standard animal feeding house at Northwest A&F University (Shanxi, China). Prior to the experiment, all goats were allowed free access to a control diet containing a forage to concentrate ratio of 65:35 for 2 weeks. Ingredients and chemical composition of the experimental diets were shown in Supplementary Table S1. After dietary adaptation, goats were randomly assigned to two groups, goats in the control group ($n = 10$) fed with a low concentrate (LC) diet containing 65% forage and 35% mixed concentrate for 19 weeks. High-grain long-term group (HL) five goats received a high-grain diet containing 65% mixed concentrate and 35% forage for 19 weeks. After 13 weeks, five goats from LC control group were randomly assigned to the high-grain short-term (HS) group received the same diet as in HL group for 6 weeks including 2 weeks dietary adaptation. All goats were fed daily at 08:00 and 18:00, respectively.

Samples Collection and Assay

At the end of the experiment, goats were slaughtered after overnight fasting. All goats were killed with neck vein injections of xylazine [0.5 mg (kg body weight)$^{-1}$; Xylosol; Ogris Pharma, Wels, Austria] and pentobarbital [50 mg (kg body weight)$^{-1}$; Release; WDT, Garbsen, Germany]. Blood samples were taken using heparin-containing vacuum tubes from jugular vein. Blood was centrifuged at 1,000 × g for 15 min at 4°C, extracted plasma in EP and stored at −20°C. Immediately after slaughter, the abdominal cavity was opened by midline incision, after that the rumen was carefully removed. The rumen was opened from the dorsal side and rumen fluid was collected and strained through four layers of cheesecloth and kept on ice until processing. The rumen tissue was then washed by PBS, collected and threw into liquid nitrogen immediately, then stored at −80°C until next process.

The rumen fluid samples were briefly centrifuged at 10,000 × *g* for 45 min at 4°C and the supernatant was aspirated gently to prevent its mixing with the pellet and passed through a disposable 0.22 μm LPS-free filter. The filtrate was collected in a sterile glass tube (previously heated at 180°C for 4 h) and heated at 100°C for 30 min. The ruminal LPS was detected by Chromogenic End-point Tachypleus Amebocyte Lysate Assay Kit (Chinese Horseshoe Crab Reagent Manufactory Co. Ltd, Xiamen, China) strictly following the manufacturer's instructions. Histamine was detected with the enzyme linked immunosorbent assay kit (Shengxing Company, Nanjing, China) and strictly following the manufacturer's instructions. The second portion of each rumen fluid sample was centrifuged at 3,000 × *g* for 15 min at 4°C immediately after collection and the supernatant was collected to analyze VFAs concentrations. The level of MCP, UN, NH3-N, lactate, and ruminal capase-3 enzyme activity was detected with the commercial assay kits (Jiancheng Company, Nanjing, China) strictly following the manufacturer's instructions, respectively.

RNA Isolation, cDNA Synthesis and Real-Time PCR

Total RNA was extracted from rumen samples with Trizol Reagent (Takara, Dalian, China). Concentration and quality of total RNA were monitored by NanoDrop ND-1000 Spectrophotometer (Thermo, USA). Then, total RNA was treated with RNase-Free DNase (M6101, Promega, USA) and reverse transcribed. Two microliter of diluted cDNA (1:40, vol/vol) was used for real-time PCR, which was performed in Mx3000P (Stratagene, USA). GAPDH was not affected by the experimental factors, and was chosen as the reference gene. All primers used in this study were synthesized by Generay Company (Shanghai, China). The method of $2^{-\Delta\Delta Ct}$ was used to analyze the real-time PCR results and gene mRNA levels were expressed as the fold change relative to the mean value of control group. Primers sequences are listed in Supplementary Table S3.

Metabolite Profiling of the Ruminal Fluid

The rumen fluid was thawed at room temperature, and centrifuged at 3,000 × *g* for 15 min at 4°C. Two hundred microliter of the supernatant was removed to 1.5 mL centrifuge tube, mixed with 10 μL dichlorobenzene alanine, then centrifuged at 13,000 × *g* for 15 min at 4°C, the supernatant removed to 200 μL tube. Twenty milligram methoxyammonium hydrochloride dissolved into 1 mL of pyridine was added to the sample after drying in 30 μL, and vortexed until completely dissolved, placed in 37°C incubator 90 min, then added 30 μL BSTFA at 70°C oven for 1 h.

Using Agilent7890A / 5975C GC-MS analysis of GC-MS platform for metabolomics sample data acquisition. Capillary column is the Agilent J&W Scientific's HP-5 ms (30 m × 0.25 mm × 0.25 μm). Instrument parameters set as follows: inlet temperature of 280° C, EI ion source temperature of 230°C, quadrupole temperature 150°C, high purity helium (purity greater than 99.999%) as a carrier gas, splitless into Injection volume 1.0 μL. Temperature program: initial temperature of 80°C, maintaining the 2 min, 10°C/min speed was raised to 320°C, and maintained 6 min. Full scan mode using mass spectrometry, mass spectrometry in the range of 50–550 (m/z). Random sequence of consecutive samples analyzed, to avoid the impact due to signal fluctuations caused by the instrument.

XCMS package of R software was used to treat LC/MS data, and then the EXCEL2007 software was used to delete the impurity peaks. Finally, two-dimension data matrix data was obtained. The matrix was created through the SIMCA-P software (Version 13.0) to analyze PCA, PLS-DA and the loading plot in the end.

DNA Extraction and 16S rDNA Gene Amplicon Pyrosequencing

Two-milliliter ruminal fluid was used for DNA extraction by PowerFecal® DNA Isolation Kit (Mobio) strictly according to the manufacturer's instructions. DNA samples were stored at −80°C for further processing. DNA purity was verified through agar gel electrophoresis, and was diluted to the concentration of 1.0 ng/μL. DNA was used as template to amplify the 16S V3-V4 region using specific primers with Barcode. The efficient hi-fi PCR enzyme and the Phusion® High-Fidelity PCR Master Mix with GC Buffer (New England Biolabs) were added to insure the amplification efficiency and accuracy. The production of PCR was verified by electrophoresis in 1.5% agarose gel, and then recycled by gel extraction kit (qiagen). The database was made with TruSeq® DNA PCR-Free Sample preparation Kit, and then used HiSeq2500 PE250 to sequencing.

Accordance to the Barcode and PCR primers sequences, raw tags were got by the FLASH (V1.2.7[1]. Further, high quality clean tags were obtained through strict filtering processing by the Qiime (V1.7.0[2]). The UCHIME Algorithm[3] and gold database[4] were used to detect chimeric sequences, removed them and acquired the effective tags. Effective tags were clustered to the Operational Taxonomic Units (OTUs) by Uparse (v7.0.1001[5]). The most abundant sequence within each OTU was designated as the representative sequence. The RDP classifier (Version2.2[6]) and GreenGene database[7] were used to species annotation. PyNAST (Version 1.2) and Green Gene "Core Set" data information in the database were used for multiple sequence alignment to get the representative sequences' phylogenetic relationship. Qiime (Version 1.7.0) was used to calculate the alpha diversity and the beta diversity. The unweighted PCoA analysis was carried out by the R (Version 2.15.3) with the WGCNA, stats and ggplot2 packages. V3–V4 region can detect the bacteria and archaea (Berg

[1] http://ccb.jhu.edu/software/FLASH/

[2] http://qiime.org/scripts/split_libraries_fastq.html

[3] http://www.drive5.com/usearch/manual/uchime_algo.html

[4] http://drive5.com/uchime/uchime_download.html

[5] http://drive5.com/uparse/

[6] http://sourceforge.net/projects/rdp-classifier/

[7] http://greengenes.lbl.gov/cgi-bin/nph-index.cgi

et al., 2012; Michelsen et al., 2014). We analyzed the bacteria and archaea, respectively.

Statistical Analysis

Data are presented as means ± SEM. The data were tested for normal distribution and analyzed by Student's unpaired *t*-test using SPSS software packages (SPSS version 19.0 for Windows; SPSS Inc., Chicago, IL, USA). Data were considered statistically significant when $P < 0.05$. The numbers of replicates used for statistics were noted in the figures. The correlation was made by the corrplot package of the R software.

RESULTS AND DISCUSSION

Alteration of VFAs and Abnormal Fermented Products after Feeding a HC Diet

Feeding a HC diet to lactating goats induced abnormal fermentation in rumen. Although there were no significant alternations of VFA concentrations in ruminal fluid among LC, HS, and HL groups, most of them showed a decrease in HC fed goats compared to LC. The ratio of acetate/propionate (A/P) was also decreased in HC goats and reached a significant decrease in HL group ($P = 0.04$) compared to LC (**Table 1**), indicating the alternation of fermentation type in rumen. Ruminal NH3-N, UN, phosphorus, and LPS concentrations were significantly increased in HS and HL goats ($P < 0.05$; **Table 1**). Compared to LC, ruminal lactate (LA) and MCP, as well as plasma histamine, LA, and UN concentrations also increased in animals fed with HC diet, particularly in the HL group ($P < 0.05$; **Table 1**).

Although VFAs concentrations in ruminal fluids were generally decreased in HC fed goats compared to LC goats, the total amount of VFA was not markedly affected by the HC diet, which was consistent with the earlier studies (Sun et al., 2015). It was reported that a lower proportion of acetate to propionate was caused by a lower fermentation of cellulose in rumen (Ribeiro et al., 2005). In this study, a significant decrease of the ratio of acetate to propionate in HL goats may indicate a lower cellulose fermentation in rumen. Moreover, increasing the proportion of butyrate in the current study was consistent with previous studies *in vitro* (Vallimont et al., 2004; Ribeiro et al., 2005) and *in vivo* (Martel et al., 2011; Oba et al., 2015). Unaffected production of branched-chain VFA (isobutyrate and isovalerate) was consistent with earlier studies (Li et al., 2014). It is widely accepted that large amount of lactate leads to an acute ruminal acidosis, and lactate is one of major products in a rapid fermentation process (Nagaraja and Titgemeyer, 2007). Lactate can cross the rumen wall and be dissolved in blood to leading an increase of plasma lactate as observed in the HS and the HL groups.

Absorption and utilization of nitrogen is promoted in lactating goat by feeding a HC diet. Our results demonstrated that the concentration of NH3-N was remarkably increased in the HS and HL groups compared to the LC (**Table 1**), which was consistent

TABLE 1 | Concentrations of metabolites in ruminal fluid and plasma.

Measure	LC	HS	HL
In ruminal fluid			
Acetate (mM)	8.2352 ± 0.7522	5.8716 ± 0.9132	5.7813 ± 1.1244
Propionate (mM)	3.5036 ± 0.2896	2.8031 ± 0.5507	2.8426 ± 0.5261
Isobutyrate (mM)	0.3511 ± 0.0330	0.3227 ± 0.0445	0.3341 ± 0.0304
Butyrate (mM)	2.3261 ± 0.0658	2.400 ± 0.5426	2.3593 ± 0.4137
Isovalerate (mM)	0.4291 ± 0.0445	0.3901 ± 0.0601	0.4142 ± 0.0398
Valerate (mM)	0.2360 ± 0.0124	0.2480 ± 0.0348	0.2702 ± 0.0473
Total VFA (mM)	15.1951 ± 1.0374	12.1353 ± 2.0902	12.1216 ± 2.1649
Acetate/Propionate	2.3595 ± 0.1251^{a}	2.1647 ± 0.1444^{ab}	2.0338 ± 0.0740^{b}
Lactate (mM)	0.6684 ± 0.1276^{b}	0.8394 ± 0.1074^{ab}	1.0925 ± 0.0818^{a}
NH3-N (mM)	9.6200 ± 1.2525^{b}	16.9000 ± 1.5161^{a}	19.1743 ± 1.2010^{a}
MCP (mg/mL)	5.1521 ± 0.0880^{b}	5.8364 ± 0.3110^{ab}	5.7729 ± 0.1168^{a}
UN (mM)	56.8734 ± 8.1717^{b}	93.5949 ± 6.0290^{a}	99.8101 ± 7.7480^{a}
Phosphorus (mM)	3.5474 ± 0.5678^{b}	5.9605 ± 0.0750^{a}	6.0517 ± 0.0948^{a}
Histamine (ng/mL)	122.3936 ± 8.7446	98.4652 ± 12.6695	112.1426 ± 9.9324
LPS (EU/mL)	$95663.3081 \pm 10939.1852^{b}$	$130604.5340 \pm 5279.7075^{a}$	$127962.8463 \pm 4932.3793^{a}$
In plasma			
Histamine (ng/mL)	185.5518 ± 32.6975^{b}	$280.8482 \pm 29.6947^{ab}$	309.8263 ± 27.3023^{a}
Phosphorus (mM)	0.2375 ± 0.0097	0.3099 ± 0.0846	0.2675 ± 0.0710
Lactate (mM)	6.7154 ± 0.2644^{b}	7.7365 ± 0.3434^{a}	7.5701 ± 0.6118^{ab}
UN (mM)	8.7595 ± 0.4860^{b}	9.9114 ± 0.8405^{ab}	10.3888 ± 0.4660^{a}
Urine UN(mM)	56.038 ± 5.4666	52.7089 ± 6.6299	53.7342 ± 4.3704

Values are expressed as the means ± SEM. Value with different small letter superscripts mean significant difference compared to LC ($P < 0.05$), and with the same or no letter superscripts mean no significant difference ($P > 0.05$). VFA, volatile fatty acids; NH3-N, Ammonia-nitrogen; MCP, microbial crude protein; UN, urine nitrogen; LPS, lipopolysaccharide.

with the previous studies (DeFrain et al., 2004; Ariko et al., 2015). Increased nitrogen suggests a more intense proteolysis (Bunthof et al., 2001). We did not find significant changes of UN in urine among the control, the HS, and the HL groups, indicating that urea might be utilized in other metabolic pathways. We found that ruminal phosphate concentration was significantly increased in the HS and the HL groups, compared to the control group (**Table 1**). However, feeding a HC diet did not change the level of phosphate in blood. Therefore, we speculated that excessive amount of ruminal phosphate was eliminated from the body through urine or feces, which will throw a potential risk to environment through methane emission.

It's well known that some metabolic diseases occurred in ruminants including acidotic rumenitis (Underwood, 1992) and laminitis (Nilsson, 1963), have been found to directly correlate with the level of endogenous histamine. In the Groot's (1998) study, histamine altered rumen epithelial barrier function by trans-epithelial electrolyte transportation and increased the passive permeability. In this study, the level of histamine in plasma was markedly increased in the HS and HL groups compared to control. High level of endogenous histamine and endogenous LPS pose a high risk to induce metabolic related diseases in lactating ruminants after digested a HC diet, as observed in the farm practice.

Alteration of Ruminal Flora

In total, 910,696 reads were obtained for the bacterial 16S rRNA genes by pyrosequencing analysis. After screening these gene sequences with strict criteria, 702,001 valid sequences were obtained, accounting for 86.62% of the raw reads. The common shared numbers among three groups were shown in Venn diagram (**Figure 1A**). The HS group had the highest number of unique sequences (220 OTUs), followed by the LC group (151 OTUs) and the HL group (55 OTUs). Additionally, there were 1139 OTUs (around 56% of total OTUs) shared among three groups. At phylum level, Firmicutes was the most abundant bacteria, with an average relative abundance of 55.10% (**Figure 1B**). Bacteroidetes were the second type with an average relative abundance of 37.82% (**Figure 1B**). We observed a notable phylum-wide shift in the Cyanobacteria and Verrucomicrobia induced by the HC diet. The Cyanobacteria abundance was significantly decreased in the HS group compared to the LC group, while Verrucomicrobia was significantly decreased in the HL group compared to the HS (**Figure 1B**).

The diversity of bacteria in rumen fluid was present by Shannon index, Chao value, and ACE index. Shannon index was markedly reduced in HS ($P = 0.02$) and HL ($P < 0.01$) groups compared to LC control group (**Figure 2A**). Chao1 values were observed in the HS group ($P < 0.01$; **Figure 2B**). ACE values were significantly decreased in the HS ($P < 0.01$) and HL ($P = 0.02$) groups (**Figure 2C**). Taken above together, feeding a HC diet induces a dramatic decrease of the diversity of ruminal bacteria particularly in long-term feeding group.

Previous studies demonstrated that feeding a HC diet for 4~6 weeks altered ruminal microbiota, which commonly caused a negative impact on the bio-diverse ecosystem (Mao et al., 2016). We found that PCoA (**Figure 2D**) can notably show the significant differences of ruminal bacterial composition among control, HS, and HL groups, which demonstrated that feeding a diet shifted from 35% concentrate to 65% proportion for 4 weeks changed the bacterial composition abruptly and unstably. However, after 19 weeks adaptation microbiota compositions maintained stability, but exhibited a lower richness and diversity than LC control group. The PCoA with unweighted UniFrac distances demonstrated that samples from LC group were clearly separated from HS and the HL groups. Moreover, the intragroup variation in HS goats was obviously presented by PCoA analysis indicating an unstable composition of ruminal microbiota.

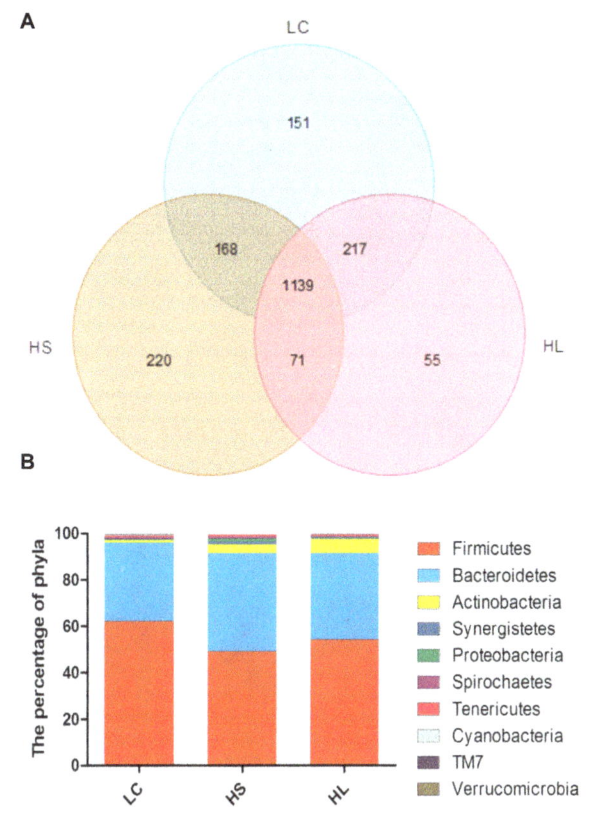

FIGURE 1 | (A) Venn diagram demonstrates overlap of bacterial OTUs at 3% dissimilarity level for low concentrate (LC), high-grain short-term (HS) and High-grain long-term group (HL) group. **(B)** Effects of length of concentrate to forage diets on the compositions of microbial phylum (as a percentage of the total sequence).

Ruminants fed with HC diets are associated with an altered rumen microbiota. **Supplementary Figure S1** shows that at genus level, there is 10 genus markedly changed among top 50 abundance OTUs by HC diet ($P < 0.10$). The abundance of *Bulleidia*, *Paracoccus*, *Pseudoramibacter_Eubacterium*, *Atopobium*, and *Selenomonas* were increased through the HC diets (**Supplementary Figure S1**). However, the level of *Oscillospira*, *YRC22*, *Bacteroides*, *CF231*, and *Akkermansia* were significantly decreased in the HC group (**Supplementary Figure S1**). In Mao et al. (2016) study, the population of *Prevotella*, *Papillibacter*, *Lysinibacillus*, *Thalassospira*,

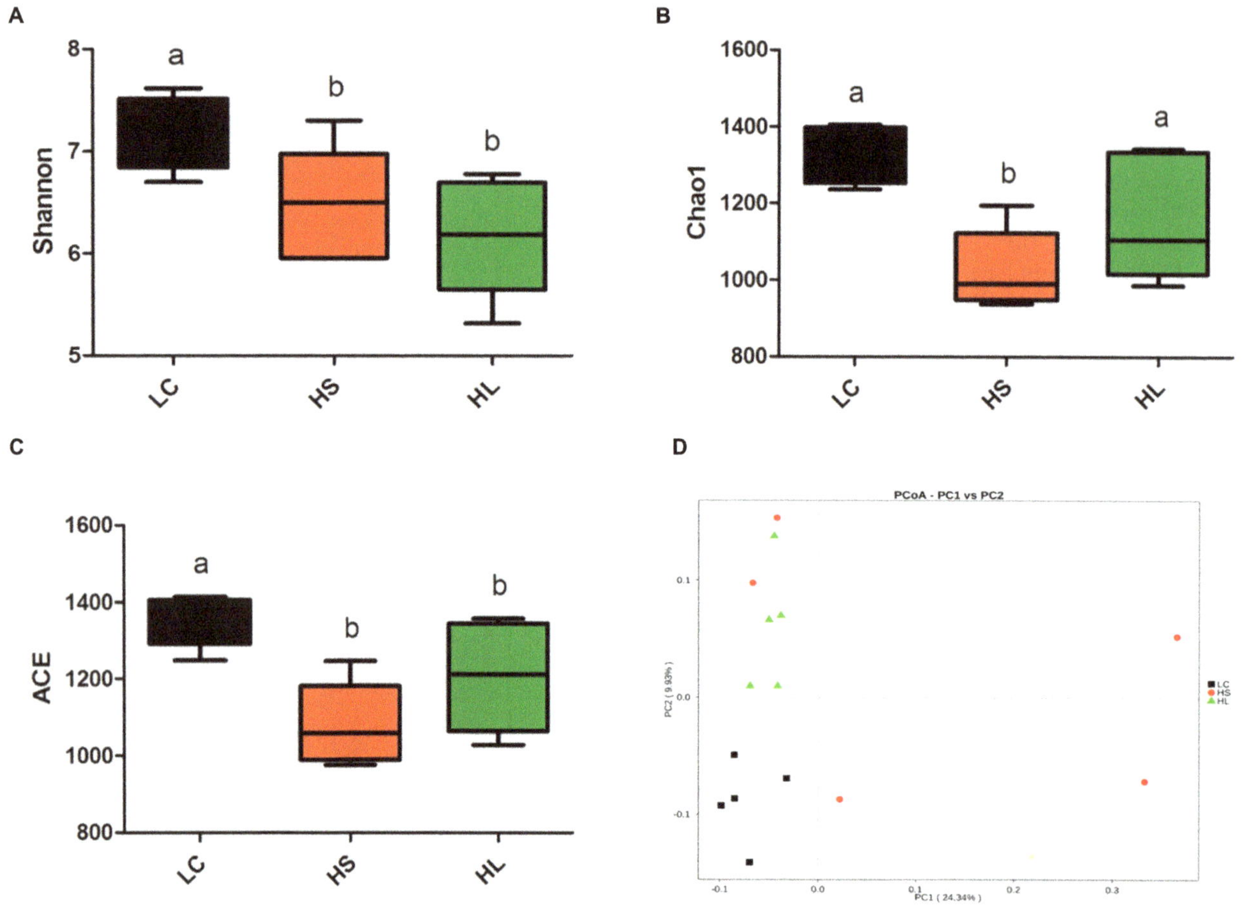

FIGURE 2 | Changes in ruminal microbial richness and diversity as a result of LC, HS and HL. (A) The bacterial richness estimated by the Shannon index. **(B)** The bacterial diversity estimated by the Chao1. **(C)** The bacterial diversity estimated by the ACE. **(D)** Unweighted principal coordinate analysis (PCOA) of rumen bacteria microbiota.

Succiniclasticum, as well as some unclassified bacteria were decreased in the HC groups, while the abundance of *Butyrivibrio, Mogibacterium, Acetitomaculum,* and unclassified *Anaerolineaceae* were increased by the HC diet. The changes in the percentage of bacterial are approximately 20~40% in Mao et al. (2016) study. The alterations of the bacteria were approximately 4%. The alteration of ruminal flora caused by the origin of animals and different management (Petri et al., 2013). High abundance of OTUs alteration could replace other OTUs that take over the analogous functions (Taxis et al., 2015), and geographical differences could also possibly affect the composition of bacteria.

The abundance of *Selenomonas, Atopobium, Bulleidia, Paracoccus,* and *Pseudoramibacter_Eubacterium* were increased. *Selenomonas* is classified into two subspecies (subsp): subsp. *lactilytica* with a capacity to utilize lactate, and incapable of utilizing lactate subsp. *ruminantium* (Asanuma et al., 2015). However, *Selenomonas ruminantium* is a representative nitrate and nitrite reducing ruminal bacterium (Asanuma et al., 2002). It is reported that an increase in nitrate and nitrite in ruminal contents induced the abundance increase of *Selenomonas ruminantium* (Asanuma et al., 2015). The increase of lactate and "N" nutrients can contribute to the increase of *Selenomonas.* As a Gram-positive anaerobic bacterium, *Atopobium* utilizes sugars and plays an important role in developing SARA (Harmsen et al., 2000; Mao et al., 2013). In this study, we found that the population of *Atopobium* was increased by feeding the HC diet, as reported in previous studies (Mao et al., 2013). It is consistently found that feeding a HC diet is likely to induce SARA disorders and other related metabolic diseases in ruminants, particularly in lactating dairy cows (Mao et al., 2016).

We found that the level of *Bacteroides, Oscillospira, Akkermansia, CF231,* and *YRC22* were decreased by the HC diet. *Bacteroides* is one of ureolytic bacteria in rumen (Slyter et al., 1968) and can produce succinic acid (Davies et al., 2007). Zhou et al. (2012) reported that the reduction potential can inhibit the growth of *Bacteroides.* In the present study, UN was considerably increased, possibly contributing to the decrease of *Bacteroides.* As a Gram-positive bacterium, *Oscillospira* is the first described bacterium (Chatton and Pérard, 1913) involved in the degradation of plant cell wall (Yanagita et al., 2003). Mackie et al. (2003) found that the abundance of *Oscillospira* in rumen

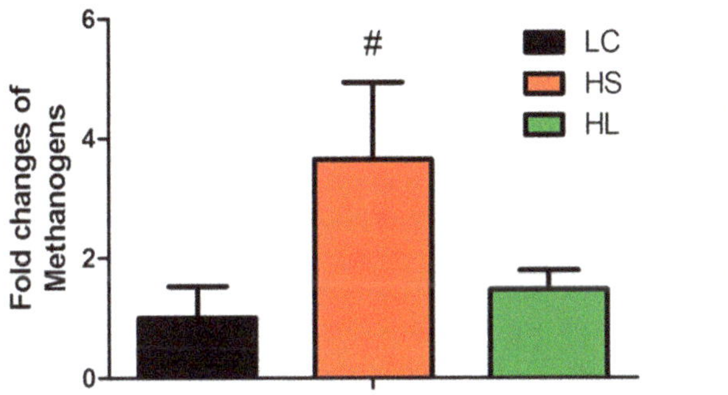

FIGURE 3 | Methanogens change. Methanogens change in three groups. Value with # superscripts means $P < 0.1$ compared to LC, and without # superscript means $P > 0.1$.

was diet-dependent and reaching the maximum level after feeding fresh-forage diets. *Akkermansia* is involved in mucosa development, as well as maintenance of intestinal integrity by utilizing mucin and antimicrobial (Everard et al., 2013). Moreover, the decrease of *CF231* and *YRC22* in ruminal fluids may be caused by higher level of nitrate (Zhao et al., 2015).

It's very important to note that the abundance of methanogens has a trend to increase in the HS group compared to LC (**Figure 3**). Mao et al. (2016) reported that SARA ruminants produced a higher level of methanogens than control healthy counterparts. Inconsistently, in Hook et al. (2011) study four non-lactating Holstein dairy cows fed with a HC diet for 3 weeks did not affect the methanogen density in the rumen. We found that the level of methanogens exhibited an increased tendency in the HS group. As the prevalent greenhouse gas, methane is highly produced by ruminants and has a potentially harmful effect on the environment.

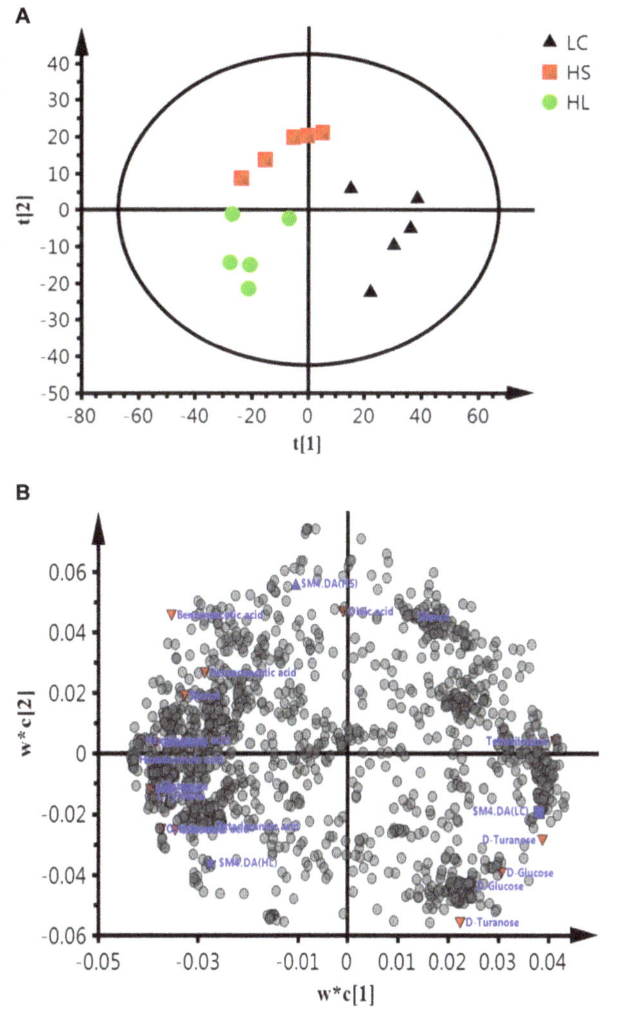

FIGURE 4 | Partial least squares discriminant analysis (PLS-DA) based on the rumen compounds data. (A) PLS-DA analysis scores plots discriminating among the rumen fluid LC, HS and HL. The different color represents different group. **(B)** Loading plot of all the commonly detected compounds projected into the PLS-DA model. The most important compounds responsible for the discrimination are labeled and colored in red.

Shift of Metabolomics in Ruminal Fluid

Gas chromatograph mass spectrometer chromatograms of the ruminal fluid metabolites were displayed in **Supplementary Figure S2**. Numbers of visible peaks were separated by GC-MS analysis. After pairwise comparison, 31, 30, and 16 differentiated metabolites with VIP > 1 between two groups was presented in Supplementary Table S2. These metabolites altered by HC diet are involved in multiple biochemical processes in the rumen, such as gluconeogenesis. PLS-DA was used to identify the key compounds responsible for the score differences. There was an obvious separation between the score plots created by the first two components and three groups of samples cluster (**Figure 4A**). Each dot represents an observation sample, and the distance between two dots represents the similarity of the sample's metabolite composition. We can observe the LC group cluster presented in the right portion; HS and HL groups were presented in the upper and lower portion of the left, respectively.

In order to verify the differentiated metabolites among LC, HS, and HL groups, a PLS-DA loading plot was created (**Figure 4B**). The variable importance in the projection (VIP) combining with the student's *t*-test (*t*-test) *p* values were used to identify which compounds was the most significant contribution in discriminating among the ruminal compounds of groups. Those compounds that were responsible (VIP > 1) for the significant difference between LC-HS, LC-HL and HS-HL were selected. Finally, 11 compounds were commonly differentiated among three groups (**Table 2**), and three of them including D-glucose, D-turanose and tetrasiloxane were significantly decreased, whereas eight metabolites including benzeneacetic acid, glutamine, hexadecanoic acid, L-tyrosine, malic acid, octadecanoic acid, oleic acid, and silanol were increased by feeding the HC diet. Silanol and tetrasiloxane may be derived from the forage, because they do not exist in any metabolic pathways. In this study, the relative density of phenylacetate acid in rumen was higher in the HS and the HL groups, which was consistent with the previous findings (Mao

TABLE 2 | Candidate ruminal compounds that significantly different among three groups.

Item	LC vs. HS			LC vs. HL		
	VIP	*P*	Fold change	VIP	*P*	Fold change
Phenylacetate acid	1.2442	0.0114	0.9493	1.1084	0.0155	0.7775
Hexadecanoic acid	1.5586	0.0000	0.3944	1.3527	0.0004	0.4954
Octadecanoic acid	1.1061	0.0332	0.4656	1.0265	0.0304	1.1477
Oleic acid	1.4849	0.0003	0.8938	1.2445	0.0033	1.7250
Glutamine	1.2610	0.0097	0.4141	1.3523	0.0004	0.5153
L-tyrosine	1.2843	0.0077	0.8296	1.3728	0.0003	1.2852
Malic acid	1.2673	0.0091	0.7655	1.3785	0.0002	1.1701
D-Glucose	1.3008	0.0065	−0.8998	1.0012	0.0364	−0.1499
D-Turanose	1.5266	0.0001	−1.1963	1.2900	0.0016	−0.9936

VIP, variable importance in the projection.

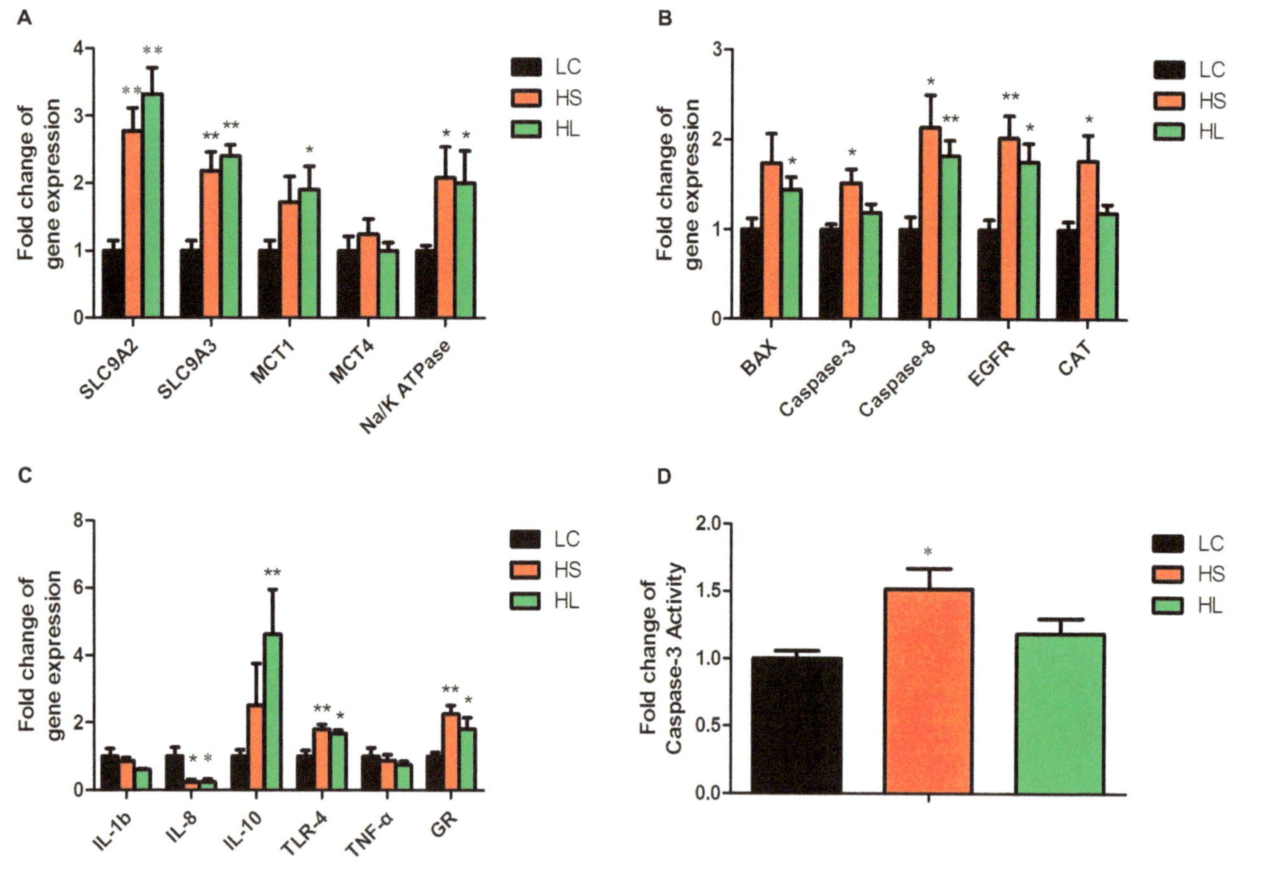

FIGURE 5 | Gene expressions in rumen. GAPDH was used as the reference gene for the gene expressions. **(A)** Genes involved in substrates transport. **(B)** Genes involved in apoptosis, proliferation and antioxidant process. **(C)** Genes involved in inflammation response. Value with different small letter superscripts mean significant difference ($P < 0.05$), and with the same or no letter superscripts mean no significant difference ($P > 0.05$). **(D)** Effects of length of concentration to forage diets on concentration of caspase-3 activity.

et al., 2016). In rumen, phenylacetate was synthesized by activity of ruminal microbiota by degrading plant constituents (Chesson et al., 1999). It has been reported that phenylacetate acid could be beneficial to the rumen bacteria (Kristensen, 1974). Our findings reveal a correlated response for phenylacetate and the proportion of some ruminal bacteria, strongly positive correlated with *Oscillospira* and *Akkermansia,* but negatively correlated with *Paracoccus.* The above results indicate that these microbes may be affected by phenylacetate, or collaboratively synthesize phenylacetate.

Feeding a HC diet to dairy goats dramatically reduced the level of D-glucose and D-turanose, which are both utilized by microbes

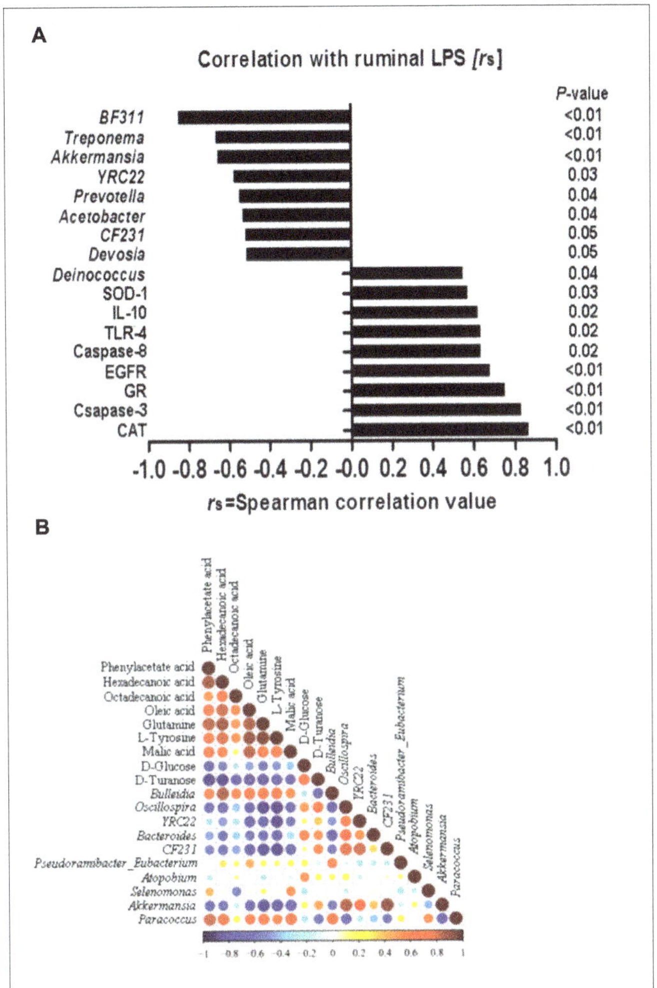

FIGURE 6 | (A) Correlation among the ruminal Lipopolysaccharide (LPS), bacteria, genes and the potential marker compounds. **(B)** Correlation analysis of 100 most abundant OTUs and the genes with LPS, only the significant ($P < 0.05$) correlations are depicted.

in rumen (Evans and Martin, 2000). After absorption, propionate is utilized for synthesis glucose in liver. More utilization and less synthesis finally led to a decrease of D-glucose. D-Turanose can be metabolized by the bacteria as well, such as *Elusimicrobium minutum* (Geissinger et al., 2009) and *Bacillus* (Seo et al., 2013). Two amino acids, glutamine and L-tyrosine were identified in ruminal fluids. The greater concentration of glutamine in the HL group suggested more proteins degraded after feeding a HC diet. Both glutamine and L-tyrosine exhibited strong correlations with microbes including *Bulleidia, Oscillospira, YRC22, CF231,* and *Akkermansia*. It's reasonable to speculate that these microbes may play an important role in producing or utilizing these two amino acids. In the current study, the amount of malic acid in the rumen was dramatically increased in the HS and HL groups compared to the control group, and a greater concentration of oleic acid and octadecanoic acid was also found in HC-fed goats. Malic acid can promote the utilization of lactate by *Selenomonas ruminantium* and then can prevent the decrease of ruminal pH (Castillo et al., 2004). Moreover, malic acid can increase the production of propionate (Khampa et al., 2006) and butyrate (Liu et al., 2009), and regulate the activity of some types of ruminal bacteria (Liu et al., 2009). The increase of octadecanoic acid comes from the hydrogenation of oleic acid, and the abundance of *Bulleidia, Oscillospira, CF231* and *Paracoccus* is strongly correlated with the level of hexadecanoic acid and oleic acid.

Changes of Functional Genes Expression in Host Cells

In the present study, the results showed that the functional genes expressions in host epithelial tissues changed greatly by feeding a HC diet. The expression of VFAs transport genes including sodium/hydrogen exchanger 2 (*SLC9A2*), sodium/hydrogen exchanger 3 (*SLC9A3*), and sodium-potassium adenosine triphosphatase (*Na/K ATPase*) in ruminal epithelium was significantly increased by feeding HC diet (**Figure 5A**). It's well known that intracellular pH is regulated in a certain physiological range, which is very important within cell homeostasis (Counillon and Pouysségur, 2000). Kiela et al. (2001) has reported that epidermal growth factor receptor (*EGFR*) can promote SLC9A2 expression. Consistently, we also found a significant increase of EGFR mRNA expression in ruminal epithelial tissues in HC-fed goats (**Figure 5B**). Monocarboxylate transporter 1 (*MCT1*) and 4 (*MCT4*) are involved in the transmembrane transport (Kirat et al., 2006; Connor et al., 2010). Taken together, the altered genes expression in the epithelial tissues indicate the changes of physiological functions including immunity, substrates transportation, as well as the homeostasis of the cross-talk between ruminal microbiota and the host proceeding the HC diet to lactating dairy goats.

Caspase-8 and the downstream effector caspase-3 (Strater et al., 1997), as well as B-cell lymphoma (*Bcl-2*) family (Nagata, 1997) were involved in the apoptotic process. In the present study, caspase-8 and *BCL-2* associated X (*BAX*) mRNA expression were significantly increased in the HS and the HL groups compared to LC group (**Figure 5B**). Caspase-3 mRNA and enzyme activity (**Figure 5D**) were not significantly changed in the HL group, however, we found a strong positive correlation of caspase-3 and -8 with ruminal LPS levels. As an antioxidant enzyme, catalase (CAT) gene expression in epithelial tissues was significantly increased in the HS goats (**Figure 5B**), and showed a strong positive correlation with ruminal LPS level (**Figure 6A**). Inflammation is a major host defense reaction. The cytokines play a key role in the initiation, maintenance, and termination of the inflammatory reaction. Tumor necrosis factor-α (*TNF-α*), interleukin-1β (*IL-1β*), interleukin-6 (*IL-6*) and 8 (*IL-8*) are pro-inflammatory, while interleukin-10 (*IL-10*) is anti-inflammatory (Wojdasiewicz et al., 2014). The expression of IL-8 mRNA was markedly decreased, whereas the expression of IL-10 and glucocorticoid (GR) gene was dramatically increased in the host epithelial tissues after feeding a HC diet (**Figure 5C**). It's well known that LPS stimulates the expression of pro-inflammatory cytokines via TLR4 signal pathway (Lu et al., 2008). However, most of cytokines expression were not altered in the HL group, which probably indicates the high tolerance of the host to ruminal endotoxin gradually, and eventually attenuate LPS responses (Ametaj et al., 2009).

Correlation between Ruminal LPS, Genes Expression and Microbiome

A correlation analysis of the 100 most abundant OTUs and genes expression with ruminal LPS levels is shown in **Figure 6A**. LPS is a structural component of the cell wall of Gram-negative bacteria, recognized by TLR-4, and activates the immune response (Gruys et al., 2005). The increase of LPS always accompanies the reduction of Gram-negative bacteria (Khafipour et al., 2009). In this study, there are nine OTUs among the 100 most abundant OTUs, which is substantially correlated with LPS level ($p < 0.05$). All of these nine OTUs are Gram-negative bacteria. Moreover, our results showed that the level of ruminal LPS also depicted a strong correlation with epithelial apoptosis, proliferation, inflammation, and anti-oxidative stress in the host rumen tissues ($r > 0.6, p < 0.05$).

Correlation between Ruminal Microbiome and Metabolome

Correlation analysis within or between the bacteria and metabolite was conducted to investigate the potential co-occurrences as displayed with a correlation matrix (**Figure 6B**). In this context, 66 correlations (34.73%) were strong positively or negatively correlated ($|r| \geq 0.6$, $p < 0.05$). There were 31 correlations strong positively or negatively correlated ($|r| \geq 0.6$, $p < 0.05$) between OTUs and metabolite. Hexadecanoic acid and *Bulleidia* had the strongest positive correlation ($r = 0.84$, $p < 0.01$), and the strongest negative correlation were found between L-tyrosine and *Oscillospira* ($r = -0.879, p < 0.01$) in the bacteria and metabolite.

CONCLUSION

Feeding a HC diet to lactating ruminants induces abnormal fermentation, metabolic perturbations, and microbiota dysbiosis with reduced bacterial richness and diversity. The microbiota dysbiosis in the HS group is more severe than that in the HL group presented by PCoA analysis. Microbiota dysbiosis and the abnormal products, particularly LPS and histamine, in rumen or circulating system induced by a HC diet is largely associated with functional genes expression in ruminal epithelial tissues. Therefore, the homeostasis of ruminal microbial ecosystem is a vital point for keeping host cells physiological functions, animal welfare, and even the healthy environment.

AUTHOR CONTRIBUTIONS

Performed experiments: CH, YG, JT, PT, ST, YL, and RC. Analyzed data: CH, JT, and ST. Conceived and designed experiments: YG, YN, RZ, RC, PT, and JT. Wrote the paper: CH and JT.

FUNDING

This work was supported by the National Nature Science Foundation of China (project no. 31272470) and National Basic Research Program of China (project no. 2011CB100802) and the Program for New Century Excellent Talents in University (NCET-13-0862) and a project funded by the Priority Academic Program Development of Jiangsu Higher Education Institutions (PAPD).

REFERENCES

Ametaj, B., Koenig, K., Dunn, S., Yang, W., Zebeli, Q., and Beauchemin, K. (2009). Backgrounding and finishing diets are associated with inflammatory responses in feedlot steers. *J. Anim. Sci.* 87, 1314–1320. doi: 10.2527/jas.2008-1196

Ariko, T., Kass, M., Henno, M., Fievez, V., Kärt, O., Kaart, T., et al. (2015). The effect of replacing barley with glycerol in the diet of dairy cows on rumen parameters and milk fatty acid profile. *Anim. Feed Sci. Technol.* 209, 69–78. doi: 10.1016/j.anifeedsci.2015.08.004

Asanuma, N., Iwamoto, M., Kawato, M., and Hino, T. (2002). Numbers of nitrate-reducing bacteria in the rumen as estimated by competitive polymerase chain reaction. *Anim. Sci. J.* 73, 199–205. doi: 10.1046/j.1344-3941.2002.00028.x

Asanuma, N., Yokoyama, S., and Hino, T. (2015). Effects of nitrate addition to a diet on fermentation and microbial populations in the rumen of goats, with special reference to Selenomonas ruminantium having the ability to reduce nitrate and nitrite. *Anim. Sci. J.* 86, 378–384. doi: 10.1111/asj.12307

Bal, M. A., Coors, J. G., and Shaver, R. D. (1997). Impact of the maturity of corn for use as silage in the diets of dairy cows on intake, digestion, and milk production. *J. Dairy Sci.* 80, 2497–2503. doi: 10.3168/jds.S0022-0302(97)76202-7

Berg, J., Brandt, K. K., Al-Soud, W. A., Holm, P. E., Hansen, L. H., Sørensen, S. J., et al. (2012). Selection for Cu-tolerant bacterial communities with altered composition, but unaltered richness, via long-term Cu exposure. *Appl. Environ. Microbiol.* 78, 7438–7446. doi: 10.1128/AEM.01071-12

Boerman, J. P., Potts, S. B., VandeHaar, M. J., Allen, M. S., and Lock, A. L. (2015). Milk production responses to a change in dietary starch concentration vary by production level in dairy cattle. *J. Dairy Sci.* 98, 4698–4706. doi: 10.3168/jds.2014-8999

Bunthof, C. J., Bloemen, K., Breeuwer, P., Rombouts, F. M., and Abee, T. (2001). Flow cytometric assessment of viability of lactic acid bacteria. *Appl. Environ. Microbiol.* 67, 2326–2335. doi: 10.1128/AEM.67.5.2326-2335.2001

Castillo, C., Benedito, J., Méndez, J., Pereira, V., Lopez-Alonso, M., Miranda, M., et al. (2004). Organic acids as a substitute for monensin in diets for beef cattle. *Anim. Feed Sci. Technol.* 115, 101–116. doi: 10.1016/j.anifeedsci.2004.02.001

Chatton, E., and Pérard, C. (1913). Schizophytes du caecum du cobaye. I. *Oscillospira guilliermondi* n.g., n.sp. *C. R. Sceances Soc. Biol.* 74, 1159–1162.

Chesson, A., Provan, G. J., Russell, W. R., Scobbie, L., Richardson, A. J., and Stewart, C. (1999). Hydroxycinnamic acids in the digestive tract of livestock and humans. *J. Sci. Food Agric.* 79, 373–378. doi: 10.1002/(SICI)1097-0010(19990301)79:3<373::AID-JSFA257>3.0.CO;2-6

Connor, E., Li, R., Baldwin, R., and Li, C. (2010). Gene expression in the digestive tissues of ruminants and their relationships with feeding and digestive processes. *Animal* 4, 993–1007. doi: 10.1017/S1751731109991285

Counillon, L., and Pouysségur, J. (2000). The expanding family of eucaryotic Na+/H+ exchangers. *J. Biol. Chem.* 275, 1–4. doi: 10.1074/jbc.275.1.1

Davies, D., Theodorou, M., and Newbold, J. (2007). Biotransformation and fermentation - exploiting the rumen and silo. *IGER Innov.* 11, 37–41.

DeFrain, J. M., Hippen, A. R., Kalscheur, K. F., and Jardon, P. W. (2004). Feeding glycerol to transition dairy cows: effects on blood metabolites and lactation performance. *J. Dairy Sci.* 87, 4195–4206. doi: 10.3168/jds.S0022-0302(04)73564-X

Evans, J. D., and Martin, S. A. (2000). Effects of thymol on ruminal microorganisms. *Curr. Microbiol.* 41, 336–340. doi: 10.1007/s002840010145

Everard, A., Belzer, C., Geurts, L., Ouwerkerk, J. P., Druart, C., Bindels, L. B., et al. (2013). Cross-talk between *Akkermansia muciniphila* and intestinal epithelium controls diet-induced obesity. *Proc. Natl. Acad. Sci. U.S.A.* 110, 9066–9071. doi: 10.1073/pnas.1219451110

Gabel, G., Aschenbach, J. R., and Muller, F. (2002). Transfer of energy substrates across the ruminal epithelium: implications and limitations. *Anim. Health Res. Rev.* 3, 15–30. doi: 10.1079/AHRR200237

Geissinger, O., Herlemann, D. P., Mörschel, E., Maier, U. G., and Brune, A. (2009). The ultramicrobacterium "*Elusimicrobium minutum*" gen. nov., sp. nov., the first cultivated representative of the termite group 1 phylum. *Appl. Environ. Microbiol.* 75, 2831–2840. doi: 10.1128/AEM.02697-08

Groot, J. (1998). Correlation between electrophysiological phenomena and transport of macromolecules in intestinal epithelium. *Vet. Q.* 20, 45–49. doi: 10.1080/01652176.1998.9694968

Gruys, E., Toussaint, M. J., Niewold, T. A., and Koopmans, S. J. (2005). Acute phase reaction and acute phase proteins. *J. Zhejiang Univ. Sci. B* 6, 1045–1056. doi: 10.1631/jzus.2005.B1045

Harmsen, H. J., Wildeboer-Veloo, A. C., Grijpstra, J., Knol, J., Degener, J. E., and Welling, G. W. (2000). Development of 16S rRNA-based probes for the *Coriobacterium* group and the atopobium cluster and their application for enumeration of Coriobacteriaceaein human feces from volunteers of different age groups. *Appl. Environ. Microbiol.* 66, 4523–4527. doi: 10.1128/AEM.66.10.4523-4527.2000

Hook, S. E., Steele, M. A., Northwood, K. S., Wright, A.-D. G., and McBride, B. W. (2011). Impact of high-concentrate feeding and low ruminal pH on methanogens and protozoa in the rumen of dairy cows. *Microb. Ecol.* 62, 94–105. doi: 10.1007/s00248-011-9881-0

Hristov, A. N., Callaway, T. R., Lee, C., and Dowd, S. E. (2012). Rumen bacterial, archaeal, and fungal diversity of dairy cows in response to ingestion of lauric or myristic acid. *J. Anim. Sci.* 90, 4449–4457. doi: 10.2527/jas.2011-4624

Jami, E., Israel, A., Kotser, A., and Mizrahi, I. (2013). Exploring the bovine rumen bacterial community from birth to adulthood. *ISME J.* 7, 1069–1079. doi: 10.1038/ismej.2013.2

Johnson, K. A., and Johnson, D. E. (1995). Methane emissions from cattle. *J. Anim. Sci.* 73, 2483–2492. doi: 10.2527/1995.7382483x

Khafipour, E., Li, S., Plaizier, J. C., and Krause, D. O. (2009). Rumen microbiome composition determined using two nutritional models of subacute ruminal acidosis. *Appl. Environ. Microbiol.* 75, 7115–7124. doi: 10.1128/aem.00739-09

Khampa, S., Wanapat, M., Wachirapakorn, C., Nontaso, N., Wattiaux, M., and Rowlison, P. (2006). Effect of levels of sodium DL-malate supplementation on ruminal fermentation efficiency of concentrates containing high levels of cassava chip in dairy steers. *Asian Australas. J. Anim. Sci.* 19:368. doi: 10.5713/ajas.2006.368

Kiela, P. R., Hines, E. R., Collins, J. F., and Ghishan, F. K. (2001). Regulation of the rat NHE3 gene promoter by sodium butyrate. *Am. J. Physiol. Gastrointest. Liver Physiol.* 281, G947–G956.

Kirat, D., Masuoka, J., Hayashi, H., Iwano, H., Yokota, H., Taniyama, H., et al. (2006). Monocarboxylate transporter 1 (MCT1) plays a direct role in short-chain fatty acids absorption in caprine rumen. *J. Physiol.* 576, 635–647. doi: 10.1113/jphysiol.2006.115931

Klevenhusen, F., Hollmann, M., Podstatzky-Lichtenstein, L., Krametter-Frotscher, R., Aschenbach, J. R., and Zebeli, Q. (2013). Feeding barley grain-rich diets altered electrophysiological properties and permeability of the ruminal wall in a goat model. *J. Dairy Sci.* 96, 2293–2302. doi: 10.3168/jds.2012-6187

Koike, S., Handa, Y., Goto, H., Sakai, K., Miyagawa, E., Matsui, H., et al. (2010). Molecular monitoring and isolation of previously uncultured bacterial strains from the sheep rumen. *Appl. Environ. Microbiol.* 76, 1887–1894. doi: 10.1128/aem.02606-09

Kristensen, S. (1974). Ruminal bioshynthesis of aromatic amino acids from arylacetic acids, glucose, shikimic acid and phenol. *Br. J. Nutr.* 31, 357–365. doi: 10.1079/BJN19740044

Li, F., Yang, X. J., Cao, Y. C., Li, S. X., Yao, J. H., Li, Z. J., et al. (2014). Effects of dietary effective fiber to rumen degradable starch ratios on the risk of sub-acute ruminal acidosis and rumen content fatty acids composition in dairy goat. *Anim. Feed Sci. Technol.* 189, 54–62. doi: 10.1016/j.anifeedsci.2013.12.011

Liu, Q., Wang, C., Yang, W., Dong, Q., Dong, K., Huang, Y., et al. (2009). Effects of malic acid on rumen fermentation, urinary excretion of purine derivatives and feed digestibility in steers. *Animal* 3, 32–39. doi: 10.1017/S1751731108003361

Lu, Y. C., Yeh, W. C., and Ohashi, P. S. (2008). LPS/TLR4 signal transduction pathway. *Cytokine* 42, 145–151. doi: 10.1016/j.cyto.2008.01.006

Mackie, R. I., Aminov, R. I., Hu, W., Klieve, A. V., Ouwerkerk, D., Sundset, M. A., et al. (2003). Ecology of uncultivated *Oscillospira* species in the rumen of cattle, sheep, and reindeer as assessed by microscopy and molecular approaches. *Appl. Environ. Microbiol.* 69, 6808–6815. doi: 10.1128/AEM.69.11.6808-6815.2003

Mao, S., Zhang, R., Wang, D., and Zhu, W. (2013). Impact of subacute ruminal acidosis (SARA) adaptation on rumen microbiota in dairy cattle using pyrosequencing. *Anaerobe* 24, 12–19. doi: 10.1016/j.anaerobe.2013.08.003

Mao, S. Y., Huo, W. J., and Zhu, W. Y. (2016). Microbiome-metabolome analysis reveals unhealthy alterations in the composition and metabolism of ruminal microbiota with increasing dietary grain in a goat model. *Environ. Microbiol.* 18, 525–541. doi: 10.1111/1462-2920.12724

Martel, C. A., Titgemeyer, E., Mamedova, L., and Bradford, B. (2011). Dietary molasses increases ruminal pH and enhances ruminal biohydrogenation during milk fat depression. *J. Dairy Sci.* 94, 3995–4004. doi: 10.3168/jds.2011-4178

Michelsen, C. F., Pedas, P., Glaring, M. A., Schjoerring, J. K., and Stougaard, P. (2014). Bacterial diversity in Greenlandic soils as affected by potato cropping and inorganic versus organic fertilization. *Polar Biol.* 37, 61–71. doi: 10.1007/s00300-013-1410-9

Mikelis, C. M., Simaan, M., Ando, K., Fukuhara, S., Sakurai, A., Amornphimoltham, P., et al. (2015). RhoA and ROCK mediate histamine-induced vascular leakage and anaphylactic shock. *Nat. Commun.* 6:6725. doi: 10.1038/ncomms7725

Nagaraja, T. G., and Titgemeyer, E. C. (2007). Ruminal acidosis in beef cattle: the current microbiological and nutritional outlook. *J. Dairy Sci.* 90(Suppl. 1), E17–E38. doi: 10.3168/jds.2006-478

Nagata, S. (1997). Apoptosis by death factor. *Cell* 88, 355–365. doi: 10.1016/S0092-8674(00)81874-7

Nilsson, S. A. (1963). Clinical, morphological, and experimental studies of laminitis in cattle. *Acta Vet. Scand.* 4(Suppl. 1):304.

Oba, M., Mewis, J., and Zhining, Z. (2015). Effects of ruminal doses of sucrose, lactose, and corn starch on ruminal fermentation and expression of genes in ruminal epithelial cells. *J. Dairy Sci.* 98, 586–594. doi: 10.3168/jds.2014-8697

Petri, R., Schwaiger, T., Penner, G., Beauchemin, K., Forster, R., McKinnon, J., et al. (2013). Changes in the rumen epimural bacterial diversity of beef cattle as affected by diet and induced ruminal acidosis. *Appl. Environ. Microbiol.* 79, 3744–3755. doi: 10.1128/AEM.03983-12

Pitta, D. W., Kumar, S., Vecchiarelli, B., Shirley, D. J., Bittinger, K., Baker, L. D., et al. (2014). Temporal dynamics in the ruminal microbiome of dairy cows during the transition period. *J. Anim. Sci.* 92, 4014–4022. doi: 10.2527/jas.2014-7621

Poulsen, M., Schwab, C., Jensen, B. B., Engberg, R. M., Spang, A., Canibe, N., et al. (2013). Methylotrophic methanogenic Thermoplasmata implicated in reduced methane emissions from bovine rumen. *Nat. Commun.* 4:1428. doi: 10.1038/ncomms2432

Ribeiro, C., Karnati, S., and Eastridge, M. (2005). Biohydrogenation of fatty acids and digestibility of fresh alfalfa or alfalfa hay plus sucrose in continuous culture. *J. Dairy Sci.* 88, 4007–4017. doi: 10.3168/jds.S0022-0302(05)73087-3

Russell, J. B., O'Connor, J. D., Fox, D. G., Van Soest, P. J., and Sniffen, C. J. (1992). A net carbohydrate and protein system for evaluating cattle diets: I. Ruminal fermentation. *J. Anim. Sci.* 70, 3551–3561.

Saleem, F., Ametaj, B., Bouatra, S., Mandal, R., Zebeli, Q., Dunn, S., et al. (2012). A metabolomics approach to uncover the effects of grain diets on rumen health in dairy cows. *J. Dairy Sci.* 95, 6606–6623. doi: 10.3168/jds.2012-5403

Seo, J., Park, T., Kwon, I., Piao, M., Lee, C., and Ha, J. K. (2013). Characterization of cellulolytic and xylanolytic enzymes of *Bacillus licheniformis* JK7 isolated from the rumen of a native Korean goat. *Asian Australas. J. Anim. Sci.* . 26, 50–58. doi: 10.5713/ajas.2012.12506

Slyter, L., Oltjen, R., Kern, D., and Weaver, J. (1968). Microbial species including ureolytic bacteria from the rumen of cattle fed purified diets. *J. Nutr.* 94, 185–192.

Soriano, F. D., Polan, C. E., and Miller, C. N. (2000). Milk production and composition, rumen fermentation parameters, and grazing behavior of dairy cows supplemented with different forms and amounts of corn grain. *J. Dairy Sci.* 83, 1520–1529. doi: 10.3168/jds.S0022-0302(00)75024-7

Strater, J., Wellisch, I., Riedl, S., Walczak, H., Koretz, K., Tandara, A., et al. (1997). CD95 (APO-1/Fas)-mediated apoptosis in colon epithelial cells: a possible role in ulcerative colitis. *Gastroenterology* 113, 160–167. doi: 10.1016/S0016-5085(97)70091-X

Sun, X., Wang, Y., Chen, B., and Zhao, X. (2015). Partially replacing cornstarch in a high-concentrate diet with sucrose inhibited the ruminal trans-10 biohydrogenation pathway in vitro by changing populations of specific bacteria. *J. Anim. Sci. Biotechnol.* 6:57. doi: 10.1186/s40104-015-0051-y

Tao, S., Duanmu, Y., Dong, H., Ni, Y., Chen, J., Shen, X., et al. (2014a). High concentrate diet induced mucosal injuries by enhancing epithelial apoptosis and inflammatory response in the hindgut of goats. *PLoS ONE* 9:e111596. doi: 10.1371/journal.pone.0111596

Tao, S., Duanmu, Y., Dong, H., Tian, J., Ni, Y., and Zhao, R. (2014b). A high-concentrate diet induced colonic epithelial barrier disruption is associated with the activating of cell apoptosis in lactating goats. *BMC Vet. Res.* 10:235. doi: 10.1186/s12917-014-0235-2

Taxis, T. M., Wolff, S., Gregg, S. J., Minton, N. O., Zhang, C., Dai, J., et al. (2015). The players may change but the game remains: network analyses of ruminal microbiomes suggest taxonomic differences mask functional similarity. *Nucleic Acids Res.* 43, 9600–9612. doi: 10.1093/nar/gkv973

Underwood, W. (1992). Rumen lactic acidosis: 1. Epidemiology and pathophysiology. *Comp. Contin. Educ. Pract. Vet.* 14, 1127–1133.

Vallimont, J., Bargo, F., Cassidy, T., Luchini, N., Broderick, G., and Varga, G. (2004). Effects of replacing dietary starch with sucrose on ruminal fermentation and nitrogen metabolism in continuous culture. *J. Dairy Sci.* 87, 4221–4229. doi: 10.3168/jds.S0022-0302(04)73567-5

Wojdasiewicz, P., Poniatowski, L. A., and Szukiewicz, D. (2014). The role of inflammatory and anti-inflammatory cytokines in the pathogenesis of osteoarthritis. *Mediat. Inflamm.* 2014:561459. doi: 10.1155/2014/561459

Yanagita, K., Manome, A., Meng, X.-Y., Hanada, S., Kanagawa, T., Tsuchida, T., et al. (2003). Flow cytometric sorting, phylogenetic analysis and in situ detection of *Oscillospira* guillermondii, a large, morphologically conspicuous but uncultured ruminal bacterium. *Int. J. Syst. Evol. Microbiol.* 53, 1609–1614. doi: 10.1099/ijs.0.02541-0

Zebeli, Q., and Metzler-Zebeli, B. U. (2012). Interplay between rumen digestive disorders and diet-induced inflammation in dairy cattle. *Res. Vet. Sci.* 93, 1099–1108. doi: 10.1016/j.rvsc.2012.02.004

Zhao, L., Meng, Q., Ren, L., Liu, W., Zhang, X., Huo, Y., et al. (2015). Effects of nitrate addition on rumen fermentation, bacterial biodiversity and abundance. *Asian Australas. J. Anim. Sci.* 28, 1433–1441. doi: 10.5713/ajas.15.0091

Zhou, Z., Yu, Z., and Meng, Q. (2012). Effects of nitrate on methane production, fermentation, and microbial populations in in vitro ruminal cultures. *Bioresour. Technol.* 103, 173–179. doi: 10.1016/j.biortech.2011.10.013

15

IL-36γ Augments Host Defense and Immune Responses in Human Female Reproductive Tract Epithelial Cells

*Sean M. Winkle, Andrea L. Throop and Melissa M. Herbst-Kralovetz**

Department of Basic Medical Sciences, College of Medicine–Phoenix, University of Arizona, Phoenix, AZ, USA

Edited by:
*Diana Bahia,
Universidade Federal de Minas Gerais, Brazil*

Reviewed by:
*Hridayesh Prakash,
University of Hyderabad, India
Jieliang Li,
Temple University, USA*

***Correspondence:**
*Melissa M. Herbst-Kralovetz
mherbst1@email.arizona.edu*

IL-36γ is a proinflamatory cytokine which belongs to the IL-1 family of cytokines. It is expressed in the skin and by epithelial cells (ECs) lining lung and gut tissue. We used human 3-D organotypic cells, that recapitulate either *in vivo* human vaginal or cervical tissue, to explore the possible role of IL-36γ in host defense against pathogens in the human female reproductive tract (FRT). EC were exposed to compounds derived from virus or bacterial sources and induction and regulation of IL-36γ and its receptor was determined. Polyinosinic-polycytidylic acid (poly I:C), flagellin, and synthetic lipoprotein (FSL-1) significantly induced expression of IL-36γ in a dose-dependent manner, and appeared to be TLR-dependent. Recombinant IL-36γ treatment resulted in self-amplification of IL-36γ and its receptor (IL-36R) via increased gene expression, and promoted other inflammatory signaling pathways. This is the first report to demonstrate that the IL-36 receptor and IL-36γ are present in the human FRT EC and that they are differentially induced by microbial products at this site. We conclude that IL-36γ is a driver for epithelial and immune activation following microbial insult and, as such, may play a critical role in host defense in the FRT.

Keywords: IL-1 family, IL-36γ, human epithelial cells, antimicrobial peptides, innate immunity, IL-36 receptor, microbial products, inflammatory mediators

INTRODUCTION

The epithelium of the human FRT plays host to a wide variety of commensal bacteria and acts as a first responder to invading microbes and changes in the microbial milieu. The human FRT can be divided into two distinct regions that differ in cellular structure and function. The lower FRT epithelium is composed of the vagina and ectocervix and is lined with a stratified squamous epithelium built to resist physical and mechanical stress. The upper FRT epithelium is made up of a single layer of columnar EC and includes the endocervix, endometrium, fallopian tubes, and the uterus (Horne et al., 2008). Site-specific differences in the magnitude and pattern of immune response have been investigated in the vagina and endocervix (Quayle, 2002; Pudney et al., 2005; Herbst-Kralovetz et al., 2008).

Abbreviations: ECs, epithelial cells; FRT, female reproductive tract; FSL-1, fibroblast stimulating lipopeptide-1; IL-36, interleukin-36; IL-36R, IL-36 receptor, KSFM, keratinocyte serum-free medium; PBST, PBS-0.05%Tween-20; qRT-PCR, quantitative real-time polymerase chain reaction; rIL-36γ, recombinant IL-36γ, STI, sexually transmitted infection.

To better understand immune function of the human FRT at these two sites, we investigated a newly characterized IL-1 family member identified to play a role in chronic inflammation and disease at other mucosal sites (Dunn et al., 2001; Gabay and Towne, 2015). The IL-36 family is a group of cytokines whose members are known for their role in mediating inflammation and host defense (Dunn et al., 2001; Gresnigt and van de Veerdonk, 2013; Gresnigt et al., 2013). While IL-36 family members have been shown to be expressed and secreted by the epithelium lining the lungs, gut and skin, this family, to our knowledge, has not previously been investigated in the human FRT (Chustz et al., 2011; Johnston et al., 2011; Gabay and Towne, 2015; Medina-Contreras et al., 2016). There are three isoforms that comprise the IL-36 family; IL-36α (IL-1F7), IL-36β (IL-1F8), and IL-36γ (IL-1F9) and these cytokines are expressed in a more tissue-restricted fashion relative to IL-1 (Dunn et al., 2001; Gresnigt and van de Veerdonk, 2013). Of the three isoforms, IL-36γ has the highest expression levels in damaged or infected epithelium that lines the skin and lungs suggesting that it plays a role in epithelial immune response (Berglof et al., 2003; Lian et al., 2012; Gabay and Towne, 2015). As tissue becomes diseased, IL-36γ promotes inflammation by binding to the IL-36 receptor (IL-36R) and stimulating the production of a number of cytokines, chemokines, adhesion molecules, antimicrobial peptides, and pro-inflammatory mediators (Towne et al., 2004; Vigne et al., 2011; Gresnigt and van de Veerdonk, 2013). Triggering IL-36R is necessary for IL-36γ to activate pathways that lead to induction of pro-inflammatory cytokine expression via NF-κB and MAPKs pathways (Chustz et al., 2011; Towne et al., 2011).

Vaginal and endocervical EC express pattern recognition receptors (PRRs) that can be triggered by sensing microbial products associated with bacteria and viruses resulting in an innate immune response (Herbst-Kralovetz et al., 2008). EC of the lower FRT respond to microbial products by secreting cytokines, chemokines, and antimicrobial peptides (Hjelm et al., 2010; Radtke et al., 2012; Doerflinger et al., 2014). Toll-like receptors (TLRs) expressed in FRT EC recognize microbe-associated molecular patterns (MAMPs) that correspond to highly conserved microbial products. Recent studies have shown that TLR agonists including polyinosinic-polycytidylic acid [poly (I:C); viral product] and flagellin (bacterial product) upregulate IL-36γ production in human skin keratinocytes, but have differential effects on IL-36γ secretion (Lian et al., 2012). Similar to IL-1β and IL-18, the IL-36 family members require processing to gain full bioactivity, the enzyme required is yet to be identified (Towne et al., 2011; Gabay and Towne, 2015). While both microbial products induce IL-36γ expression, only poly (I:C) has been shown to induce secretion (Lian et al., 2012). As FRT EC express TLRs 2, 3, 5, and 6 at the highest levels (Herbst-Kralovetz and Pyles, 2006), we tested poly (I:C; TLR3 agonist), flagellin (TLR5 agonist), and FSL-1 (bacterial product; TLR2/6 agonist) for induction and secretion of IL-36γ in the human FRT.

To investigate the function and regulation of IL-36γ in the FRT, we used our well-characterized 3-D human vaginal and endocervical EC models (Barrila et al., 2010; Hjelm et al., 2010; Radtke et al., 2012; Doerflinger et al., 2014). These models faithfully recapitulate many of the physiologic traits of FRT EC, including cellular architecture, adhesion, interaction, and immune function (Hjelm et al., 2010; Radtke et al., 2012; McGowin et al., 2013; Doerflinger et al., 2014). Furthermore, these two models can be used to evaluate site-specific differences in the magnitude and/or pattern of immune response at these sites. FRT EC models were exposed to microbial products to measure innate inflammatory responses in the FRT via induction and secretion of IL-36γ. In addition, we investigated the autocrine function of IL-36γ in these 3-D FRT human EC by treating with rIL-36γ. This is the first report to demonstrate that IL-36 receptor and IL-36γ are both expressed by ECs lining the human lower FRT and important in regulating host defense and immune responses in a TLR and site-specific manner.

MATERIALS AND METHODS

Culturing 3-D Human Vaginal and Endocervical Epithelial Cells

Human endocervical (A2) ECs were added to a suspension of collagen-coated dextran microcarrier beads (Sigma-Aldrich, St. Louis, MO, USA) in KSFM (Life Technologies, Grand Island, NY, USA) supplemented with 5 ng/ml human recombinant epidermal growth factor, 500 μg/ml bovine pituitary extract, 44 μg/ml $CaCl_2$ (Sigma-Aldrich), and 100 μg/ml primocin (InvivoGen, San Diego, CA, USA) as previously described (Radtke et al., 2012). Human vaginal (V19I) EC were added to a suspension of collagen-coated dextran microcarrier beads in a 1:1 mixture of supplemented KSFM and EpiLife medium (Life Technologies) as previously described (Hjelm et al., 2010). Cell and bead suspensions were transferred into a slow turning lateral vessel bioreactor and incubated at 37°C as previously described (Hjelm et al., 2010; Radtke and Herbst-Kralovetz, 2012; Radtke et al., 2012). After 28 days of growth, differentiated endocervical and vaginal aggregates were quantified using a Countess machine (Life Technologies) and viability was measured by trypan blue exclusion.

Human Cervical and Vaginal Epithelial Tissue Total RNA and Protein

Human cervical and vaginal total protein samples and human cervical and vaginal total RNA samples were purchased from BioChain (Newark, CA, USA). According to BioChain, human vaginal total protein and total RNA samples were acquired from a post-menopausal Caucasian female. Human cervical total protein samples were acquired from a pre-menopausal Asian female and human cervical total RNA samples were acquired from a pre-menopausal Caucasian female (BioChain). As per Biochain, all analyzed tissues were determined to be non-diseased per the clinical report and were screened by serology for HIV, HBV, HCV, and HTLV. cDNA was synthesized from total RNA (1 μg) cDNA by reverse transcription (iScript cDNA Synthesis Kit, Bio-Rad, Hercules, CA, USA) prior to qRT-PCR assays.

TLR Agonist and Cytokine Treatment

For all experimental treatments, 3-D aggregates were transferred into 24-well plates (1×10^5–5×10^5 cells per ml). Both 3-D vaginal EC and 3-D endocervical EC were treated with poly (I:C; InvivoGen) at 1, 25, and 100 μg/ml, flagellin (InvivoGen) at 0.5 and 5 μg/ml, FSL-1 (InvivoGen) at 0.01 and 0.1 μg/ml, or recombinant IL-36γ (PeproTech, Rocky Hill, NJ, USA) at 1, 10, 100, and 500 ng/ml for 24 h at 37°C. Additional wells were treated with sterile Dulbecco's PBS to provide a negative control. Following cell culture treatment, supernatants were harvested and stored at −20°C until analysis. Cell pellets were stored at −20°C in lysis buffer for RNA or protein extraction.

Protein Extraction and Quantification

For protein extraction, aggregates were pelleted and stored at −20°C in 1 ml of RIPA buffer [150 mM NaCl (Thermo Fisher Scientific), 50 mM Tris-HCL pH 8.0 (Life Technologies), 1 mM EDTA (Life Technologies), 1% NP-40 (Sigma-Aldrich), 1% sodium deoxycholate (Sigma-Aldrich), 0.1% SDS (Sigma-Aldrich), 1 mM PMSF (Sigma-Aldrich), and 5 μg/ml leupeptin (Sigma-Aldrich)]. Frozen cell pellets were thawed on ice. Once thawed, cells were incubated for 10 min on ice with intermittent mixing by inversion. Next, cells were sheared with a 23-gage needle connected to a 1 ml syringe. After incubating on ice for 20 min, cells were pelleted at 10,000 × *g* and supernatant was removed. A Bradford assay was performed in a microtiter plate to determine total protein concentration of the extracted cell pellet and culture supernatant. Absorbance was read at 595 nm on a Biotek ELx800 Microplate Reader (BioTek, Winooski, VT, USA) and experimental values were compared to the calculated standard curve to acquire total protein concentration.

ELISA Quantification of Intracellular and Secreted Human IL-36γ

High binding microtiter plates (Corning, Tewksbury, MA, USA) were coated with 50 μl/well of monoclonal rat anti-human IL-36γ antibody (R&D Systems, Minneapolis, MN, USA) at 2 μg/ml in PBS and incubated overnight at 4°C. The microtiter plates were washed three times with PBS-0.05%Tween-20 (PBST) then 50 μl of twofold serial dilutions of recombinant human IL-36γ (rIL-36γ, PeproTech) was added in duplicate to generate a standard curve. Experimental samples were added to each well in duplicate and all samples were incubated for 2 h at 37°C. The microtiter plates were then washed three times with PBST then biotinylated polyclonal goat anti-human IL-36γ detection antibody (R&D Systems) diluted at 2 μg/ml with 1% skim milk in PBST was added to each well and incubated for 2 h at 37°C. Following the incubation, the plates were washed with PBST three times. After washing, 50 μl streptavidin-HRP (R&D Systems) was added at a 1:250 dilution in PBST to each well and incubated for 1 h at 37°C. The plates were washed three times with PBST and then developed by addition of 50 μl tetramethylbenzidine substrate solution (Thermo Fisher Scientific, Waltham, MA, USA) to each well and incubated in the dark for up to 30 min at room temperature. The colorimetric reaction was stopped by addition of 50 μl/well of 1 M phosphoric acid and absorbance read at 450 nm on a Biotek ELx800 Microplate Reader (BioTek). Results were reported in fold as compared to PBS treated cell extracts.

Human IL-36γ Western Blot Analysis

Cell culture supernatants and cell pellet extracts were boiled for 10 min in 2× SDS buffer (6% SDS, 25 mM Tris-HCL pH 6.5, 10% glycerol, 0.1 M DTT, 20 μg/ml bromophenol blue). Total protein (30 μg) was loaded into 4–15% polyacrylamide Mini-PROTEAN TGX precast gels (Bio-Rad). After proteins were separated by SDS-PAGE, gels were transferred to polyvinylidene diflouride membranes (Life Technologies) using a dry blotting system (iBlot, Life Technologies). Levels of IL-36γ were determined using biotinylated goat anti-human IL-36γ diluted to 4 μg/ml in PBST with 1% dry milk, followed by streptavidin-HRP diluted 1:250 (R&D Systems). Levels of β-tubulin were examined using mouse anti-β-tubulin (Santa Cruz, Biotechnology, Dallas, TX, USA) diluted 1:1000 with horseradish peroxidase labeled goat anti-mouse (Santa Cruz Biotechnology) as a secondary antibody. Membranes were developed using ECL substrate (Life Technologies).

Quantification of Human Cytokines and Chemokines by Multiplex Analysis

Supernatants from 3-D vaginal and endocervical EC aggregates treated with rIL-36γ as described above were collected cytokine secretion levels were quantified. Cytokine concentrations were determined using a custom four-plex human cytokine kit containing IL-1B, IL-6, CCL20, and TNFα (EMD Millipore, Billerica, MA, USA) using the manufacturer's protocol. The data were collected using a Bio-Plex 200 System with Bio-Plex 5.0 Manager software (Bio-Rad).

RNA Extraction and Quantitative Real-Time PCR Analysis

RNA was extracted from 3-D endocervical and 3-D vaginal EC using the Qiagen RNeasy kit following the manufacturer's instructions (Qiagen, Valencia, CA, USA). cDNA was synthesized from 1 μg RNA by reverse transcription (iScript cDNA Synthesis Kit, Bio-Rad) and analyzed by qRT-PCR. qRT-PCR was performed with an Applied Biosystems 7500 Fast Real Time PCR System (Life Technologies) using customized primers purchased from IDT (Integrated DNA Technologies, Coralville, IA, USA) and iTAQ Universal SYBR Green Supermix (Bio-Rad). The following primers were used in this study: IL-1β forward, 5′-ACAGATGAAGTGCTCCTTCCA-3′ and reverse 5′-GTCGGAGATTCGTAGCTGGAT-3′ (Stordeur et al., 2002), HE4 forward 5′-CGGCTTCACCCTAGTCTCAG-3′ and reverse 5′-AAAGGGAGAAGCTGTGGTCA-3′ (Drapkin et al., 2005), IL 36γ forward 5′-CCCAGTCACTGTTGCTGTTA-3′ and reverse 5′-CAGTCTTGGCACGGTAGAAA-3′, IL-36R forward 5′-GCTGGAGTGTCCACAGCATA-3′ and reverse 5′-GCGATAAGCCCTCCTATCAA-3′ (Mutamba et al., 2012). IL-6, IL-8, HBD2, HBD4, and SLPI primers were previously described (Radtke et al., 2012; Doerflinger et al., 2014). Relative transcript levels were determined using a GAPDH housekeeping gene transcript and are reported as fold relative to negative control unless otherwise noted.

Statistical Analysis

All of the data for this study was generated using three independent batches of 3-D endocervical or vaginal EC aggregates in three independent experiments, and an average of all three of these experimental results are presented. An unpaired two-tailed Student t-test with Welch's correction was performed to determine statistical significance. GraphPad Prism v5.0 software was used for statistical analysis (GraphPad, San Diego, CA, USA). Levels of significance are reported as follows; $^{\wedge}P < 0.05$; $^{*}P < 0.01$; $^{**}P < 0.001$.

RESULTS

Human Tissue and 3-D Vaginal and Endocervical EC Express IL-36γ and IL-36R

IL-36γ has been previously shown to be expressed in skin and ECs lining mucosal tissue (Jensen, 2010; Chustz et al., 2011; Gabay and Towne, 2015), however, this cytokine has not been shown to be expressed previously in the FRT. Human IL-36γ levels were measured in both 3-D vaginal and endocervical EC by qRT-PCR analysis. Untreated 3-D vaginal and endocervical EC, and human cervical and vaginal tissue were analyzed by qRT-PCR assays targeting IL-36γ and IL-36R (**Figure 1A**). Basal expression of IL-36γ was significantly ($P < 0.001$) higher in 3-D endocervical EC compared to 3-D vaginal EC. Human 3-D vaginal and endocervical EC, as well as, human vaginal and cervical tissue expressed IL-36γ at levels much higher than IL-36R (**Figure 1A**). This is the first report to characterize IL-36R and IL-36γ in the EC lining the human FRT, furthermore, these results demonstrate our model's ability to serve as a tool to further study IL-36 function.

Western blots were performed to confirm production/ synthesis of human IL-36γ in EC of the FRT. Cell lysates from human 3-D vaginal and endocervical EC models and cell lysates from human cervical and vaginal tissue, were analyzed by Western blot (**Figure 1B**). We found that IL-36γ was constitutively produced in human 3-D vaginal and endocervical EC and vaginal and endocervical tissue. Basal synthesis of IL-36R was lower than the detection limits of Western blot analysis (data not shown) in both PBS treated and TLR agonist stimulated FRT EC cells. TNFα, a cytokine previously shown to induce IL-36γ (Friedrich et al., 2014), was used as a positive control to test if IL-36γ is inducible in the FRT EC. Three-D vaginal and endocervical EC were treated with TNFα for 24 h. Following TNFα treatment, IL-36γ gene expression levels in 3-D vaginal and endocervical EC were significantly ($P < 0.001$) increased 32-fold, and 19.6-fold respectively (**Figure 1C**). The 3-D vaginal EC and tissue exhibited higher baseline levels of IL-36γ and IL-36R relative to 3-D endocervical EC and tissue and TNFα treatment induced higher levels of IL-36γ expression in 3-D vaginal EC relative to the endocervical EC.

Microbial Products Differentially Induce IL-36γ Expression and Secretion in EC of the Human FRT

Microbial products elicit potent immune responses in EC of the FRT through ligation with TLRs. We have previously shown that TLR agonists: poly(I:C; viral product and TLR3 agonist), flagellin (bacterial product and TLR5 agonist), and

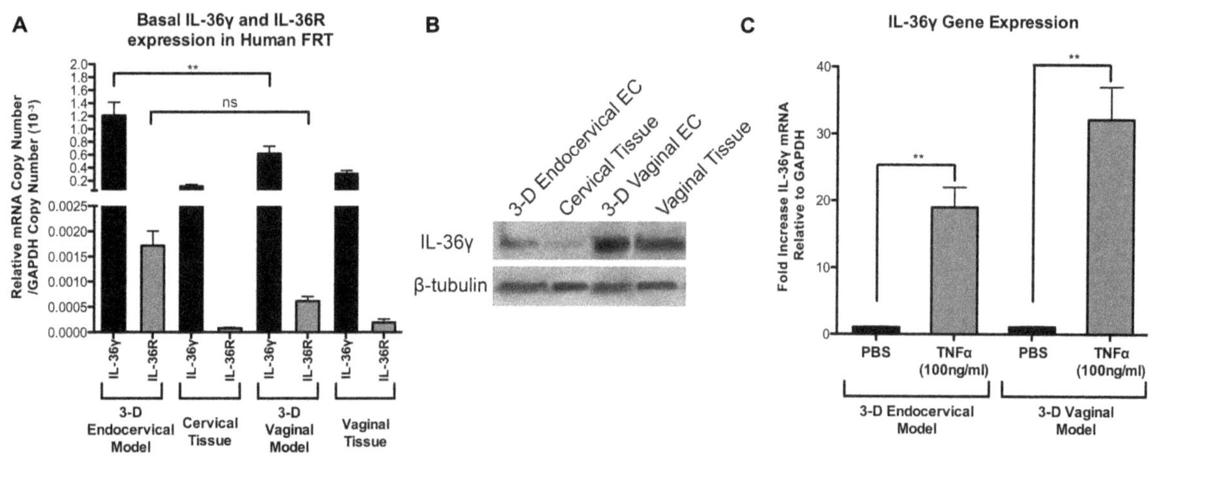

FIGURE 1 | Human 3-D vaginal and cervical epithelial cells and tissue express IL-36γ and IL-36R. (A) Expression of IL-36γ and IL-36R in 3-D human vaginal and endocervical cell cultures, and human vaginal and cervical tissue. cDNA from cell cultures and tissue were analyzed by qRT-PCR and mRNA copy numbers were normalized against GAPDH. **(B)** Cell lysates from 3-D vaginal and endocervical cell cultures, vaginal tissue and cervical tissue were evaluated by Western blot for expression of IL-36γ. β-tubulin was used as a loading control. 30 μg total protein from each lysate was loaded. **(C)** TNFα (100 ng/ml) treated 3-D cultures of vaginal and endocervical epithelial cells expressed IL-36γ. mRNA levels were normalized against GAPDH and expression of IL-36γ was reported as fold change compared to PBS treated samples. Data are shown as mean ± SD from three independent experiments each performed in duplicate. $^{\wedge}P < 0.05$; $^{*}P < 0.01$; $^{**}P < 0.001$; unpaired two-tailed Student t-test with Welch's correction.

FSL-1 (bacterial product and TLR2/6 agonist) induce cytokine expression and secretion in primary and immortalized vaginal and endocervical EC grown as monolayers and in 3-D (Herbst-Kralovetz et al., 2008; Hjelm et al., 2010; Radtke et al., 2012; Doerflinger et al., 2014). These studies have also shown that TLRs 2, 3, 5, and 6 are abundantly expressed in these cell lines and these TLRs remain functional and responsive when cultured under rotating conditions (Herbst-Kralovetz et al., 2008; Hjelm et al., 2010; Radtke and Herbst-Kralovetz, 2012; Radtke et al., 2012). Maximal cytokine induction was detected 24 h after treatment of microbial products in these studies, hence we focused on that time point herein (Herbst-Kralovetz et al., 2008). Human 3-D vaginal and endocervical EC were treated with a panel of microbial products for 24 h, and IL-36γ expression levels were quantified by qRT-PCR analysis (**Figure 2**). Following poly (I:C), FSL-1, and flagellin treatment did not result in cellular adherence to beads as observed by light microscopy and cellular viability was not decreased by more than 10% as measured by trypan blue exclusion (data not shown). IL-36γ expression increased following treatment with each microbial product in a concentration dependent manner. We observed that poly (I:C)

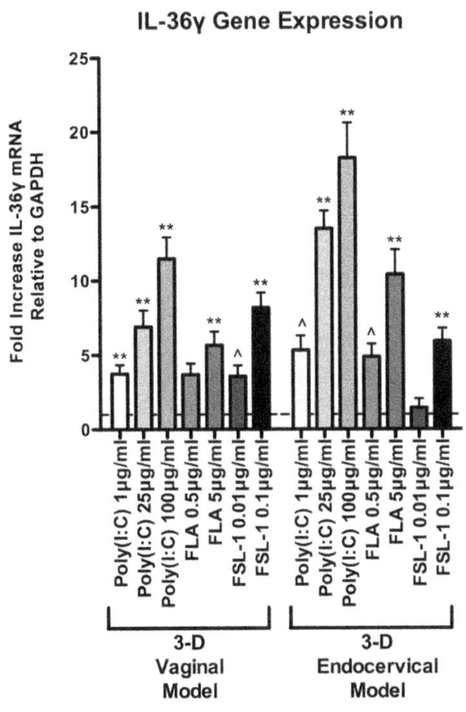

FIGURE 2 | Differential IL-36γ induction in 3-D human vaginal and endocervical epithelial cells treated with microbial products. Gene expression analysis of 3-D vaginal and endocervical EC. Cell cultures were exposed to microbial products [poly (I:C); 1, 25, and 100 μg/ml, flagellin (FLA); 0.5 and 5 μg/ml, and fibroblast stimulating lipopeptide-1 (FSL-1); 0.01 and 0.1 μg/ml] for 24 h. Expression of IL-36γ was determined by qRT-PCR and reported relative to GAPDH as fold change compared to PBS treated samples. Horizontal dashed line indicates IL-36γ expression in PBS treated cells. Data are shown as mean ± SD from three independent experiments each performed in duplicate. ^$P < 0.05$; *$P < 0.01$; **$P < 0.001$; unpaired two-tailed Student *t*-test with Welch's correction.

induced the highest amount of IL-36γ gene expression in both cell types (**Figure 2**). In 3-D vaginal EC IL-36γ expression was induced by treatment with poly (I:C; 100 μg/ml), flagellin (5 μg/ml), and FSL-1 (0.1 μg/ml) by 10.2-, 5.1-, and 8.9-fold, respectively. Endocervical EC treated with poly (I:C), flagellin, and FSL-1 also resulted in significant increases ($P < 0.001$) in IL-36γ expression levels at 20.2-, 11.7-, and 7-fold, respectively. Poly (I:C) and flagellin induced higher levels of IL-36γ expression in 3-D endocervical EC relative to vaginal EC ($P < 0.001$).

Since IL-36γ mRNA levels increased in a dose dependent manner, we performed an IL-36γ targeted ELISA to quantify protein levels following treatment with these microbial products. After treatment with the poly (I:C), flagellin, FSL-1, and PBS, cell lysates and cellular supernatants were collected and assayed for IL-36γ by ELISA. Poly (I:C), flagellin, and FSL-1 treatments induced IL-36γ production in a dose dependent manner in both 3-D vaginal and 3-D endocervical EC when compared to PBS treated controls (**Figure 3B**). All of the microbial products tested induced an increase in IL-36γ mRNA in 3-D FRT EC, however, the magnitude of protein production and secretion levels were microbial product specific. Treatment with higher concentrations of poly (I:C) resulted in higher levels of secreted IL-36γ in 3-D vaginal and endocervical EC (**Figures 3A,B**) when measured by ELISA. Flagellin induced a significant increase ($P < 0.01$) in intracellular IL-36γ production in vaginal and endocervical ($P < 0.01$) EC. Poly (I:C) induced significant levels of IL-36γ secretion by vaginal (1 μg/ml; $P < 0.01$, 25 μg/ml; $P < 0.001$, and 100 μg/ml; $P < 0.001$) and endocervical (1 μg/ml; [not significant (NS)] 25 μg/ml; $P < 0.001$, and 100 μg/ml; $P < 0.001$) EC. FSL-1 induction of intracellular IL-36γ production significantly increased secretion of IL-36γ in both vaginal (0.01 μg/ml; $P < 0.01$, 0.01 μg/ml; $P < 0.001$) and endocervical (0.01 μg/ml; $P < 0.01$, 0.01 μg/ml; $P < 0.001$) EC. Less IL-36γ secretion occurred following treatment with flagellin relative to control treated cells.

IL-36γ probed Western blots were performed to verify ELISA results. FRT EC were treated with poly (I:C), flagellin, and FSL-1 in increasing concentrations for 24 h. Both the supernatant and extracted cell pellets of TLR agonist treated FRT EC were analyzed to measure intracellular and secreted IL-36γ levels. Patterns of IL-36γ protein synthesis detected by Western blot mirrored the quantitative data generated by ELISA, supporting the protein production profiles of FRT EC (Supplementary Figure S1). Poly (I:C) and FSL-1 treatment resulted in the highest induction of secreted IL-36γ levels relative to untreated wells by ELISA and Western blot, whereas flagellin treatment induced significantly higher intracellular levels of IL-36γ (**Figure 3** and Supplementary Figure S1). As such, there was an inverse relationship between IL-36γ induction by these microbial products.

Recombinant IL-36γ Induces an Autocrine IL-36γ Loop in Human 3-D Vaginal and Endocervical EC

IL-36γ has been shown to induce expression of itself, thereby exhibiting an autocrine signaling loop in keratinocytes (Lian

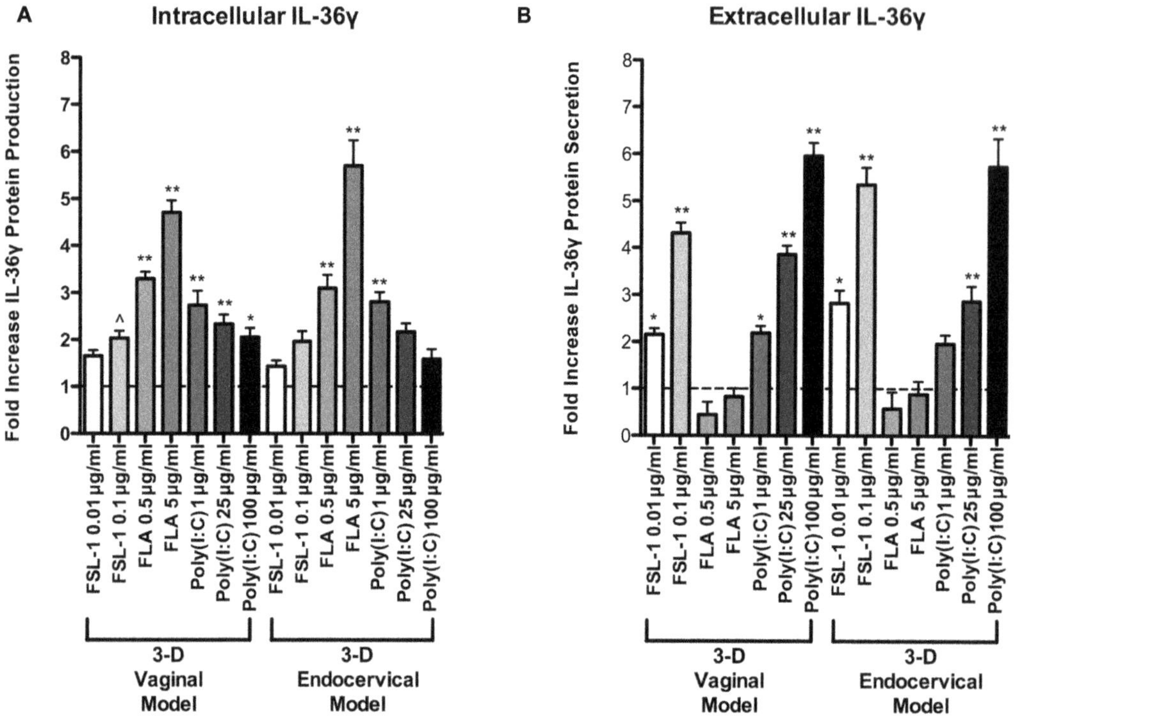

FIGURE 3 | Extracellular and intracellular localization of IL-36γ in epithelial cells of the human female reproductive tract following treatment with microbial products. Three-D vaginal and endocervical EC were treated with PBS, flagellin (FLA; 5 or 0.5 μg ml^{-1}), FSL-1 (0.1 or 0.01 μg ml^{-1}), or poly (I:C; 100, 25, or 1 μg ml^{-1}) for 24 h. **(A)** Cell culture lysates and **(B)** supernatants were assayed by ELISA to quantify IL-36γ production and secretion. Levels of IL-36γ were determined and reported as fold change relative to PBS treated samples. Horizontal dashed line indicates IL-36γ expression in PBS treated cells. Data are shown as mean ± SD from three independent experiments each performed in duplicate. ^$P < 0.05$; *$P < 0.01$; **$P < 0.001$; unpaired 2-tailed Student *t*-test with Welch's correction.

et al., 2012). Three-D vaginal and endocervical EC were treated with increasing doses of rIL-36γ (1, 10, 100, and 500 ng/ml) for 24 h. qRT-PCR assays targeting IL-36γ and IL-36R was performed following treatment with rIL-36γ. We observed a significant increase in IL-36γ gene expression in both 3-D vaginal (10 ng/ml; $P < 0.001$, 100 ng/ml; $P < 0.001$, 500 ng/ml; $P < 0.001$) and endocervical EC (10 ng/ml; $P < 0.01$, 100 ng/ml; $P < 0.001$, 500 ng/ml $P < 0.001$; **Figure 4A**). A similar pattern in IL-36R expression in 3-D vaginal [10 ng/ml; 4.2-fold ($P < 0.01$), 100 ng/ml; 7.9-fold ($P < 0.001$), 500 ng/ml; 10.4-fold ($P < 0.001$)] and endocervical EC [10 ng/ml; 8.3-fold ($P < 0.05$), 100 ng/ml; 13.9-fold ($P < 0.01$), 500 ng/ml; 15.7-fold ($P < 0.001$; **Figure 4B**)] was observed, albeit at lower levels. The fold increases in both IL-36γ and IL-36R were dependent on rIL-36γ treatment concentration and expression levels were higher in the 3-D endocervical EC compared to the vaginal EC.

Pro-inflammatory Cytokine and Antimicrobial Peptide Production in 3-D FRT EC is Increased Following Recombinant IL-36γ Treatment

We have shown that microbial products induce IL-36γ and influence its secretion. To examine the role IL-36γ plays in triggering inflammatory signaling pathways in FRT EC and to determine if it is concentration dependent, we treated 3-D endocervical and vaginal EC with rIL-36γ. Treatment with rIL-36γ induced antimicrobial peptide, pro-inflammatory cytokine, and chemokine production in a dose dependent manner (**Figures 5** and **6**). IL-8 (**Figure 5A**) mRNA levels were significantly increased ($P < 0.001$) in both cell models following 100 and 500 ng/ml rIL-36γ treatments. rIL-36γ treatment (500 ng/ml) resulted in a significant difference ($P < 0.001$) in CCL20 production (**Figure 5B**) and 3-D vaginal EC CCL20 levels were significantly higher ($P < 0.001$) at 500 ng/ml compared to the 3-D endocervical EC. Antimicrobial peptides including HE4 (**Figure 5C**), SLPI (**Figure 5D**), HBD-2 (**Figure 5E**), and HBD-4 (**Figure 5F**) were significantly increased in a dose dependent manner following rIL-36γ treatments. Treatment with rIL-36γ at 100 and 500 ng/ml resulted in significant increases in HE4 in 3-D vaginal ($P < 0.001$ and $P < 0.01$) and endocervical ($P < 0.05$ and $P < 0.001$). SLPI mRNA levels increased at similar magnitudes following rIL-36γ treatments at 100 and 500 ng/ml in 3-D vaginal ($P < 0.001$) and endocervical ($P < 0.05$) EC.

Of these AMP, two notable site-specific differences were observed in the induction HBD-2 and HBD-4 in 3-D FRT EC following treatment with100 and 500 ng/ml of rIL-36γ. Fold values for 3-D vaginal cells were higher for HBD-2 and HBD-4 relative to endocervical EC. HBD-2 increased 12.7-fold

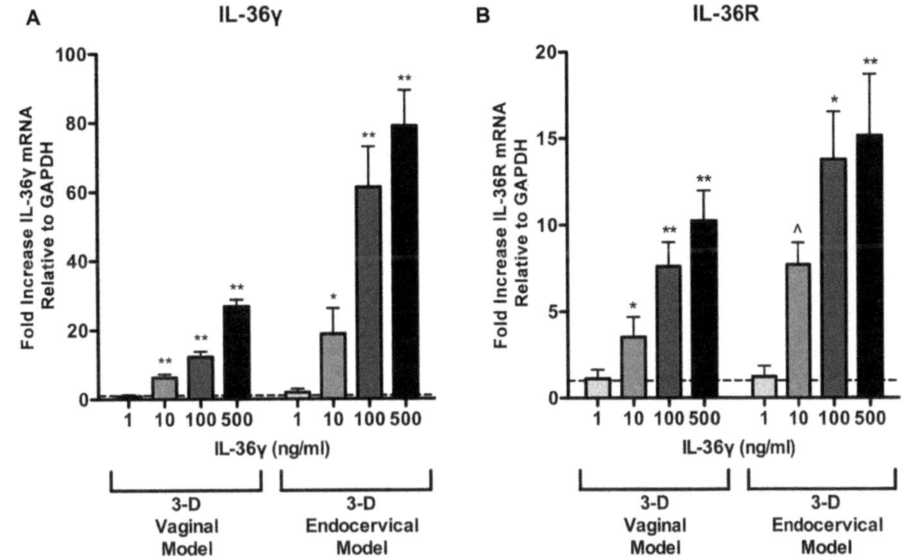

FIGURE 4 | Induction of IL-36γ and IL-36R expression following recombinant IL-36γ treatment. 3-D vaginal and endocervical EC were treated with increasing doses of recombinant IL-36γ (1, 10, 100, and 500 ng/ml) for 24 h. **(A)** IL-36γ and **(B)** IL-36R expression was determined by qRT-PCR and reported relative to GAPDH as fold change compared to PBS treated samples. Horizontal dashed line indicates IL-36γ expression in PBS treated cells. Data are shown as mean ± SD from three independent experiments each performed in duplicate. ^$P < 0.05$; *$P < 0.01$; **$P < 0.001$; unpaired two-tailed Student *t*-test with Welch's correction.

($P < 0.001$) and 39.4-fold ($P < 0.001$) and HBD-4 mRNA was increased 24.6-fold ($P < 0.001$) and 55.4-fold ($P < 0.001$) in 3-D vaginal EC, respectively. rIL-36γ treated 3-D endocervical EC increased 6.9- and 18-fold at 100 and 500 ng/ml, respectively, for HBD-2 at 15.6-fold ($P < 0.001$) and 31.8-fold ($P < 0.001$) for the HBD-4 gene. HBD 2 and HBD-4 expression was significantly higher in 3-D vaginal EC than 3-D endocervical EC following rIL-36γ treatments at 100 and 500 ng/ml ($P < 0.05$ and $P < 0.01$ respectively). SLPI and HE4, were the only AMP assayed that resulted in more robust response in endocervical EC when compared to vaginal EC, but these site specific differences in expression were not significant. IL-8 gene expression levels were similar in both cell types. CCL20 expression levels varied in the two cell types depending on the concentration of rIL-36γ.

We have shown that rIL-36γ increases gene expression of chemokines and AMP. To quantify levels of additional proinflammatory cytokines and chemokines we assayed cell supernatants with a multiplex bead-based immunodetection assay. Secretion of IL-1β, IL-6, CCL20, and TNFα was induced in a dose dependent fashion following 24 h treatments with rIL-36γ (**Figure 6**). CCL20 secretion was significantly ($P < 0.01$) increased following treatment with rIL-36γ at 100 and 500 ng/ml in both 3-D vaginal and endocervical EC. IL-1β secretion was significantly ($P < 0.01$) increased following rIL-36γ treatments at 500 ng/ml in 3-D vaginal EC. rIL-36γ treatments at 100 and 500 ng/ml resulted in significant increases in IL-1β secretion ($P < 0.05$) in 3-D endocervical EC as well. Secretion of IL-6 was significantly ($P < 0.01$) increased in 3-D vaginal EC following rIL-36γ treatments of 100 and 500 ng/ml. IL-6 and TNFα secretion was induced in 3-D endocervical EC, however, the magnitude of secretion was higher in 3-D vaginal EC. Secreted IL-6 increased from 86.48 to 552.9 pg/ml (*NS*) for 100 ng/ml rIL-36γ and 951.5 pg/ml ($P < 0.01$) for 500 ng/ml rIL-36γ. TNFα levels from 3-D endocervical EC increased only slightly following rIL-36γ treatments and the increase was not significant. Overall, pro-inflammatory cytokines, chemokines, and AMP were significantly induced in 3-D vaginal and cervical EC following treatment with rIL-36γ. The pattern of cytokine and AMP induction was similar for vaginal and cervical cells and was also dose-dependent, but the magnitude was higher in the 3-D vaginal EC model relative to the 3-D cervical model for most immune mediators evaluated.

DISCUSSION

This is the first report demonstrating a role for IL-36γ in host defense and amplification of the mucosal immune response by ECs lining the FRT. We showed that in the lower FRT, IL-36γ stimulates the innate immune response by induction of pro-inflammatory cytokines, chemokines, and AMP, thereby promoting mucosal inflammation. This study provides the foundation for future studies that aim to elucidate the mechanisms of IL-36 family members in the FRT during homeostasis, STI, disease pathogenesis, and reproductive sequelae.

Recent literature on IL-36γ has been primarily focused on chronic inflammatory processes that include psoriasis and chronic obstructive pulmonary disease (COPD), where IL-36γ was found to be abundantly expressed in the skin and the mucosal ECs of the lung (Li et al., 2014; Gabay and Towne, 2015) and result in sustained inflammation (Chustz et al., 2011; Gabay

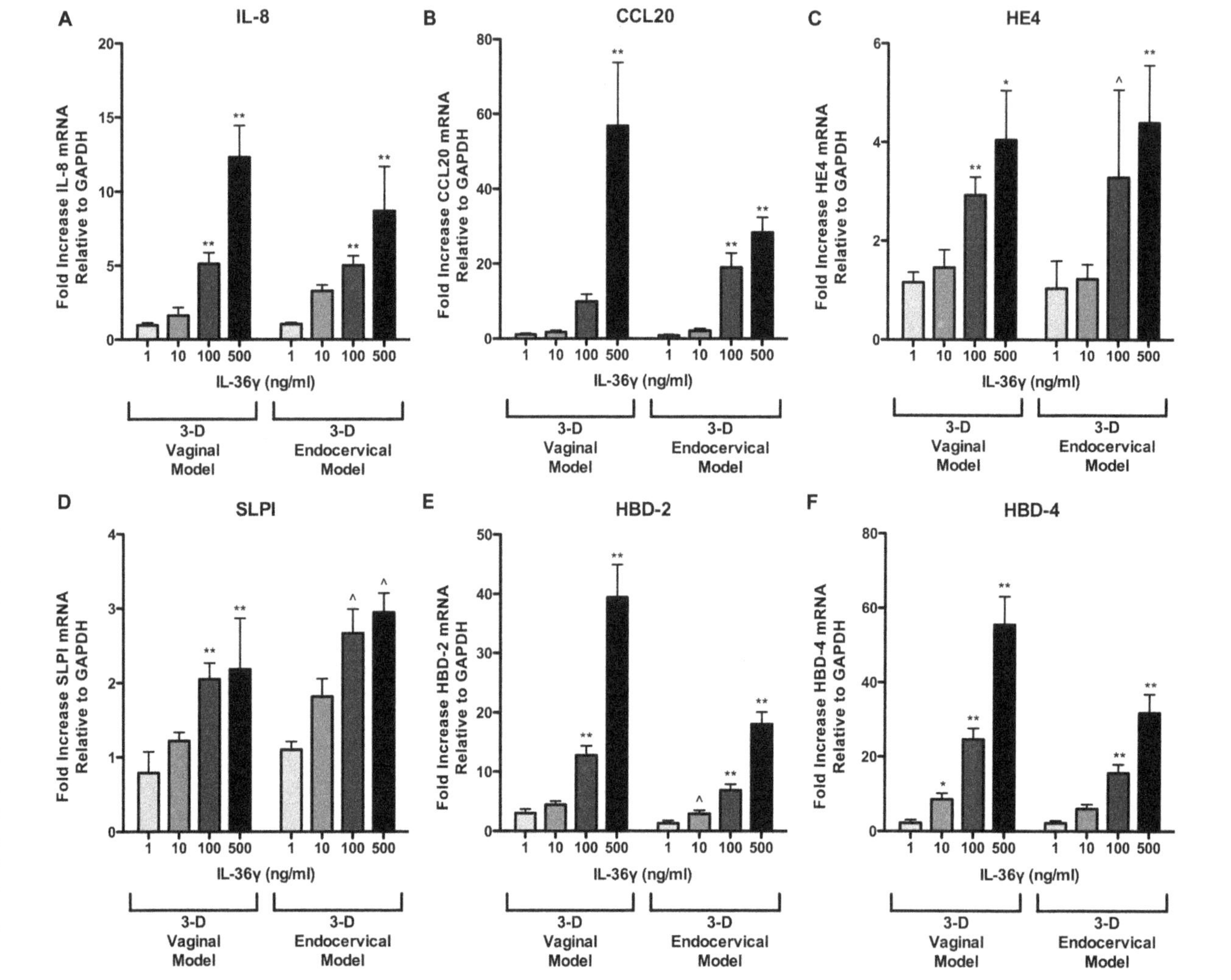

FIGURE 5 | Recombinant IL-36γ treatment induces antimicrobial peptide expression in 3-D human vaginal and endocervical epithelial cells. 3-D vaginal and endocervical cell cultures were treated with increasing doses of recombinant IL-36γ (1, 10, 100, or 500 ng/ml) for 24 h. Relative expression levels of **(A)** IL-8, **(B)** CCL20, **(C)** HE4, **(D)** SLPI, **(E)** HBD-2, and **(F)** HBD-4 were determined by qRT-PCR and reported relative to GAPDH as fold change compared to PBS treated cells Data are shown as mean ± SD from three independent experiments each performed in duplicate. ^$P < 0.05$; *$P < 0.01$; **$P < 0.001$; unpaired two-tailed Student *t*-test with Welch's correction.

and Towne, 2015). However, there have been limited studies investigating the role of IL-36γ in host defense and infection at mucosal sites (Vos et al., 2005; Gresnigt et al., 2013; Segueni et al., 2015).

Herein we report that IL-36γ is expressed in human cervical and vaginal ECs and tissue. A limitation to the study evaluating IL-36γ and IL-36R in vaginal tissue, is the lack of availability of premenopausal non-diseased vaginal tissue. However, our data demonstrated that premenopausal cervical tissue and post-menopausal vaginal tissue both express IL-36γ and IL-36R. Future studies will be required to determine the impact of sex hormones on this cytokine and associated receptor expression. Using our well-characterized human 3-D cervical and vaginal EC models we demonstrate that these cells respond to rIL-36γ exposure in a site-specific manner. Overall, 3-D vaginal EC expressed and secreted higher levels of these innate immune mediators following rIL-36γ treatments relative to 3-D endocervical EC, however, this response did not seem to correspond to IL-36R expression (**Figure 1**).

IL-36γ increases and prolongs inflammation through interaction with TNFα, a known inducer and inductee of IL-36γ (Towne and Sims, 2012). TNFα is expressed in excess in psoriatic lesions (Cuchacovich et al., 2008) leading to upregulation of IL-36γ. IL-36γ mRNA and protein have been shown to be induced in keratinocytes following treatment with TNFα (Kumar et al., 2000). Consistent with findings in bronchial ECs (Chustz et al., 2011), IL-36γ levels were increased following TNFα treatment in 3-D FRT EC, demonstrating that expression of IL-36γ is inducible in the human FRT. TNFα and IL-36γ participate in a mutually reinforcing gene expression loop, initiated by IL-36γ, TNFα, as well as other cytokines (Friedrich et al., 2014). Increased levels of TNFα have been shown to result

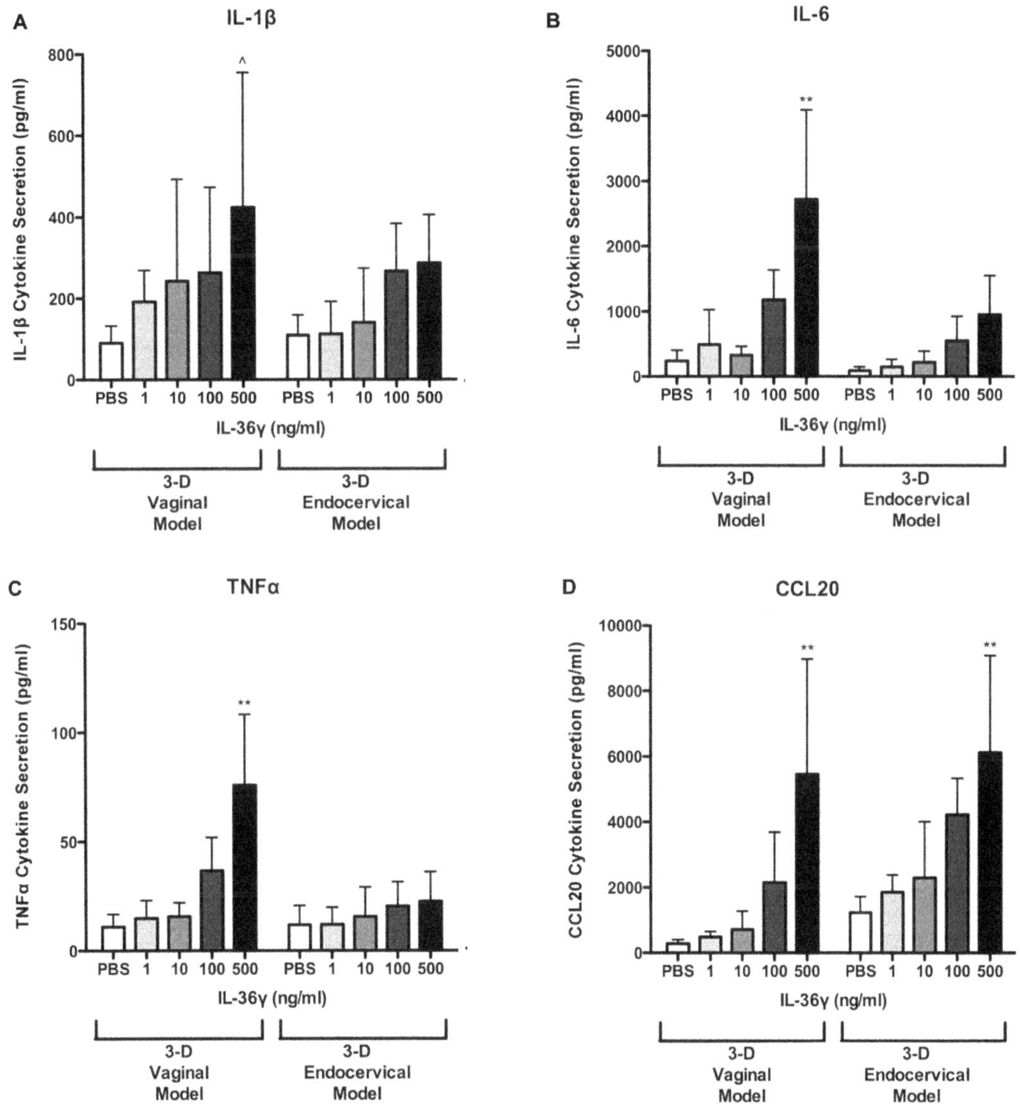

FIGURE 6 | 3-D human vaginal and endocervical epithelial cells secrete increased levels of pro-inflammatory cytokines and chemokines in response to treatment with recombinant IL-36γ. 3-D vaginal and endocervical cell cultures were treated with PBS or increasing doses of recombinant IL-36γ (1, 10, 100, or 500 ng/ml) for 24 h. Expression levels of **(A)** IL-1β, **(B)** IL-6, **(C)** TNFα, and chemokine **(D)** CCL20 were determined by multiplex analysis and reported in pg/ml. Data are shown as mean ± SD from three independent experiments each performed in duplicate. ^$P < 0.05$; *$P < 0.01$; **$P < 0.001$; unpaired two-tailed Student *t*-test with Welch's correction.

from microbial product exposure and pathogenic insults in the FRT (Jarvis et al., 1999; Agrawal et al., 2008; Hjelm et al., 2010; Radtke et al., 2012; Krzyzowska et al., 2013; Doerflinger et al., 2014) and as such, IL-36γ could be bolstering the inflammatory cascade in the FRT.

Poly (I:C), FSL-1, and flagellin treatments each resulted in unique IL-36γ expression and secretion patterns in our 3-D FRT EC models. TLR2, 3, 5, and 6 are the most abundantly expressed and functional in the FRT and our 3-D human FRT models (Herbst-Kralovetz et al., 2008; Hjelm et al., 2010; Radtke and Herbst-Kralovetz, 2012; Radtke et al., 2012). Flagellin targets TLR5, which activates the pro-inflammatory NF-κB pathway in parallel to apoptotic caspase cascades (Zeng et al., 2006). FSL-1 signals through TLR2/6, which also leads to the pro-inflammatory NF-κB pathway, but also triggers the MAPK pathway leading to possible apoptosis through MyD88 and caspase 8 (Aliprantis et al., 2000). Increased levels of IL-36γ were detected in 3-D FRT EC cell culture supernatant following FSL-1 treatment, but not following flagellin treatment. The mechanism of secretion is unclear as IL-36γ lacks a conventional signal sequence (Towne et al., 2011), and further work must be performed to determine IL-36γ secretion mechanisms in FRT EC. Poly (I:C) triggers TLR3 in the endosome, activating inflammation through the NF-κB pathway. Poly (I:C) has also been shown to induce FRT EC pro-inflammatory cytokine secretion, recruiting immune cells that activate cellular defense mechanisms, limiting viral replication until an adaptive immune response is generated (Schaefer et al., 2005; Herbst-Kralovetz and Pyles, 2006). It is unclear whether poly (I:C) induces IL-36γ secretion via a signaling sequence or via cell death, but it is possible that IL-36γ acts as an alarmin, alerting surrounding cells of danger (Lian et al., 2012).

In our study, levels of extracellular IL-36γ significantly increased and intracellular IL-36γ decreased following high dose treatments with poly (I:C). In contrast, flagellin treatment in FRT EC resulted in increased intracellular IL-36γ protein with corresponding supernatants containing low levels of extracellular IL-36γ. These results are similar to IL-36γ expression patterns observed in a study by Lian et al. in keratinocytes (Lian et al., 2012). These microbial product-specific induction and secretion relationships suggest that IL-36γ influences TLR dependent cellular defense responses unique to different forms of microbial attack. Further studies are required, but our data suggest that IL-36γ secretion may play a role in antimicrobial host defense and regulating tissue specific immune responses to infection in the FRT by amplifying the mucosal immune response.

IL-36γ activates the release of cytokines and chemokines from surrounding cells, leading to inflammatory host defense against invading pathogens. Prolonged inflammation in the FRT may also act as a warning sign of infection, indicating that IL-36γ may act as a marker for FRT disease or inflammation. A recent proteomics report demonstrated that cervicovaginal lavages collected from women with bacterial vaginosis exhibited elevated levels of IL-36α and IL-36 receptor antagonist relative to controls, further supporting a role for IL-36 family members in the FRT (Borgdorff et al., 2015). Both 3-D vaginal and endocervical EC responded to rIL-36γ treatment by upregulating gene expression of IL-6, IL-8, IL-1β, and TNFα. Each of these immune mediators has been found to be elevated in human cervicovaginal lavages collected from women with STI (Trifonova et al., 2007; Kyongo et al., 2015), as well as in the cell lines used in our 3-D FRT EC cultures following STI exposure (Herbst-Kralovetz et al., 2008; Hjelm et al., 2010; Radtke and Herbst-Kralovetz, 2012; McGowin et al., 2013). However, preliminary studies have not shown significant changes in TLR expression following rIL-36γ treatment (data not shown). We found that treating the 3-D FRT EC with rIL-36γ induced expression of IL-36R and cellular IL-36γ. A study by Friedrich et al. with keratinocytes defined this autocrine function as a self-amplifying loop involving IL-17 (Friedrich et al., 2014). IL-36γ has also been shown to induce CCL20, a Th17 cytokine, in lung fibroblasts due to asthma and COPD (Chustz et al., 2011) and in dermatoses via LL-37 (Li et al., 2014). rIL-36γ induced upregulation of IL-36γ and IL-36R mRNA as well as CCL20 mRNA and protein in 3-D FRT EC models suggesting the induction of an IL-36γ amplification loop following TLR activation. The increased secretion of CCL20 and its impact on Th17 cell recruitment and activation may play a role in host defense against FRT insult and is a future area of investigation.

AMP are present in the FRT during homeostasis and production is increased following infection in the FRT, as detected in cervicovaginal lavages (Novak et al., 2007; Keller et al., 2012; Madan et al., 2012; Pellett Madan et al., 2015). AMP induction by IL-36γ has been observed in keratinocytes and bronchial ECs (Chustz et al., 2011; Lian et al., 2012; Li et al., 2014). We investigated a panel of AMP known to be expressed in the lower FRT (Yarbrough et al., 2015) for induction by rIL-36γ exposure. Our data shows that each of these AMP (CCL20, HE4, SLPI, HBD-2, and HBD-4) were induced by treatment with rIL-36γ, further promoting inflammation in FRT EC in a dose-dependent fashion. CCL20, HBD-2, and HBD-4 upregulation was more robust in 3-D vaginal EC relative to cervical EC. Following pathological insult, serine protease inhibitors HE4 and SLPI have an inverse relationship, for example in bacterial vaginosis, HE4 is elevated, whereas SLPI is decreased (Mitchell et al., 2008). Interestingly, rIL-36γ treatment induced HE4 and SLPI expression in 3-D FRT models, similar to IL-1 treatment (Bingle et al., 2006). Further investigation is needed to determine if AMP induction by IL-36γ enhances host defense or results in damaging inflammation.

In the FRT, IL-36γ may function as a key regulator of mucosal inflammation. This is the first study to report the induction and regulation of IL-36γ, IL-36R, and related immune mediators in the human FRT following microbial product exposure. Both IL-36γ and IL-36R were induced by rIL-36γ in a dose dependent fashion, as well as a number of AMPs, pro-inflammatory cytokines, and chemokines in the FRT. The IL-36 family has been associated with the pathogenesis of several inflammatory diseases. Our data suggest that IL-36γ could play a critical role in initiating, promoting and sustaining epithelial-mediated inflammation to microbial products in the FRT. As such, future work will be aimed at studying the role of the IL-36 family in the context of the FRT in terms of host response to vaginal microbiota, STI pathogens, and other reproductive disorders.

AUTHOR CONTRIBUTIONS

MH-K contributed to the conception and design of the experiments. SW and AT acquired the data. SW, AT, and MH-K all contributed to the analysis and interpretation of the data, drafting the manuscript and revising it critically for intellectual content and approve of the final version of the manuscript.

FUNDING

This work was supported by the National Institutes of Health and NIAID Grant 1R15AI113457-01A1 (to MH-K).

ACKNOWLEDGMENT

The authors would like to acknowledge Dr. Pawel Laniewski for critical review of the manuscript.

REFERENCES

Agrawal, T., Vats, V., Wallace, P. K., Salhan, S., and Mittal, A. (2008). Role of cervical dendritic cell subsets, co-stimulatory molecules, cytokine secretion profile and beta-estradiol in development of sequalae to *Chlamydia trachomatis* infection. *Reprod. Biol. Endocrinol.* 6:46. doi: 10.1186/1477-7827-6-46

Aliprantis, A. O., Yang, R. B., Weiss, D. S., Godowski, P., and Zychlinsky, A. (2000). The apoptotic signaling pathway activated by Toll-like receptor-2. *EMBO J.* 19, 3325–3336. doi: 10.1093/emboj/19.13.3325

Barrila, J., Radtke, A. L., Crabbe, A., Sarker, S. F., Herbst-Kralovetz, M. M., Ott, C. M., et al. (2010). Organotypic 3D cell culture models: using the rotating wall vessel to study host-pathogen interactions. *Nat. Rev. Microbiol.* 8, 791–801. doi: 10.1038/nrmicro2423

Berglof, E., Andre, R., Renshaw, B. R., Allan, S. M., Lawrence, C. B., Rothwell, N. J., et al. (2003). IL-1Rrp2 expression and IL-1F9 (IL-1H1) actions in brain cells. *J. Neuroimmunol.* 139, 36–43. doi: 10.1016/S0165-5728(03)00130-9

Bingle, L., Cross, S. S., High, A. S., Wallace, W. A., Rassl, D., Yuan, G., et al. (2006). WFDC2 (HE4): a potential role in the innate immunity of the oral cavity and respiratory tract and the development of adenocarcinomas of the lung. *Respir. Res.* 7:61. doi: 10.1186/1465-9921-7-61

Borgdorff, H., Gautam, R., Armstrong, S. D., Xia, D., Ndayisaba, G. F., van Teijlingen, N. H., et al. (2015). Cervicovaginal microbiome dysbiosis is associated with proteome changes related to alterations of the cervicovaginal mucosal barrier. *Mucosal Immunol.* 9, 621–633. doi: 10.1038/mi.2015.86

Chustz, R. T., Nagarkar, D. R., Poposki, J. A., Favoreto, S. Jr., Avila, P. C., Schleimer, R. P., et al. (2011). Regulation and function of the IL-1 family cytokine IL-1F9 in human bronchial epithelial cells. *Am. J. Respir. Cell Mol. Biol.* 45, 145–153. doi: 10.1165/rcmb.2010-0075OC

Cuchacovich, R., Espinoza, C. G., Virk, Z., and Espinoza, L. R. (2008). Biologic therapy (TNF-alpha antagonists)-induced psoriasis: a cytokine imbalance between TNF-alpha and IFN-alpha? *J. Clin. Rheumatol.* 14, 353–356. doi: 10.1097/RHU.0b013e318190dd88

Doerflinger, S. Y., Throop, A. L., and Herbst-Kralovetz, M. M. (2014). Bacteria in the vaginal microbiome alter the innate immune response and barrier properties of the human vaginal epithelia in a species-specific manner. *J. Infect. Dis.* 209, 1989–1999. doi: 10.1093/infdis/jiu004

Drapkin, R., von Horsten, H. H., Lin, Y., Mok, S. C., Crum, C. P., Welch, W. R., et al. (2005). Human epididymis protein 4 (HE4) is a secreted glycoprotein that is overexpressed by serous and endometrioid ovarian carcinomas. *Cancer Res.* 65, 2162–2169. doi: 10.1158/0008-5472.CAN-04-3924

Dunn, E., Sims, J. E., Nicklin, M. J., and O'Neill, L. A. (2001). Annotating genes with potential roles in the immune system: six new members of the IL-1 family. *Trends Immunol.* 22, 533–536. doi: 10.1016/S1471-4906(01)02034-8

Friedrich, M., Tillack, C., Wollenberg, A., Schauber, J., and Brand, S. (2014). IL-36gamma sustains a proinflammatory self-amplifying loop with IL-17C in anti-TNF-induced psoriasiform skin lesions of patients with Crohn's disease. *Inflamm. Bowel Dis.* 20, 1891–1901. doi: 10.1097/MIB.0000000000000198

Gabay, C., and Towne, J. E. (2015). Regulation and function of interleukin-36 cytokines in homeostasis and pathological conditions. *J. Leukoc. Biol.* 97, 645–652. doi: 10.1189/jlb.3RI1014-495R

Gresnigt, M. S., Rosler, B., Jacobs, C. W., Becker, K. L., Joosten, L. A., van der Meer, J. W., et al. (2013). The IL-36 receptor pathway regulates *Aspergillus fumigatus*-induced Th1 and Th17 responses. *Eur. J. Immunol.* 43, 416–426. doi: 10.1002/eji.201242711

Gresnigt, M. S., and van de Veerdonk, F. L. (2013). Biology of IL-36 cytokines and their role in disease. *Semin. Immunol.* 25, 458–465. doi: 10.1016/j.smim.2013.11.003

Herbst-Kralovetz, M. M., and Pyles, R. B. (2006). Quantification of poly(I:C)-mediated protection against genital herpes simplex virus type 2 infection. *J. Virol.* 80, 9988–9997. doi: 10.1128/JVI.01099-06

Herbst-Kralovetz, M. M., Quayle, A. J., Ficarra, M., Greene, S., Rose, W. A. II, Chesson, R., et al. (2008). Quantification and comparison of toll-like receptor expression and responsiveness in primary and immortalized human female lower genital tract epithelia. *Am. J. Reprod. Immunol.* 59, 212–224. doi: 10.1111/j.1600-0897.2007.00566.x

Hjelm, B. E., Berta, A. N., Nickerson, C. A., Arntzen, C. J., and Herbst-Kralovetz, M. M. (2010). Development and characterization of a three-dimensional organotypic human vaginal epithelial cell model. *Biol. Reprod.* 82, 617–627. doi: 10.1095/biolreprod.109.080408

Horne, A. W., Stock, S. J., and King, A. E. (2008). Innate immunity and disorders of the female reproductive tract. *Reproduction* 135, 739–749. doi: 10.1530/REP-07-0564

Jarvis, G. A., Li, J., and Swanson, K. V. (1999). Invasion of human mucosal epithelial cells by *Neisseria gonorrhoeae* upregulates expression of intercellular adhesion molecule 1 (ICAM-1). *Infect. Immun.* 67, 1149–1156.

Jensen, L. E. (2010). Targeting the IL-1 family members in skin inflammation. *Curr. Opin. Investig. Drugs* 11, 1211–1220.

Johnston, A., Xing, X., Guzman, A. M., Riblett, M., Loyd, C. M., Ward, N. L., et al. (2011). IL-1F5, -F6, -F8, and -F9: a novel IL-1 family signaling system that is active in psoriasis and promotes keratinocyte antimicrobial peptide expression. *J. Immunol.* 186, 2613–2622. doi: 10.4049/jimmunol.1003162

Keller, M. J., Madan, R. P., Shust, G., Carpenter, C. A., Torres, N. M., Cho, S., et al. (2012). Changes in the soluble mucosal immune environment during genital herpes outbreaks. *J. Acquir. Immune Defic. Syndr.* 61, 194–202. doi: 10.1097/QAI.0b013e31826867ae

Krzyzowska, M., Baska, P., Orlowski, P., Zdanowski, R., Winnicka, A., Eriksson, K., et al. (2013). HSV-2 regulates monocyte inflammatory response via the Fas/FasL pathway. *PLoS ONE* 8:e70308. doi: 10.1371/journal.pone.0070308

Kumar, S., McDonnell, P. C., Lehr, R., Tierney, L., Tzimas, M. N., Griswold, D. E., et al. (2000). Identification and initial characterization of four novel members of the interleukin-1 family. *J. Biol. Chem.* 275, 10308–10314. doi: 10.1074/jbc.M000449200

Kyongo, J. K., Crucitti, T., Menten, J., Hardy, L., Cools, P., Michiels, J., et al. (2015). Cross-sectional analysis of selected genital tract immunological markers and molecular vaginal microbiota in sub-saharan African women, with relevance to HIV risk and prevention. *Clin. Vaccine Immunol.* 22, 526–538. doi: 10.1128/CVI.00762-14

Li, N., Yamasaki, K., Saito, R., Fukushi-Takahashi, S., Shimada-Omori, R., Asano, M., et al. (2014). Alarmin function of cathelicidin antimicrobial peptide LL37 through IL-36gamma induction in human epidermal keratinocytes. *J. Immunol.* 193, 5140–5148. doi: 10.4049/jimmunol.1302574

Lian, L. H., Milora, K. A., Manupipatpong, K. K., and Jensen, L. E. (2012). The double-stranded RNA analogue polyinosinic-polycytidylic acid induces keratinocyte pyroptosis and release of IL-36gamma. *J. Invest. Dermatol.* 132, 1346–1353. doi: 10.1038/jid.2011.482

Madan, R. P., Carpenter, C., Fiedler, T., Kalyoussef, S., McAndrew, T. C., Viswanathan, S., et al. (2012). Altered biomarkers of mucosal immunity and reduced vaginal Lactobacillus concentrations in sexually active female adolescents. *PLoS ONE* 7:e40415. doi: 10.1371/journal.pone.0040415

McGowin, C. L., Radtke, A. L., Abraham, K., Martin, D. H., and Herbst-Kralovetz, M. (2013). *Mycoplasma genitalium* infection activates cellular host defense and inflammation pathways in a 3-dimensional human endocervical epithelial cell model. *J. Infect. Dis.* 207, 1857–1868. doi: 10.1093/infdis/jit101

Medina-Contreras, O., Harusato, A., Nishio, H., Flannigan, K. L., Ngo, V., Leoni, G., et al. (2016). Cutting Edge: IL-36 receptor promotes resolution of intestinal damage. *J. Immunol.* 196, 34–38. doi: 10.4049/jimmunol.1501312

Mitchell, C. M., Balkus, J., Agnew, K. J., Cohn, S., Luque, A., Lawler, R., et al. (2008). Bacterial vaginosis, not HIV, is primarily responsible for increased vaginal concentrations of proinflammatory cytokines. *AIDS Res. Hum. Retroviruses* 24, 667–671. doi: 10.1089/aid.2008.0268

Mutamba, S., Allison, A., Mahida, Y., Barrow, P., and Foster, N. (2012). Expression of IL-1Rrp2 by human myelomonocytic cells is unique to DCs and facilitates DC maturation by IL-1F8 and IL-1F9. *Eur. J. Immunol.* 42, 607–617. doi: 10.1002/eji.201142035

Novak, R. M., Donoval, B. A., Graham, P. J., Boksa, L. A., Spear, G., Hershow, R. C., et al. (2007). Cervicovaginal levels of lactoferrin, secretory leukocyte protease inhibitor, and RANTES and the effects of coexisting vaginoses in human immunodeficiency virus (HIV)-seronegative women with a high risk of heterosexual acquisition of HIV infection. *Clin. Vaccine Immunol.* 14, 1102–1107. doi: 10.1128/CVI.00386-06

Pellett Madan, R., Masson, L., Tugetman, J., Werner, L., Grobler, A., Mlisana, K., et al. (2015). Innate antibacterial activity in female genital tract secretions is associated with increased risk of HIV acquisition. *AIDS Res. Hum. Retroviruses* 31, 1153–1159. doi: 10.1089/AID.2015.0011

Pudney, J., Quayle, A. J., and Anderson, D. J. (2005). Immunological microenvironments in the human vagina and cervix: mediators of cellular immunity are concentrated in the cervical transformation zone. *Biol. Reprod.* 73, 1253–1263. doi: 10.1095/biolreprod.105.043133

Quayle, A. J. (2002). The innate and early immune response to pathogen challenge in the female genital tract and the pivotal role of epithelial cells. *J. Reprod. Immunol.* 57, 61–79. doi: 10.1016/S0165-0378(02)00019-0

Radtke, A. L., and Herbst-Kralovetz, M. M. (2012). Culturing and applications of rotating wall vessel bioreactor derived 3D epithelial cell models. *J. Vis. Exp.* 62:3868. doi: 10.3791/3868

Radtke, A. L., Quayle, A. J., and Herbst-Kralovetz, M. M. (2012). Microbial products alter the expression of membrane-associated mucin and antimicrobial peptides in a three-dimensional human endocervical epithelial cell model. *Biol. Reprod.* 87:132. doi: 10.1095/biolreprod.112.103366

Schaefer, T. M., Fahey, J. V., Wright, J. A., and Wira, C. R. (2005). Innate immunity in the human female reproductive tract: antiviral response of uterine epithelial cells to the TLR3 agonist poly(I:C). *J. Immunol.* 174, 992–1002. doi: 10.4049/jimmunol.174.2.992

Segueni, N., Vigne, S., Palmer, G., Bourigault, M. L., Olleros, M. L., Vesin, D., et al. (2015). Limited contribution of IL-36 versus IL-1 and TNF pathways in host response to mycobacterial infection. *PLoS ONE* 10:e0126058. doi: 10.1371/journal.pone.0126058

Stordeur, P., Poulin, L. F., Craciun, L., Zhou, L., Schandene, L., de Lavareille, A., et al. (2002). Cytokine mRNA quantification by real-time PCR. *J. Immunol. Methods* 259, 55–64. doi: 10.1016/S0022-1759(01)00489-6

Towne, J. E., Garka, K. E., Renshaw, B. R., Virca, G. D., and Sims, J. E. (2004). Interleukin (IL)-1F6, IL-1F8, and IL-1F9 signal through IL-1Rrp2 and IL-1RAcP to activate the pathway leading to NF-kappaB and MAPKs. *J. Biol. Chem.* 279, 13677–13688. doi: 10.1074/jbc.M400117200

Towne, J. E., Renshaw, B. R., Douangpanya, J., Lipsky, B. P., Shen, M., Gabel, C. A., et al. (2011). Interleukin-36 (IL-36) ligands require processing for full agonist (IL-36alpha, IL-36beta, and IL-36gamma) or antagonist (IL-36Ra) activity. *J. Biol. Chem.* 286, 42594–42602. doi: 10.1074/jbc.M111.267922

Towne, J. E., and Sims, J. E. (2012). IL-36 in psoriasis. *Curr. Opin. Pharmacol.* 12, 486–490. doi: 10.1016/j.coph.2012.02.009

Trifonova, R. T., Bajpai, M., Pasicznyk, J. M., Chandra, N., Doncel, G. F., and Fichorova, R. N. (2007). Biomarkers of leukocyte traffic and activation in the vaginal mucosa. *Biomarkers* 12, 608–622. doi: 10.1080/135475007016 00670

Vigne, S., Palmer, G., Lamacchia, C., Martin, P., Talabot-Ayer, D., Rodriguez, E., et al. (2011). IL-36R ligands are potent regulators of dendritic and T cells. *Blood* 118, 5813–5823. doi: 10.1182/blood-2011-05-356873

Vos, J. B., van Sterkenburg, M. A., Rabe, K. F., Schalkwijk, J., Hiemstra, P. S., and Datson, N. A. (2005). Transcriptional response of bronchial epithelial cells to *Pseudomonas aeruginosa*: identification of early mediators of host defense. *Physiol. Genom.* 21, 324–336. doi: 10.1152/physiolgenomics. .00289.2004

Yarbrough, V. L., Winkle, S., and Herbst-Kralovetz, M. M. (2015). Antimicrobial peptides in the female reproductive tract: a critical component of the mucosal immune barrier with physiological and clinical implications. *Hum. Reprod. Update* 21, 353–377. doi: 10.1093/humupd/dmu065

Zeng, H., Wu, H., Sloane, V., Jones, R., Yu, Y., Lin, P., et al. (2006). Flagellin/TLR5 responses in epithelia reveal intertwined activation of inflammatory and apoptotic pathways. *Am. J. Physiol. Gastrointest. Liver Physiol.* 290, G96–G108. doi: 10.1152/ajpgi.00273.2005

Histoplasma capsulatum-Induced Cytokine Secretion in Lung Epithelial Cells is Dependent on Host Integrins, Src-Family Kinase Activation and Membrane Raft Recruitment

*Paloma K. Maza and Erika Suzuki**

Department of Microbiology, Immunology and Parasitology, Escola Paulista de Medicina, Universidade Federal de São Paulo, São Paulo, Brazil

Edited by:
Olivier Dussurget, University Paris Diderot, Institut Pasteur, INSERM, INRA, France

Reviewed by:
Birgit Strobl, University of Veterinary Medicine Vienna, Austria
Jessica Lynn Humann, Florida A&M University, USA

***Correspondence:**
Erika Suzuki
erika.suzuki@unifesp.br

Histoplasma capsulatum var. *capsulatum* is a dimorphic fungus that causes histoplasmosis, a human systemic mycosis with worldwide distribution. In the present work, we demonstrate that *H. capsulatum* yeasts are able to induce cytokine secretion by the human lung epithelial cell line A549 in integrin- and Src-family kinase (SFK)-dependent manners. This conclusion is supported by small interfering RNA (siRNA) directed to α3 and α5 integrins, and PP2, an inhibitor of SFK activation. siRNA and PP2 reduced IL-6 and IL-8 secretion in *H. capsulatum*-infected A549 cell cultures. In addition, α3 and α5 integrins from A549 cells were capable of associating with *H. capsulatum* yeasts, and this fungus promotes recruitment of these integrins and SFKs to A549 cell membrane rafts. Corroborating this finding, membrane raft disruption with the cholesterol-chelator methyl-β-cyclodextrin reduced the levels of integrins and SFKs in these cell membrane domains. Finally, pretreatment of A549 cells with the cholesterol-binding compound, and also a membrane raft disruptor, filipin, significantly reduced IL-6 and IL-8 levels in A549-*H.capsulatum* cultures. Taken together, these results indicate that *H. capsulatum* yeasts induce secretion of IL-6 and IL-8 in human lung epithelial cells by interacting with α3 and α5 integrins, recruiting these integrins to membrane rafts, and promoting SFK activation.

Keywords: *Histoplasma capsulatum*, fungi, epithelial cells, cytokine, integrin, membrane rafts, Src-family kinases

INTRODUCTION

Histoplasmosis is a human systemic mycosis caused by the fungal pathogen *Histoplasma capsulatum*. This fungus presents two varieties: *H. capsulatum* var. *capsulatum* and *H. capsulatum* var. *duboisii* that are etiological agents of the classical and the African histoplasmoses, respectively. Classical histoplasmosis is widely distributed in the Americas. In the United States, highly endemic areas include the Mississippi and Ohio River valleys. This mycosis also occurs in countries of Central and South America, and in Brazil, outbreaks of histoplasmosis have been reported after exposure to *H. capsulatum* fragments (Martins et al., 2003; Guimarães et al., 2006; Oliveira et al., 2006; Rocha-Silva et al., 2014).

Histoplasma capsulatum is a thermally dimorphic fungus, which is found in soil, caves, and abandoned constructions that are enriched in bat or bird excrements (Smith and Kauffman, 2012). Infection with *H. capsulatum* occurs by inhaling microconidia or mycelial fragments which then settle in the host's lungs and convert to yeast forms (Mihu and Nosanchuk, 2012). Some fungi are internalized by phagocytes and are able to survive and multiply within macrophages, allowing, in this manner, dissemination of *H. capsulatum* to several organs through the bloodstream or lymphatic system (Mihu and Nosanchuk, 2012; Adenis et al., 2014).

The severity of this mycosis depends on the number of inhaled fungal particles and the immune status of the host. In immunocompetent individuals, a small inoculum can cause asymptomatic infection or acute pulmonary histoplasmosis. Individuals with pre-existing lung diseases, such as emphysema, may develop chronic pulmonary histoplasmosis, and immunocompromised patients may present disseminated histoplasmosis (Smith and Kauffman, 2012). Histoplasmosis is responsible for low rates of morbidity and mortality among immunocompetent patients. However, among immunocompromised patients, morbidity, and mortality of this mycosis have increased mostly due to HIV (Adenis et al., 2014). Histoplasmosis is an AIDS-defining illness, and some authors consider HIV-associated histoplasmosis a neglected disease in South America. Unhappily, these cases are often confused with tuberculosis or pneumocystosis (Nacher et al., 2013). In Brazil, in the state of Ceará, a study of a 4-years period (2006–2010) reported 208 cases of histoplasmosis in HIV-positive patients. Histoplasmosis was the first indicator of AIDS in about 39% of the cases. About 80% of these patients were not being treated with highly active antiretroviral therapy (HAART) at the moment of histoplasmosis diagnosis, and about 42% of these patients died (Brilhante et al., 2012).

Besides acting as a structural barrier, several research groups have demonstrated the importance of epithelial cells in modulating the immune system in various body tissues. In the lungs, for example, type II pneumocytes are among the cells that form the alveolar epithelium, and they are involved not only in surfactant production and repair of alveoli after a lung injury, but also in the immune response against particles and inhaled microorganisms (Mason, 2006). To participate in the host's innate immunity, airway epithelial cells produce a wide range of inflammatory mediators, such as growth factors, cytokines, and chemokines, that promote recruitment and activation of immune cells to the sites of infection (Suzuki et al., 2008; Proud and Leigh, 2011).

Recently, our group has demonstrated that the human fungal pathogen *Paracoccidioides brasiliensis* induces interleukin (IL)-6 and IL-8 secretion by the human lung epithelial cell line A549. This cytokine secretion was dependent on activation of some host cell signaling kinases, such as ERK 1/2 (extracellular signal-regulated kinase 1/2), p38 MAPK (p38 mitogen-activated protein kinase) and PKC δ (protein kinase C δ; Maza et al., 2012; Alcantara et al., 2015). Later, we demonstrated that integrins are one type of receptor involved in the secretion of IL-6 and IL-8 (Barros et al., 2016). In addition, *P. brasiliensis* promoted an increase of integrin expression in these epithelial cells, and clustering of α3 and α5 integrins into host membrane rafts was also observed in the presence of this fungus (Barros et al., 2016).

Integrins are heterodimeric transmembrane glycoproteins consisting of α and β subunits. In mammals, 18 α and 8 β integrin subunits non-covalently dimerize to form 24 different receptors. Integrins, a major class of receptors involved in cell adhesion to other cells or to extracellular matrix, are able to bind to a wide variety of ligands, including adhesive proteins present on the surface of pathogens. Therefore, in this manner, several pathogens hijack host cell signaling to invade and survive in the host, leading to the establishment of an infection (Hauck et al., 2012).

Integrins may be recruited and clustered into membrane rafts (Leitinger and Hogg, 2002; Wang et al., 2013). These cell membrane structures are dynamic nanoscale domains enriched in sterols, sphingolipids, and specific proteins. Upon stimulation, membrane rafts coalesce to form larger platforms, compartmentalizing and activating cell signaling (Simons and Sampaio, 2011). Some pathogens, such as *Listeria monocytogenes*, *Toxoplasma gondii*, and herpes simplex virus, recruit membrane rafts for host cell invasion (Seveau et al., 2004; Gianni et al., 2010; Cruz et al., 2014). Previously, we verified that *P. brasiliensis* is able to recruit these cell domains for host cell adhesion, and also, cytokine secretion (Maza et al., 2008; Barros et al., 2016).

Engagement of integrins may activate Src-family kinases (SFK), which are non-receptor tyrosine kinases that participate in regulating several cellular events, such as cell growth, division, differentiation and survival (Engen et al., 2008; Ingley, 2008; Okada, 2012). In response to pathogens or other stimuli, some studies have also shown SFK involvement in cytokine secretion by epithelial cells (Ren et al., 2005; Kannan et al., 2006; Lin et al., 2006; Bentley et al., 2007; Eucker et al., 2014). For example, Eucker et al. (2014) recently demonstrated that *Campylobacter jejuni* promotes IL-8 secretion by the INT 407 human intestinal epithelial cells, and that this is triggered in response to engagement of β1 integrins and activation of focal adhesion kinase (FAK), paxillin and Src. Our group previously observed that *P. brasiliensis* also induces SFK activation and recruitment of these signaling molecules to A549 cell membrane rafts (Maza et al., 2008), but the participation of integrins in SFK activation and the role of these kinases in *P. brasiliensis* epithelial cell infection were not determined yet.

Regarding host innate immunity and infection mechanisms by human pathogenic fungi, such as *P. brasiliensis* or *H. capsulatum*, most of the studies were performed with cells of the myeloid lineage, which include macrophages, neutrophils and dendritic cells. However, since the first reports describing the secretion of cytokines by epithelial cells in the 1990s (Stadnyk, 1994), various research groups have demonstrated the importance of these cells in modulating the host immune system. Cleaver et al. (2014), for example, recently showed that mice that inhaled Toll-like receptor (TLR) agonists were protected against lethal pneumonia. This protection persisted even after the reduction or depletion of neutrophils, alveolar macrophages, dendritic cells, mature lymphocytes, or natural killer cells. Moreover, airway epithelial cells treated with TLR agonists were able to kill pathogenic bacteria. Thus, the authors concluded that lung epithelial cells are important for pulmonary antimicrobial defense, and for patients with impaired leukocyte-mediated immunity, epithelial cells may be critical for the antimicrobial action (Cleaver et al., 2014).

As studies about interaction of epithelial cells with *H. capsulatum* are still incipient, in the present study, we aimed to analyze the secretion of inflammatory cytokines by A549 epithelial cells during infection with *H. capsulatum* yeasts. To the best of our knowledge, this is the first report describing the role of α3 and α5 integrins, SFK activation, and membrane rafts in *H. capsulatum*-inducible cytokine secretion.

MATERIALS AND METHODS

Fungal Growth Conditions

Histoplasma capsulatum, strain 496, was kindly provided by Dr. Olga F. Gompertz, Escola Paulista de Medicina, Universidade Federal de São Paulo, São Paulo, Brazil. Yeast forms were grown 5–7 days at 37°C, 100 rpm, in BHI medium (Brain Heart Infusion, Becton, Dickinson and Company, USA) as described elsewhere (Toledo et al., 2001).

Histoplasma capsulatum yeasts, grown for 3 days, were washed three times with Dulbecco's Modified Eagle's Medium (DMEM; Sigma, USA) and used for interaction assays with A549 cells.

A549 Cell Culture

Human lung epithelial cell line A549 was grown in DMEM supplemented with 10% fetal bovine serum (FBS; Vitrocell Embriolife, Brazil), 10 mM HEPES, 100 U/ml penicillin, and 100 μg/ml streptomycin (complete DMEM) at 37°C, 5% CO_2.

Analysis of Cytokine Levels in Culture Supernatants of A549 Cells during Incubation with *H. capsulatum*

1.8×10^4 A549 cells were cultured in 24-well plates with complete DMEM. After 72 h, A549 cells were maintained overnight in FBS-free DMEM (FBS starvation). Next, A549 cells were incubated with 2.0×10^6 *H. capsulatum* yeasts [multiplicity of infection (MOI) of 8 yeasts per A549 cell] for 5, 16, or 24 h. After incubation with *H. capsulatum*, culture supernatants were collected and centrifuged at 1300 × *g* to remove fungi. IL-6, IL-8, and IL-10 levels in these supernatants were determined using DuoSet® ELISA Kits (R&D Systems), according to manufacturer's instructions.

In some experiments, after FBS starvation, A549 cells were incubated for 2 h in FBS-free DMEM containing 0.1, 1, or 10 μM PP2 (an inhibitor of SFK activation, Calbiochem, USA), 1 μg/ml filipin (a cholesterol-binding compound that disrupts membrane rafts, Sigma, USA), or 0.025% or 0.05% DMSO (DMSO concentrations used as vehicle for PP2 and filipin, respectively). And then, *H. capsulatum* yeasts were added to the cultures and incubated for 16 h. IL-6 and IL-8 levels in these culture supernatants were determined as described above.

Analysis of Integrin Expression in A549 Cells during Incubation with *H. capsulatum*

1.0×10^5 A549 cells were cultured in 6-well plates with complete DMEM. After 48 h, A549 cells were incubated overnight in FBS-free DMEM. Next, A549 cells were incubated with 6.5×10^6 *H. capsulatum* yeasts (MOI of 8 yeasts per A549 cell) for different periods of time. After incubation with fungi, A549 cells were washed three times with phosphate buffered saline (PBS, 10 mM sodium phosphate buffer, pH 7.2, containing 150 mM NaCl), harvested with a cell scraper, and lysed with TNE (25 mM Tris buffer, pH 7.5, with 150 mM NaCl, 5 mM EDTA pH 7.5) containing 1% Brij® 98 (Sigma, USA) and a mixture of inhibitors (IMix – 5 mM Na_3VO_4, 100 μM leupeptin, 125 μg/ml aprotinin, 1 mM AEBSF, all inhibitors were purchased from Sigma, USA) for 30 min at 4°C. Protein content in samples was measured using a Micro BCA™ Protein Assay Kit (Thermo Scientific, USA) according to manufacturer's instructions.

Ten micrograms of protein were loaded per well of SDS-PAGE gels, and expression of integrins was evaluated by Western blot.

Association of A549 Cell Integrins with *H. capsulatum* Yeasts

1.12×10^6 A549 cells were cultured in 150-mm plates with complete DMEM. After 72 h, A549 cells were maintained overnight in FBS-free DMEM. A549 cells were washed three times with PBS, harvested with cell scraper, and lysed with lysis buffer (50 mM Tris buffer, pH 7.2, containing 150 mM NaCl, 1 mM $CaCl_2$, 1% Brij® 98 and IMix) for 30 min at 4°C.

After protein quantification of A549 cell lysates, 1 mg protein in 500 μl was incubated with 5.0×10^8 *H. capsulatum* yeasts at 4°C with gentle shaking. As controls, fungi were incubated with A549 cell-free lysis buffer. After overnight incubation, samples were centrifuged, and supernatants containing *H. capsulatum*-unassociated proteins were collected. Fungi (pellet) were washed five times with 500 μl of lysis buffer. After each washing step, the supernatant containing *H. capsulatum*-unassociated proteins was collected ($U_{Fractions}$). Next, *H. capsulatum*-associated proteins ($A_{Fraction}$) were eluted with 60 μl of sample buffer (125 mM Tris-HCl, pH 6.8; 4% sodium dodecyl sulfate; 20% glycerol; 0.05% bromophenol blue), boiled for 5 min, and centrifuged. Then, aliquots of A549 cell lysates, *H. capsulatum*-associated ($A_{Fraction}$) and unassociated ($U_{Fractions}$) proteins, and A549 cell-free lysis buffer, that was incubated with this fungus, were submitted to SDS-PAGE. Integrins and caspase-3 were analyzed by Western blot using antibodies anti-α3 and α5 integrins, and caspase-3.

Silencing of Integrins in A549 Cells by Small Interfering RNA (siRNA)

2.0×10^5 A549 cells were cultured in 6-well plates with DMEM supplemented with 10% FBS and 10 mM HEPES in the absence of antibiotics. After 24 h, A549 cells were washed three times with DMEM supplemented with 1% FBS, and then, transfected with Lipofectamine® RNAiMAX (Life Technologies, USA) and Silencer® Select Pre-designed α3 or α5 integrin siRNA (#s7543 and #s7549, Life Technologies, USA) at a final concentration of 10 nM. Silencer® Select Negative Control No. 1 siRNA (Life Technologies, USA) was used as negative control. After 24 h, A549 cells were washed three times with DMEM, and then, incubated with 6.5×10^6 *H. capsulatum* yeasts (MOI of 8 yeasts per A549 cell). After 16 h, culture supernatants were collected for IL-6 and IL-8 analysis. Concomitantly, A549 cells were washed three times with PBS, harvested with a cell scraper, and lysed with TNE containing 1% Brij® 98 and IMix for 30 min at 4°C. Silencing of α3 and α5 integrins was analyzed by Western blot.

Analysis of SFK Activation during the Interaction of A549 Cells with *H. capsulatum*

1.0×10^5 A549 cells were cultured in 6-well plates with complete DMEM. After 48 h, A549 cells were maintained overnight in FBS-free DMEM to decrease basal phosphorylation of SFK. Next, A549 cells were incubated with 6.5×10^6 *H. capsulatum* yeasts (MOI of 8 yeasts per A549 cell) for 15, 30, 60, 120, or 180 min. After incubation with fungi, A549 cells were washed three times with PBS containing 1 mM Na_3VO_4 (PBS/SV), harvested with a cell scraper, and lysed with TNE containing 1% Brij® 98 and IMix for 30 min at 4°C.

In some experiments, A549 cells were transfected with α3 or α5 integrin siRNA, maintained overnight in FBS-free DMEM, and then, incubated with *H. capsulatum* yeasts for 3 h. After incubation with fungi, A549 cells were washed, harvested, and lysed as described above.

Twenty micrograms of protein were loaded per well of SDS-PAGE gels, and SFK activation was evaluated by Western blot.

Detergent-Resistant Membrane Isolation

1.12×10^6 A549 cells were cultured in 150-mm plates with complete DMEM. After 72 h, A549 cells were maintained overnight in FBS-free DMEM. Next, A549 cells were incubated with 1.5×10^8 *H. capsulatum* yeasts (MOI of 8 yeasts per A549 cell). After 3 h, A549 cells were washed three times with PBS containing 1 mM Na_3VO_4 (PBS/SV), and detergent-resistant membranes (DRMs) were isolated as previously described (Maza et al., 2008). Briefly, harvested cells were lysed with TNE containing 1% Brij® 98 and IMix for 30 min at 4°C. Next, after Dounce-homogenization, the lysate was centrifuged at $1300 \times g$, for 7 min, and the supernatant obtained, termed post-nuclear fraction, was subjected to protein quantification. Same amount of protein was submitted to sucrose gradient, and centrifuged at $260{,}800 \times g$ at 4°C, using the SW 41 Ti rotor Beckman Coulter. After 16 h, 12 fractions of 1 ml each were collected and numbered from top to bottom. Aliquots of DRM fractions (fractions 4–6) and non-DRM fractions (fractions 10–12) were submitted to SDS-PAGE and analyzed by Western blot.

In some experiments, the membrane cholesterol of the cell homogenate was removed as described elsewhere (Maza et al., 2008). Briefly, A549 cells were incubated with *H. capsulatum* yeasts for 3 h then washed, harvested, and centrifuged, producing a cell pellet that was subsequently incubated with 10 mM methyl-β-cyclodextrin (MβCD) in TNE containing IMix for 30 min at 37°C with gentle shaking. Control experiments were performed in the absence of MβCD. Then, Brij® 98 was added to a final concentration of 1%, and after 30 min at 4°C, DRMs were isolated as described above. Aliquots of DRM fractions (fractions 4–6) were submitted to SDS-PAGE, and analyzed by Western blot.

SDS-PAGE and Western Blot

Aliquots, each containing the same amount of protein, were loaded onto 10% SDS-PAGE gels and then transferred to PVDF membranes. Next, membranes were blocked with 5% non-fat dry milk (Cell Signaling, USA) in TBST (200 mM Tris buffer, pH 8.0, containing 150 mM NaCl and 0.1% Tween® 20) at room temperature. After 1 h, membranes were incubated with 1% BSA in TBST containing the primary antibodies: mouse anti-α3 integrin (1:1000, sc-374242, Santa Cruz, CA, USA), rabbit anti-α5 integrin (1:1000, #4705, Cell Signaling, USA), mouse anti-glyceraldehyde 3-phosphate dehydrogenase (GAPDH; 1:10000, Cat.No.39-8600, Invitrogen, USA), rabbit anti-caspase-3 (1:1000, #9665, Cell Signaling, USA), rabbit anti-Phospho (P)-SFK (Tyr^{416}; 1:1000, #2101, Cell Signaling, USA), rabbit anti-cSrc/SFK (1:1000, sc-18, Santa Cruz, CA, USA), or rabbit anti-caveolin-1 (1:2500, sc-894, Santa Cruz, CA, USA). After overnight incubation at 4°C, membranes were incubated for 1 h at room temperature with HRP-conjugated anti-rabbit (1:2000, #7074, Cell Signaling, USA), or anti-mouse antibodies (1:2000, A-10668, Invitrogen, USA) diluted in 1% BSA in TBST. After each step, membranes were washed three times with TBST.

Reactive proteins were detected using SuperSignal™ West Pico Chemiluminescent Substrate (Thermo Scientific, USA) and were documented with a Uvitec Cambridge System (UVITEC, UK). In some experiments, after immunoblotting with anti-P-SFK, membranes were stripped using Restore Western Blot Stripping Buffer (Thermo Scientific, USA), and reprobed with anti-cSrc/SFK (1:1000, sc-18, Santa Cruz, CA, USA). For protein quantification, densitometric analyses were performed using Scion Image (Scion Corporation, USA).

Cell Viability and Statistical Significance

The viabilities of A549 cell and *H. capsulatum* were measured by MTT (3-[4,5-dimethylthiazol-2-yl]-2,5-diphenyltetrazolium bromide) assay as described previously (Maza et al., 2008).

Briefly, to determine A549 cell viability, cells were incubated in the presence or absence of 0.1, 1, or 10 μM PP2, 1 μg/ml filipin, or 0.025 or 0.05% DMSO for 2 h and then with *H. capsulatum* yeasts for 16 h. After incubation with fungi, A549 cells were washed three times with DMEM without phenol red and incubated using the same medium containing 0.5 mg/ml MTT (Life Technologies, USA) for 2 h. The medium was removed, formazan was solubilized with DMSO, and absorbance was determined at 540 nm with a microplate reader.

To determine fungal viability, after incubation with PP2, filipin or DMSO for 16 h, *H. capsulatum* yeasts were washed three times with DMEM without phenol red and incubated with 0.5 mg/ml MTT as described above.

Statistical significance was evaluated using Student's *t*-test. $p < 0.01$ or $p < 0.05$ was considered significant.

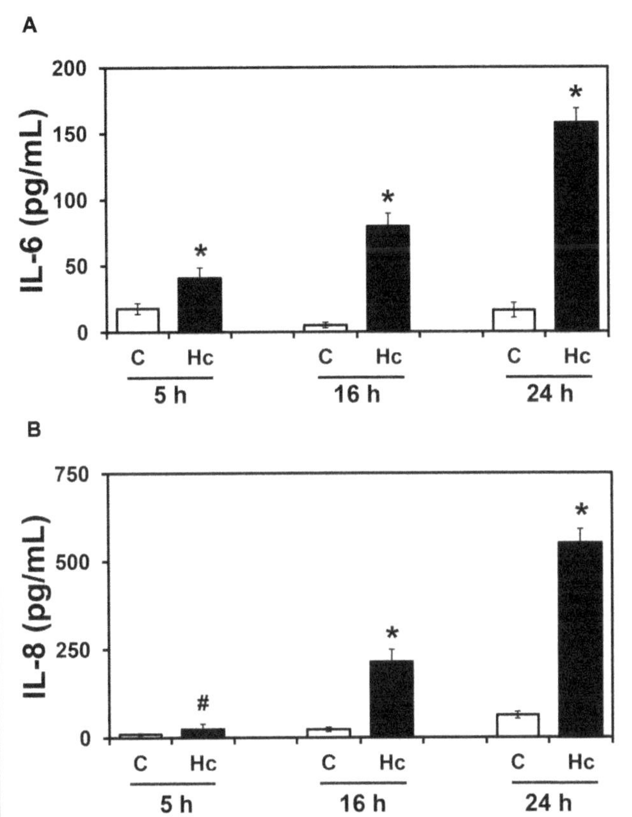

FIGURE 1 | IL-6 (A) and IL-8 (B) levels in culture supernatants of A549 cells during interaction with *H. capsulatum*. A549 cells were incubated in the absence (C) or presence (Hc) of *H. capsulatum* yeasts for 5, 16 or 24 h. Culture supernatants were collected, and IL-6 and IL-8 levels were determined by ELISA. Values represent the mean of triplicate experiments ± the standard deviation. *$p < 0.01$ and #$p < 0.05$ when compared to C of the related time period. Similar results were obtained from three independent experiments.

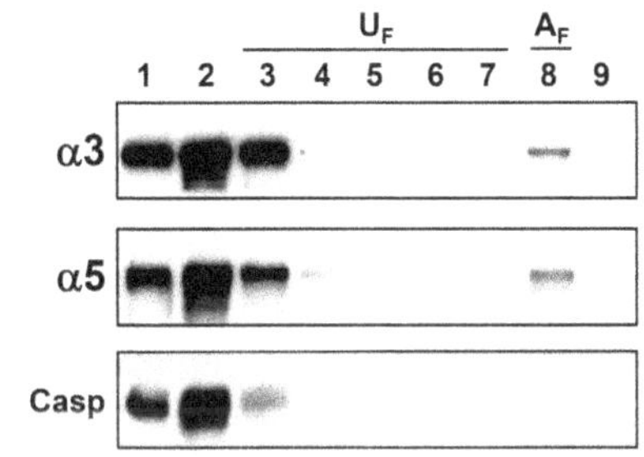

FIGURE 2 | Association of α3 and α5 integrins from A549 cells with *H. capsulatum* yeasts. A549 cell lysate was incubated with *H. capsulatum* yeasts overnight at 4°C. Then, samples were centrifuged, and supernatants were collected. Fungi were washed five times to remove unassociated proteins. Next, *H. capsulatum*-associated proteins were eluted with sample buffer. Some fungi were also incubated with A549 cell-free lysis buffer. Lane 1, A549 cell lysate. Lane 2, *H. capsulatum*-unassociated proteins. Lanes 3–7, washing steps (1–5) containing *H. capsulatum*-unassociated proteins (U_F). Lane 8, associated A549 cell proteins with *H. capsulatum* (A_F). Lane 9, *H. capsulatum* yeasts incubated with A549 cell-free lysis buffer. Proteins were analyzed by Western blot using antibodies anti-α3 and -α5 integrins. Caspase-3 (Casp) was used as a negative control of association. Blots are representative of two independent experiments.

RESULTS

Cytokine Secretion by A549 Cells during Interaction with *H. capsulatum*

To verify whether *H. capsulatum* yeasts induce secretion of inflammatory cytokines by epithelial cells, the human lung epithelial cell line A549 was incubated with *H. capsulatum* yeasts for different periods of time (5-24 h), and levels of IL-6, IL-8, and IL-10 in these culture supernatants were determined by ELISA. IL-6 and IL-8 levels significantly increased in a time-dependent manner (**Figure 1**). Regarding IL-6, after 5, 16 and 24 h of A549-*H. capsulatum* interaction, this cytokine levels increased 2.3-, 15.9- and 9.8-fold over basal levels, respectively. Under the same conditions, IL-8 levels increased 2.4-, 9.3- and 8.8-fold for 5, 16 and 24 h time periods, respectively. On the other hand, the anti-inflammatory cytokine IL-10 was undetectable in these culture supernatants (data not shown).

By MTT assay, we verified that cell viability was unaffected when A549 cells were incubated with *H. capsulatum* yeasts for 5–24 h (Supplementary Table 1). Despite this result, as major differences of cytokine levels were observed at 16 h, following experiments were performed within this period.

Association of α3 and α5 Integrins from Epithelial Cells with *H. capsulatum* Yeasts

First, to determine whether *H. capsulatum* is able to modulate the expression of α3 and α5 integrins, A549 cells were incubated with this fungus for different periods of time (0.5–16 h). By Western blot, it was observed that *H. capsulatum* yeasts were not able to increase the expression levels of these integrins (Supplementary Figure 1).

Next, to evaluate whether α3 or α5 integrin from epithelial cells interacts with *H. capsulatum*, A549 cell lysate was incubated with yeast forms of this fungus. Associated A549 cell proteins with *H. capsulatum* ($A_{Fraction}$ - A_F) and unassociated proteins from washing steps ($U_{Fractions}$ - U_F) were analyzed by Western blot. α3 and α5 integrins from A549 cells, associated with yeasts, were eluted and recovered in $A_{Fraction}$ (**Figure 2**, lane 8). As expected, caspase-3 (a cytoplasmatic protein of A549 cell) was not recovered in $A_{Fraction}$ (**Figure 2**, lane 8), indicating that there is no interaction between this cytoplasmatic protein and *H. capsulatum*. In addition, we verified that unassociated proteins were removed efficiently by washing steps (**Figure 2**, lanes 3–7, $U_{Fractions}$). To demonstrate that detected proteins by Western blot were derived from A549 cells, and not from *H. capsulatum*, fungi were also incubated with A549 cell-free lysis buffer. Incubation with this lysis buffer did not extract *H. capsulatum* proteins that could be recognized by the antibodies anti-α3 and anti-α5 integrins (**Figure 2**, lane 9). Therefore, these results indicate that α3 and α5 integrins from A549 cells were associated with *H. capsulatum* yeasts.

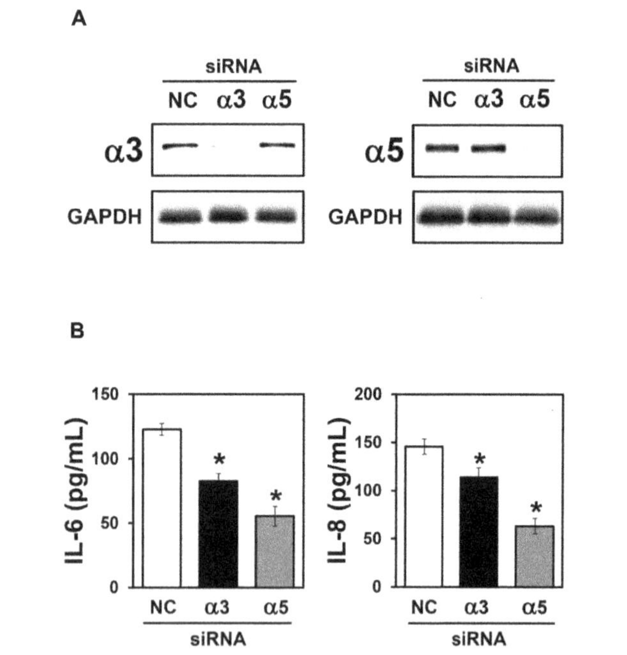

FIGURE 3 | Effect of α3 or α5 integrin silencing on IL-6 and IL-8 secretion by A549 cells during interaction with *H. capsulatum*. A549 cells were transfected with α3 or α5 integrin-directed siRNA or with Negative Control (NC) siRNA for 24 h, and then, incubated with *H. capsulatum* yeasts for 16 h. After incubation with fungi, culture supernatants were collected for determination of IL-6 and IL-8 levels, and A549 cells were harvested, lysed, and analyzed by Western blot. **(A)** Silencing of α3 and α5 integrins was confirmed by Western blot. GAPDH was used as protein loading control. Blots are representative of three independent experiments. **(B)** IL-6 and IL-8 levels in culture supernatants were determined by ELISA. Values represent the mean of triplicate experiments ± the standard deviation. $^{*}p < 0.01$ when compared to NC siRNA. Similar results were obtained from three independent experiments.

Involvement of α3 and α5 Integrins in Cytokine Secretion by A549 Cells during Interaction with *H. capsulatum*

Integrins are able to modulate cytokine secretion in various cell types, including epithelial cells (Lubin et al., 2003; Schmid et al., 2004; Gianni and Campadelli-Fiume, 2014). Therefore, to determine the importance of α3 and α5 integrins on IL-6 and IL-8 secretion by A549 cells, during interaction with *H. capsulatum*, silencing of these integrins was performed by using small interfering RNA (siRNA). First, by Western blot (**Figure 3A**) and densitometric analysis (data not shown), we verified that α3 and α5 integrin-directed siRNAs reduced the expression of these receptors by 89 and 87%, respectively, when compared to A549 cells transfected with negative control siRNA.

Next, after incubation with *H. capsulatum* yeasts, we analyzed IL-6 and IL-8 levels in culture supernatants of A549 cells transfected with negative control, α3 or α5 integrin-directed siRNA. By ELISA, it was verified that α3 integrin-directed siRNA was able to reduce significantly 32.7% of IL-6 and 21.9% of IL-8 levels when compared to A549 cells transfected with negative control siRNA (**Figure 3B**). Reduction of IL-6 and IL-8 levels was even more pronounced when α5 integrin-directed siRNA was used (55.0% for IL-6 and 56.7% for IL-8; **Figure 3B**). It was also verified that, in cultures of A549 cells transfected with both integrins (α3 and α5)-directed siRNAs, the decrease of IL-6 and IL-8 levels was the same as observed for A549 cells transfected only with α5-directed integrin siRNA (data not shown). Taken together, these results indicate that α3 and α5 integrins are involved in IL-6 and IL-8 secretion during A549-*H. capsulatum* interaction.

Activation of Src-Family Kinases (SFK) in A549 Cells during Interaction with *H. capsulatum*

Some pathogens exploit integrins for host cell adhesion and invasion, triggering activation of several signaling molecules, including downstream tyrosine kinases such as SFKs (Scibelli et al., 2007; Ulanova et al., 2009; Hauck et al., 2012). Therefore, in order to verify whether A549-*H. capsulatum* interaction induces SFK activation, A549 cells were first incubated with *H. capsulatum* yeasts for different periods of time (15–180 min), and then, levels of SFK phosphorylated at Tyr^{416} (P-SFK) were analyzed by Western blot. As shown in **Figure 4A**, *H. capsulatum* induced SFK activation as early as 15 min, increasing up to 9.1-fold over basal levels after 3 h of A549 cell-*H. capsulatum* interaction.

Involvement of SFK Activation in Cytokine Secretion by A549 Cells during Interaction with *H. capsulatum*

PP2 (an inhibitor of SFK activation) was used to determine the role of SFK activation in cytokine secretion during A549-*H. capsulatum* interaction. By ELISA, cytokine levels were evaluated in A549 cell-*H. capsulatum* cultures, it was verified that PP2 decreased in a dose-dependent manner IL-6 and IL-8 levels up to 60.3 and 55.1%, respectively (**Figure 4B**). The IC_{50} values of PP2 for IL-6 and IL-8 levels were 3.9 and 7.6 μM, respectively. Taken together, these results indicate that *H. capsulatum* promotes IL-6 and IL-8 secretion in A549 cells in an SFK activation-dependent manner.

A549 cell and fungal viabilities were verified by MTT assay in these experiments. More than 97.0% of A549 cells were viable in the presence of *H. capsulatum* yeasts and PP2, and no morphological changes were observed (Supplementary Table 2). *H. capsulatum* yeasts were viable in the presence of different concentrations of PP2 (Supplementary Table 3).

Role of α3 and α5 Integrins in SFK Activation in A549 Cells during Interaction with *H. capsulatum*

A549 cells were transfected with α3 or α5 integrin-directed siRNA, and then, incubated with *H. capsulatum* in order to evaluate the importance of these cell receptors on SFK activation. By Western blot and densitometric analyses, it was verified that α3 integrin-directed siRNA reduced SFK activation by 57% up to 64% when compared to A549 cells transfected with negative

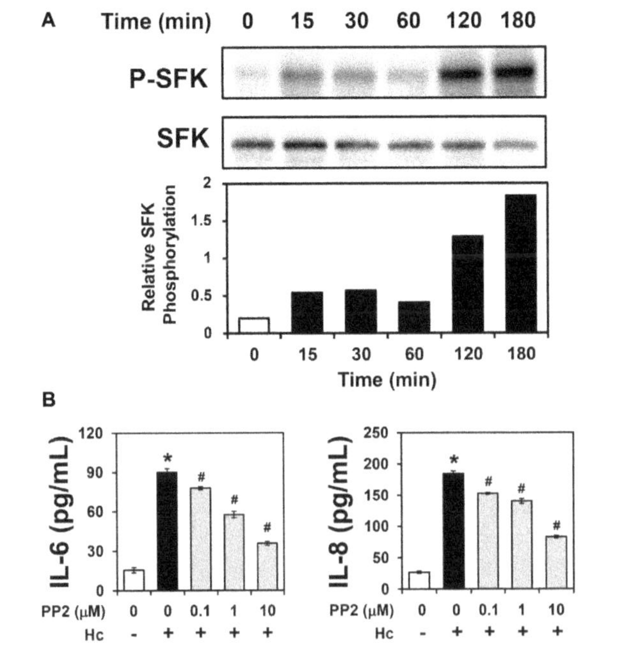

FIGURE 4 | Activation of SFKs in A549 cells during interaction with *H. capsulatum*, and effect of PP2 on IL-6 and IL-8 secretion by A549 cells during interaction with *H. capsulatum*. (A) A549 cells were incubated with *H. capsulatum* yeasts for 15, 30, 60, 120 or 180 min, and then P-SFK (Tyr^{416}) was analyzed by Western blot. SFK was used as protein loading control. Relative SFK phosphorylation was determined by densitometric analysis of bands obtained by Western blot, and values represent the ratio of the intensity of P-SFK band divided by the corresponding intensity of SFK band. Blots are representative of three independent experiments. **(B)** A549 cells were incubated for 2 h in the absence or presence of 0.1, 1, or 10 μM PP2 (an inhibitor of SFK activation), and then, with *H. capsulatum* yeasts (Hc) for 16 h. After incubation with fungi, culture supernatants were collected, and IL-6 and IL-8 levels were determined by ELISA. To analyze basal cytokine levels, A549 cells were incubated in the absence of PP2 and *H. capsulatum* (PP2 0/Hc −). Values represent the mean of triplicate experiments ± the standard deviation. *$p < 0.01$ when compared to A549 cells incubated in the absence of PP2 and *H. capsulatum* (PP2 0/Hc −). #$p < 0.01$ when compared to A549 cells incubated with *H. capsulatum* in the absence of PP2 (PP2 0/Hc +). Similar results were obtained from three independent experiments.

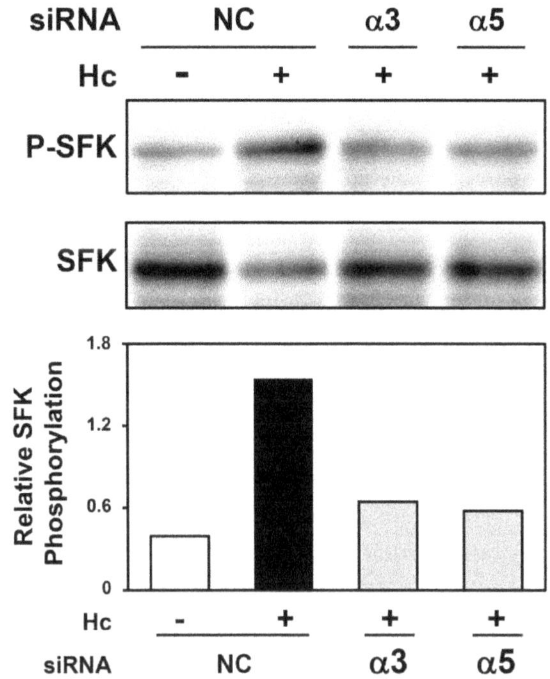

FIGURE 5 | Effect of α3 or α5 integrin silencing on SFK activation in A549 cells during interaction with *H. capsulatum*. A549 cells were transfected with α3 or α5 integrin-directed siRNA or with Negative Control (NC) siRNA for 24 h, and then, incubated with *H. capsulatum* yeasts for 3 h. After incubation with fungi, A549 cells were harvested, lysed, and then P-SFK (Tyr^{416}) was analyzed by Western blot. SFK was used as protein loading control. Relative SFK phosphorylation was determined by densitometric analysis of the bands obtained by Western blot, and values represent the ratio of the intensity of P-SFK band divided by the corresponding intensity of SFK band.

control siRNA, and incubated with fungi. In addition, SFK activation was decreased by 42% up to 62% when A549 cells were transfected with α5 integrin-directed siRNA (**Figure 5** and Supplementary Figure 2). These results indicate that the interaction between *H. capsulatum* and α3 and α5 integrins is important for SFK activation in A549 epithelial cells. Silencing of these integrins under the same culture conditions was confirmed by Western blot (data not shown).

Localization of α3 and α5 Integrins and SFKs in Epithelial Cell Membranes Rafts during A549 Cell-*H. capsulatum* Interaction

As previous reports have demonstrated that integrins and/or SFKs may be recruited to membrane rafts under a particular stimulation (Leitinger and Hogg, 2002; Maza et al., 2008; Wang et al., 2013), we investigated whether *H. capsulatum* promotes the recruitment of these proteins to A549 cell membrane rafts. To study this event, DRMs, which contain membrane rafts, were isolated by sucrose gradient/ultracentrifugation method.

After A549-*H. capsulatum* interaction, it was verified the dislodgment of α3 and α5 integrins from non-DRMs (fractions 10–12) to DRMs (fractions 4–6; **Figure 6**). Regarding SFK activation, incubation of epithelial cells with this fungus promoted an increase of P-SFK (Tyr^{416}) levels into DRMs (fractions 4–6; **Figure 6**). As expected, caveolin-1, a marker for DRM isolation efficiency, was observed mostly in A549 cell DRMs (fractions 4–6) of both conditions, i.e., A549 cells incubated in the absence (Control) or presence of *H. capsulatum* (**Figure 6**).

Membrane rafts are enriched in cholesterol, therefore cholesterol dependence in the recruitment of integrins and SFK to DRMs was analyzed. For this, cholesterol of A549 cell homogenate was depleted with methyl-β-cyclodextrin (MβCD). Treatment with MβCD reduced the levels of α3 and α5 integrins, P-SFK and SFK in DRMs (fractions 4–6; **Figure 7**). Taken together, these results indicate that *H. capsulatum*

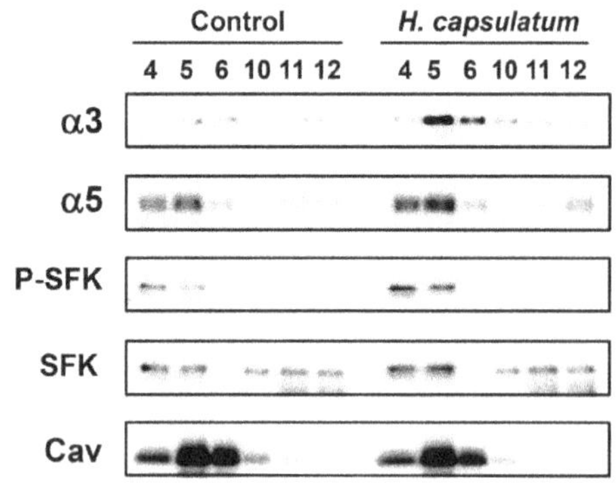

FIGURE 6 | Localization of SFKs, α3 and α5 integrins in A549 cell DRMs during interaction with *H. capsulatum*. A549 cells were incubated in the absence (Control) or presence of *H. capsulatum* yeasts. After 3 h, A549 cell DRMs were isolated by using a sucrose gradient and ultracentrifugation. Aliquots of DRM fractions (4–6) and non-DRM fractions (10–12) were submitted to SDS-PAGE. α3 and α5 integrins, P-SFK (Tyr^{416}), SFK and caveolin-1 were analyzed by Western blot. Caveolin-1 (Cav) was used as a marker of DRM isolation efficiency. Blots are representative of three independent experiments.

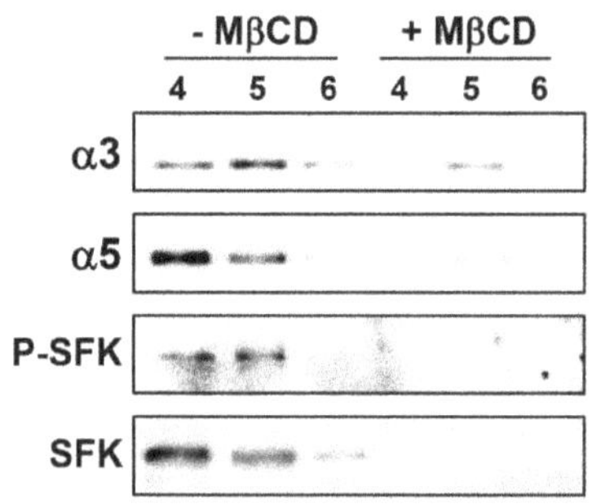

FIGURE 7 | Effect of membrane raft disruption on the localization of SFKs, α3 and α5 integrins in A549 cell DRMs during interaction with *H. capsulatum*. After incubation with *H. capsulatum* yeasts for 3 h, A549 cells were harvested, lysed and membrane cholesterol of cell homogenate was removed by methyl-β-cyclodextrin (+ MβCD). Control in the absence of MβCD (− MβCD) was also performed. Then, A549 cell DRMs were isolated by using a sucrose gradient and ultracentrifugation. Aliquots of DRM fractions (4–6) were submitted to SDS-PAGE. α3 and α5 integrins, P-SFK (Tyr^{416}) and SFK were analyzed by Western blot. Blots are representative of two independent experiments.

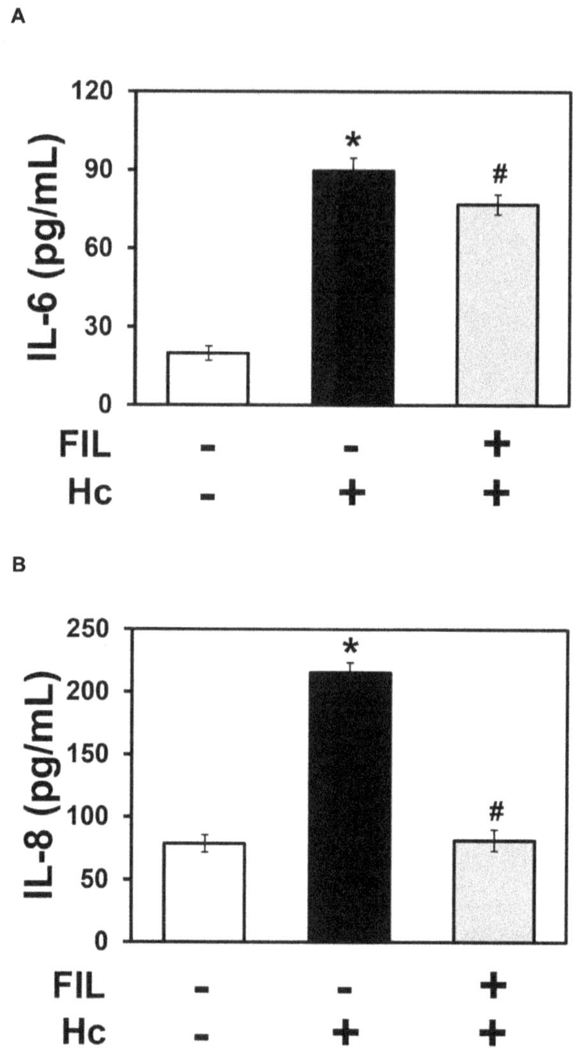

FIGURE 8 | Effect of filipin on IL-6 and IL-8 secretion by A549 cells during interaction with *H. capsulatum*. A549 cells were incubated in the absence or presence of 1 μg/ml filipin (FIL, a cholesterol-binding compound that disrupts membrane rafts). After 2 h, *H. capsulatum* yeasts (Hc) were added to these cultures. After 16 h, culture supernatants were collected, and IL-6 **(A)** and IL-8 **(B)** levels were determined by ELISA. To analyze basal cytokine levels, A549 cells were incubated in the absence of filipin and *H. capsulatum* (FIL −/Hc −).Values represent the mean of triplicate experiments ± the standard deviation. *$p < 0.01$ when compared to A549 cells incubated in the absence of filipin and *H. capsulatum* (FIL −/Hc −). #$p < 0.01$ when compared to A549 cells incubated with *H. capsulatum*, in the absence of filipin (FIL −/Hc +). Similar results were obtained from three independent experiments.

induces cholesterol-dependent recruitment of these proteins to membrane rafts in A549 cells.

Effect of Membrane Raft Disruption on IL-6 and IL-8 Secretion by A549 Cells during Interaction with *H. capsulatum*

The cholesterol-binding compound filipin disrupts membrane rafts, and it was used to evaluate the importance of these domains on *H. capsulatum*-inducible cytokine secretion by A549 cells. By ELISA, it was verified that 1 μg/ml filipin significantly decreased IL-6 and IL-8 levels by 14 and 62%, respectively, when compared to A549 cell cultures incubated with fungi, in the absence of filipin (**Figure 8**). This result indicates, especially for IL-8, that membrane raft recruitment is important for cytokine secretion by A549 epithelial cells during interaction with *H. capsulatum*.

A549 cell and fungal viability was verified by MTT assay. A549 cells were viable in the presence of *H. capsulatum* yeasts and filipin, and no morphological changes were observed (Supplementary Table 4). *H. capsulatum* yeasts were also viable in the presence of filipin (Supplementary Table 5).

DISCUSSION

Most of the studies about the interaction between pathogenic fungi and host cells are performed using macrophages, dendritic cells, or neutrophils. However, over the last two decades, several groups have demonstrated the ability of epithelial cells to produce inflammatory mediators. In this manner, studies about the epithelial cell mechanisms involved in pathogen-inducible cytokine secretion are important for understanding the role of these cells in the host's innate immune defense. To the best of our knowledge, this is the first report showing that *H. capsulatum* yeasts are able to promote cytokine release by pulmonary epithelial cells and, more importantly, this event is dependent on α3 and α5 integrins, SFK activation, and membrane rafts clustering.

In this work, it was demonstrated that *H. capsulatum* yeasts promote secretion of the inflammatory cytokines IL-6 and IL-8 by A549 epithelial cells, but not the anti-inflammatory cytokine IL-10. This result was expected because we recently verified that the fungal pathogen *P. brasiliensis* also promotes IL-6 and IL-8 secretion by A549 cells (Maza et al., 2012). Furthermore, other fungi induce cytokine secretion during epithelial cell–fungus interaction. *Aspergillus fumigatus* and *Cryptococcus neoformans*, for example, induce IL-8 secretion by human bronchial epithelial cells (BEAS-2B), while *Candida albicans* promotes IL-6 secretion by human oral epithelial cells (TR146), but not IL-8 (Balloy et al., 2008; Guillot et al., 2008; Moyes et al., 2011), demonstrating that the cytokine profile depends on the pathogen and epithelial cell type studied.

Several groups have demonstrated that host cell integrins are exploited by different pathogens. These microorganisms express molecules that bind directly to integrins, or to extracellular matrix proteins which then associate with integrins. Both ways lead pathogens to manipulate host signaling pathways and subvert cell processes in order to survive and proliferate (Scibelli et al., 2007). Furthermore, pathogen adhesion to host cell surface is essential for establishing infection, and several microorganisms are able to adhere to integrins. In addition, integrins are able to modulate cytokine secretion in different cell types, including epithelial cells (Lubin et al., 2003; Schmid et al., 2004; Gianni and Campadelli-Fiume, 2014).

In this work, it was verified whether integrins are involved in the secretion of IL-6 and IL-8, promoted by *H. capsulatum*, in A549 cells. By siRNA, silencing of α3 and α5 integrins led to IL-6 and IL-8 level reduction in A549-*H. capsulatum* cultures. Therefore, some of the mechanisms by which these integrins induce this cytokine secretion were investigated. First, we verified that *H. capsulatum* interacts with α3 and α5 integrins in A549 cells, because these integrins, present in A549 cell lysates, associated with *H. capsulatum* yeasts. Despite this result, at the moment, we are not certain whether *H. capsulatum* interacts with α3 and α5 integrins directly, or indirectly, by binding to an extracellular matrix ligand. There is a high probability that this interaction is indirect, because *H. capsulatum* yeasts are able to bind to murine laminin that is recognized by α3β1 integrin (McMahon et al., 1995; Kikkawa et al., 1998; Tagliari et al., 2012). This hypothesis is under current investigation in our laboratory.

In addition, pathogens interact with integrins and may lead to cytokine secretion by several mechanisms. For example, the adhesin A of the bacterium *Yersinia enterocolitica* (YadA) mediates adhesion and invasion of HeLa cells and promotes IL-8 secretion by engaging Rho GTPases, MAPKs and NF-κB activation. All these events are dependent on β1 integrin, because blocking antibodies against this integrin reduced IL-8 production and host cell adhesion (Schmid et al., 2004). The periodontopathogen *Treponema denticola* (Td92) is another example that interacts with integrins. Interaction of a surface protein of this bacterium with α5β1 integrin promotes NLRP3 inflammasome activation, IL-1β secretion, and NF-κB signaling pathway (Jun et al., 2012).

Besides integrins, membrane rafts and SFKs may also be involved in cytokine secretion promoted by different pathogens. For example, Cheon et al. (2008) and Im et al. (2009) showed that treatment of A549 cells with nystatin reduces IL-8 secretion promoted by bacterial flagellin and peptidoglycan, indicating the importance of membrane rafts on this cytokine secretion. Regarding SFKs, Kannan et al. (2006) demonstrated that *Pseudomonas aeruginosa* infection promotes Lyn (a member of SFK) activation in A549 cells. In addition, the authors showed that Lyn is involved in cytokine secretion, because PP2 reduced IL-1β levels in A549-*P. aeruginosa* supernatants. In this work, using different approaches, we show that *H. capsulatum* yeasts lead to IL-6 and IL-8 secretion by associating with α3 and α5 integrin, recruiting these integrins to A549 cell membrane rafts, and then activating SFKs.

Comparing the results of our previous work with *P. brasiliensis* (Barros et al., 2016), and those obtained with *H. capsulatum*, we observed some differences in the importance of α3 and α5 integrins in IL-6 and IL-8 secretion by A549 cells. First, although both fungi are able to interact with α3 and α5 integrins, *H. capsulatum* infection did not alter these integrin levels in A549 cells, while *P. brasiliensis* was able to promote an increase of α3 and α5 integrins. Regarding integrin involvement in cytokine secretion, major differences were seen when A549 cells were transfected with α3 integrin siRNA and IL-8 levels were measured. α3 integrin seems to be critical for IL-8 secretion promoted by *P. brasiliensis*, but not by *H. capsulatum*. Therefore, together, these results indicate that each fungal pathogen interacts differently with these epithelial cells, promoting cytokine secretion by several mechanisms.

In addition, besides integrins, several receptors may be involved in cytokine secretion. Some studies have shown the cooperation between integrins and TLRs (Marre et al., 2010; Gianni and Campadelli-Fiume, 2014). For example, Lerman et al. (2014) related the importance of α3β1 integrin expression on TLR-induced cytokine production by neutrophils during sepsis. The authors showed that integrin deletion reduced IL-6 and IL-10 secretion by neutrophils stimulated with Pam_3CSK_4 (a TLR2/1 specific stimulus), indicating that α3β1 integrin cooperates with TLR2-induced cytokine responses. Therefore, the involvement of

TLRs and other receptors in integrin-mediated cytokine secretion during interaction with pathogenic fungi will also be studied in our laboratory.

AUTHOR CONTRIBUTIONS

PM and ES designed the project and experiments, analyzed the data and wrote the manuscript. PM performed all the experiments.

FUNDING

This study was supported by the Brazilian agencies: Fundação de Amparo à Pesquisa do Estado de São Paulo (FAPESP – Grants # 2011/22773-6 and 2012/11792-2), Conselho Nacional de Desenvolvimento Científico e Tecnológico (CNPq), and Coordenação de Aperfeiçoamento de Pessoal de Nível Superior (CAPES).

REFERENCES

Adenis, A. A., Aznar, C., and Couppié, P. (2014). Histoplasmosis in HIV-infected patients: a review of new developments and remaining gaps. *Curr. Trop Med. Rep.* 1, 119–128.

Alcantara, C., Maza, P. K., Barros, B. C., and Suzuki, E. (2015). Role of protein kinase C in cytokine secretion by lung epithelial cells during infection with *Paracoccidioides brasiliensis*. *Pathog. Dis.* 73:ftv045. doi: 10.1093/femspd/ftv045

Balloy, V., Sallenave, J. M., Wu, Y., Touqui, L., Latgé, J. P., Si-Tahar, M., et al. (2008). *Aspergillus fumigatus*-induced interleukin-8 synthesis by respiratory epithelial cells is controlled by the phosphatidylinositol 3-kinase, p38 MAPK, and ERK1/2 pathways and not by the toll-like receptor-MyD88 pathway. *J. Biol. Chem.* 283, 30513–30521. doi: 10.1074/jbc.M803149200

Barros, B. C., Maza, P. K., Alcantara, C., and Suzuki, E. (2016). *Paracoccidioides brasiliensis* induces recruitment of α3 and α5 integrins into epithelial cell membrane rafts, leading to cytokine secretion. *Microbes Infect.* 18, 68–77. doi: 10.1016/j.micinf.2015.09.003

Bentley, J. K., Newcomb, D. C., Goldsmith, A. M., Jia, Y., Sajjan, U. S., and Hershenson, M. B. (2007). Rhinovirus activates interleukin-8 expression via a Src/p110 β phosphatidylinositol 3-kinase/Akt pathway in human airway epithelial cells. *J. Virol.* 81, 1186–1194. doi: 10.1128/JVI.02309-06

Brilhante, R. S., Fechine, M. A., Mesquita, J. R., Cordeiro, R. A., Rocha, M. F., Monteiro, A. J., et al. (2012). Histoplasmosis in HIV-positive patients in Ceará, Brazil: clinical-laboratory aspects and in vitro antifungal susceptibility of *Histoplasma capsulatum* isolates. *Trans. R. Soc. Trop. Med. Hyg.* 106, 484–488. doi: 10.1016/j.trstmh.2012.05.003

Cheon, I. S., Woo, S. S., Kang, S. S., Im, J., Yun, C. H., Chung, D. K., et al. (2008). Peptidoglycan-mediated IL-8 expression in human alveolar type II epithelial cells requires lipid raft formation and MAPK activation. *Mol. Immunol.* 45, 1665–1673. doi: 10.1016/j.molimm.2007.10.001

Cleaver, J. O., You, D., Michaud, D. R., Pruneda, F. A., Juarez, M. M., Zhang, J., et al. (2014). Lung epithelial cells are essential effectors of inducible resistance to pneumonia. *Mucosal Immunol.* 7, 78–88. doi: 10.1038/mi.2013.26

Cruz, K. D., Cruz, T. A., de Moraes, G., Paredes-Santos, T. C., Attias, M., and de Souza, W. (2014). Disruption of lipid rafts interferes with the interaction of *Toxoplasma gondii* with macrophages and epithelial cells. *Biomed. Res. Int.* 2014:687835. doi: 10.1155/2014/687835

Engen, J. R., Wales, T. E., Hochrein, J. M., Meyn, M. A., Banu Ozkan, S., Bahar, I., et al. (2008). Structure and dynamic regulation of Src-family kinases. *Cell Mol. Life Sci.* 65, 3058–3073. doi: 10.1007/s00018-008-8122-2

Eucker, T. P., Samuelson, D. R., Hunzicker-Dunn, M., and Konkel, M. E. (2014). The focal complex of epithelial cells provides a signalling platform for interleukin-8 induction in response to bacterial pathogens. *Cell Microbiol.* 16, 1441–1455. doi: 10.1111/cmi.12305

Gianni, T., and Campadelli-Fiume, G. (2014). The epithelial αvβ3-integrin boosts the MYD88-dependent TLR2 signaling in response to viral and bacterial components. *PLoS Pathog.* 10:e1004477. doi: 10.1371/journal.ppat.1004477

Gianni, T., Gatta, V., and Campadelli-Fiume, G. (2010). αVβ3-integrin routes herpes simplex virus to an entry pathway dependent on cholesterol-rich lipid rafts and dynamin2. *Proc. Natl. Acad. Sci. U.S.A.* 107, 22260–22265. doi: 10.1073/pnas.1014923108

Guillot, L., Carroll, S. F., Badawy, M., and Qureshi, S. T. (2008). Cryptococcus neoformans induces IL-8 secretion and CXCL1 expression by human bronchial epithelial cells. *Respir. Res.* 9:9. doi: 10.1186/1465-9921-9-9

Guimarães, A. J., Nosanchuk, J. D., and Zancopé-Oliveira, R. M. (2006). Diagnosis of histoplasmosis. *Braz. J. Microbiol.* 37, 1–13. doi: 10.1590/S1517-83822006000100001

Hauck, C. R., Borisova, M., and Muenzner, P. (2012). Exploitation of integrin function by pathogenic microbes. *Curr. Opin. Cell Biol.* 24, 637–644. doi: 10.1016/j.ceb.2012.07.004

Im, J., Jeon, J. H., Cho, M. K., Woo, S. S., Kang, S. S., Yun, C. H., et al. (2009). Induction of IL-8 expression by bacterial flagellin is mediated through lipid raft formation and intracellular TLR5 activation in A549 cells. *Mol. Immunol.* 47, 614–622. doi: 10.1016/j.molimm.2009.09.004

Ingley, E. (2008). Src family kinases: regulation of their activities, levels and identification of new pathways. *Biochim. Biophys. Acta* 1784, 56–65. doi: 10.1016/j.bbapap.2007.08.012

Jun, H. K., Lee, S. H., Lee, H. R., and Choi, B. K. (2012). Integrin α5β1 activates the NLRP3 inflammasome by direct interaction with a bacterial surface protein. *Immunity* 36, 755–768. doi: 10.1016/j.immuni.2012.05.002

Kannan, S., Audet, A., Knittel, J., Mullegama, S., Gao, G. F., and Wu, M. (2006). Src kinase Lyn is crucial for *Pseudomonas aeruginosa* internalization into lung cells. *Eur. J. Immunol.* 36, 1739–1752. doi: 10.1002/eji.200635973

Kikkawa, Y., Sanzen, N., and Sekiguchi, K. (1998). Isolation and characterization of laminin-10/11 secreted by human lung carcinoma cells. *laminin*-10/11 mediates cell adhesion through integrin α3β1. *J. Biol. Chem.* 273, 15854–15859. doi: 10.1074/jbc.273.25.15854

Leitinger, B., and Hogg, N. (2002). The involvement of lipid rafts in the regulation of integrin function. *J. Cell Sci.* 115, 963–972.

Lerman, Y. V., Lim, K., Hyun, Y. M., Falkner, K. L., Yang, H., Pietropaoli, A. P., et al. (2014). Sepsis lethality via exacerbated tissue infiltration and TLR-induced cytokine production by neutrophils is integrin α3β1-dependent. *Blood* 124, 3515–3523. doi: 10.1182/blood-2014-01-552943

Lin, C. H., Cheng, H. W., Hsu, M. J., Chen, M. C., Lin, C. C., and Chen, B. C. (2006). c-Src mediates thrombin-induced NF-κB activation and IL-8/CXCL8 expression in lung epithelial cells. *J. Immunol.* 177, 3427–3438. doi: 10.4049/jimmunol.177.5.3427

Lubin, F. D., Segal, M., and McGee, D. W. (2003). Regulation of epithelial cell cytokine responses by the α3β1 integrin. *Immunology* 108, 204–210. doi: 10.1046/j.1365-2567.2003.01577.x

Marre, M. L., Petnicki-Ocwieja, T., DeFrancesco, A. S., Darcy, C. T., and Hu, L. T. (2010). Human integrin α3β1 regulates TLR2 recognition of lipopeptides from endosomal compartments. *PLoS ONE* 5:e12871. doi: 10.1371/journal.pone.0012871

Martins, E. M. L., Marchiori, E., Damato, S. D., Pozes, A. S., Silva, A. C. G., and Dalston, M. (2003). *Histoplasmose pulmonar* aguda: relato de uma microepidemia. *Radiol. Brasil.* 36, 147–151. doi: 10.1590/S0100-39842003000300005

Mason, R. J. (2006). Biology of alveolar type II cells. *Respirology* 11(Suppl.), S12–S15. doi: 10.1111/j.1440-1843.2006.00800.x

Maza, P. K., Oliveira, P., Toledo, M. S., Paula, D. M. B., Takahashi, H. K., Straus, A. H., et al. (2012). *Paracoccidioides brasiliensis* induces secretion of IL-6 and IL-8 by lung epithelial cells. Modulation of host cytokine levels by fungal proteases. *Microbes Infect.* 14, 1077–1085. doi: 10.1016/j.micinf.2012.05.016

Maza, P. K., Straus, A. H., Toledo, M. S., Takahashi, H. K., and Suzuki, E. (2008). Interaction of epithelial cell membrane rafts with *Paracoccidioides brasiliensis* leads to fungal adhesion and Src-family kinase activation. *Microbes Infect.* 10, 540–547. doi: 10.1016/j.micinf.2008.02.004

McMahon, J. P., Wheat, J., Sobel, M. E., Pasula, R., Downing, J. F., and Martin, W. J. (1995). Murine laminin binds to *Histoplasma capsulatum*. A possible mechanism of dissemination. *J. Clin. Invest.* 96, 1010–1017. doi: 10.1172/JCI118086

Mihu, M. R., and Nosanchuk, J. D. (2012). *Histoplasma virulence* and host responses. *Int. J. Microbiol.* 2012:268123. doi: 10.1155/2012/268123

Moyes, D. L., Murciano, C., Runglall, M., Islam, A., Thavaraj, S., and Naglik, J. R. (2011). *Candida albicans* yeast and hyphae are discriminated by MAPK signaling in vaginal epithelial cells. *PLoS ONE* 6:e26580. doi: 10.1371/journal.pone.0026580

Nacher, M., Adenis, A., and Mc Donald, S. (2013). Disseminated histoplasmosis in HIV-infected patients in South America: a neglected killer continues on its rampage. *PLoS Negl. Trop Dis.* 7:e2319. doi: 10.1371/journal.pntd.0002319

Okada, M. (2012). Regulation of the SRC family kinases by Csk. *Int. J. Biol. Sci.* 8, 1385–1397. doi: 10.7150/ijbs.5141

Oliveira, F. D. M., Unis, G., and Severo, L. C. (2006). Microepidemia de histoplasmose em Blumenau, Santa Catarina. *J. Brasil. Pneumol.* 32, 375–378. doi: 10.1590/S1806-37132006000400018

Proud, D., and Leigh, R. (2011). Epithelial cells and airway diseases. *Immunol. Rev.* 242, 186–204. doi: 10.1111/j.1600-065X.2011.01033.x

Ren, L., Blanchette, J. B., White, L. R., Clark, S. A., Heffner, D. J., Tibbles, L. A., et al. (2005). Soluble fibronectin induces chemokine gene expression in renal tubular epithelial cells. *Kidney Int.* 68, 2111–2120. doi: 10.1111/j.1523-1755.2005.00667.x

Rocha-Silva, F., Figueiredo, S. M., Silveira, T. T., Assunção, C. B., Campolina, S. S., Pena-Barbosa, J. P., et al. (2014). Histoplasmosis outbreak in Tamboril cave-Minas Gerais state. *Brazil. Med. Mycol. Case Rep.* 4, 1–4. doi: 10.1016/j.mmcr.2013.12.002

Schmid, Y., Grassl, G. A., Bühler, O. T., Skurnik, M., Autenrieth, I. B., and Bohn, E. (2004). *Yersinia enterocolitica* adhesin a induces production of interleukin-8 in epithelial cells. *Infect. Immun.* 72, 6780–6789. doi: 10.1128/IAI.72.12.6780-6789.2004

Scibelli, A., Roperto, S., Manna, L., Pavone, L. M., Tafuri, S., Della Morte, R., et al. (2007). Engagement of integrins as a cellular route of invasion by bacterial pathogens. *Vet. J.* 173, 482–491. doi: 10.1016/j.tvjl.2006.01.010

Seveau, S., Bierne, H., Giroux, S., Prévost, M. C., and Cossart, P. (2004). Role of lipid rafts in E-cadherin- and HGF-R/Met-mediated entry of *Listeria monocytogenes* into host cells. *J. Cell Biol.* 166, 743–753. doi: 10.1083/jcb.200406078

Simons, K., and Sampaio, J. L. (2011). Membrane organization and lipid rafts. *Cold Spring Harb. Perspect. Biol.* 3:a004697. doi: 10.1101/cshperspect.a004697

Smith, J. A., and Kauffman, C. A. (2012). Pulmonary fungal infections. *Respirology* 17, 913–926. doi: 10.1111/j.1440-1843.2012.02150.x

Stadnyk, A. W. (1994). Cytokine production by epithelial cells. *FASEB J.* 8, 1041–1047.

Suzuki, T., Chow, C. W., and Downey, G. P. (2008). Role of innate immune cells and their products in lung immunopathology. *Int. J. Biochem. Cell Biol.* 40, 1348–1361. doi: 10.1016/j.biocel.2008.01.003

Tagliari, L., Toledo, M. S., Lacerda, T. G., Suzuki, E., Straus, A. H., and Takahashi, H. K. (2012). Membrane microdomain components of *Histoplasma capsulatum* yeast forms, and their role in alveolar macrophage infectivity. *Biochim. Biophys. Acta Biomemb.* 1818, 458–466. doi: 10.1016/j.bbamem.2011.12.008

Toledo, M. S., Levery, S. B., Suzuki, E., Straus, A. H., and Takahashi, H. K. (2001). Characterization of cerebrosides from the thermally dimorphic mycopathogen *Histoplasma capsulatum*: expression of 2-hydroxy fatty N-acyl (E)-Delta(3)-unsaturation correlates with the yeast-mycelium phase transition. *Glycobiology* 11, 113–124. doi: 10.1093/glycob/11.2.113

Ulanova, M., Gravelle, S., and Barnes, R. (2009). The role of epithelial integrin receptors in recognition of pulmonary pathogens. *J. Innate Immun.* 1, 4–17. doi: 10.1159/000141865

Wang, R., Bi, J., Ampah, K. K., Zhang, C., Li, Z., Jiao, Y., et al. (2013). Lipid raft regulates the initial spreading of melanoma A375 cells by modulating β1 integrin clustering. *Int. J. Biochem. Cell Biol.* 45, 1679–1689. doi: 10.1016/j.biocel.2013.04.031

Candida albicans: The Ability to Invade Epithelial Cells and Survive under Oxidative Stress is Unlinked to Hyphal Length

Paloma K. Maza[1†], *Alexis Bonfim-Melo*[1†], *Ana C. B. Padovan*[2,3], *Renato A. Mortara*[1], *Cristina M. Orikaza*[1], *Lilian M. Damas Ramos*[4], *Tauany R. Moura*[5], *Frederico M. Soriani*[5], *Ricardo S. Almeida*[4], *Erika Suzuki*[1] and *Diana Bahia*[1,5*]

[1] Disciplina de Parasitologia, Departamento de Microbiologia, Imunologia e Parasitologia, Escola Paulista de Medicina, Universidade Federal de São Paulo, São Paulo, Brazil, [2] Laboratório Especial de Micologia, Disciplina de Infectologia, Departamento de Medicina, Escola Paulista de Medicina, Universidade Federal de São Paulo, São Paulo, Brazil, [3] Departamento de Microbiologia e Imunologia, Universidade Federal de Alfenas, Alfenas, Brazil, [4] Laboratório de Micologia Médica e Microbiologia Bucal, Departamento de Microbiologia, Universidade Estadual de Londrina, Londrina, Brazil, [5] Departamento de Biologia Geral, Instituto de Ciências Biológicas, Universidade Federal de Minas Gerais, Belo Horizonte, Brazil

Edited by:
Alexandre Morrot, Federal University of Rio de Janeiro, Brazil

Reviewed by:
Hridayesh Prakash, University of Hyderabad, India
Manuel Vilanova, Instituto de Ciências Biomédicas Abel Salazar (ICBAS), Universidade do Porto, Portugal

***Correspondence:**
Diana Bahia
dianabahia@hotmail.com

† The authors wish it to be known that, in their opinion, the first two authors should be regarded as joint First Authors.

In its hyphal form, *Candida albicans* invades epithelial and endothelial cells by two distinct mechanisms: active penetration and induced endocytosis. The latter is dependent on a reorganization of the host cytoskeleton (actin/cortactin recruitment), whilst active penetration does not rely on the host's cellular machinery. The first obstacle for the fungus to reach deep tissues is the epithelial barrier and this interaction is crucial for commensal growth, fungal pathogenicity and host defense. This study aimed to characterize *in vitro* epithelial HeLa cell invasion by four different isolates of *C. albicans* with distinct clinical backgrounds, including a *C. albicans* SC5314 reference strain. All isolates invaded HeLa cells, recruited actin and cortactin, and induced the phosphorylation of both Src-family kinases (SFK) and cortactin. Curiously, L3881 isolated from blood culture of a patient exhibited the highest resistance to oxidative stress, although this isolate showed reduced hyphal length and displayed the lowest cell damage and invasion rates. Collectively, these data suggest that the ability of *C. albicans* to invade HeLa cells, and to reach and adapt to the host's blood, including resistance to oxidative stress, may be independent of hyphal length.

Keywords: *Candida albicans*, hyphae, HeLa cells, actin, cortactin, cell signaling

INTRODUCTION

Candida albicans is a polymorphic opportunistic human pathogen that takes on the most common morphologies of yeast, pseudohyphae, and true hyphae. All morphology types are found during infection; however, hyphae are generally the invasive form of *C. albicans* (Jacobsen et al., 2012). *C. albicans* frequently causes superficial infections in mildly immunocompromised patients, but in severely immunocompromised or critically ill patients, this fungus may cause life-threatening, disseminated disease. The *Candida* genus is associated with approximately 80% of nosocomial fungal infections, representing the major cause of fungemia showing a high mortality rate (40–60%) (Perlroth et al., 2007; Doi et al., 2016).

For most microbial pathogens, adherence to epithelial surfaces is the first step to initiating an infection, and is even a prerequisite for their survival and eventual distribution. After attachment, many pathogens invade host cells (Filler and Sheppard, 2006). With regard to *C. albicans,* this fungus uses various mechanisms of pathogenicity: escape from host immune responses, morphological change (yeast-to-hypha transition) and invasion, which is supported by hyphae-associated factors such as adhesion molecules and the secretion of hydrolytic enzymes (Wilson et al., 2009).

In order to infect mucosal tissues, gain access to blood and subsequently escape from the bloodstream, *C. albicans* must interact with epithelial and endothelial cells (Davis, 2009). The invasion of non-professional phagocytic host cells by *C. albicans* can occur through two mechanisms: induced endocytosis and active penetration. The latter mechanism does not rely on the host's cellular machinery, but possibly on physical pressure applied by the advancing hyphal tip and the secretion of extracellular hydrolases (Wilson et al., 2009).

In contrast, invasion via induced endocytosis is dependent on host cell actin microfilament dynamics. *C. albicans* stimulates epithelial cells to produce pseudopodia-like structures that surround the fungus and cause its uptake. In endocytosis, induced proteases of *C. albicans* have little or no role (Frank and Hostetter, 2007). This highly effective invasive mechanism is predominantly triggered by the interaction between a fungal adhesin (Als3) (Phan et al., 2007) and either human E-cadherin or the epidermal growth factor receptor 2 (Her2) on epithelial cells (Zhu et al., 2012). Such interactions stimulate actin rearrangement in host cells and the subsequent internalization of fungal cells (Phan et al., 2007; Zhu et al., 2012; Yang et al., 2014). In addition, *C. albicans* exploits clathrin-dependent endocytosis, which requires dynamin and cortactin, to invade human epithelial cells (Moreno-Ruiz et al., 2009). Cortactin is involved in actin-related cellular processes ranging from lamellipodium protrusion and extracellular matrix degradation to the uptake of intracellular pathogens such as bacteria and parasites (Chen et al., 2003; Bonfim-Melo et al., 2015). Cortactin interacts with the Arp2/3 complex, newly formed actin filaments and a variety of actin-binding/regulation proteins (Daly, 2004); it is also regulated by phosphorylation (Martinez-Quiles et al., 2004). Oral and vaginal epithelial cells recognize both the yeast and hyphal forms of *C. albicans.* As a consequence, they activate cellular signaling mechanisms, including mitogen-activated protein kinase (MAPK), nuclear factor kappa-light-chain-enhancer of activated B cells (NF-κB), and phosphatidylinositol 3-kinase (PI3K) pathways (Moyes et al., 2010, 2014, 2015; Naglik et al., 2014).

With regard to induced endocytosis, our group has previously shown that extracellular amastigote (EAs) infective forms of *Trypanosoma cruzi* seem to share with *C. albicans* a similar pathway for entry into the host cell. Both events occur via an actin-dependent mechanism and involve cortactin (Moreno-Ruiz et al., 2009; Bonfim-Melo et al., 2015). In addition, *T. cruzi* recruited actin and cortactin to sites of adherence and invasion into HeLa cells and also selectively induced cortactin phosphorylation during cell invasion (Bonfim-Melo et al., 2015). However, possible cortactin phosphorylation in response to *C. albicans* invasion into non-phagocytic cells, such as HeLa cells, has never been assessed. We propose that epithelial cells represent the first obstacle for *Candida* invasion which probably will determinate the fate of this infection. We would like to consider invasion in a broad spectrum which includes crossing the endothelial barrier, reaching and adapting to blood.

In this study, we verified the ability of different *Candida albicans* isolates (the reference strain SC5314 and three *C. albicans* clinical isolates) to invade HeLa epithelial cells. First, we analyzed hyphal growth, cell internalization and damage rates and resistance to oxidative stress. Furthermore, we investigated HeLa cell cortactin/actin recruitment and Src-family kinases (SFK)/cortactin phosphorylation/activation in response to these isolates. Our data showed that all isolates are capable of invading HeLa cells, recruiting actin/cortactin and inducing SFK and cortactin phosphorylation. Interestingly, L3881 isolate, that has a natural *HWP1* mutation, the shortest hyphae and the lowest internalization and damage rates, was able to reach the blood of a patient and to survive under nutritional and oxidative stress, indicating that the latter phenomenon was independent of hyphal length or damage/invasion rate.

MATERIALS AND METHODS

Candida Strains

C. albicans isolates were obtained from various sources: 997,5 g was isolated from the blood of a deceased patient (Chaves et al., 2012), L3881 (LEMI #L3881, Padovan et al., 2009) was isolated from blood of a patient who had candidemia which was cured clinically, and L3837 was an isolate from the oral cavity (oropharynx, LEMI# L3837, non-hematogenic). *C. albicans* SC5314 is a reference strain. The isolates were kindly provided by LEMI (Laboratorio Especial de Micologia, Universidade Federal de São Paulo, Brazil). *C. albicans* isolates were grown at 30°C on YPD (1% [w/v] yeast extract; 2% [w/v] peptone, 2% [w/v] dextrose) medium plates. Before performing all experiments, a single colony of *C. albicans* isolates was inoculated in YPD liquid media and incubated in an orbital shaker at 30°C for 24 h. Then, *C. albicans* cells were washed with phosphate buffered saline, pH 7.2 (PBS), harvested by centrifugation, and counted in a hemocytometer chamber.

Cell Culture

HeLa cells (human epithelial cells from a fatal cervical adenocarcinoma) were obtained from the Instituto Adolfo Lutz (São Paulo, SP, Brazil) and maintained in complete RPMI medium with 10% fetal bovine serum (FBS) supplemented with antibiotics (10 μg/mL streptomycin and 100 U/mL penicillin; Sigma-Aldrich, St. Louis, MO, USA) in a humidified atmosphere at 37°C and 5% CO_2.

Antibodies

Rabbit anti-pT202/Y204 ERK1/2 (#9101S) and rabbit anti-pY416 SFK (#2101S) were obtained from Cell Signaling, Beverly, MA, USA. Rabbit anti-pY466 cortactin (PA114115) was obtained from Thermo Scientific, Waltham, MA, USA. Mouse monoclonal anti-β-actin (AC-74, A2228) was obtained from Sigma-Aldrich. Mouse monoclonal anti-cortactin (A-4; sc-55578) was from Santa Cruz Biotechnology (Santa Cruz County, CA, USA). The secondary antibody, goat anti-mouse AlexaFluor® 488 (A11001), was from Invitrogen, Carlsbad, CA, USA. Goat anti-rabbit IgG peroxidase (A6154) and goat anti-mouse IgG peroxidase (A4416) were obtained from Sigma-Aldrich.

Hyphal Growth Assay

C. albicans blastospores (3×10^5) were seeded in complete RPMI medium at 37°C and 5% CO_2 for 2 h on glass coverslips in a 24-well plate. *C. albicans* cells were fixed with 4% paraformaldehyde, washed with phosphate buffered saline, pH 7.2 (PBS), and analyzed under a BX51 microscope (Olympus, Tokyo, Japan) with a 60× objective. Random images from 10 fields were acquired by a DP71 camera (Olympus) and hyphal lengths were measured using ImageProPlus 6.2 software (Media Cybernetics, Rockville, MD, USA). Data were treated and plotted using Office Excel and Prism 5 (GraphPad, San Diego, CA, USA). Experiments were performed in triplicate.

Cell Internalization Assay

To evaluate cell internalization by *C. albicans* strains, HeLa cells were cultured in complete RPMI medium on glass coverslips treated with acetic acid until they form confluent monolayers (80–100% confluence). After washing with PBS, monolayers were seeded with 2×10^5 *C. albicans* blastospores in RPMI 1640 medium without FBS for 2 h at 37°C in 5% CO_2. Coverslips were fixed with 4% paraformaldehyde for 30 min. To stain *C. albicans* cells localized only to the outside of epithelial cells, coverslips were incubated with 25 μg/mL Concanavalin A-fluorescein conjugate (Invitrogen) in PBS for 45 min. After permeabilization with 0.1% Triton X-100 for 15 min, fungal cells localized outside and inside epithelial cells were stained with 10 μg/mL Calcofluor White (Sigma-Aldrich) in 0.1 M Tris hydrochloride (pH 9.0) for 20 min. After each step, HeLa cells were washed three times with PBS. Coverslips were mounted on slides with 50% glycerol and analyzed under a fluorescence microscope (Zeiss, Jena, Germany; Axio Observer Research microscope). This method allowed differentiation between hyphae of the fungus that were located in the external medium and hyphae that invaded the host cell (intracellular) (Almeida et al., 2008). Three experiments in duplicate were analyzed and 10 randomly chosen fields per coverslip were evaluated.

Epithelial Cell Damage Assay

To evaluate the cell damage caused by *C. albicans* strains, HeLa cells were incubated in complete RPMI medium on glass coverslips treated with acetic acid until they form confluent monolayers (80–100% confluence). Afterwards, 1 mL PBS was added and removed from each well to wash the cells. Finally, the monolayers were co-incubated with 10^6 *C. albicans* blastospores in 1 mL RPMI 1640 medium without FBS for 8 h at 37°C and 5% CO_2. To quantify cell damage, supernatants were removed and the amount of lactate dehydrogenase (LDH) released into the medium was determined using LDH Cytotoxicity Assay Kit (Cayman Chemical, Ann Arbor, MI, USA) following manufacturer's instructions. Samples containing only epithelial cells or fungal cells alone were used as controls.

Phenotypic Characterization

C. albicans blastospores of the SC5314 reference strain and L3881 (10^2–10^6) were seeded in YPD medium plates containing different stressor agents: 200 mM $CaCl_2$; 2 mM $FeSO_4$; 200 mM $MgCl_2$; 200 mM NaCl; 3 mM ethylenediaminetetraacetic acid (EDTA); 0.125 μg/mL 4-nitroquinoline N-oxide (4NQO); 50 μM camptothecin (CPT); 0.005% methyl methanesulfonate (MMS); 5 mM hydrogen peroxide (H_2O_2); and 1.25 mM paraquat (Pqt). Plates were incubated for 24–48 h at 30°C.

Host Molecules Recruitment, Immunofluorescence, and Confocal Microscopy

HeLa cells were cultured in complete RPMI medium on glass coverslips treated with acetic acid until they form confluent monolayers (80–100% confluence). After washing with PBS, HeLa cells were seeded with 10^5 *C. albicans* blastospores in RPMI 1640 medium without FBS for 2 h at 37°C in 5% CO_2. Coverslips were washed with PBS, fixed with 3.5% paraformaldehyde for 15 min at room temperature and sequentially incubated with Calcofluor White (10 μg/mL), phalloidin-TRITC (10 ng/mL), anti-cortactin (20 μg/mL) and anti-mouse AlexaFluor® 488 (20 μg/mL) secondary antibodies diluted in PGNS solution (PBS pH 7.2 with 0.1% gelatin [Sigma-Aldrich], 0.1% sodium azide [Sigma-Aldrich] and 0.2% saponin [Sigma-Aldrich]) at room temperature in an humidified chamber for 1 h. Coverslips were mounted on slides in pH 9.0 buffered glycerol solution with 9 mM *p*-phenylenediamine. Then, HeLa-*C. albicans* interactions were evaluated and imaged in a TCS SP5 II Tandem Scanner (Leica, Wetzlar, Germany) confocal microscope with a 63 × 1.40 NA objective. Image processing and analysis were performed with Imaris v.7.4.2 software (Bitplane, Belfast, Ireland) or ImageJ (https://imagej.nih.gov/ij/).

Signaling and Western Blot Assays

Western blots were performed essentially as described by Bonfim-Melo et al. (2015). HeLa cells were seeded with 10^8 *C. albicans* blastospores and incubated for various time intervals (ranging from 0 to 180 min) at 37°C in 5% CO_2. Protein extracts were quantified using Micro BCA Protein Assay Kit (Thermo Scientific), according to the manufacturer's instruction.

Statistical Analysis

All results are presented as the mean ± standard deviation (s.d.). Differences were considered statistically significant when $p < 0.05$, as determined by one-way ANOVA and Tukey's multiple comparison post-test. Analyses and graphs were created using Prism 5 (GraphPad, San Diego, CA, USA) software.

RESULTS

Hyphal Growth, Internalization, and Damage Rates

Hwp1 (hyphal wall protein 1) is a well-characterized *C. albicans* cell surface protein, expressed only on hyphae, which mediates binding to epithelial cells. The isolate L3881 is a *HWP1-2* homozygous strain that showed reduced *HWP1* expression, biofilm production, and hyphae formation (Padovan et al., 2009). In fact, we verified that *C. albicans* L3881 produced the shortest hyphae and in HeLa cells produced the lowest internalization and damage rates when compared to other strains (**Figures 1A–C**). The L3837 oral strain showed similar hyphal lengths to the SC5314 reference strain (**Figures 1A,B**). HeLa cell internalization and cellular damage rates were lower for L3837 than for SC5314 (**Figures 1C,D**). The 997,5 g isolate also displayed a similar hyphal length when compared to the reference isolate, SC5314. However, 997,5 g displayed the highest cell internalization and cell damage rates (**Figures 1C,D**) in this study. The characteristics of all four *Candida* species are summarized in the Supplementary Table 1 (Supplementary Material).

The L3881 Isolate Showed the Poorest Internalization Fitness but with a High Resistance to Oxidative and Nutritional Stress

To explore possible physiological alterations, we tested the SC5314 reference strain and the L3881 blood isolate and analyzed their ability to deal with different kinds of stresses (**Figure 1E**). SC5314 and L3881 strains demonstrated similar sensitivities in the presence of ions such as calcium, magnesium and iron, and during osmotic stress (NaCl). On the other hand, in the presence of ethylenediaminetetraacetic acid (EDTA), a chelator of bivalent ions, L3881 isolate displayed a more resistant phenotype, indicating lower nutritional requirements needed for maintaining growth. In the presence of DNA damage or stress agents of replication (4-nitro-quinoline-N-oxide [4NQO]; camptothecin [CPT] and methylmethanesulfonate [MMS]), both strains showed similar phenotypes in response to functional mechanisms of DNA repair. SC5314 and L3881 isolates were also tested in the presence of oxidative stress inducers such as H_2O_2 and Paraquat (Pqt). Paraquat is a potent oxidative stress inducer because it greatly increases reactive oxygen species (ROS) production and inhibits the regeneration of reducing equivalents and compounds necessary for maintaining the redox status. Surprisingly, L3881 the *HWP1-2* homozygous strain with reduced *HWP1* expression with shortest hyphae and the lowest internalization and damage rates in HeLa cells was found to be more resistant to oxidative stress (**Figure 1E**).

Different Clinical Isolates of *C. albicans* Invade HeLa Cells and Recruit Actin and Cortactin

After observing distinct levels of cell cytotoxicity and invasiveness promoted by different isolates, we then evaluated whether host actin and cortactin could be recruited by these isolates and also be differentially modulated. By staining hyphae, inside and outside HeLa cells, with Calcofluor White and by using Concanavalin A to stain hyphae that were not internalized, we tested the ability of the *C. albicans* SC5314 reference strain to invade HeLa cells. **Figure 2A** clearly shows the invasion of HeLa cells by *C. albicans* hyphae. **Figure 2B** shows that cortactin was recruited to host cell ruffles by *C. albicans* SC5314 with a similar distribution to that of F-actin recruitment. Actually, all *C. albicans* isolates recruited cortactin and actin, and both molecules colocalized at sites of hyphae internalization (**Figure 2C**).

Biphasic Activation of ERK Kinase Pathway and Induction of Cortactin-Phosphorylated SFK

We first tested ERK 1/2 (one of the MAP kinase components) signaling in response to *C. albicans* SC5314 invasion in HeLa cells at several time points (**Figure 3**). As previously shown with TR146 oral epithelial cells by Moyes et al. (2010), *C. albicans* SC5314 also induced a biphasic MAP kinase response in another cell type, i.e., HeLa cells (**Figure 3A**) in a characteristic pattern: an increasing activation within the first 15 min (3–15 min), followed by a rapid decline (30 min), and increased activation again beginning at the 75 min time point, ending in a further decline of activation after 120 min (**Figure 3A**, 180 min). The initial activation was observed directly after contact with yeast cells, possibly due to the recognition of fungal cell wall components; the second activation phase corresponds to hyphae formation. However, such biphasic activation has not been observed with the other *Candida* isolates (not shown). We also tested cortactin phosphorylation at Y^{466} (phosphorylated site by SFKs) under the same conditions (**Figure 3B**). Cortactin phosphorylation at Y^{466} started to increase after 30–45 min and was sustained until 75 min of incubation, after which the signal declined.

Src-Family Kinases (SFKs) and Cortactin Signaling during *C. albicans* Invasion into HeLa Cells

Once our experimental approaches using HeLa cells showed that *C. albicans* SC5314 induced a biphasic pattern of ERK signaling similar to the one obtained by Moyes et al. (2010) but in a different cell type, we evaluated SFK and cortactin signaling at selected time points in HeLa cells. **Figures 3C,D** shows that all *C. albicans* strains induced, at different levels, both SFK and cortactin phosphorylation in HeLa cells.

Regarding cortactin, 997,5 g, oral L3837 and blood L3881 strains also activated cortactin phosphorylation by Src at the Y^{466} residue; this phosphorylation was sustained for at least 180 min of *Candida*–HeLa cell contact. Lipid rafts have no impact in the epithelial cell invasion of all *Candida* isolates (Supplementary Figure 1A). Conversely, SFK inhibition negatively modulated all *Candida albicans* isolates invasion to HeLa cells (Supplementary Figure 1B).

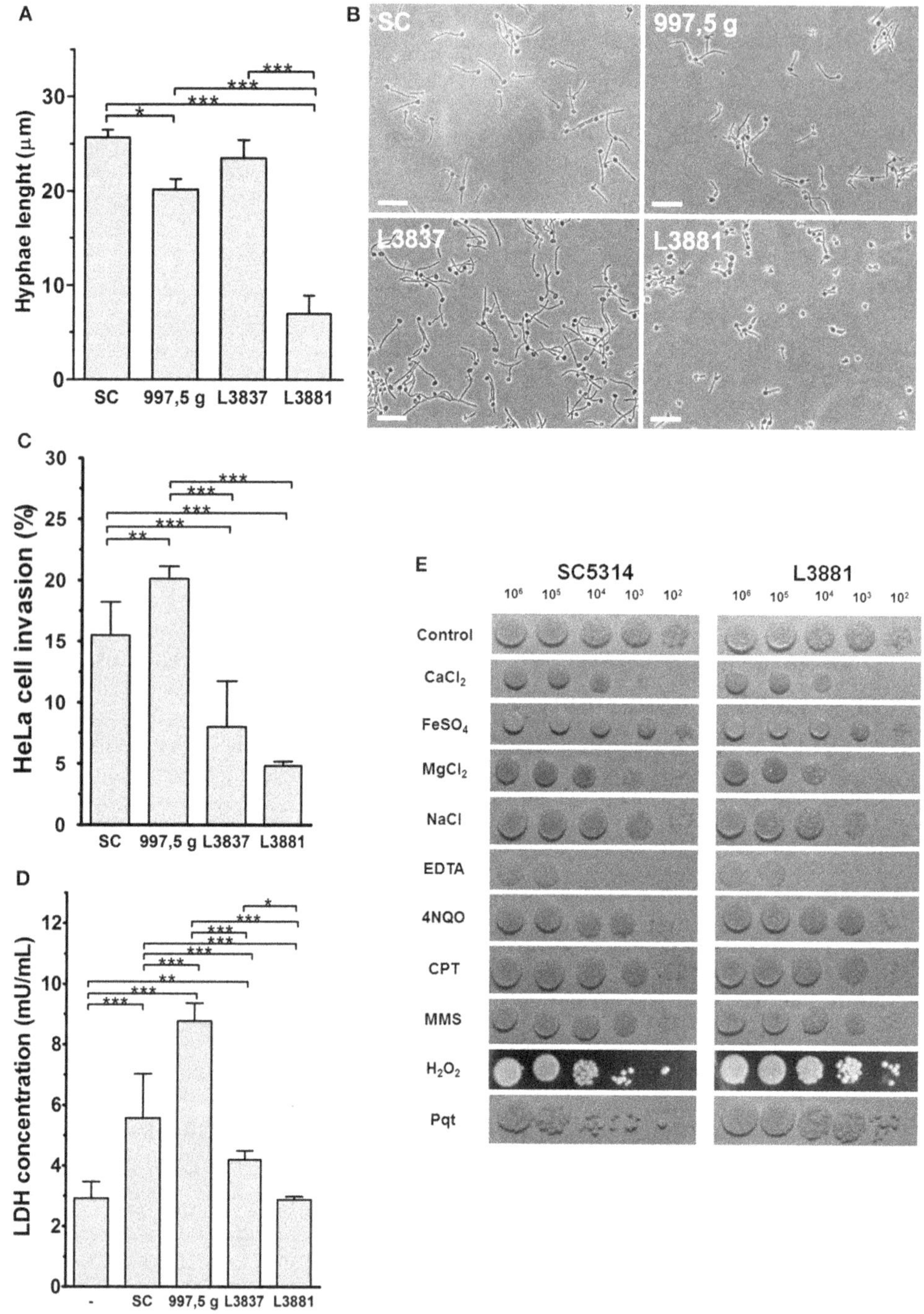

FIGURE 1 | Characterization of HeLa epithelial cell invasion by four *C. albicans* isolates. **(A,B)** The blood isolate L3881 displayed a smaller hyphal length. Blastospores from SC5314, 997,5 g, L3881 and L3837 isolates were seeded on glass coverslips for 2 h at 37°C and 5% CO_2, washed, fixed and their hyphal length measured under optical microscopy. **(A)** Hyphal length of all isolates used (mean ± s.d., $^{*}p < 0.05$ and $^{***}p < 0.001$). **(B)** Representative fields of each isolate observed under optical microscopy, showing the overall aspect and size of hyphal length. Bar = 20 μm. **(C,D)** The 997,5 g and L3881 blood isolates displayed the highest and lowest invasion index and cytotoxicity for HeLa cells, respectively. **(C)** Blastospores from all isolates were seeded with HeLa cells for 2 h at 37°C and 5% CO_2, washed, fixed, stained for intracellular/extracellular hyphae discrimination and the cell invasion indexes evaluated under epifluorescence microscopy. The 997,5 g and SC5314 isolates displayed higher invasion indexes compared to the other isolates (L3837 and L3881; mean ± s.d. $^{**}p < 0.01$ and $^{***}p < 0.001$). **(D)** For cytotoxicity assays, HeLa cells were seeded with blastospores for 8 h at 37°C and 5% CO_2, washed and the amount of lactate dehydrogenase in culture supernatants was quantified by a commercial kit. L3881 did not induce toxicity in HeLa cells while the 997,5 g strain induced significantly higher cytotoxicity compared to other isolates (mean ± s.d., $^{*}p < 0.05$, $^{**}p < 0.01$, and $^{***}p < 0.001$). **(E)** L3881 showed more plasticity under stress. 10^2–10^6 blastospores of the SC5314 reference strain and L3881 blood isolate were seeded on plates containing different stressor agents for 24–48 h at 30°C. Fungi were cultured in the presence of 200 mM $CaCl_2$; 2 mM $FeSO_4$; 200 mM $MgCl_2$; 200 mM NaCl; 3 mM ethylenediaminetetraacetic acid (EDTA); 0.125 μg/mL 4-nitroquinoline N-oxide (4NQO); 50 μM camptothecin (CPT); 0.005% methyl methanesulfonate (MMS); 5 mM hydrogen peroxide (H_2O_2); and 1.25 mM paraquat (Pqt). All assays are representative of triplicate experiments.

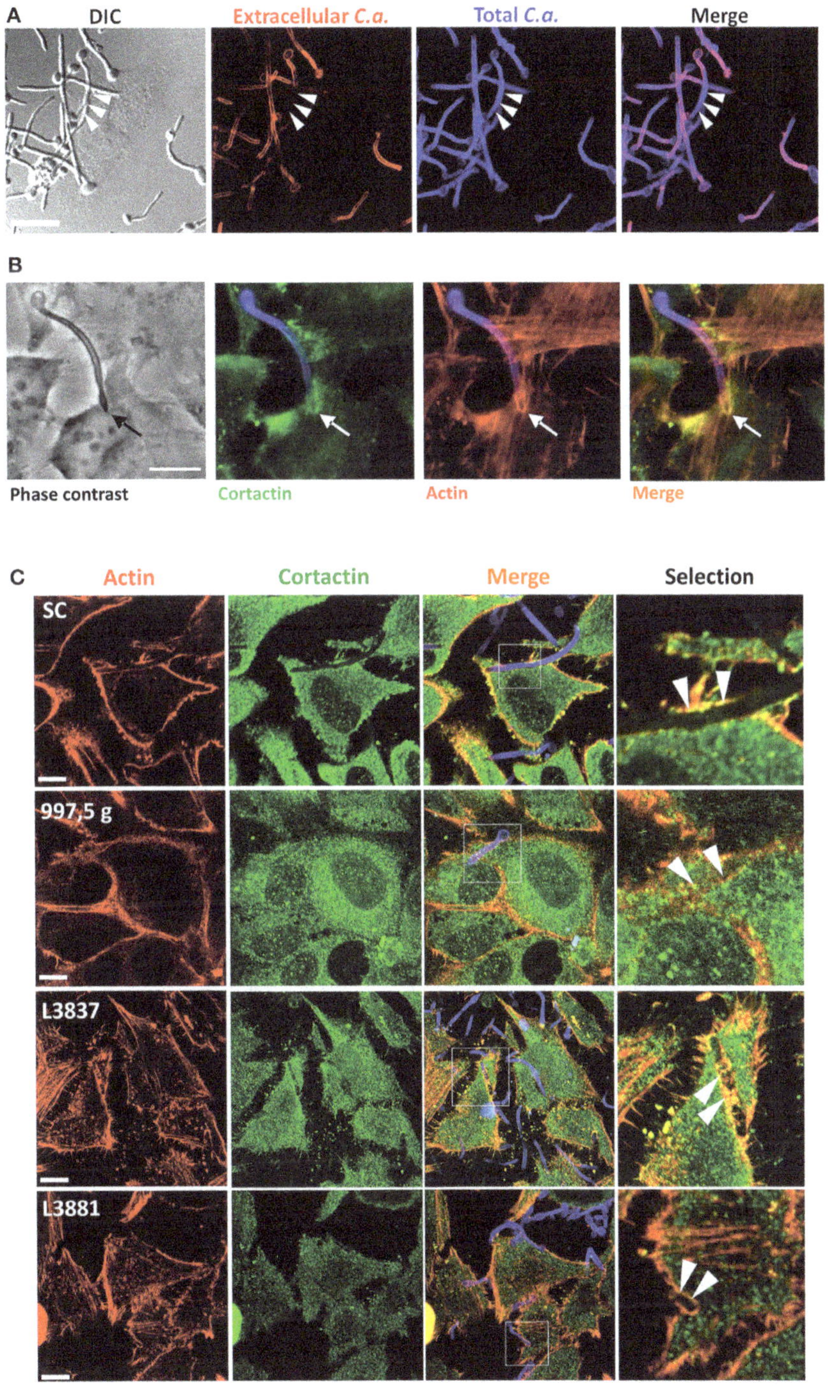

FIGURE 2 | The hyphae of all four *C. albicans* isolates induced the recruitment and colocalization of actin and cortactin. **(A)** *C. albicans* SC5314 strain invaded HeLa cells. *C. albicans* SC5314 blastospores were seeded with HeLa cells for 2 h at 37°C and 5% CO_2. Extracellular hyphae were stained with Concanavalin A–rhodamine (red), and, after permeabilization, extra- and intracellular hyphae were stained with Calcofluor White (blue). Bar = 10 μm. The arrows show hyphae internalization. **(B)** Interaction of HeLa cells with *C. albicans* SC5314 led to the recruitment and colocalization of actin and cortactin. *C. albicans* SC5314 blastospores were seeded for 2 h with HeLa cells and stained with phalloidin-TRITC (actin: red-orange), Calcofluor white (fungal cell wall: blue) and anti-cortactin antibody followed by Alexa Fluor-488® conjugated secondary antibody (cortactin: green). Epifluorescence microscopy. Bar: 10 μm. The arrows clearly show the recruitment and colocalization of both molecules at the hyphae invasion site in HeLa cells. **(C)** The four *C. albicans* isolates recruit actin and cortactin. Blastospores from SC5314, 997,5 g, L3881 and L3837 isolates were seeded with HeLa cells in RPMI medium for 2 h at 37°C and 5% CO_2, washed, fixed, immunostained for cortactin, labeled with phalloidin for host actin and evaluated by confocal microscopy. Invading hyphae from all four isolates were able to induce the recruitment and colocalization of actin (red) and cortactin (green) to the proximity of the hyphal invasion sites (arrowheads). Bar = 20 μm.

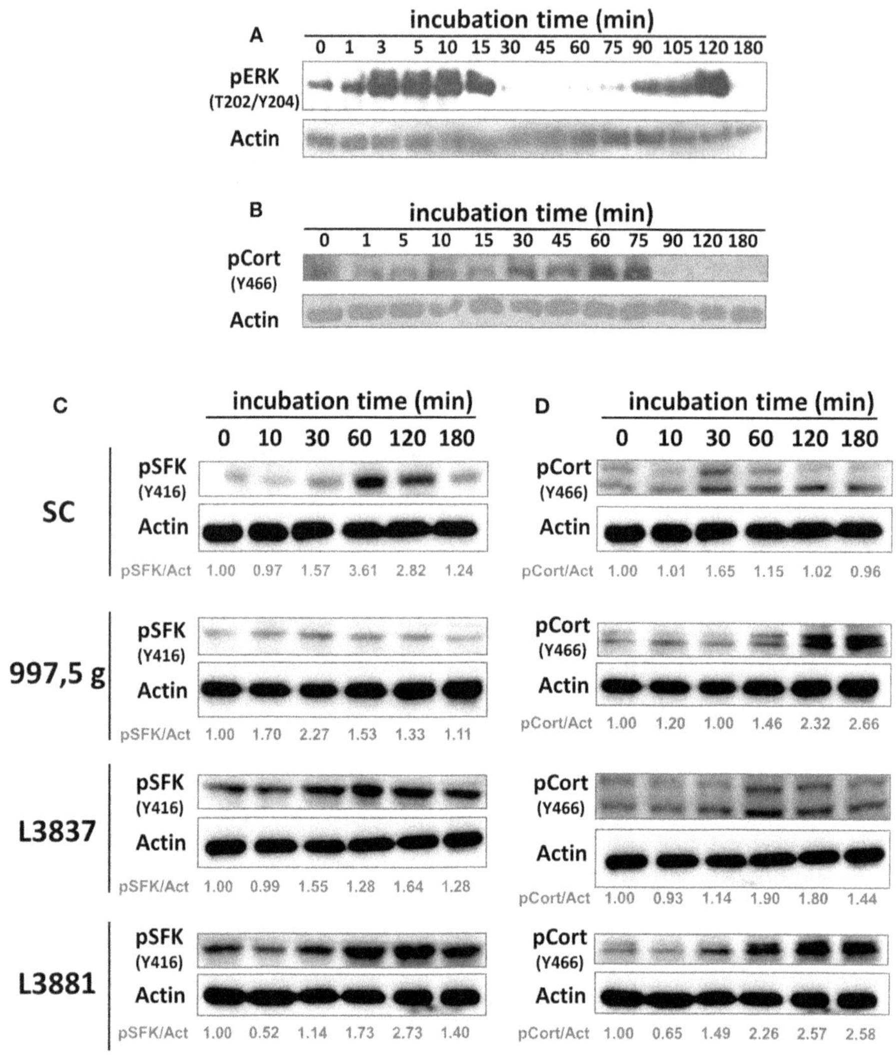

FIGURE 3 | *C. albicans* SC5314 strain induced host cell biphasic ERK 1/2 activation and cortactin phosphorylation by SFKs. HeLa cells were incubated with blastospores from the SC5314 strain for the indicated time points. After incubation, HeLa cells were harvested, lysed, and their protein lysates separated by sodium dodecyl sulfate polyacrylamide gel electrophoresis (SDS-PAGE) and analyzed by Western blot using the indicated antibodies. The SC5314 strain induced **(A)** a biphasic activation of host ERK 1/2 (the other isolates did not, not shown) and also **(B)** the phosphorylation of cortactin (pCort [Y466]) by SFKs. These are representative observations from at least three independent experiments. SC5314, 997,5 g, oral L3837 and blood L3881 isolates induced host cell signaling activation. HeLa cells were incubated with blastospores from four *C. albicans* isolates for the indicated time points. After incubation, HeLa cells were harvested, lysed, and SFK and cortactin activation were analyzed by Western blot using the indicated antibodies. Cortactin, SFKs and actin have different molecular weights (80–85, 60, and 45 kDa, respectively), thus the nitrocellulose membrane was cut into three horizontal strips, blocked and each strip was incubated with anti-pY466 cortactin, anti-pY416 SFK or anti-actin. **(C)** pSFK (Y416) and **(D)** pCort (Y466), cortactin phosphorylated by Src at the Y^{466} residue. Actin was used for protein loading normalization. Actin labeling is duplicated in **(C,D)**. pSFK/Act = densitometric rate of pSFK over actin. pCort/Act = densitometric rate of pCort over actin. These are representative observations from at least three independent experiments.

DISCUSSION

Candida albicans is an opportunistic human pathogen that can cause superficial and hematogenous infections. The latter result in substantial morbidity and mortality and are a common, costly problem in healthcare settings. To access the systemic circulation, *C. albicans* has to invade both epithelial and endothelial barriers. For epithelial cells, two mechanisms have been described their invasion by *C. albicans*: active penetration and induced, actin-driven endocytosis. In this study, we have demonstrated that *C. albicans* strains—the reference strain SC5314 and three isolates with different clinical backgrounds—can invade HeLa cells and recruit cortactin and actin to sites of hyphae internalization, suggesting that at least part of the invasion process is mediated by induced endocytosis—which is dependent on dynamic microfilaments of the host—independently of the source of the clinical isolate. All four strains induced

SFK/cortactin phosphorylation/activation, but to varying levels.

Invasive Ability, Cortactin/Actin Recruitment, Induction of Src/Cortactin Phosphorylation and Survival Under Oxidative Stress by *C. albicans* Are Independent of Hyphal Length or Damage/Internalization Rates

In this study, all isolates of *C. albicans* from varying clinical backgrounds were able to invade and damage HeLa cells to different levels, and induced the accumulation of cortactin and actin around *C. albicans* hyphae that were in the process of being internalized by HeLa cells. Similar effects were observed for the reference strain, SC5314, in other epithelial cell types (Moreno-Ruiz et al., 2009). In addition, the invasive ability of *C. albicans* is apparently not related to the clinical background of the isolate.

It is noteworthy that 997,5 g strain—isolated from the blood of a patient who died within 24 h (unpublished results)—exhibited the greatest ability of all isolates tested to both internalize cells and cause damage *in vitro*.

Conversely, *C. albicans* L3881—isolated from the blood of a patient who had experienced candidemia but had undergone a clinical cure—expressed the *HWP1-2* allele and showed very reduced levels of *HWP1* mRNA expression (Padovan et al., 2009). In addition to displaying the shortest hyphae, as expected, *C. albicans* L3881 also showed the lowest rate of damage and internalization. Despite this, *C. albicans* L3881 was able to reach blood vessels, invade HeLa cells, recruit actin/cortactin, activate Src and cortactin by phosphorylation and, interestingly, survive under the pressure of nutritional and oxidative stress (i.e., ROS).

It is worth mentioning that many groups have previously shown the importance of hyphae length using *HWP1* mutants. For instance, Tsuchimori et al. (2000) have shown that a *HWP1* deletion mutant forms shorter hyphae and has decreased endothelial cell adherence, uptake and damage. Moreover, Wächtler et al. (2011) have examined a number of mutants for adhesion, internalization and damage of oral epithelial cells and have shown that a mutant with a HWP1 deletion has significant reduction in its epithelial cell adhesion and damage, but not internalization, which already indicates that hyphal length is likely to be important, but not an obstacle to cellular invasion. Indeed, in all cases, the deletions have been constructed artificially and the cell types used were different from this present study.

As a host defense mechanism, the production of ROS is a marked characteristic of phagocytic cells, which use such oxygen intermediates to kill invading pathogens. On one hand, *C. albicans* is relatively resistant to ROS, tolerating up to 20 mmol/L hydrogen peroxide (H_2O_2) under some conditions (Brown et al., 2014). Given that adaptation to oxidative stress is essential for pathogenicity, one can conclude that the inactivation of detoxifying enzymes, such as superoxide dismutases by *C. albicans* lead to severe attenuations in virulence and viability inside immune cells. Upon interaction with pathogens, phagocytes rapidly produce ROS. The ability to counteract oxidative attack by the patient would presumably aid fungal adaptation and invasiveness through the host. In the present study, L3881 isolate was resistant to oxidative stress and could reach the blood vessels of the patient, despite displaying a *HWP1* natural deletion and the shorter hyphae. It is an open question whether L3881 subverts the innate immune response by mediating the efficient detoxification of host-derived ROS, which ultimately allows it to reach deep tissues such as blood vessels.

Moreover, L3881 was also more resistant to EDTA than the SC5314 reference strain. This evidence indicates that the L3881 blood isolate has developed plasticity to deal with challenges and maintain its growth under different environmental conditions.

Cortactin possesses a series of C-terminal tyrosines that are heavily Src-phosphorylated and implicated in regulating actin remodeling during cell motility (Head et al., 2003). Phosphorylation of cortactin by Src greatly enhanced Arp2/3 complex-mediated actin polymerization (Tehrani et al., 2007). This event required Nck, an adaptor that links phosphorylated cortactin with neuronal WASp (N-WASp) and WASp-interacting protein (WIP) to activate the Arp2/3 complex, enhancing polymerization. Src phosphorylation of cortactin is important for actin polymerization during host cell invasion by some bacteria (*Shigella flexneri* and *Staphylococcus aureus*) or parasites (*Cryptosporidium parvum*) (Chen et al., 2003; Bougnères et al., 2004; Agerer et al., 2005).

The *C. albicans* reference strain and clinical isolates activated Src phosphorylation and consequently Src phosphorylation of cortactin (Y466) during HeLa cell invasion, thus pointing to a concerted series of events involving phosphorylation, and culminating in actin remodeling and actin-driven *C. albicans* invasion. Actin-driven invasion is a process that can be interrupted by both the up- and down-regulation of actin polymerization, thus drawing attention to the fact that actin polymerization must be carefully controlled to generate productive results; excessive assembly can bring the network to a halt (Haglund and Welch, 2011).

The tyrosine SFKs play a fundamental role in a wide variety of cellular processes including morphogenesis and proliferation, phagocytosis and host–pathogen interactions. In this study, it became clear that SFKs are important for *C. albicans* invasion, which has been evidenced by the decrease of internalization rate of all *Candida* isolates into HeLa cells when using a SFK inhibitor. SFKs normally drive actin-dependent rearrangements required for cellular migration and may be used by the fungus to promote cell invasion in the same way of several bacteria (Mounier et al., 2009). Upon pathogen-cell adhesion, SFKs may associate to and activate other molecules such as FAK to regulate several actin-related processes, including phosphorylation of key substrates such as paxillin and cortactin (Ferreira et al., 2016). It has also been shown that activation of host cell SFK is correlated to the actin dependent internalization of bacterial pathogens (Pizarro-Cerdá et al., 2016). Src-dependent remodeling of the actin cytoskeleton has been also reported for viruses and parasites, suggesting that such signaling corresponds to a common feature that controls actin dynamics at the membrane (Mitra and Schlaepfer, 2006; Munter et al., 2006).

As far as we know this is the first time that the inhibition of Src-familiy kinases has been shown to hinder host cell invasion by *C. albicans*, pointing to a key role for SFK in driving actin polymerization and consequently the internalization of *C. albicans*.

Several pathogens have been described to enter professional phagocyte cells via cholesterol-enriched membrane lipid raft microdomains. In fact, *C. albicans* uptake by human monocyte has been strongly impaired by the cholesterol extracting agent methyl-β-cyclodextrin in the study by de Turris et al. (2015). It is worth mentioning that HeLa cells are not professional phagocytes as monocytes or macrophages. This is probably why the disruption of lipid rafts in HeLa cells did not impact on *C. albicans* internalization.

Collectively, these results suggest that *C. albicans* invasion of HeLa epithelial cells—and also the ability to reach and adapt to live in blood—may be independent of fungal adherence and hyphal length, or even the ability to damage cells, since L3881, isolated from a patient's blood, has been able to both invade cells and, recruit actin/cortactin, activate SFKs and cortactin, survive under nutritional and oxidative stress and reach blood tissues, despite its shorter hyphae.

Final Consideration—Building a Niche

When discussing *C. albicans* invasion, it is known that to infect mucosal tissues, gain access to blood and escape blood flow, *C. albicans* must interact with both epithelial and endothelial cells, respectively (Davis, 2009). In superficial infections (skin and mucosa), this fungus only reaches blood vessels when a mechanical disruption (wound or cut) to epithelial tissue occurs.

Interestingly, oropharyngeal candidiasis is prevalent in HIV-positive patients and, although rare (especially in the terminally ill), HIV-positive patients can develop systemic candidiasis. Regarding the mechanisms involved in *Candida* penetration of epithelial cells, what is known from *in vitro* experiments is that active penetration is the predominant form of invasion, while induced endocytosis seems to occur in the initial phases of tissue colonization (Wächtler et al., 2012).

The fungus uses a variety of strategies in order to survive; other types of interactions may exist *in vivo* that we have not yet seen *in vitro*. In systemic infections, it is believed that the gastrointestinal tract is a major source of *Candida* invasion in addition to catheters or surgery, for example (Perlroth et al., 2007). The basic pH of gastrointestinal tissue, its 37°C temperature, ample contact with epithelial cells and an iron-restricted environment are strong inducers of hyphae formation, i.e., this fungus always forms hyphae when it meets its host (Almeida et al., 2008). However, in infected organs, hyphae, yeasts and pseudohyphae can be observed by histological analysis, probably because yeasts are part of the proliferation process (the more yeasts present, the more hyphae formed). However, the importance of pseudohyphae remains unknown.

Hematogenous infections caused by pathogens of the *Candida* genus have increased heavily in Latin America (Doi et al., 2016) mainly because modern medical treatments are becoming more invasive and more people are being treated with immunosuppressive drugs. Hematogenous candidiasis is a silent disease. Often, clinicians are only able to detect the infectious agent when the patient already has an irreversible condition. The mortality rate for hematogenous candidiasis is 40–60%, which are devastating numbers. However, drugs commonly used to control the infection cause pathogen resistance. Epithelial cells play a key role in many human–microbiota and/or pathogens interactions. Interactions of these cells with *C. albicans* have been shown to be critical to host and fungal cell responses (Moyes et al., 2015). Molecules of both *Candida* and the host are important for *Candida* internalization into epithelial cells, which is the first step to further dissemination leading to hematogenic infections. A better elucidation of these factors may significantly contribute to the control of systemic infections in hospitals, such as the rational development of new compounds against the essential components triggered during the invasion of host cells by the fungus.

AUTHOR CONTRIBUTIONS

Conception of the study: DB. Designed the experiments: PM, AB, AP, FS, RA, ES, and DB. Performed the experiments: PM, AB, CO, LR, TM. Interpretation of the results and data analysis: PM, AB, AP, RM, FS, RA, ES, and DB. Contributed reagents/materials/analysis tools: AP, RM, FS, RA, ES, and DB. Wrote the manuscript: PM, AB, and DB. All authors read and approved the final manuscript.

FUNDING

The authors wish to gratefully acknowledge the financial support of CNPq (477867/2010-3), FAPESP (2011/22773-6; 2015/25652-6) and FAPEMIG (APQ-01674-14).

ACKNOWLEDGMENTS

The authors are indebted to Prof. Dr. Bernhard Hube (Department of Microbial Pathogenicity Mechanisms, Leibniz Institute for Natural Product Research and Infection Biology—Hans Knoell Institute Jena, Germany) for his critical revision of the manuscript and valuable comments. DB, ES, and RM are each the recipients of CNPq fellowships. The authors would also like to thank BioMed Proofreading (Cleveland, OH, USA).

REFERENCES

Agerer, F., Lux, S., Michel, A., Rohde, M., Ohlsen, K., and Hauck, C. R. (2005). Cellular invasion by *Staphylococcus aureus* reveals a functional link between focal adhesion kinase and cortactin in integrin-mediated internalization. *J. Cell Sci.* 118, 2189–2200. doi: 10.1242/jcs.02328

Almeida, R. S., Brunke, S., Albrecht, A., Thewes, S., Laue, M., Edwards, J. E., et al. (2008). The hyphal-associated adhesin and invasin Als3 of *Candida albicans* mediates iron acquisition from host ferritin. *PLoS Pathog.* 4:e1000217. doi: 10.1371/journal.ppat.1000217

Bonfim-Melo, A., Zanetti, B. F., Ferreira, É. R., Vandoninck, S., Han, S. W., Van Lint, J., et al. (2015). *Trypanosoma cruzi* extracellular amastigotes trigger the protein kinase D1-cortactin-actin pathway during cell invasion. *Cell Microbiol.* 17, 1797–1810. doi: 10.1111/cmi.12472

Bougnères, L., Girardin, S. E., Weed, S. A., Karginov, A. V., Olivo-Marin, J. C., Parsons, J. T., et al. (2004). Cortactin and Crk cooperate to trigger actin polymerization during *Shigella* invasion of epithelial cells. *J. Cell Biol.* 166, 225–235. doi: 10.1083/jcb.200402073

Brown, A. J., Budge, S., Kaloriti, D., Tillmann, A., Jacobsen, M. D., Yin, Z., et al. (2014). Stress adaptation in a pathogenic fungus. *J. Exp. Biol.* 217(Pt 1), 144–155. doi: 10.1242/jeb.088930

Chaves, G. M., Santos, F. P., and Colombo, A. L. (2012). The persistence of multifocal colonization by a single ABC genotype of *Candida albicans* may predict the transition from commensalism to infection. *Mem. Inst. Oswaldo Cruz.* 107, 198–204. doi: 10.1590/S0074-02762012000200008

Chen, X. M., Huang, B. Q., Splinter, P. L., Cao, H., Zhu, G., McNiven, M. A., et al. (2003). *Cryptosporidium parvum* invasion of biliary epithelia requires host cell tyrosine phosphorylation of cortactin via c-Src. *Gastroenterology* 125, 216–228. doi: 10.1016/S0016-5085(03)00662-0

Daly, R. J. (2004). Cortactin signalling and dynamic actin networks. *Biochem. J.* 15, 13–25. doi: 10.1042/BJ20040737

Davis, D. A. (2009). How human pathogenic fungi sense and adapt to pH, the link to virulence. *Curr. Opin. Microbiol.* 12, 365–370. doi: 10.1016/j.mib.2009.05.006

de Turris, V., Teloni, R., Chiani, P., Bromuro, C., Mariotti, S., Pardini, M., et al. (2015). *Candida albicans* targets a lipid raft/dectin-1 platform to enter human monocytes and induce antigen specific T cell responses. *PLoS ONE* 10:e0142531. doi: 10.1371/journal.pone.0142531

Doi, A. M., Pignatari, A. C., Edmond, M. B., Marra, A. R., Camargo, L. F., Siqueira, R. A., et al. (2016). Epidemiology and microbiologic characterization of nosocomial candidemia from a Brazilian National Surveillance program. *PLoS ONE* 11:e0146909. doi: 10.1371/journal.pone.0146909

Ferreira, É. R., Horjales, E., Bonfim-Melo, A., Cortez, C., da Silva, C. V., De Groote, M., et al. (2016). Unique behavior of *Trypanosoma cruzi* mevalonate kinase: a conserved glycosomal enzyme involved in host cell invasion and signaling. *Sci. Rep.* 6:24610. doi: 10.1038/srep24610

Filler, S. G., and Sheppard, D. C. (2006). Fungal invasion of normally non-phagocytic host cells. *PLoS Pathog.* 2:e129. doi: 10.1371/journal.ppat.0020129

Frank, C. F., and Hostetter, M. K. (2007). Cleavage of E-cadherin, a mechanism for disruption of the intestinal epithelial barrier by *Candida albicans*. *Transl. Res.* 149, 211–222. doi: 10.1016/j.trsl.2006.11.006

Gillum, A. M., Tsay, E. Y., and Kirsch, D. R. (1984). Isolation of the *Candida albicans* gene for orotidine-5V-phosphate decarboxylase by complementation of *S. cerevisiae* ura3 and *E. coli* pyrF mutations. *Mol. Gen. Genet.* 198, 179–182. doi: 10.1007/BF00328721

Haglund, C. M., and Welch, M. D. (2011). Pathogens and polymers, Microbe-host interactions illuminate the cytoskeleton. *J. Cell Biol.* 195, 7–17. doi: 10.1083/jcb.201103148

Head, J. A., Jiang, D., Li, M., Zorn, L. J., Schaefer, E. M., Parsons, J. T., et al. (2003). Cortactin tyrosine phosphorylation requires Rac1 activity and association with the cortical actin cytoskeleton. *Mol. Biol. Cell* 14, 3216–3229. doi: 10.1091/mbc.e02-11-0753

Jacobsen, I. D., Wilson, D., Wächtler, B., Brunke, S., Naglik, J. R., and Hube, B. (2012). *Candida albicans* dimorphism as a therapeutic target. *Expert Rev. Anti Infect. Ther.* 10, 85–93. doi: 10.1586/eri.11.152

Martinez-Quiles, N., Ho, H. Y., Kirschner, M. W., Ramesh, N., and Geha, R. S. (2004). Erk/Src phosphorylation of cortactin acts as a switch on-switch off mechanism that controls its ability to activate N-WASP. *Mol. Cell. Biol.* 24, 5269–5280. doi: 10.1128/MCB.24.12.5269-5280.2004

Mitra, S. K., and Schlaepfer, D. D. (2006). Integrin-regulated FAK-Src signaling in normal and cancer cells. *Curr. Opin. Cell Biol.* 18, 516–523. doi: 10.1016/j.ceb.2006.08.011

Moreno-Ruiz, E., Galán-Díez, M., Zhu, W., Fernández-Ruiz, E., d'Enfert, C., Filler, S. G., et al. (2009). *Candida albicans* internalization by host cells is mediated by a clathrin-dependent mechanism. *Cell. Microbiol.* 11, 1179–1189. doi: 10.1111/j.1462-5822.2009.01319.x

Mounier, J., Popoff, M. R., Enninga, J., Frame, M. C., Sansonetti, P. J., and Van Nhieu, G. T. (2009). The IpaC carboxyterminal effector domain mediates src-dependent actin polymerization during shigella invasion of epithelial cells. *PLoS Pathog.* 5:e1000271. doi: 10.1371/journal.ppat.1000271

Moyes, D. L., Richardson, J. P., and Naglik, J. R. (2015). *Candida albicans*-epithelial interactions and pathogenicity mechanisms, scratching the surface. *Virulence* 6, 338–346. doi: 10.1080/21505594.2015.1012981

Moyes, D. L., Runglall, M., Murciano, C., Shen, C., Nayar, D., Thavaraj, S., et al. (2010). A biphasic innate immune MAPK response discriminates between the yeast and hyphal forms of *Candida albicans* in epithelial cells. *Cell Host Microbe.* 8, 225–235. doi: 10.1016/j.chom.2010.08.002

Moyes, D. L., Shen, C., Murciano, C., Runglall, M., Richardson, J. P., Arno, M., et al. (2014). Protection against epithelial damage during *Candida albicans* infection is mediated by PI3K/Akt and mammalian target of rapamycin signaling. *J. Infect. Dis.* 209, 1816–1826. doi: 10.1093/infdis/jit824

Munter, S., Way, M., and Frischknecht, F. (2006). Signaling during pathogen infection. *Sci. STKE* 2006:re5.S. doi: 10.1126/stke.3352006re5

Naglik, J. R., Richardson, J. P., and Moyes, D. L. (2014). *Candida albicans* pathogenicity and epithelial immunity. *PLoS Pathog.* 10:e1004257. doi: 10.1371/journal.ppat.1004257

Padovan, A. C., Chaves, G. M., Colombo, A. L., and Briones, M. R. (2009). A novel allele of HWP1, isolated from a clinical strain of *Candida albicans* with defective hyphal growth and biofilm formation, has deletions of Gln/Pro and Ser/Thr repeats involved in cellular adhesion. *Med. Mycol.* 47, 824–835. doi: 10.3109/13693780802669574

Perlroth, J., Choi, B., and Spellberg, B. (2007). Nosocomial fungal infections, epidemiology, diagnosis, and treatment. *Med. Mycol.* 45, 321–346. doi: 10.1080/13693780701218689

Phan, Q. T., Myers, C. L., Fu, Y., Sheppard, D. C., Yeaman, M. R., Welch, W. H., et al. (2007). Als3 is a *Candida albicans* invasin that binds to cadherins and induces endocytosis by host cells. *PLoS Biol.* 5:e64. doi: 10.1371/journal.pbio.0050064

Pizarro-Cerdá, J., Charbit, A., Enninga, J., Lafont, F., and Cossart, P. (2016). Manipulation of host membranes by the bacterial pathogens Listeria, Francisella, Shigella and Yersinia. *Semin. Cell Dev. Biol.* 60, 155–167. doi: 10.1016/j.semcdb.2016.07.019

Tehrani, S., Tomasevic, N., Weed, S., Sakowicz, R., and Cooper, J. A. (2007). Src phosphorylation of cortactin enhances actin assembly. *Proc. Natl. Acad. Sci. U.S.A.* 104, 11933–11938. doi: 10.1073/pnas.0701077104

Tsuchimori, N., Sharkey, L. L., Fonzi, W. A., French, S. W., Edwards, J. E. Jr., and Filler, S. G. (2000). Reduced virulence of HWP1-deficient mutants of *Candida albicans* and their interactions with host cells. *Infect. Immun.* 68, 1997–2002. doi: 10.1128/IAI.68.4.1997-2002.2000

Wächtler, B., Citiulo, F., Jablonowski, N., Förster, S., Dalle, F., Schaller, et al. (2012). *Candida albicans*-epithelial interactions, dissecting the roles of active penetration, induced endocytosis and host factors on the infection process. *PLoS ONE* 7:e36952. doi: 10.1371/journal.pone.0036952

Wächtler, B., Wilson, D., Haedicke, K., Dalle, F., and Hube, B. (2011). From attachment to damage: defined genes of *Candida albicans* mediate adhesion, invasion and damage during interaction with oral epithelial cells. *PLoS ONE* 6:e17046. doi: 10.1371/journal.pone.0017046

Wilson, D., Thewes, S., Zakikhany, K., Fradin, C., Albrecht, A., Almeida, R., et al. (2009). Identifying infection-associated genes of *Candida albicans* in the postgenomic era. *FEMS Yeast Res.* 9, 688–700. doi: 10.1111/j.1567-1364.2009.00524.x

Yang, W., Yan, L., Wu, C., Zhao, X., and Tang, J. (2014). Fungal invasion of epithelial cells. *Microbiol. Res.* 169, 803–810. doi: 10.1016/j.micres.2014.02.013

Zhu, W., Phan, Q. T., Boontheung, P., Solis, N. V., Loo, J. A., and Filler, S. G. (2012). EGFR and HER2 receptor kinase signaling mediate epithelial cell invasion by *Candida albicans* during oropharyngeal infection. *Proc. Natl. Acad. Sci. U.S.A.* 109, 14194–14199. doi: 10.1073/pnas.1117676109

18

A Critical Role for Hemolysin in *Vibrio fluvialis*-Induced IL-1β Secretion Mediated by the NLRP3 Inflammasome in Macrophages

Liqiong Song[1†], Yuanming Huang[1†], Meng Zhao[1], Zhihao Wang[1], Shujing Wang[1], Hui Sun[1], Biao Kan[1], Guangxun Meng[2‡], Weili Liang[1,2‡] and Zhihong Ren[1,2*‡]*

[1] State Key Laboratory for Infectious Disease Prevention and Control, Collaborative Innovation Center for Diagnosis and Treatment of Infectious Diseases, National Institute for Communicable Disease Control and Prevention – Chinese Center for Disease Control and Prevention, Beijing, China, [2] Unit of Innate Immunity, Key Laboratory of Molecular Virology and Immunology, Institute Pasteur of Shanghai – Chinese Academy of Sciences, Shanghai, China

Edited by:
Diana Bahia, Universidade Federal de São Paulo, Brazil

Reviewed by:
Birgit Strobl, University of Veterinary Medicine Vienna, Austria
Yukihiro Akeda, Osaka University, Japan

***Correspondence:**
Zhihong Ren and Weili Liang, State Key Laboratory for Infectious Disease Prevention and Control, Collaborative Innovation Center for Diagnosis and Treatment of Infectious Diseases, National Institute for Communicable Disease Control and Prevention – Chinese Center for Disease Control and Prevention, 155 Changbai Road, Changping, Beijing 102206, China
renzhihong@icdc.cn; liangweili@icdc.cn

† These authors have contributed equally to this work.
‡ These authors shared senior authorship.

Vibrio fluvialis causes human diarrhea, but the pathogenesis is not well-studied. We hypothesized that *V. fluvialis*-secreted hemolysin (VFH) may induce IL-1β secretion through the activation of the NLRP3 inflammasome and contribute to the pathogenicity of *V. fluvialis.* To examine this possibility, we constructed VFH mutant and complement strains and demonstrated that *V. fluvialis*-induced IL-1β production and cytotoxicity in human monocytic THP-1 cells and mouse macrophages is attributed to VFH. To evaluate the role of VFH *in vivo*, we infected adult C57BL/6 mice intraperitoneally and suckling C57/B6 mice orally with various strains. The mice treated with 10^8 CFU wild-type *V. fluvialis* or cell-free supernatant containing VFH induced significantly higher IL-1β production in peritoneal lavage fluid or in colon compared with those infected with the mutant strain, while no effect on TNF and IL-6 production was observed at day 5 or 24 h post-infection. VFH contributed to pathological changes and IL-1β release independent of colonization of *V. fluvialis* in the colon. VFH has no effect on the synthesis of pro-IL-1β, but rather it triggers the processing of pro-IL-1β into IL-1β. Furthermore, using deficient mouse strains, we verified that *V. fluvialis*-induced IL-1β is mediated through activation of Caspase-1 and the NLRP3 inflammasome *ex vivo.* Confocal microscopy suggests that VFH contributes to cathepsin B release. Furthermore, *V. fluvialis*-induced IL-1β secretion requires potassium (K^+) efflux and reactive oxygen species production. Our results provide new evidence for the role of VFH in the activation of the NLRP3 inflammasome and pathogenesis in response to *V. fluvialis* infection.

Summary Sentence: *Vibrio fluvialis*-secreted hemolysin induces IL-1β secretion through the activation of the NLRP3 inflammasome and contributes to the pathogenicity of *V. fluvialis*.

Keywords: *Vibrio fluvialis*, hemolysin, IL-1β, NLRP3 inflammasomes, macrophages

---

Abbreviations: ASC, adapter molecule; *Asc*$^{-/-}$, Asc-deficient; BMMs, mouse bone marrow derived macrophages; *Caspase-1*$^{-/-}$, Caspase-1-deficient; i.p., intraperitoneally; LDH, lactate dehydrogenase; MOI, multiplicity of infection; NLRs, nod-like receptors; *Nlrp3*$^{-/-}$, Nlrp3-deficient; NLRP3, NLR family pyrin domain containing 3; PLF, peritoneal lavage fluid; pUC-*vfh*, Δ*vfh V. fluvialis* with complement plasmid; ROS, reactive oxygen species; VFH, *V. fluvialis*-secreted hemolysin; Δ*vfh*, isogenic *vfh* mutant; WT, wild-type.

Introduction

Vibrio fluvialis causes mild to moderate dehydration, vomiting, fever, abdominal pain, diarrhea and it can be isolated from human diarrheal feces and aquatic environments (Igbinosa and Okoh, 2010; Ramamurthy et al., 2014). *V. fluvialis* infection has become an increasing public health hazard worldwide and frequently occurs in countries where the raw seafood is largely consumed (Ramamurthy et al., 2014). *V. fluvialis* accounted for 10% of *Vibrio*-caused clinical cases along the Gulf Coast in U.S. (Levine and Griffin, 1993). In China, information about the etiological characteristics of *V. fluvialis* and its epidemiology of infection are limited due to complexity in the identification and less attention in the pathogen surveillance (Liang et al., 2013).

It has been reported that *V. fluvialis* elicits intestinal fluid when fed to suckling mice and produces an array of virulence factors and toxins (Lockwood et al., 1982; Kothary et al., 2003). Cell-free culture filtrates of *V. fluvialis* strains are able to evoke distinct cytotoxic and vacuolization effects on HeLa cells (Chakraborty et al., 2005). Quorum sensing in *V. fluvialis* positively regulates production of an extracellular protease and hemolysin and affects cytotoxic activity against epithelial tissue cultures (Wang et al., 2013). Of note, the *V. fluvialis*-caused diarrhea with presence of blood is different from cholera. Thus far, the underlying mechanisms of inflammatory bloody diarrhea caused by *V. fluvialis* have not been fully defined.

The innate immune system recognizes microbial infections through a vast array of pathogen-associated molecular patterns (Takeuchi and Akira, 2010). After recognition of bacterial components, different TLRs activate signaling via the adapter MyD88, leading to the activation of the NF-κB signaling pathway. IL-1β, together with TNF-α and IL-6, are thought to be important proinflammatory mediators in initiating and maintaining the inflammatory response to pathogen and disease development during infection. IL-1β is expressed as a pro-form, which requires proteolytic cleavage for maturation. The cysteine protease Caspase-1 mediates the proteolytic processing and secretion of mature IL-1β (Fink and Cookson, 2005).

Caspase-1 is activated within inflammasomes, multiprotein complexes that also contain NLRs and the ASC. NLRP3 is essential for Caspase-1 activation in response to a variety of microbial molecules, necrotic cells and endogenous danger-associated molecules (Gurcel et al., 2006; Mariathasan et al., 2006; Kanneganti et al., 2007). Numerous pathogens activate the NLRP3 inflammasome through bacterial toxins, such as VFH (Toma et al., 2010), hemolysin of *Staphylococcus aureus* (Munoz-Planillo et al., 2009), pneumolysin (McNeela et al., 2010; Witzenrath et al., 2011), cytotoxins from *Aeromonas hydrophila* (McCoy et al., 2010). However, little is known about the host immune response to *V. fluvialis* infection. We hypothesized that upon *V. fluvialis* infection, the pore-forming VFH may induce the activation of NLRP3 inflammasome, leading to inflammatory response.

Here, we demonstrate that VFH induces cytotoxicity and the secretion of IL-1β in response to *V. fluvialis* infection in macrophages. VFH mediates the activation of the NLRP3 inflammasome and contributes to inflammatory pathology in the colon of suckling mouse orally infected with *V. fluvialis*. We further show that VFH-induced NLRP3 activation requires ROS production, cathepsin B release, and K^+ efflux.

Materials and Methods

Bacterial Strains

Wild-type *V. fluvialis* strain 85003 used in this study was isolated from patient in China with diarrhea. Its genome sequence is available in the Sequence Read Archive under accession no. SRX397301 (Lu et al., 2014). The Δ*vfh* was constructed as previously described using allele replacement strategies (Wang et al., 2013). To construct the pUC-*vfh*, a chromosomal DNA fragment comprising the *vfh* open reading frame and its promoter sequence was amplified with the primer pair 5′-CGG AAT TCT AAG ATC ATG TCT GAA TGT-3′/5′-CGG GAT CCC GAC TGA GTT CAG CTC TCA C-3′. The amplicon was cloned as an *EcoRI-BamHI* fragment in pUC18 and further confirmed by DNA sequencing. The resultant plasmid, pUC-*vfh*, was introduced into *V. fluvialis* Δ*vfh* by electroporation (Marcus et al., 1990). Hemolytic phenotypes were examined using Columbia blood agar and 2% sheep erythrocytes. Unless noted otherwise, all strains were grown with aeration in brain heart infusion broth at 37°C.

Mice and Cell Culture

C57BL/6 WT mice were obtained from Beijing Vital River Laboratory Animal Technology Co. Ltd. *Asc*$^{-/-}$ and *Nlrp3*$^{-/-}$) mice have been described previously (Mariathasan et al., 2004; Mao et al., 2014). *Caspase-1*$^{-/-}$ mice were obtained from the Jackson Laboratory and crossed onto the C57BL/6 genetic background for 10 generations. These mice are also deficient for functional Caspase-11 (Kayagaki et al., 2011). All animal studies were performed in accordance with protocols approved by the Welfare & Ethical Inspection in Animal Experimentation Committee at the Chinese CDC. BMMs were prepared from the femurs and tibias of the above mice and cultured for 6 days in RPMI 1640 supplemented with 10% heat-inactivated fetal bovine serum (Life Technologies, Grand Island, NY, USA), recombinant macrophage colony-stimulating factor (20 ng/ml; R&D), 25 mmol/L HEPES, 2 mmol/L glutamine, 100 KU/L penicillin, and 100 mg/L streptomycin (all from Gibco Invitrogen, Grand Island, NY, USA). Peritoneal exudate cells were obtained by peritoneal lavage and enriched for macrophages using the method of Kumagai et al. (1979). Macrophages prepared in these methods were at least 90% pure, as assessed by expression of F4/80 using flow cytometry. The human monocytic cell line THP-1 and mouse macrophage cell line IC-21 were purchased from ATCC (TIB-202 and TIB-186). THP-1 cells were allowed to differentiate to macrophages by incubating for 48 h in the presence of 1 nM phorbol myristate acetate. All cells in this study were cultured in RPMI 1640 at a maximum concentration of 1×10^6 cells/ml.

In Vivo Infection

Five-days-old C57BL/6 suckling mice were infected orally with WT *V. fluvialis*, Δ*vfh V. fluvialis*, or pUC-*vfh* at 10^8 CFU in 100 μl sterile PBS. Control mice were given PBS by gavage. Mice were sacrificed 24 h post-infection (p.i.), and the ileum and colon were homogenized or collected for staining. Diluted homogenates were plated on LB plates containing 50 μg/ml streptomycin to quantify the CFU of *V. fluvialis* colonization. A portion of the colon was fixed in 10% formaldehyde, paraffin embedded, cut into 6-mm sections, stained with hematoxylin and eosin and observed under optical microscopy. A blinded grading was performed semi-quantitatively by an outside pathologist to assess the relative degree of pathology changes. The scoring of pathology slides was assessed as follows: 0, no inflammation; 1^+, mild inflammatory cell infiltrate with cuffing around vessels; 2^+, increased inflammation with hyperemia; 3^+, severe inflammation involving vascular dilation; and 4^+, submucosal edema.

For cytokine analysis, Four-weeks-old WT C57BL/6 mice were inoculated intraperitoneally (i.p.) with 10^7, 10^8, or 10^9 CFU of different bacteria in 200 μl PBS or PBS alone for control mice. Mice were sacrificed at day 5 p.i. Cytokine production in PLF was measured by ELISA using commercial kits (BD Biosciences, San Jose, CA, USA).

Ex Vivo Bacterial Infection, Lactate Dehydrogenase Assay, ELISA, and ATP Assessment

Various cells were incubated with WT, Δ*vfh*, or Δ*vfh* containing pUC-*vfh* or control vector pUC18 strains of *V. fluvialis* at a MOI of 50 per cell for 3 h without LPS priming. For some experiments, the BMMs were pre-incubated for 2 h with CA-074 Me (C5857, Calbiochem, San Diego, CA, USA), Z-YVAD-FMK (Alexis Biochemicals, Lörrach, Germany), oATP (Merck, Darmstadt, Germany), *N*-acetyl-L-cysteine (NAC; A7250 Sigma-Aldrich), or KCl at the indicated concentrations. At the indicated time points, LDH activity in the culture supernatants was measured using a Cytotox96 Kit (Promega, Madison, WI, USA). The concentrations of cytokines in cell-free supernatants were quantified using commercial ELISA kits (BD Biosciences, San Jose, CA, USA). The release of ATP from the BMMs was monitored using a bioluminescence assay kit (Molecular Probes).

Immunoblotting

Cell-free supernatants were concentrated using Amicon Ultra-4 10K Centrifugal Filter Devices (Millipore, Bedford, MA, USA). Cell extracts and concentrated supernatants were analyzed by immunoblotting. Antibodies specific for IL-1β (no. sc-52012; Santa Cruz, CA, USA), Caspase-1 (no. sc-56036), or GAPDH (no. sc-137179; Santa Cruz, CA, USA) and fluorescence-labeled secondary antibody (IRDye 800-labeled anti-rabbit IgG; 611-132-002; Rockland, Gilbertsville, PA, USA) were used, and the proteins were detected using an Odyssey Infrared Imaging System (LI-COR, Lincoln, NE, USA).

Confocal Laser Scanning Microscopy

Mouse BMMs were left untreated or treated with bacteria. After 3 h, the extracellular bacteria were removed and the cells were fixed with 3% paraformaldehyde. The cells were then permeabilized and stained as described (Zhang et al., 2012).

Statistical Analysis

Statistical analysis was performed using one-way ANOVA with Newman–Keuls post-testing. The correlation between ATP and IL-1β concentrations in the supernatants of BMMs co-cultured with *V. fluvialis* was assessed using the Pearson's test and linear regression. Values of $P < 0.05$ were considered significant.

Results

V. fluvialis Colonizes Mouse Ileum and Colon and Induces IL-1β Release

To determine whether *V. fluvialis* can infect and productively colonize mice, we challenged suckling mice orally with different doses of *V. fluvialis* for 24 h. Our results show that *V. fluvialis* colonized in both ileum and colon in a dose-dependent manner (**Figure 1A**).

To assess the macrophage response to *V. fluvialis*, we treated mouse BMMs with *V. fluvialis* at different MOI. *V. fluvialis* induced IL-1β secretion and cytotoxicity in a dose-dependent manner, with peak levels of IL-1β production at 50 MOI, which caused 20% cytotoxicity (**Figures 1B,C**). At this MOI, *V. fluvialis* also promoted a time-dependent increase in IL-1β production, which peaked at 3 h p.i., as well as an increase in cytotoxicity (**Figures 1D,E**).

To assess the proinflammatory response induced by *V. fluvialis in vivo*, 4-weeks-old C57BL/6 mice were injected i.p. with 10^7 or 10^8 CFU of *V. fluvialis* and the cytokines levels in PLF were measured by ELISA at day 5 p.i. The results showed that the mice injected with *V. fluvialis* induced higher levels of IL-1β secretion, though the levels of TNF-α and IL-6 did not change significantly (**Figure 1F**). These results verify that *V. fluvialis* can colonize the mouse colon and specifically activate IL-1β.

VFH Plays a Critical Role in *V. fluvialis*-Induced IL-1β Release and Cytotoxicity *Ex Vivo*

To determine which component of *V. fluvialis* mediates *V. fluvialis*-induced IL-1β release and cytotoxicity in various cells, we constructed an isogenic mutant (Δ*vfh*) and complement strain of VFH (pUC-*vfh*). WT and pUC-*vfh* strains induced significantly higher levels of IL-1β and cytotoxicity compared with the Δ*vfh* strains in human THP-1 cells, mouse IC-21 cells, mouse peritoneal macrophages (PMs), and BMMs at 3 h p.i. (**Figures 2A,B**). In contrast, VFH did not contribute to the IL-6 and TNF-α production induced by *V. fluvialis* (**Figure 2C**).

To confirm that the secreted form of VFH can trigger IL-1β release and cytotoxicity, we treated BMMs with different supernatants of overnight cultures from WT, Δ*vfh*, and pUC-*vfh* strains for 3 h. The supernatant of WT and pUC-*vfh* induced significantly higher levels of IL-1β production and cytotoxicity in BMMs than the supernatant of Δ*vfh* (**Figures 2D,E**),

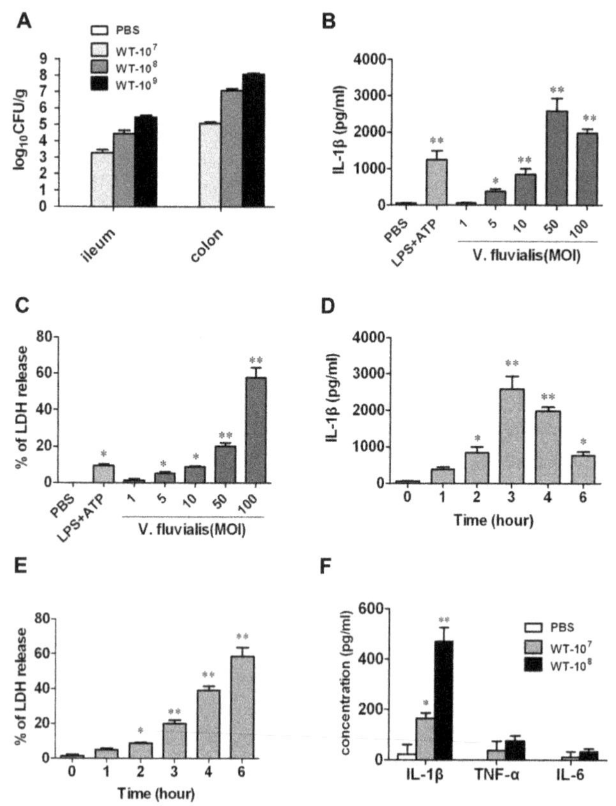

FIGURE 1 | *Vibrio fluvialis* colonizes the mouse colon and induces IL-1β production and cytotoxicity in BMMs. (A) Five-days-old C57BL/6 sucking mice were inoculated orally with 10^7, 10^8, or 10^9 CFU of *V. fluvialis,* and control mice were given PBS by gavage. The ileum and colon were collected and homogenized after 24 h, and bacterial colonization was determined by plating diluted homogenates on LB plates containing 50 μg/ml streptomycin ($n = 6$ mice per group). **(B)** BMMs (1×10^6) were infected with *V. fluvialis* at the indicated MOI. PBS was added as a negative control and LPS (100 ng/ml) + ATP (5 mM) was added as a positive control. After 3 h treatment, IL-1β production in culture supernatants was analyzed by ELISA. **(C)** BMMs (1×10^6) were infected with *V. fluvialis* at different doses. The release of LDH was measured as an indicator of cytotoxicity. **(D)** BMMs (1×10^6) were treated with *V. fluvialis* (MOI, 50) and supernatants were collected at different time points to detect IL-1β by ELISA. **(E)** BMMs (1×10^6) were treated with *V. fluvialis* (MOI, 50) and supernatants were collected at different time points to assess LDH for cytotoxicity. **(F)** Four-weeks-old C57BL/6 mice were injected i.p. with 10^7 or 10^8 CFU of *V. fluvialis,* and at day 5 p.i. PLF was harvested for IL-1β, TNF-α and IL-6 measurement by ELISA ($n = 6$ mice per group). ELISA and cytotoxicity data represent mean ± SD of six independent mice. $^*P < 0.05$; $^{**}P < 0.01$.

though the production of TNF-α and IL-6 was comparable for all supernatants (**Figure 2F**). These results suggest that VFH has an essential and specific role in the release of IL-1β and cytotoxicity in human and mouse monocyte/macrophage cell lines and primary mouse macrophages.

VFH Plays a Critical Role in *V. fluvialis*-Induced IL-1β Release and Cytotoxicity *In Vivo*

To evaluate the role of VFH in inflammatory response, bacterial colonization, and pathology change induced by *V. fluvialis in vivo*, we infected adult mice i.p. with 10^8 CFU WT, Δ*vfh,* and pUC-*vfh* strains and measured the cytokines in PLF at day 5 p.i. The mice injected with WT and pUC-*vfh* strains induced significantly more IL-1β than those injected with Δ*vfh* (**Figure 3A**). The VFH-dependent IL-1β induction was confirmed in colon using a suckling mouse model (**Figure 3B**). In contrast to the results for IL-1β, no difference among the strains was observed for TNF-α (**Figures 3C,D**) or IL-6 (**Figures 3E,F**).

To verify the *in vivo* function of soluble VHF, we inoculated suckling mice with supernatants from overnight cultures of various strains. The supernatant from WT and pUC-*vfh* compared with Δ*vfh* induced significantly higher levels of IL-1β release in colon without affecting TNF and IL-6 production (**Figure 3G**). Furthermore, we did not observe a difference in bacterial colonization in the colon after mice were challenged with three strains (**Figure 3H**), suggesting that VFH contributes to IL-1β release independent of colonization of *V. fluvialis.* Taken together, these findings confirm that VFH secreted by *V. fluvialis* specifically mediates IL-1β release upon *V. fluvialis* challenge.

VFH of *V. fluvialis* Induces Histopathological Lesions in the Colon

To characterize the inflammatory histopathology in colon caused by *V. fluvialis* infection, we inoculated five-days-old sucking mice with various strains or cell-free supernatant from overnight culture with various strains by gavage. At 24 h p.i., there were no lesions or obvious abnormalities in the negative control mice treated with PBS (**Figure 4A**); however, WT *V. fluvialis* caused submucosal edema, vascular dilation, hyperemia, and inflammatory cell infiltration as indicated by red arrow (**Figure 4B**). Mice infected with Δ*vfh* showed similar pathology to the control mice (**Figure 4C**), whereas the pUC-*vfh* strain induced severe lesions including obvious epithelial shedding, glandular structure damage and inflammatory cell infiltration as pointed by arrow (**Figure 4D**). The effects of the Δ*vfh* and WT supernatants were similar to the effects of the Δ*vfh* and WT strains. (**Figures 4E,F**). The summary data showed that the mice treated with WT *V. fluvialis* or supernatants of WT culture had significantly higher colon pathology scores than did those mice treated with Δ*vfh* or supernatants of Δ*vfh* (**Figures 4G,H**). All these data verify that VFH contributes to the pathogenicity of *V. fluvialis* in the colon.

VFH Triggers IL-1β via Activation of Caspase-1 and the NLRP3 Inflammasome

To study the mechanism of VFH-triggered IL-1β, we measured mRNA expression level of IL-1β induced by different strains and found that VFH had no effect on the mRNA expression of IL-1β (**Supplementary Figure S1**). We further examined the production of inactive pro-IL-1β (p31) and mature active IL-1β (p17) in supernatants and cell lysates using immunoblotting after infection. WT and pUC-*vfh* induced significantly higher levels of active mature IL-1β in the supernatants than Δ*vfh* did, which is consistent with the ELISA results; however, all strains induced similar levels of biologically inactive pro-IL-1β in cell lysates (**Figure 5A**). Furthermore, the secretion of p10

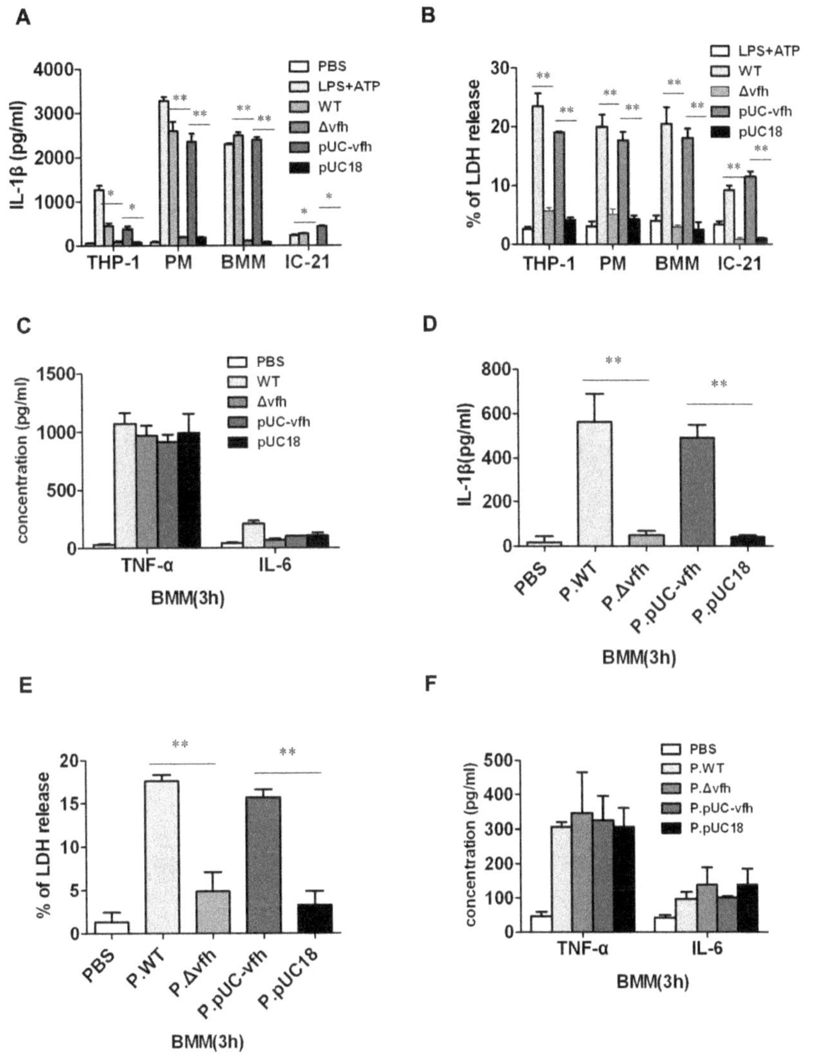

FIGURE 2 | *Vibrio fluvialis*-secreted hemolysin triggers IL-1β production and cytotoxicity *ex vivo*. (A–C) 1 × 10^6 human monocytic THP-1 cells, mouse peritoneal macrophages (PMs), mouse BMM, or mouse macrophage IC-21 cells were infected at MOI 50 for 3 h with WT *V. fluvialis* (WT), *V.* fluvialis with a mutated VFH gene (Δ*vfh*), Δ*vfh* complemented with a VFH gene (pUC-*vfh*), or Δ*vfh* complemented with an empty vector (pUC18). The levels of IL-1β **(A)**, IL-6 and TNF-α **(C)** in the supernatants were determined by ELISA, and the levels of LDH release **(B)** was quantified using the Cytotox96 Kit (Promega). **(D–F)** BMMs were treated for 3 h with PBS or supernatant from overnight culture of WT, Δ*vfh*, pUC-*vfh*, or pUC18 strains in PBS (designated as P.WT, P.Δ*vfh*, P.pUC-*vfh*, and P.pUC18). The levels of IL-1β **(D)**, IL-6 and TNF-α **(F)** in the supernatants were determined by ELISA, and the LDH release **(E)** was quantified using the Cytotox96 Kit. Results represent mean ± SD of at least three independent experiments. $^*P < 0.05$; $^{**}P < 0.01$.

subunit of Caspase-1 was evident in the supernatants of BMMs infected with WT and pUC-*vfh* but not in negative control cells or those treated with the Δ*vfh* strain (**Figure 5B**). These results verify that the induction of IL-1β by VFH occurs at the level of inflammasome processing. To verify the role of Caspase-1, we repeated the ELISA experiments in the presence of the Caspase-1 inhibitor Z-YVAD-FMK. Z-YVAD-FMK dramatically inhibited IL-1β secretion (**Figure 5C**), but not TNF-α secretion (**Figure 5D**). Furthermore, *V. fluvialis*-induced IL-1β was abolished for BMMs isolated from *Caspase-1*$^{-/-}$, *Nlrp3*$^{-/-}$, or *Asc*$^{-/-}$ mice, though the release of TNF-α was not affected (**Figures 5E,F**). Giving the fact that VFH did not obviously affect message level of IL-1β and it indeed induced Caspase-1 activation and caused more mature IL-1β secreted, we conclude that VFH may affect IL-1β production at the level of IL-1β processing or secreting.

Cathepsin B Release Contributes to *V. fluvialis*-Induced-IL-1β Secretion in Mouse BMMo

To determine whether cathepsin B is activated and released from cells upon *V. fluvialis*-induced IL-1β release, we performed immunofluorescence staining and confocal microscopy of BMMs. Uninfected cells showed punctate staining, which was obviously diminished in WT-infected cells and pUC-Δ*vfh*-infected cells, but was apparent in Δ*vfh*-infected cells. These

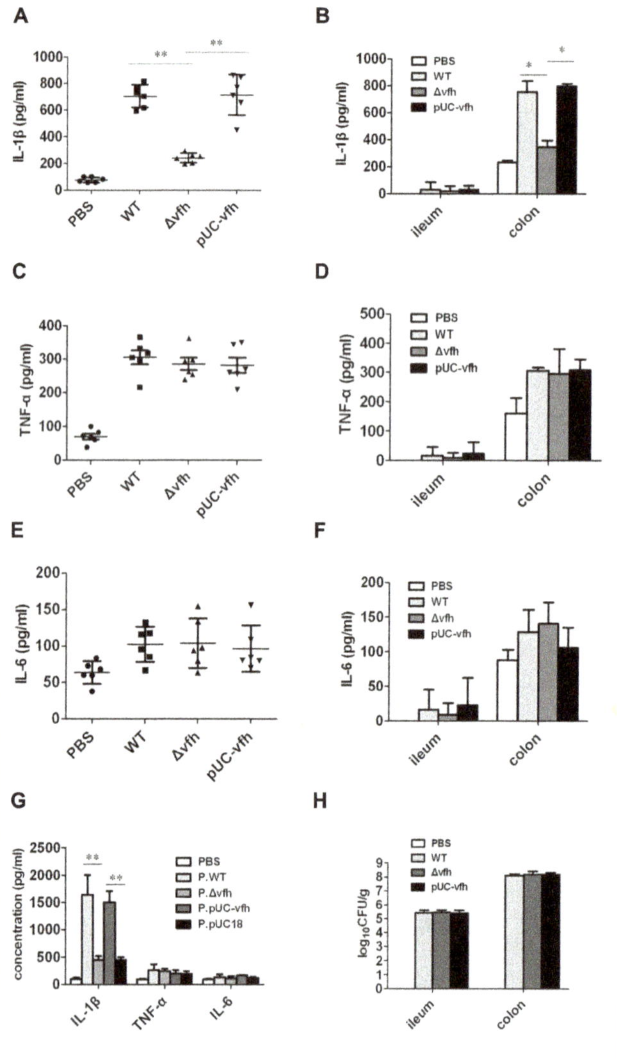

FIGURE 3 | *Vibrio fluvialis*-secreted hemolysin triggers IL-1β production and cytotoxicity *in vivo*. (A,C,E) C57BL/6 mice were injected i.p. with PBS as control or 10^8 CFU WT, Δ*vfh*, or pUC-*vfh*. Mice ($n = 6$) were sacrificed 5 days after injection, and the IL-1β **(A)**, TNF **(C)** and IL-6 **(E)** in PLF was measured by ELISA. **(B,D,F)** Five-days-old sucking mice were infected orally with PBS as a control or 10^8 CFU WT, Δ*vfh*, or pUC-*vfh*. Mice ($n = 6$) were sacrificed 24 h p.i. and the ileum and colon were homogenized. The production of IL-1β **(B)**, TNF **(D)** and IL-6 **(F)** in homogenate was assayed by ELISA. **(G)** Five-days-old sucking mice ($n = 6$) were treated by gavage with cell-free supernatant from overnight culture of various strains. The level of IL-1β, TNF, and IL-6 in homogenates of colon at 24 h p.i. were measured by ELISA. **(H)** The *V. fluvialis* colonization in ileum and colon was quantified by plating the diluted homogenates on LB plates containing 50 μg/ml streptomycin. Results represent the mean + SD of triplicates. $^*P < 0.05$; $^{**}P < 0.01$.

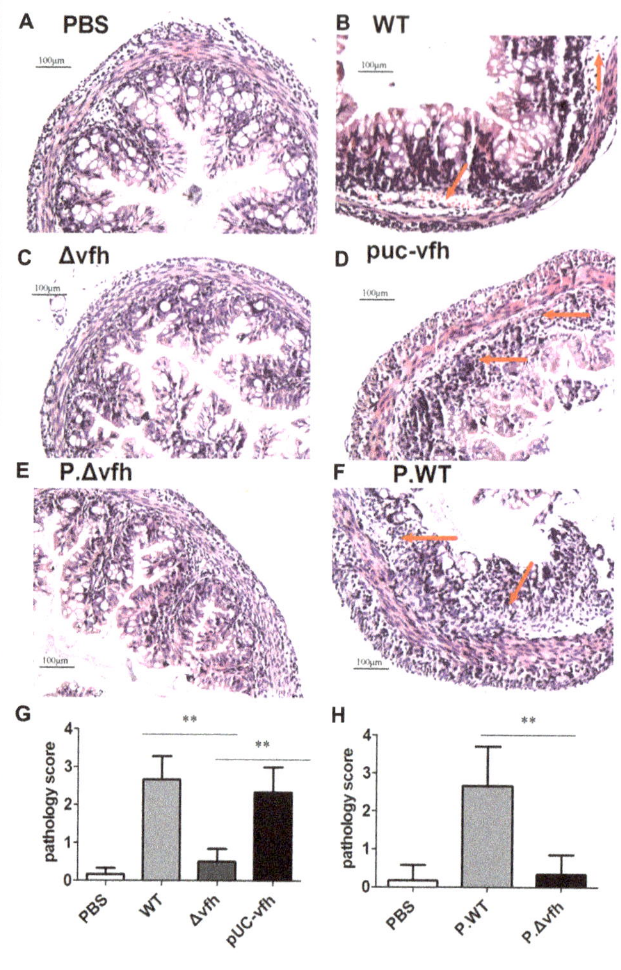

FIGURE 4 | Histopathological analysis of colon lesions. Five days C57 BL/6 sucking mice were inoculated orally with various strains or supernatant from cell-free overnight culture of various strain of *V. fluvialis*. Colons were collected 24 h p.i., and the specimens were fixed, embedded, cut, and stained with H.E. **(A)** Negative control mouse treated with PBS. **(B)** Infection with WT *V. fluvialis*. **(C)** Infection with the mutant strain Δ*vfh*. **(D)** Infection with complemented mutant strain pUC-*vfh*. **(E)** Treatment with supernatant from Δ*vfh*. **(F)** Treatment with supernatant from WT. **(G)** The pathology scoring in mice treated with WT or Δ*vfh* or pUC-*vfh* strain. **(H)** The pathology scoring in mice treated with supernatant from Δ*vfh* and WT. The result represent mean ± SD of six independent mice. $^*P < 0.05$; $^{**}P < 0.01$ and are shown as 200 magnification times, **(A–H)** 200×.

results suggest that VFH contributes to cathepsin B release from phagoendosomes to the cytosol (**Figure 6A**).

To verify these findings, we assessed the effects of the cathepsin B inhibitor CA-074Me on *V. fluvialis*-induced IL-1β secretion. CA-074Me significantly attenuated IL-1β production in BMMs infected with WT *V. fluvialis* (**Figure 6B**) but had no effect on TNF-α secretion (**Figure 6C**).

V. fluvialis-Induced Activation of the NLRP3 Inflammasome in BMMs Requires Potassium (K^+) Efflux, and ROS Production

To investigate the role of K^+ efflux in *V. fluvialis*-induced IL-1β production in BMMs, we added KCl to the cell culture medium to block K^+ efflux prior to infection with *V. fluvialis*. KCl almost completely inhibited IL-1β production (**Figure 7A**), though no effects on TNF levels were observed (**Figure 7B**). Because extracellular ATP has been shown to trigger K^+ efflux and NLRP3 activation via the ATP receptor P2X7 (Franchi et al., 2007), we sought to determine whether VFH affects extracellular ATP release. The results show that the WT and pUC-*vfh*

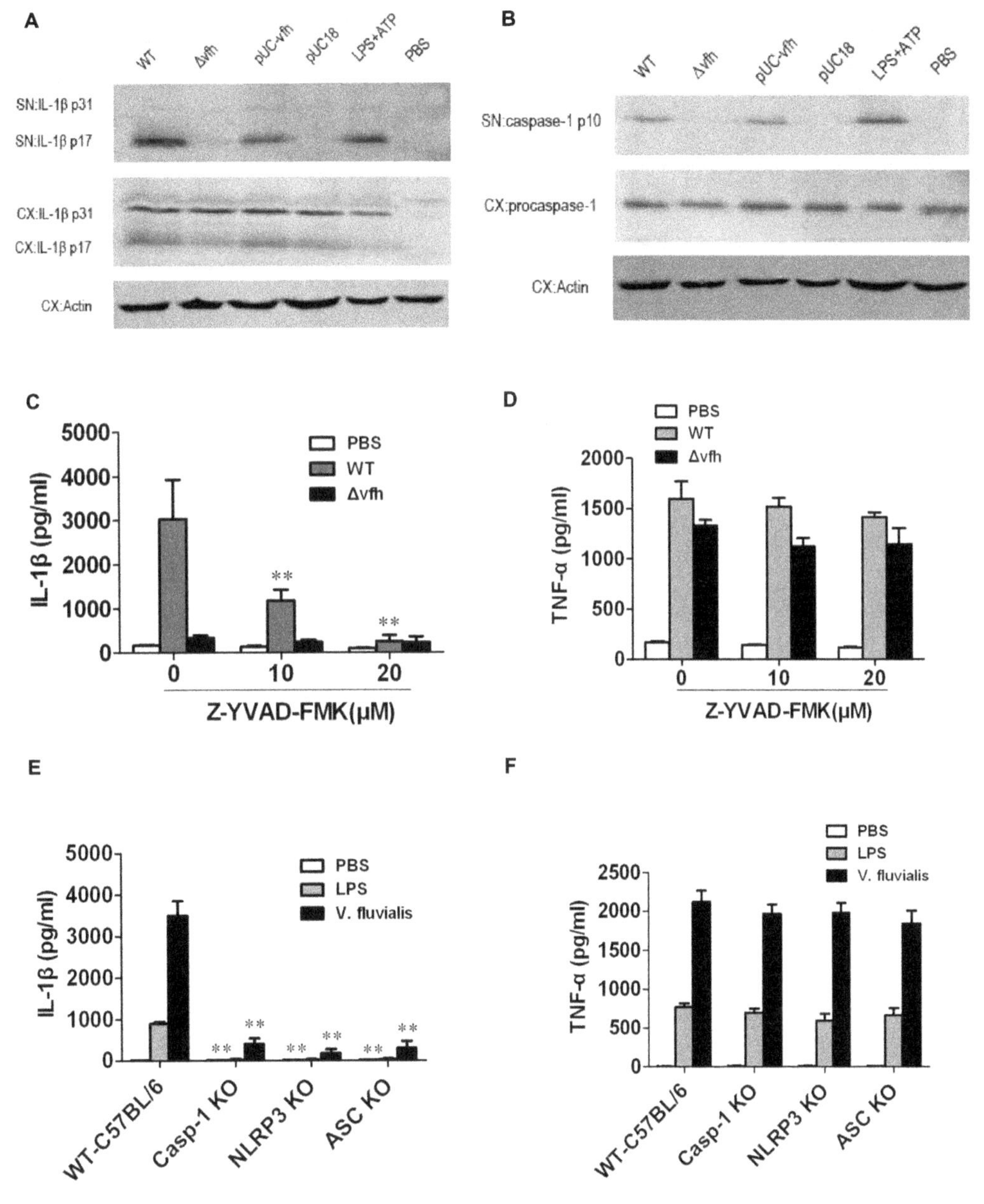

FIGURE 5 | ***Vibrio fluvialis*****-secreted hemolysin-induced IL-1β secretion is dependent on Caspase-1 and NLRP3 inflammasome activation in mouse BMM.** BMMs (1×10^6) were infected with WT, Δ*vfh*, pUC-*vfh*, or pUC18 *V. fluvialis* strains at MOI 50 for 3 h or were incubated with PBS (negative control) or LPS + ATP (positive control). **(A)** Amounts of IL-1β p17 and p31 in supernatant (SN) and cell extract (CX) were visualized by Western blotting. **(B)** Amounts of active Caspase-1 p10 and procaspase-1 and in SN and CX were visualized by Western blotting. **(C,D)** The levels of IL-1β secretion **(C)** and TNF-α secretion **(D)** after infection of BMMs in the presence of the Caspase-1 inhibitor Z-YVAD-FMK were measured by ELISA. **(E,F)** The levels of IL-1β secretion **(E)** and TNF-α secretion **(F)** were measured in BMMs isolated from WT C57 B/6 or *Caspase-1*$^{-/-}$, *Nlrp3*$^{-/-}$, or *Asc*$^{-/-}$ mouse after infection with *V. fluvialis*. Values represent the mean + SD of triplicates and are representative of three independent experiments. $^{*}P < 0.05$; $^{**}P < 0.01$.

strains induced significantly higher levels of ATP release into the supernatant than the Δ*vfh* strain did (**Figure 7C**). Furthermore, there was a significantly positive correlation between ATP release and IL-1β secretion in BMMs (**Figure 7D**). To further study whether *V. fluvialis* exerts its effect through the ATP receptor P2X7, we tested the effects of the P2X7R inhibitor, oATP, on IL-1β release. oATP significantly reduced *V. fluvialis*-induced IL-1β levels (**Figure 7E**) although it had no effect on TNF-α production (**Figure 7F**). Finally, to determine the role of ROS in *V. fluvialis*-induced NLRP3 inflammasome activation, we pretreated BMMs with the ROS inhibitor NAC. NAC impaired *V. fluvialis*-induced IL-1β release in a dose-dependent manner (**Figure 7G**). NAC also reduced the production of TNF-α to a lesser extent (**Figure 7H**), which is different from the effects of KCl and oATP. Collectively, these findings suggest that *V. fluvialis* induction of IL-1β involves K^+ efflux, extracellular ATP release, and the production of ROS.

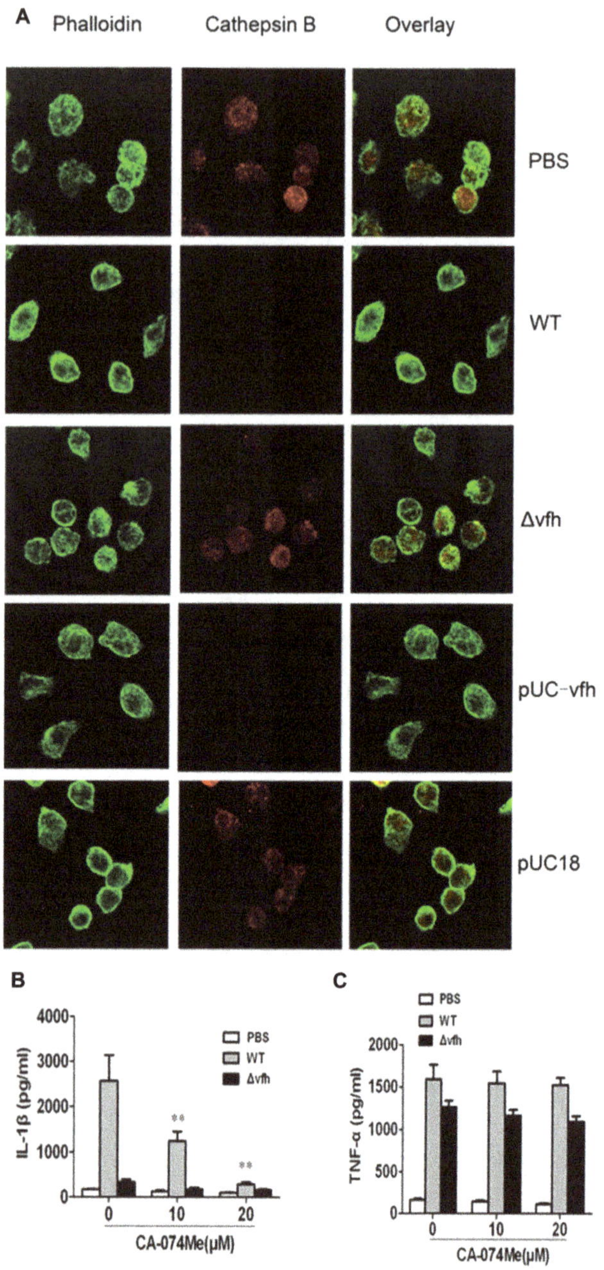

FIGURE 6 | Cathepsin B is involved in IL-1β secretion in BMMs upon *V. fluvialis* infection. (A) BMMs were left untreated or were infected with WT, Δ*vfh*, pUC-*vfh*, or pUC18 *V. fluvialis* strains for 2 h. Cathepsin B was visualized by confocal laser scanning microscopy using a specific Ab. Actin was stained with phalloidin AF488. **(B,C)** BMMs were treated with cathepsin B inhibitor CA-074Me at the indicated concentrations for 1 h before infection with WT or Δ*vfh* for 3 h. IL-1β **(B)** and TNF-α **(C)** in the supernatants were quantified by ELISA. Results represent mean ± SD of three independent experiments. $^*P < 0.05$; $^{**}P < 0.01$.

Discussion

Prior to this study, little was known regarding the mechanism of inflammatory response induced by *V. fluvialis*. Previous work identified VFH as a virulence factor of *V. fluvialis* for its role in lysing erythrocytes of various host origin and activating fluid accumulation (Kothary et al., 2003). VFH is a member of the pore-forming toxin family, which has a wide spectrum of cytocidal activity. It has been reported that bacterial hemolysins, for example, listeriolysin O of *Listeria monocytogenes* (Meixenberger et al., 2010), hemolytic cytolysin pneumolysin of *Staphylococcus pneumonia* (Witzenrath et al., 2011) and hemolysin of *S. aureus* (Kebaier et al., 2012), can induce the release of IL-1β through activation of NLRP3 inflammasome. We compared the amino sequence of these toxins and found out that the amino acid similarity between VFH and pneumolysin of *S. pneumonia* or listeriolysin O of *L. monocytogenes* or Staphylococcal α-hemolysin is 23, 27, or 24%, respectively. Of note, the amino similarity between VFH and *Vibrio cholera* hemolysin is 72%. However, it is not known whether VFH contributes to inflammation response during *V. fluvialis* infection. The presented data in this manuscript not only provides information that VFH induces IL-1β release dependent on NLRP3, but also demonstrate that VFH plays a key role in a suckling mouse model in contributing to host colon pathology and inflammatory response upon *V. fluvialis* infection, suggesting that VFH may be an important virulence factor of *V. fluvialis* causing inflammatory diarrhea in humans.

In this study, we unexpectedly found that VFH induced greater amount of IL-1β in colon but no detectable level of IL-1β in ileum (data not shown) and that IL-1β production was independent of colonization of *V. fluvialis* using the suckling mouse model. One possible explanation is that there may be some difference in the number of IL-1β-producing cells (like macrophages) or in their function of producing cytokines in response to *V. fluvialis* stimulation between ileum and colon. There is plenty of microbiota colonization in colon so the inflammatory response by innate immune cells in colon maybe functionally different from the cells in ileum. The exact mechanisms underlying the different inflammatory response induced by *V. fluvialis* between ileum and colon deserve further investigation. We agree that the inflammation triggered by VFH is a "double-edged sword" for the host. On the one hand, VFH-induced IL-1β, a key cytokine in the host's immune response, contributes to controlling of the bacteria infection. On the other hand, if too much inflammation was induced, it will cause damage to the host cells and tissues and to evade certain defense mechanisms resulting in the severe inflammatory pathology changes.

We demonstrated that VFH can specifically trigger IL-1β but not TNF-α and IL-6 secretion. IL-1β is produced in a two-step process with the first step involving the generation of the biologically inactive precursor pro–IL-1β, typically in response to TLR activation (Bauernfeind et al., 2009; Franchi et al., 2009), and the second step involving pro–IL-1β cleavage by Caspase-1 into an active cytokine. The activation of Caspase-1 is controlled by inflammasomes. Recent studies have shown that many bacterial toxins, including the hemolysins of various other bacterial species trigger NLRP3 inflammasome activation (Harder et al., 2009; Munoz-Planillo et al., 2009; McNeela et al., 2010; Meixenberger et al., 2010; Toma et al., 2010; Holzinger et al., 2012; Zhang et al.,

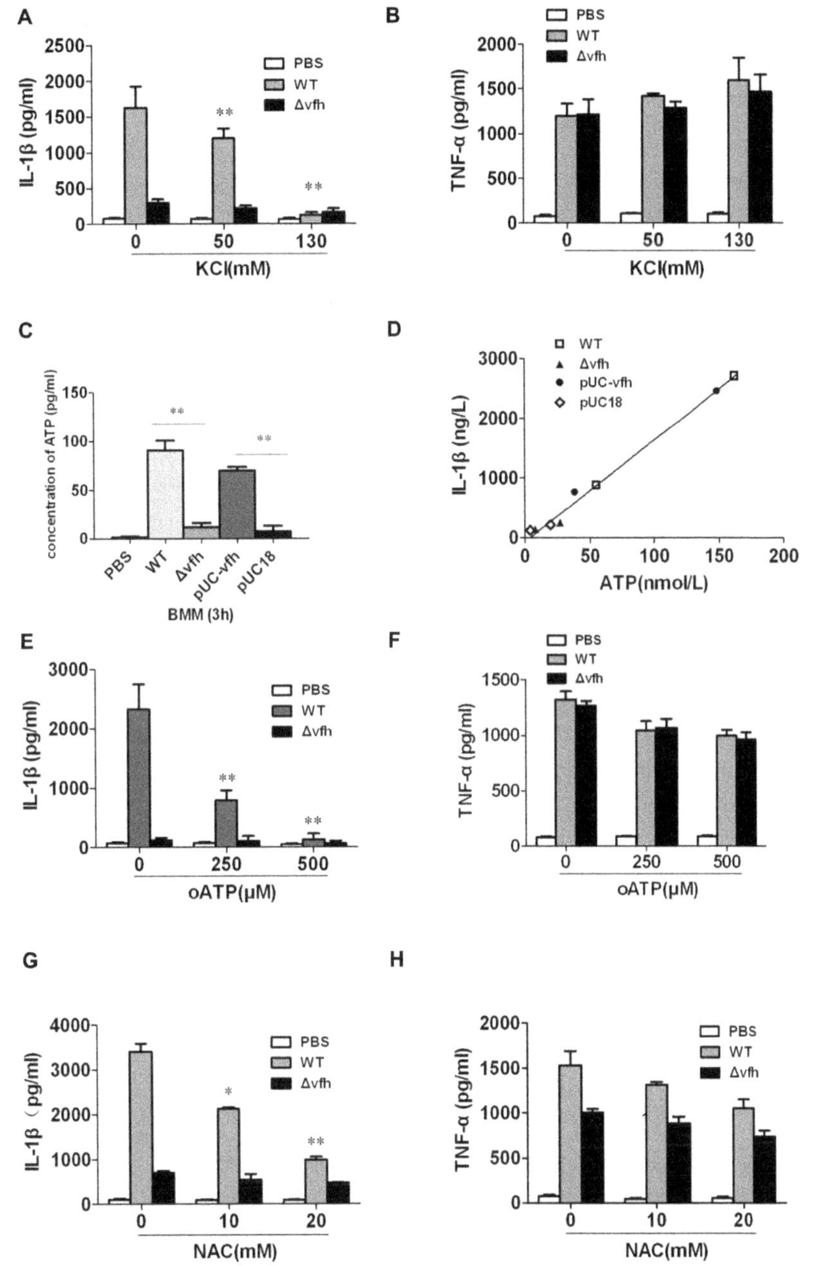

FIGURE 7 | *Vibrio fluvialis* hemolysin-induced IL-1β production in BMMs requires ATP signaling, K^+-efflux and ROS generation. (A,B) BMMs were infected with *V. fluvialis* WT or Δ*vfh* in the absence or presence of KCl. IL-1β **(A)** and TNF-α **(B)** in the supernatants were determined at 3 h p.i. using ELISA. **(C)** The release of ATP from the BMMs infected with various strains was monitored using a bioluminescence assay kit. **(D)** The correlation between extracellular ATP release and IL-1β secretion is plotted for BMMs infected with different *V. fluvialis* strains at 3 h. **(E–H)** BMMs were infected with *V. fluvialis* WT or Δ*vfh* in the absence or presence of oxidized ATP (oATP; **E,F**) or *N*-acetyl-L-cysteine (NAC; **G,H**). IL-1β and TNF-α in the supernatants were determined at 3 h p.i. using ELISA. Results represent mean ± SD of three independent experiments. $^*P < 0.05$; $^{**}P < 0.01$.

2012). Our data show that VFH did not affect the expression of IL-1β mRNA and pro-IL-1β synthesis but had an effect on activating the NLRP3 inflammasome and inducing the mature active IL-1β, therefore we propose that VFH affects IL-1β production at the level of IL-1β processing or secretion. The mature biological active IL-1β plays a critical role in host immune response. We confirmed the effect of VFH on secreted mature IL-1β and therefore have added a new member to the list of bacterial pore-forming toxins triggering IL-1β via Caspase-1 and NLRP3 inflammasome activation.

Cellular stimulation triggers ATP release and subsequently activates purinergic receptors in an autocrine and/or paracrine manner (Piccini et al., 2008), although certain non-nucleotide inflammasome activators may interact directly with purinergic receptors, which does not depend on ATP release (Babelova et al., 2009). In the present study, we demonstrated that VHF contributes to the ATP release into supernatant and that P2X7R signaling is involved in the inflammasome activation induced by *V. fluvialis*. A recent study has shown that the efflux of K^+ is responsible for the maturation of pro-IL-1β (Petrilli et al., 2007). K^+ efflux occurs upon the engagement of extracellular ATP with the P2X7R or directly through bacterial pore-forming toxins (Munoz-Planillo et al., 2013). Some bacterial pore-forming toxins were reported to cause K^+ efflux by permeabilizing the plasma membrane (Walev et al., 1993; Fiser et al., 2012), which activates Nlrp3 independently of P2X7R (Harder et al., 2009). Therefore, we propose that VFH may induce extracellular ATP release, which results in K^+ efflux via P2X7R indirectly. It has yet to be determined how K^+ efflux is induced by other classes of inflammasome activators. More recently, calcium has been shown to play a role in NLRP3 inflammasome activation (Lee et al., 2012; Rossol et al., 2012). Whether migration of calcium is required in *V. fluvialis*-induced activation of the NLRP3 inflammasome warrants further investigation.

NLR family pyrin domain containing 3 activators induce lysosomal damage, which leads to the release of cathepsin B into the cytosol. The lysosomal proteases, in turn, could either degrade a putative NLRP3 inhibitor or cleave a substrate in the cytosol that would generate a NLRP3 ligand (Halle et al., 2008). Our results demonstrate that VFH significantly contributes to cathepsin B release and that an inhibitor of cathepsin B causes a modest decrease in IL-1β production in BMMs. It remains to be determined how cathepsin B release plays its role in VFH-induced NLRP3 activation in BMMs. Furthermore, our data show that ROS production is required for *V. fluvialis*-triggered NLRP3 activation. Consistent with these results, previous studies have shown that ROS inhibitors such as DPI and NAC strikingly inhibit IL-1β and IL-8 production in mouse macrophages (Bauernfeind et al., 2011; Mao et al., 2014).

In summary, this study demonstrates that VFH plays a key role in mediating IL-1β secretion and histopathology in the colon during *V. fluvialis* infection *in vivo*. Our data also suggest that VFH-mediated activation of the NLRP3 inflammasome is critically involved in the proinflammatory response upon *V. fluvialis* infection. Several signaling pathways are involved in NLRP3 activation, though K^+ efflux appears to play a more important role in *V. fluvialis*-induced IL-1β production than ATP and cathepsin B release do in this study. These findings are helpful for understanding the role of VFH in the pathogenesis of *V. fluvialis*.

Author Contributions

LS and ZR conceived and designed the experiments. LS, YH, MZ, ZW, HS, and SW performed the experiments. LS and ZW analyzed the data. WL, BK, and GM discussed the results. LS and ZR wrote the paper.

Acknowledgments

We are grateful to Dr. Warren Strober for sharing NLRP3 deficient mice and Dr. Vishva M. Dixit for providing $Asc^{-/-}$ mice. This work was supported by grants from the National Natural Science Foundation of China (No.81371761 to ZR, No. 31170868 to GM; No. 81171640 to WL).

References

Babelova, A., Moreth, K., Tsalastra-Greul, W., Zeng-Brouwers, J., Eickelberg, O., Young, M. F., et al. (2009). Biglycan, a danger signal that activates the NLRP3 inflammasome via toll-like and P2X receptors. *J. Biol. Chem.* 284, 24035–24048. doi: 10.1074/jbc.M109.014266

Bauernfeind, F., Bartok, E., Rieger, A., Franchi, L., Nunez, G., and Hornung, V. (2011). Cutting edge: reactive oxygen species inhibitors block priming, but not activation, of the NLRP3 inflammasome. *J. Immunol.* 187, 613–617. doi: 10.4049/jimmunol.1100613

Bauernfeind, F. G., Horvath, G., Stutz, A., Alnemri, E. S., Macdonald, K., Speert, D., et al. (2009). Cutting edge: NF-kappaB activating pattern recognition and cytokine receptors license NLRP3 inflammasome activation by regulating NLRP3 expression. *J. Immunol.* 183, 787–791. doi: 10.4049/jimmunol.0901363

Chakraborty, R., Chakraborty, S., De, K., Sinha, S., Mukhopadhyay, A. K., Khanam, J., et al. (2005). Cytotoxic and cell vacuolating activity of *Vibrio fluvialis* isolated from paediatric patients with diarrhoea. *J. Med. Microbiol.* 54, 707–716. doi: 10.1099/jmm.0.45820-0

Fink, S. L., and Cookson, B. T. (2005). Apoptosis, pyroptosis, and necrosis: mechanistic description of dead and dying eukaryotic cells. *Infect. Immun.* 73, 1907–1916. doi: 10.1128/IAI.73.4.1907-1916.2005

Fiser, R., Masin, J., Bumba, L., Pospisilova, E., Fayolle, C., Basler, M., et al. (2012). Calcium influx rescues adenylate cyclase-hemolysin from rapid cell membrane removal and enables phagocyte permeabilization by toxin pores. *PLoS Pathog.* 8:e1002580. doi: 10.1371/journal.ppat.1002580

Franchi, L., Eigenbrod, T., and Nunez, G. (2009). Cutting edge: TNF-alpha mediates sensitization to ATP and silica via the NLRP3 inflammasome in the absence of microbial stimulation. *J. Immunol.* 183, 792–796. doi: 10.4049/jimmunol.0900173

Franchi, L., Kanneganti, T. D., Dubyak, G. R., and Nunez, G. (2007). Differential requirement of P2X7 receptor and intracellular K^+ for Caspase-1 activation induced by intracellular and extracellular bacteria. *J. Biol. Chem.* 282, 18810–18818. doi: 10.1074/jbc.M610762200

Gurcel, L., Abrami, L., Girardin, S., Tschopp, J., and Van Der Goot, F. G. (2006). Caspase-1 activation of lipid metabolic pathways in response to bacterial pore-forming toxins promotes cell survival. *Cell* 126, 1135–1145. doi: 10.1016/j.cell.2006.07.033

Halle, A., Hornung, V., Petzold, G. C., Stewart, C. R., Monks, B. G., Reinheckel, T., et al. (2008). The NALP3 inflammasome is involved in the innate immune response to amyloid-beta. *Nat. Immunol.* 9, 857–865. doi: 10.1038/ni.1636

Harder, J., Franchi, L., Munoz-Planillo, R., Park, J. H., Reimer, T., and Nunez, G. (2009). Activation of the Nlrp3 inflammasome by *Streptococcus pyogenes* requires streptolysin O and NF-kappa B activation but proceeds independently of TLR signaling and P2X7 receptor. *J. Immunol.* 183, 5823–5829. doi: 10.4049/jimmunol.0900444

Holzinger, D., Gieldon, L., Mysore, V., Nippe, N., Taxman, D. J., Duncan, J. A., et al. (2012). *Staphylococcus aureus* Panton-Valentine leukocidin induces an inflammatory response in human phagocytes via the NLRP3 inflammasome. *J. Leukoc. Biol.* 92, 1069–1081. doi: 10.1189/jlb.0112014

Igbinosa, E. O., and Okoh, A. I. (2010). *Vibrio fluvialis*: an unusual enteric pathogen of increasing public health concern. *Int. J. Environ. Res. Public Health* 7, 3628–3643. doi: 10.3390/ijerph7103628

Kanneganti, T. D., Lamkanfi, M., Kim, Y. G., Chen, G., Park, J. H., Franchi, L., et al. (2007). Pannexin-1-mediated recognition of bacterial molecules activates the cryopyrin inflammasome independent of Toll-like receptor signaling. *Immunity* 26, 433–443. doi: 10.1016/j.immuni.2007.03.008

Kayagaki, N., Warming, S., Lamkanfi, M., Vande Walle, L., Louie, S., Dong, J., et al. (2011). Non-canonical inflammasome activation targets Caspase-11. *Nature* 479, 117–121. doi: 10.1038/nature10558

Kebaier, C., Chamberland, R. R., Allen, I. C., Gao, X., Broglie, P. M., Hall, J. D., et al. (2012). *Staphylococcus aureus* alpha-hemolysin mediates virulence in a murine model of severe pneumonia through activation of the NLRP3 inflammasome. *J. Infect. Dis.* 205, 807–817. doi: 10.1093/infdis/jir846

Kothary, M. H., Lowman, H., Mccardell, B. A., and Tall, B. D. (2003). Purification and characterization of enterotoxigenic El Tor-like hemolysin produced by *Vibrio fluvialis*. *Infect. Immun.* 71, 3213–3220. doi: 10.1128/IAI.71.6.3213-3220.2003

Kumagai, K., Itoh, K., Hinuma, S., and Tada, M. (1979). Pretreatment of plastic Petri dishes with fetal calf serum. A simple method for macrophage isolation. *J. Immunol. Methods* 29, 17–25. doi: 10.1016/0022-1759(79)90121-2

Lee, G. S., Subramanian, N., Kim, A. I., Aksentijevich, I., Goldbach-Mansky, R., Sacks, D. B., et al. (2012). The calcium-sensing receptor regulates the NLRP3 inflammasome through Ca^{2+} and cAMP. *Nature* 492, 123–127. doi: 10.1038/nature11588

Levine, W. C., and Griffin, P. M. (1993). *Vibrio* infections on the Gulf Coast: results of first year of regional surveillance. Gulf Coast *Vibrio* Working Group. *J. Infect. Dis.* 167, 479–483.

Liang, P., Cui, X., Du, X., Kan, B., and Liang, W. (2013). The virulence phenotypes and molecular epidemiological characteristics of *Vibrio fluvialis* in China. *Gut Pathog.* 5:6. doi: 10.1186/1757-4749-5-6

Lockwood, D. E., Kreger, A. S., and Richardson, S. H. (1982). Detection of toxins produced by *Vibrio fluvialis*. *Infect. Immun.* 35, 702–708.

Lu, X., Liang, W., Wang, Y., Xu, J., Zhu, J., and Kan, B. (2014). Identification of genetic bases of *Vibrio fluvialis* species-specific biochemical pathways and potential virulence factors by comparative genomic analysis. *Appl. Environ. Microbiol.* 80, 2029–2037. doi: 10.1128/AEM.03588-13

Mao, L., Zhang, L., Li, H., Chen, W., Wang, H., Wu, S., et al. (2014). Pathogenic fungus *Microsporum canis* activates the NLRP3 inflammasome. *Infect. Immun.* 82, 882–892. doi: 10.1128/IAI.01097-13

Marcus, H., Ketley, J. M., Kaper, J. B., and Holmes, R. K. (1990). Effects of DNase production, plasmid size, and restriction barriers on transformation of *Vibrio cholerae* by electroporation and osmotic shock. *FEMS Microbiol. Lett.* 56, 149–154. doi: 10.1111/j.1574-6968.1990.tb04139.x

Mariathasan, S., Newton, K., Monack, D. M., Vucic, D., French, D. M., Lee, W. P., et al. (2004). Differential activation of the inflammasome by Caspase-1 adaptors ASC and Ipaf. *Nature* 430, 213–218. doi: 10.1038/nature02664

Mariathasan, S., Weiss, D. S., Newton, K., Mcbride, J., O'Rourke, K., Roose-Girma, M., et al. (2006). Cryopyrin activates the inflammasome in response to toxins and ATP. *Nature* 440, 228–232. doi: 10.1038/nature04515

McCoy, A. J., Koizumi, Y., Toma, C., Higa, N., Dixit, V., Taniguchi, S., et al. (2010). Cytotoxins of the human pathogen *Aeromonas hydrophila* trigger, via the NLRP3 inflammasome, Caspase-1 activation in macrophages. *Eur. J. Immunol.* 40, 2797–2803. doi: 10.1002/eji.201040490

McNeela, E. A., Burke, A., Neill, D. R., Baxter, C., Fernandes, V. E., Ferreira, D., et al. (2010). Pneumolysin activates the NLRP3 inflammasome and promotes proinflammatory cytokines independently of TLR4. *PLoS Pathog.* 6:e1001191. doi: 10.1371/journal.ppat.1001191

Meixenberger, K., Pache, F., Eitel, J., Schmeck, B., Hippenstiel, S., Slevogt, H., et al. (2010). *Listeria monocytogenes*-infected human peripheral blood mononuclear cells produce IL-1beta, depending on listeriolysin O and NLRP3. *J. Immunol.* 184, 922–930. doi: 10.4049/jimmunol.0901346

Munoz-Planillo, R., Franchi, L., Miller, L. S., and Nunez, G. (2009). A critical role for hemolysins and bacterial lipoproteins in *Staphylococcus aureus*-induced activation of the Nlrp3 inflammasome. *J. Immunol.* 183, 3942–3948. doi: 10.4049/jimmunol.0900729

Munoz-Planillo, R., Kuffa, P., Martinez-Colon, G., Smith, B. L., Rajendiran, T. M., and Nunez, G. (2013). K(+) efflux is the common trigger of NLRP3 inflammasome activation by bacterial toxins and particulate matter. *Immunity* 38, 1142–1153. doi: 10.1016/j.immuni.2013.05.016

Petrilli, V., Papin, S., Dostert, C., Mayor, A., Martinon, F., and Tschopp, J. (2007). Activation of the NALP3 inflammasome is triggered by low intracellular potassium concentration. *Cell Death Differ.* 14, 1583–1589. doi: 10.1038/sj.cdd.4402195

Piccini, A., Carta, S., Tassi, S., Lasiglie, D., Fossati, G., and Rubartelli, A. (2008). ATP is released by monocytes stimulated with pathogen-sensing receptor ligands and induces IL-1beta and IL-18 secretion in an autocrine way. *Proc. Natl. Acad. Sci. U.S.A.* 105, 8067–8072. doi: 10.1073/pnas.0709684105

Ramamurthy, T., Chowdhury, G., Pazhani, G. P., and Shinoda, S. (2014). *Vibrio fluvialis*: an emerging human pathogen. *Front. Microbiol.* 5:91. doi: 10.3389/fmicb.2014.00091

Rossol, M., Pierer, M., Raulien, N., Quandt, D., Meusch, U., Rothe, K., et al. (2012). Extracellular Ca^{2+} is a danger signal activating the NLRP3 inflammasome through G protein-coupled calcium sensing receptors. *Nat. Commun.* 3:1329. doi: 10.1038/ncomms2339

Takeuchi, O., and Akira, S. (2010). Pattern recognition receptors and inflammation. *Cell* 140, 805–820. doi: 10.1016/j.cell.2010.01.022

Toma, C., Higa, N., Koizumi, Y., Nakasone, N., Ogura, Y., Mccoy, A. J., et al. (2010). Pathogenic *Vibrio* activate NLRP3 inflammasome via cytotoxins and TLR/nucleotide-binding oligomerization domain-mediated NF-kappa B signaling. *J. Immunol.* 184, 5287–5297. doi: 10.4049/jimmunol.0903536

Walev, I., Martin, E., Jonas, D., Mohamadzadeh, M., Muller-Klieser, W., Kunz, L., et al. (1993). Staphylococcal alpha-toxin kills human keratinocytes by permeabilizing the plasma membrane for monovalent ions. *Infect. Immun.* 61, 4972–4979.

Wang, Y., Wang, H., Liang, W., Hay, A. J., Zhong, Z., Kan, B., et al. (2013). Quorum sensing regulatory cascades control *Vibrio fluvialis* pathogenesis. *J. Bacteriol.* 195, 3583–3589. doi: 10.1128/JB.00508-13

Witzenrath, M., Pache, F., Lorenz, D., Koppe, U., Gutbier, B., Tabeling, C., et al. (2011). The NLRP3 inflammasome is differentially activated by pneumolysin variants and contributes to host defense in pneumococcal pneumonia. *J. Immunol.* 187, 434–440. doi: 10.4049/jimmunol.1003143

Zhang, X., Cheng, Y., Xiong, Y., Ye, C., Zheng, H., Sun, H., et al. (2012). Enterohemorrhagic *Escherichia coli* specific enterohemolysin induced IL-1beta in human macrophages and EHEC-induced IL-1beta required activation of NLRP3 inflammasome. *PLoS ONE* 7:e50288. doi: 10.1371/journal.pone.0050288

19

Receptor-Interacting Protein Kinase-2 Inhibition by CYLD Impairs Antibacterial Immune Responses in Macrophages

Katharina Wex[1], Ursula Schmid[1], Sissy Just[1], Xu Wang[1], Rebecca Wurm[1], Michael Naumann[2], Dirk Schlüter[1,3] and Gopala Nishanth[1,3]**

[1] Institute of Medical Microbiology and Hospital Hygiene, Otto-von-Guericke University Magdeburg, Magdeburg, Germany, [2] Institute of Experimental Internal Medicine, Otto-von-Guericke University Magdeburg, Magdeburg, Germany, [3] Organ-Specific Immune Regulation, Helmholtz Centre for Infection Research, Braunschweig, Germany

Edited by:
Olivier Dussurget,
University Paris Diderot, France

Reviewed by:
Vida A. Dennis,
Alabama State University, USA
Attila Szabo,
University of Debrecen, Hungary

***Correspondence:**
Dirk Schlüter
dirk.schlueter@med.ovgu.de;
Gopala Nishanth
nishanth.gopala@med.ovgu.de

Upon infection with intracellular bacteria, nucleotide oligomerization domain protein 2 recognizes bacterial muramyl dipeptide and binds, subsequently, to receptor-interacting serine/threonine kinase 2 (RIPK2), which activates immune responses via the nuclear factor kappa-light-chain enhancer of activated B cells (NF-κB) and extracellular signal-regulated kinase (ERK) pathways. Activation of RIPK2 depends on its K63 ubiquitination by E3 ligases, whereas the deubiquitinating enzyme A20 counter regulates RIPK2 activity by cleaving K63-polyubiquitin chains from RIPK2. Here, we newly identify the deubiquitinating enzyme CYLD as a new inhibitor of RIPK2. We show that CYLD binds to and removes K63-polyubiquitin chains from RIPK2 in *Listeria monocytogenes* (Lm) infected murine bone marrow-derived macrophages. CYLD-mediated K63 deubiquitination of RIPK2 resulted in an impaired activation of both NF-κB and ERK1/2 pathways, reduced production of proinflammatory cytokines interleukin-6 (IL-6), IL-12, anti-listerial reactive oxygen species (ROS) and nitric oxide (NO), and, finally, impaired pathogen control. In turn, RIPK2 inhibition by siRNA prevented activation of NF-κB and ERK1/2 and completely abolished the protective effect of CYLD deficiency with respect to the production of IL-6, NO, ROS, and pathogen control. Noteworthy, CYLD also inhibited autophagy of *Listeria* in a RIPK2-ERK1/2-dependent manner. The protective function of CYLD deficiency was dependent on interferon gamma (IFN-γ) prestimulation of infected macrophages. Interestingly, the reduced NF-κB activation in CYLD-expressing macrophages limited the protective effect of IFN-γ by reducing NF-κB-dependent signal transducers and activators of transcription-1 (STAT1) activation. Taken together, our study identifies CYLD as an important inhibitor of RIPK2-dependent antibacterial immune responses in macrophages.

Keywords: ***Listeria monocytogenes*, receptor-interacting protein kinase-2, CYLD, macrophages, antibacterial immune responses**

INTRODUCTION

Macrophages play a crucial role in the innate immune response including sensing and phagocytosis of bacterial pathogens at the site of infection (1). In bacterial infections, phagocyte activation is augmented by type II interferon (IFN-γ) produced by natural killer (NK) and T cells (2). IFN-γ primarily signals through the Janus kinase (JAK) – signal tranducers and activators of transcription (STAT)-1 pathway and primes the secretion of proinflammatory cytokines, production of superoxide anions, generation of nitrogen and oxygen free radicals and thereby conditions macrophages as potent antibacterial effector cells (3, 4). Of note, STAT1 is also activated by type I IFNs, which can be produced by various cell types including macrophages in bacterial infection (5). However, in contrast to IFN-γ, type I IFN is not protective and may even exacerbate infection with *Listeria monocytogenes* (Lm), a facultative intracellular bacterium (6, 7).

In addition to cell surface and endosomal Toll-like receptors, cytosolic nucleotide oligomerization domain protein (NOD)-like receptors, in particular NOD2, recognize intracellular bacteria such as Lm. Activated NOD2 binds to receptor-interacting serine/threonine kinase 2 (RIPK2) (8). RIPK2 undergoes K63 polyubiquitination resulting in the activation of nuclear factor kappa-light-chain enhancer of activated B cells (NF-κB) and mitogen-activated protein kinases (MAPKs) (9–13). The activity of RIPK2 is counter-regulated by the deubiquitinating enzyme A20, which disassembles K63-linked ubiquitin chains from RIPK2 (10). NOD2-dependent RIPK2 activation is essential for an effective antibacterial response, since RIPK2-deficient macrophages have an impaired pathogen control (14–16) and RIPK2-deficient mice are highly susceptible to listeriosis (17).

Listeria monocytogenes may cause severe infections in pregnant women and immunocompromised patients (18). In mice, *Listeria* rapidly home to Kupffer cells, i.e., liver-resident macrophages (19). In addition, uninfected myeloid cell populations including macrophages, monocytes, and TNF/iNOS-producing dendritic cells are recruited to the site of infection and are crucial for the control of listeriosis (20). In macrophages, both NF-κB and MAPK induce the synthesis of protective (i) proinflammatory cytokines including IL-6 and IL-12, (ii) ROS, (iii) nitric oxide (NO), and (iv) autophagosomes (21–24). The formation of autophagosomes is either induced via the NOD2/RIPK2/extracellular signal-regulated kinase (ERK) pathway (25) or NOD2 independent by the autophagy-related protein 16-1 (ATG16L1) (26). In the autophagosomes, phagocytosed Lm are killed by ROS (27–29). In further amplification of antibacterial activity, ROS activates the STAT1 pathway leading to increased production of NO, which reacts with ROS to produce peroxynitrite, an even stronger antibacterial molecule (30).

The deubiquitinating enzyme CYLD cleaves K63-linked polyubiquitin chains from specific substrates (31), including tumor necrosis factor receptor-associated factors (TRAF)-2, TRAF6, transforming growth factor beta-activated kinase 1 (TAK1), B cell lymphoma 3 (BCL3), STAT3, nuclear factor kappa B essential modulator (NEMO), and retinoic acid inducible gene 1 (RIG-1) (32–36), and negatively regulates the activation of NF-κB, MAPKs, and type I IFN production. In infectious diseases, CYLD has a disease-specific effect and CYLD-deficient mice suffer from exacerbated *Escherichia coli* pneumonia and *Haemophilus influenzae* middle ear infection (37, 38) but are protected against lethal *Streptococcus pneumoniae* and Lm infection (36, 37).

Earlier, we have shown that macrophage-derived IL-6 induced STAT3-dependent protective fibrin production by hepatocytes in listeriosis (36). However, it remained unclear whether CYLD also directly regulated the antibacterial function of infected macrophages, in particular, the NOD2-dependent activation of NF-κB, MAPKs, and autophagy. Here, we newly identified that CYLD inhibits NOD2-dependent antibacterial activity of macrophages by K63 deubiquitination of RIPK2.

RESULTS

CYLD Impairs Production of Cytokines, ROS, and NO and Reduces Activation of NF-κB, ERK1/2, and p38MAPK in Lm-Infected BMDM

Upon infection with Lm, Cyld$^{-/-}$ BMDM showed no significant improvement in pathogen control as compared to WT BMDM (**Figure 1A**, $p < 0.05$). In good agreement with previous reports (39, 40), pretreatment of BMDM with IFN-γ, greatly augmented the listericidal activity of BMDM. Importantly, the IFN-γ-induced listericidal effect was significantly stronger in Cyld$^{-/-}$ as compared to WT BMDM. Since CYLD only regulated pathogen control upon IFN-γ stimulation, we further focused on the function of CYLD in IFN-γ-stimulated Lm-infected BMDM.

In addition to pathogen control, the production of listericidal ROS and NO as well as the cytokines IL-6 and IL-12 were significantly increased in IFN-γ-treated Lm-infected Cyld$^{-/-}$ BMDM (**Figures 1B–E**). To identify the underlying CYLD-regulated signaling pathways, the activation of NF-κB, ERK1/2, and p38MAPK were analyzed by WB in IFN-γ-stimulated Lm-infected BMDM. As displayed in **Figure 1F**, Cyld$^{-/-}$ BMDM showed enhanced activation of NF-κB as indicated by increased phosporylation of p65 and IκB-α. In addition, activation of ERK1/2 and p38MAPK were augmented in Cyld$^{-/-}$ BMDM. In contrast, stimulation with IFN-γ only was insufficient to induce activation of these signaling pathways.

CYLD Impairs Activation of STAT1

Mice with macrophage-specific deletion of STAT1 are susceptible to listeriosis (41). Since IFN-γ augments bactericidal activity and NO production of Lm-infected macrophages [**Figure 1A** (39, 40)] and the immunostimulatory activity of IFN-γ is mediated by the transcription factor STAT1, we studied the impact of CYLD on the activation of STAT1 in Lm-infected IFN-γ-stimulated BMDM. Within 30 min post infection, the phosphorylation of cytoplasmic and nuclear STAT1 (Y701) and STAT1 (S727) was increased in Cyld$^{-/-}$ as compared to WT BMDM (**Figure 2A**). To study how CYLD reduces STAT1 activation and nuclear accumulation in IFN-γ-stimulated Lm-infected BMDM, we analyzed whether CYLD might directly bind to STAT1. However, immunoprecipitation experiments showed that CYLD

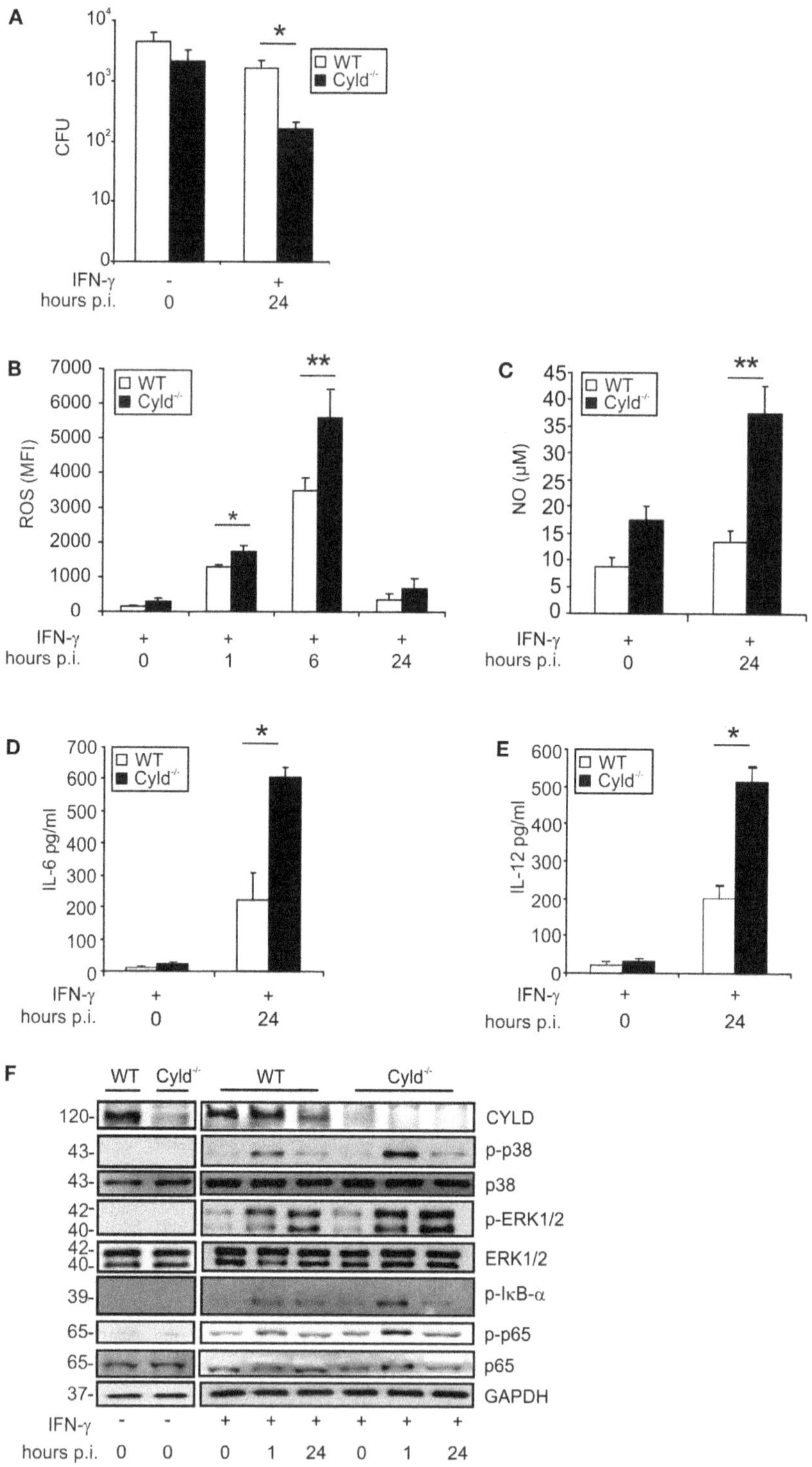

FIGURE 1 | CYLD impairs pathogen control and decreases cytokine production and activation of NF-κB and MAP kinases. WT and $Cyld^{-/-}$ BMDM (1×10^6cells/group), stimulated with IFN-γ (100 U/ml) for 24 h followed by infection with Lm (MOI 5:1). **(A)** The bacterial load was determined in 1×10^6 IFN-γ-stimulated and unstimulated Lm-infected WT and $Cyld^{-/-}$ BMDM 24 h p.i. **(B)** Levels of intracellular ROS were analyzed by flow cytometry in 1×10^6 uninfected (0 h p.i.) and infected (1, 6, 24 h p.i.) WT and $Cyld^{-/-}$ BMDM. **(C)** The amount of NO was measured in the supernatant of 1×10^6 uninfected and infected WT and $Cyld^{-/-}$ BMDM colorimetrically using a Griess assay kit. **(D,E)** IL-6 **(D)** and IL-12 **(E)** were determined in the supernatant of 1×10^6 Lm-infected WT and $Cyld^{-/-}$ BMDMs by flow cytometric bead array. **(F)** Proteins were isolated from 1×10^6 uninfected and infected WT and $Cyld^{-/-}$ BMDM at the indicated time points and stained with antibodies specific for CYLD, p-p38, p38, p-ERK1/2, ERK1/2, p-IκB-α, p-p65, and p65. GAPDH was used as loading control. In **(A–E)**, data represent the mean + SD from triplicate wells. In **(A–F)**, data from one of the three representative experiments are shown.

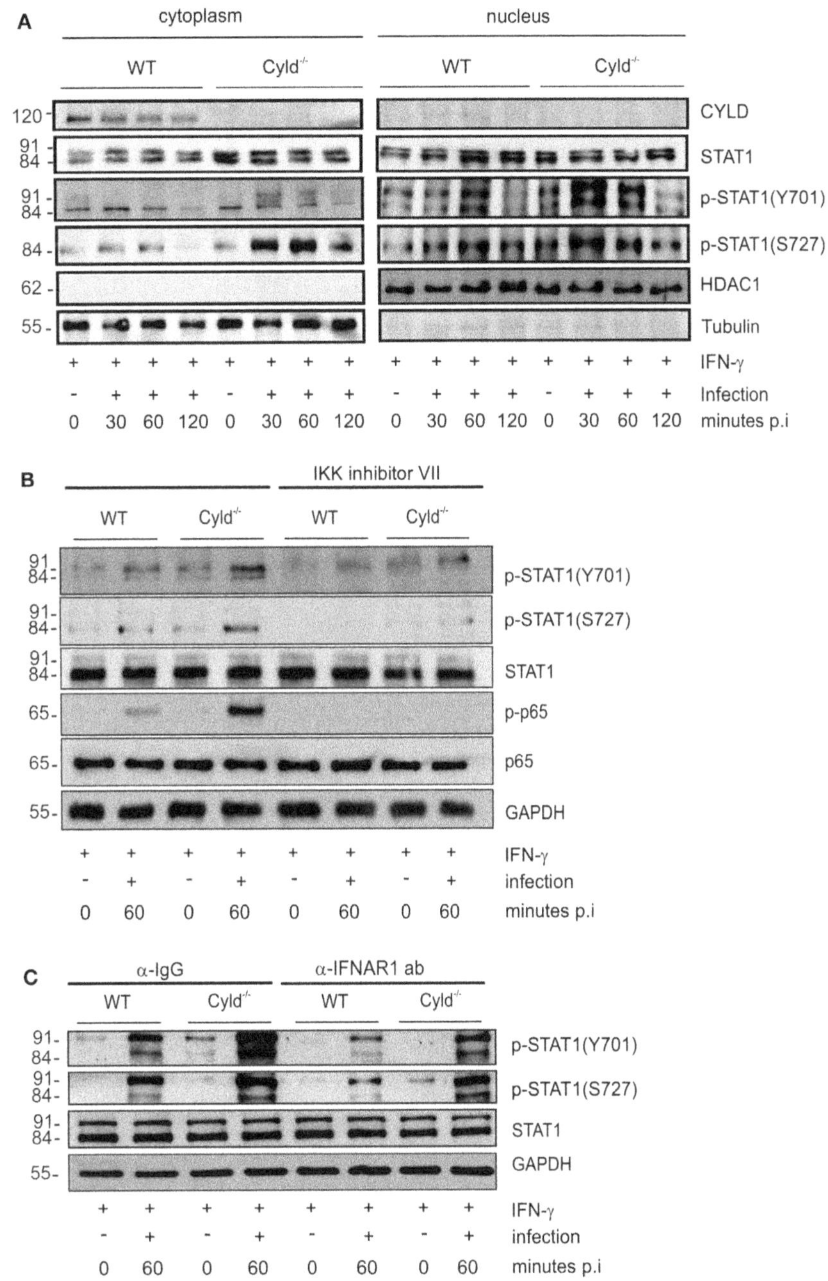

FIGURE 2 | CYLD diminishes activation of STAT1 upon stimulation with IFN-γ. (A–C) 1×10^6 WT and Cyld$^{-/-}$ BMDM were stimulated with IFN-γ (100 U/ml) for 24 h. Thereafter, the indicated groups of BMDM were additionally infected with Lm (MOI 5:1) for 24 h. Proteins were isolated from uninfected (0 min p.i.) and infected (30, 60 min p.i.) BMDM as indicated. **(A)** Proteins were isolated from the cytoplasm and nucleus of BMDM respectively at the indicated time points after infection. WBs were stained with anti-tubulin and anti-histone deacetylase (HDAC) as marker proteins for the cytoplasm and nucleus, respectively. In addition to total STAT1, phosphorylation of STAT1 at position Y701 and S727, respectively, was analyzed by WB. **(B)** BMDM were either left untreated or treated with IKK inhibitor (1 μM) beginning in parallel to IFN-γ treatment. Proteins were isolated at the indicated time points and WB analysis for p-65, p-p-65, STAT1, p-STAT1 (Y701), p-STAT1 (S727), and GAPDH was performed. **(C)** IFN-γ-stimulated, -uninfected, and -infected BMDM were either treated with control IgG or anti-IFNAR1 antibody (30 μg/ml) beginning 30 min before infection with Lm. Proteins were isolated at the indicated time points and WB analysis for p-65, p-p-65, STAT1, p-STAT1 (Y701), p-STAT1 (S727), and GAPDH was performed. In **(A–C)**, data represent one of the two independent experiments.

and STAT1 did not interact with each other (data not shown). Importantly, the inhibition of NF-κB activity by an IKK inhibitor decreased p-STAT1 to equally low levels in both Lm-infected WT and Cyld$^{-/-}$ BMDM indicating that the activation of STAT1 was dependent on the augmented *Listeria*-induced NF-κB activation (**Figure 2B**).

In listeriosis, type I IFNs, which also activate STAT1, are mainly produced by macrophages and exacerbate the diseases (5–7). Therefore, we studied the effect of type I IFNs on the CYLD-mediated suppression of STAT1 activation in Lm-infected IFN-γ-stimulated BMDM. Blocking of IFNAR1 by IFNAR1-specific antibodies in Lm-infected IFN-γ-stimulated BMDM decreased p-STAT1 (Y701) and p-STAT1 (S727) levels in Lm-infected WT and Cyld$^{-/-}$ BMDM without abolishing the strongly enhanced STAT1 activation of Cyld$^{-/-}$ BMDM (**Figure 2C**). Thus, in addition to IFN-γ, the activation of STAT1 was partially dependent on autocrine type I IFN-mediated activation of STAT1 (**Figure 2C**).

CYLD Impairs RIPK2-Mediated Activation and Antibacterial Function of BMDM

Since RIPK2 activates the canonical NF-κB and ERK pathways, upon intracellular recognition of Lm by the NOD2 receptor, we studied the impact of CYLD on RIPK2 expression in IFN-γ-stimulated Lm-infected BMDM. In IFN-γ-stimulated and Lm-infected BMDM, RIPK2 protein was increased in Cyld$^{-/-}$ as compared to WT BMDM (**Figure 3A**). To study whether increased RIPK2 levels in Cyld$^{-/-}$ BMDM led to the increased activation of NF-κB and MAPKs pathways, IFN-γ-stimulated WT and Cyld$^{-/-}$ BMDM were treated with RIPK2 and control siRNA, respectively, for 48 h prior to Lm infection. WB analysis showed an efficient knockdown of RIPK2 in RIPK2-siRNA-treated BMDM 24 h p.i., whereas treatment with control siRNA did not influence RIPK2 protein levels (**Figure 3B**). RIPK2 knockdown caused a strong decrease of p-IκB-α (**Figure 3B**) and p-ERK (**Figure 3B**) and diminished differences in the expression levels of both molecules in Cyld$^{-/-}$ and WT BMDM. In addition, RIPK2-siRNA-treatment of IFN-γ-stimulated Lm-infected BMDM significantly reduced IL-6 (**Figure 3C**), ROS (**Figure 3D**), and NO levels (**Figure 3E**) in both Cyld$^{-/-}$ and WT BMDM. Consequently, siRNA-mediated RIPK2 inhibition significantly increased colony-forming units (CFUs) in RIPK2-siRNA-treated Cyld$^{-/-}$ and WT BMDM (**Figure 3F**). Thus, the improved immune response in Lm-infected Cyld$^{-/-}$ BMDM is RIPK2 dependent.

K63 Deubiquitination of RIPK2 by CYLD

To study whether CYLD directly regulates RIPK2 in Lm-infected BMDM, we immunoprecipitated CYLD from Lm-infected IFN-γ-stimulated WT and Cyld$^{-/-}$ BMDM. WB analysis of immunoprecipitates detected CYLD and RIPK2 only in WT BMDM but not in Cyld$^{-/-}$ BMDM indicating that CYLD interacts with RIPK2 (**Figure 4A**). Transfection of Cyld$^{-/-}$ BMDM with MYC-DDK-RIPK2, HA-CYLD (CYLD WT), and catalytically inactive CYLD (CYLD C/S) followed by immunoprecipitation of DDK-RIPK2 further showed that WT CYLD reduced K63 ubiquitination of RIPK2 (**Figure 4B**). Catalytically inactive CYLD interacted with RIPK2 but failed to reduce K63 ubiquitination of RIPK2 (**Figure 4B**).

CYLD Regulates ERK1/2-Mediated Production of Antibacterial Effector Molecules, Autophagy, and Control of Lm

Since NOD2-mediated RIPK2 activation results in the activation of ERK (**Figure 1F**), we further studied the impact of CYLD on ERK1/2-dependent pathogen control and production of ROS and NO. Inhibition of ERK by pretreatment of BMDM with U0126, an ERK1/2 inhibitor, increased CFU and reduced the levels of ROS (**Figure 5A**) as well as NO (**Figure 5B**) in BMDM from both mouse strains. Whereas ERK1/2 inhibition abolished differences between WT and Cyld$^{-/-}$ mice with respect to CFU and ROS, NO levels were still significantly increased in Cyld$^{-/-}$ mice.

Autophagy is an intrinsic defense mechanism against Lm and, therefore, we further investigated whether CYLD-dependent RIPK2 and ERK1/2 activation influences the formation of autophagosomes. WB for the autophagosome marker LC3B-II indicated increased autophagosome formation in Cyld$^{-/-}$ BMDM (**Figure 5C**). Fluorescence microscopy results showed increased number of autophagosomes (green) (**Figures 5D,E**) in Cyld$^{-/-}$ BMDM accompanied by reduced Lm burden (red) (**Figure 5D**). Furthermore, the enhanced formation of autophagosome in Cyld$^{-/-}$ BMDM was abolished upon knockdown of RIPK2 by siRNA (**Figures 5D,E**) as well as inhibition of the downstream molecule ERK1/2 (**Figures 5D,E**). The reduction in autophagomes resulted in increased Lm load in BMDM from both groups (**Figure 5D**) indicating that CYLD inhibits the formation of autophagosomes by regulating RIPK2-mediated ERK1/2 expression. In accordance to the reduced production of ROS, NO, and autophagosomal killing of Lm in ERK-inhibitor-treated WT and Cyld$^{-/-}$ BMDM, inhibition of ERK resulted in significantly increased CFUs in both groups (**Figure 5F**).

DISCUSSION

The present study identified CYLD as a novel and important inhibitor of RIPK2-dependent protective immune responses in Lm-infected murine macrophages. Activation of RIPK2 by the Lm-sensing intracellular pattern-recognition receptor NOD2 is essential for the downstream stimulation of the NF-κB and MAPK pathways (42, 43), NOD2-induced immune responses in macrophages (25) and the control of listeriosis *in vivo* (17). Previous studies have shown that CYLD limited RIPK2-mediated activation of NF-κB by removing K63-polyubiquitin chains from NEMO (8, 44). In addition to NEMO, the proinflammatory function of RIPK2 is also dependent on its K63 polyubiquitination. Although multiple E3 ligases including Pellino-3, ITCH, cellular inhibitor of apoptosis (cIAP)-1, cIAP2, X-linked apoptosis of protein (XIAP) and TRAF2, TRAF5, and TRAF6 (9, 11–13, 45) perform K63 polyubiquitination of RIPK2, the only counter-regulatory RIPK2 K63-deubiquitinating enzyme identified so far is A20 (10). Here, we demonstrate that CYLD also directly binds to RIPK2 and cleaves K63-polyubiquitin chains from RIPK2 in Lm-infected macrophages resulting in a reduced activation of NF-κB, p38 MAPK, and ERK1/2 and, consequently, in an impaired control of Lm in macrophages.

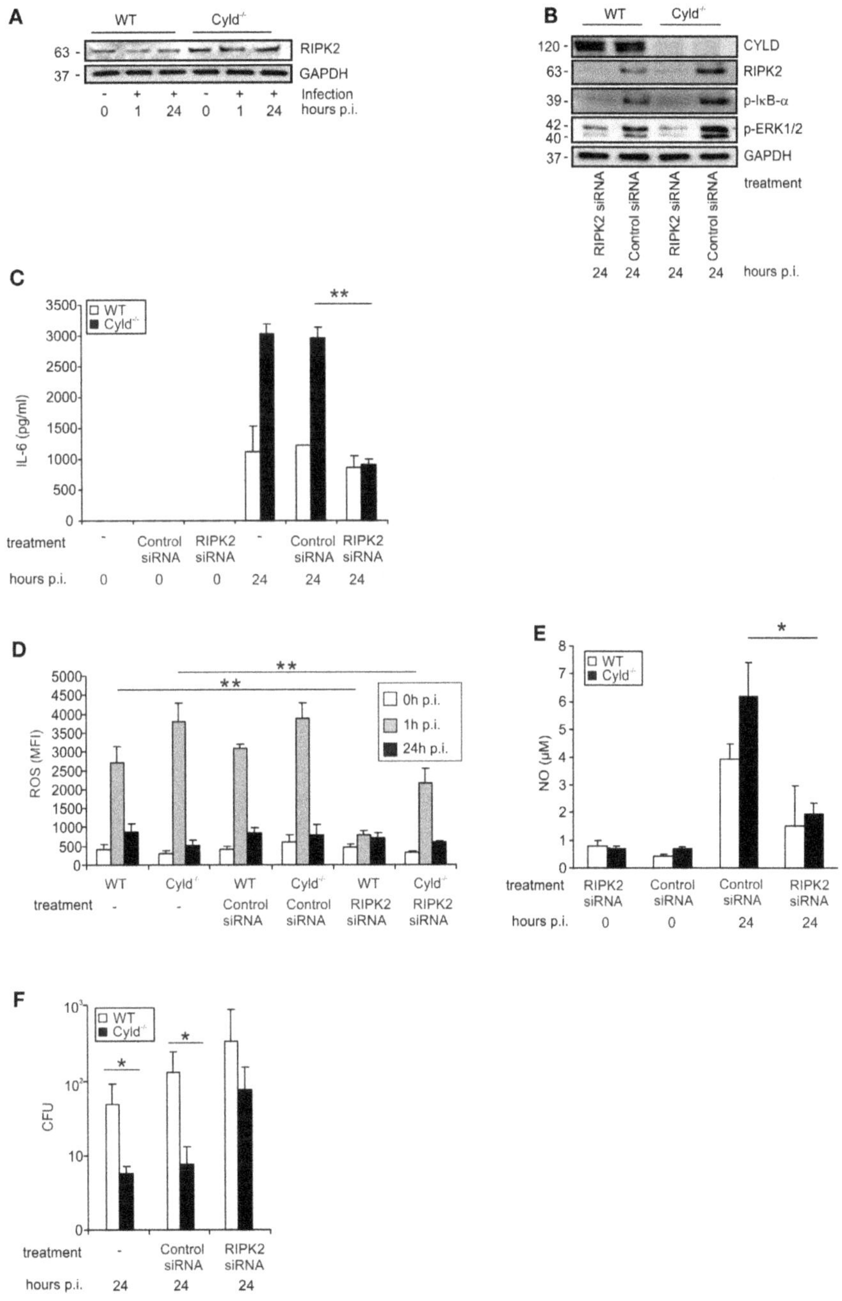

FIGURE 3 | CYLD-mediated inhibition of NF-κB and MAP kinases activation is dependent on RIPK2. WT and Cyld$^{-/-}$ BMDM were cultivated and the indicated subgroups of BMDM (1 × 10^6/well) were transfected with RIPK2 and control siRNA, respectively, beginning 48 h before stimulation with IFN-γ (100 U/ml, all groups of BMDM). After 24 h of IFN-γ-stimulation, the indicated groups of BMDM were infected with Lm (MOI 5:1) for the indicated time points or left untreated (0 h p.i.). **(A)** Proteins were isolated from indicated groups of 1 × 10^6 uninfected and Lm-infected BMDM and analyzed for RIPK2 and GAPDH protein expression by WB. In this experiment, the same proteins were used as in **Figure 1F**. **(B)** BMDMs were treated with siRNA and infected with Lm as indicated. Proteins were isolated and analyzed by WB for RIPK2, p-ERK1/2, p-IκBα, and GAPDH protein expression. **(C)** The indicated groups of RIPK2 and control siRNA-treated WT and Cyld$^{-/-}$ BMDM were infected with Lm or left uninfected. The supernatant of BMDM cultures was harvested before infection (0) or 24 h post infection (p.i.). IL-6 was determined in the supernatant by flow cytometry using cytometric bead array. Data show the mean + SD of triplicate wells per group. **(D)** BMDMs were transfected with RIPK2 and control siRNA and infected with Lm as indicated. Intracellular ROS production of 1 × 10^6 WT and Cyld$^{-/-}$ BMDM was analyzed by flow cytometry employing a ROS detection kit. BMDM were analyzed before infection (0 h p.i.) and at the time point of infection (1 h p.i. and 24 h p.i.). Data show the mean + SD of triplicate wells per group. **(E)** The supernatant of the indicated groups of RIP2K and control siRNA treated, uninfected, and infected WT and Cyld$^{-/-}$ BMDM (1 × 10^6 cells) was harvested before and 24 h p.i. NO was determined photometrically using a Griess assay kit. Data show the mean + SD of triplicate wells per group. **(F)** Twenty-four hours p.i., the bacterial load was determined in Lm-infected WT and Cyld$^{-/-}$ BMDM (1 × 10^6 cells), which were treated with RIPK2 and control siRNA as indicated. Data show the mean + SD of triplicate wells per group. In **(A–F)**, data represent one of the two independent experiments with similar results.

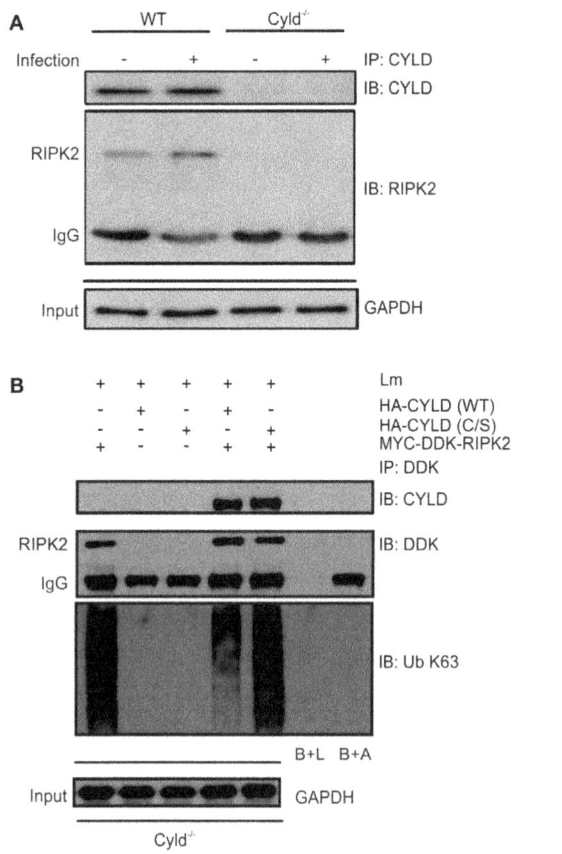

FIGURE 4 | CYLD cleaves K63-polyubiquitin chains from RIPK2. **(A)** Proteins from 1×10^6 IFN-γ stimulated, infected (MOI 5:1, 24 h after infection) and uninfected WT and Cyld$^{-/-}$ BMDM were immunoprecipitated with CYLD antibody. Immunoprecipitates were stained for CYLD and RIPK2. GAPDH was used as loading control. **(B)** Cyld$^{-/-}$ BMDM (1×10^6 cells) were transfected with MYC-DDK-RIPK2, HA-CYLD (WT), and mutant HA-CYLD (C/S), which lacks catalytic activity, as indicated. After 24 h of infection, total lysates were immunoprecipitated with anti-DDK. Immunoprecipitates were stained for the indicated proteins. Beads plus lysate without antibody before immunoprecipitation (B + L), beads plus DDK antibody without lysate (B + A), and GAPDH were used as control. In **(A,B)**, data represent one of the two independent experiments.

In addition to listeriosis, RIPK2 plays essential protective role upon infection with the intracellular bacteria *Legionella pneumophila*, *Chlamydophila pneumonia*, and *Mycobacterium tuberculosis* (15, 17, 46). In sharp contrast, RIPK2 promotes inflammation and lethality during infection with the Gram-negative bacteria *Pseudomonas aeruginosa* and *E. coli* due to excessive activation of NF-κB and MAPK (47). Interestingly, Cyld deficiency also increases susceptibility to infection with *E. coli* and *H. influenzae* due to immunopathology induced by the hyperactivation of the same signaling pathways (37, 38), whereas it protects from lethal listeriosis (36). Thus, the underlying infection determines the impact of RIPK2 and CYLD on the outcome of the disease and our identification of a direct inhibition of RIPK2 by CYLD provides a mechanistical explanation of the antagonistic effects of these two signaling molecules.

The crucial importance of the CYLD/RIPK2 interaction for the reduction of protective anti-Lm immune responses is shown by the complete abolishment of the protective effect of Cyld deficiency by RIPK2 inhibition. In the absence of RIPK2, Lm-infected WT, and Cyld$^{-/-}$ macrophages produced less ROS and NO, showed reduced autophagy and, finally, exhibited diminished control of the pathogen without any significant differences between Cyld-competent and -deficient BMDM.

CYLD regulates the NF-κB-dependent production of antibacterial ROS and NO in macrophages (36, 48). Here, we also demonstrate that CYLD also inhibits ERK1/2-dependent ROS production. However, compared to NF-κB-dependent ROS production, the effect of ERK1/2 inhibition on reduction of ROS was lower (≈50%) in both WT and Cyld$^{-/-}$ BMDM. In addition, NO production was significantly reduced by ERK1/2 inhibition in both Lm-infected WT and Cyld$^{-/-}$ BMDM. In contrast to ROS, ERK inhibition did not abolish the increased production of NO in Cyld$^{-/-}$ BMDM. Thus, both the increased NF-κB and ERK1/2 activation in Cyld$^{-/-}$ BMDM contributes to the increased ROS and NO production, although NF-κB seems to be the more potent regulator. This is also illustrated by the more than 100-fold increase of CFU in both NF-κB inhibited WT and Cyld$^{-/-}$ BMDM as compared to a <10-fold increase in macrophages with ERK1/2 inhibition (**Figure 5F**) (36).

Autophagy is an essential cell intrinsic mechanism to combat intracellular *Listeria* and RIPK2 is known to regulate autophagy via ERK activation (25). Here, we identify for the first time that CYLD also impairs autophagosome formation in Lm-infected macrophages (**Figures 5D–E**). Inhibition of RIPK2 and ERK caused a reduction in autophagosome formation accompanied by increased Lm burden in both WT and Cyld$^{-/-}$ BMDM. Thus, CYLD impairs production of antibacterial effector molecules and autophagosome formation, which all contribute to the impaired pathogen control.

The convergence of type I and type II IFN signaling on the level of STAT1 places this signaling molecule in a central position in antibacterial defense of macrophages (41).

In good agreement with previous studies (39, 40), IFN-γ prestimulation of macrophages greatly augmented their capacity to control Lm. In extension and importantly, the protective effect of Cyld deficiency on the control of Lm was completely dependent on IFN-γ prestimulation of macrophages. IFN-γ prestimulation of WT and Cyld$^{-/-}$ BMDM without infection resulted in an equally weak activation and nuclear translocation of STAT1, a critical IFN-γ-induced transcription factor (**Figure 2A**). In accordance with the CYLD-independent activation of STAT1 by IFN-γ, we could not detect a direct interaction of CYLD with STAT1. In sharp contrast, STAT1 activation was much stronger in Cyld$^{-/-}$ BMDM upon additional Lm infection. This suggests an indirect regulation of STAT1 activation by NOD2/RIPK2-dependent immune responses. In fact, we could identify that inhibition of NF-κB abolished the increased STAT1 activation in Lm-infected IFN-γ-stimulated Cyld$^{-/-}$ BMDM (**Figure 2B**). This indirect regulation of STAT1 signaling by CYLD resembles the function of the deubiquitinating enzyme A20, which also inhibits STAT1 activation indirectly by suppressing NF-κB activation (49).

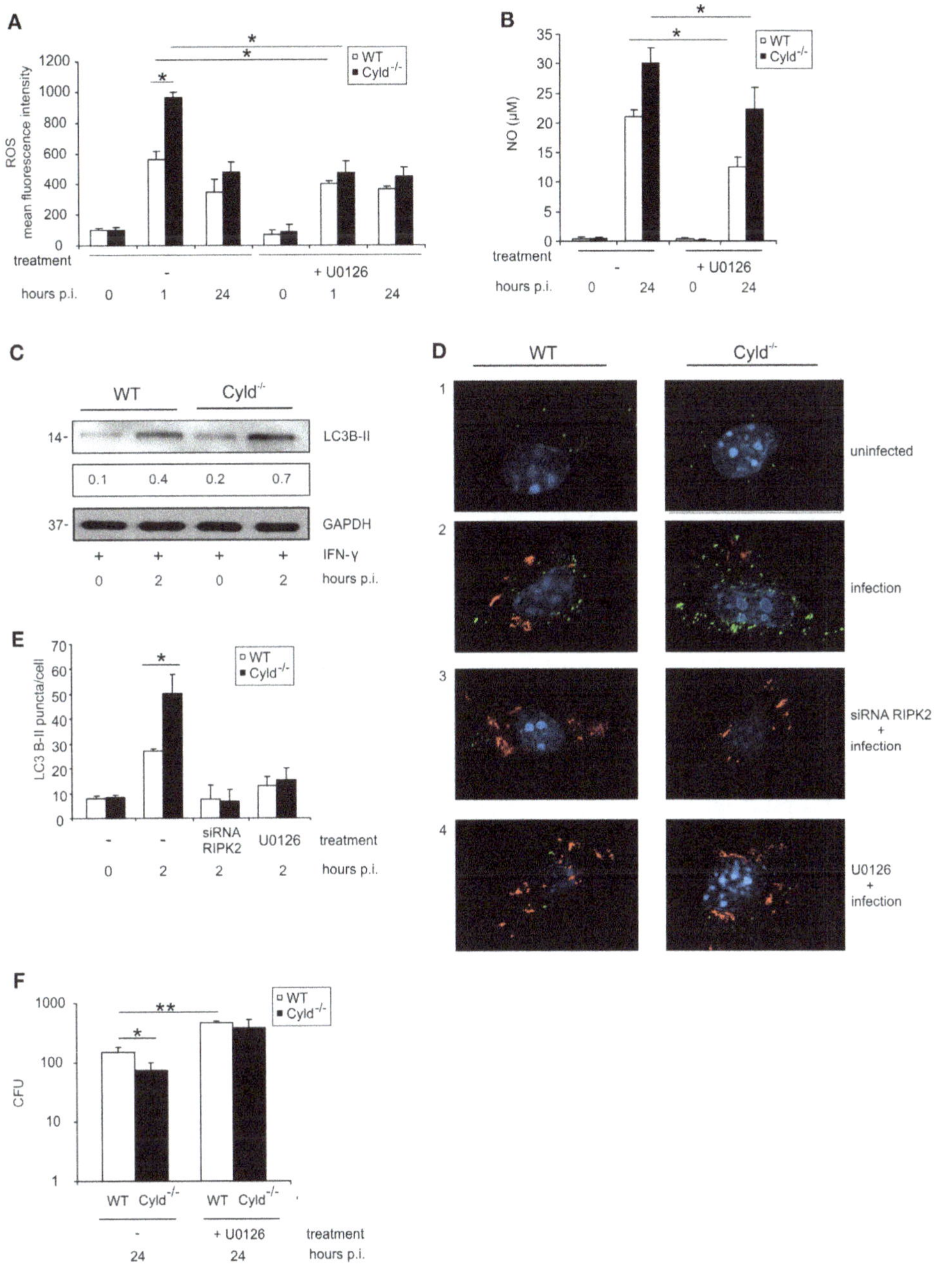

FIGURE 5 | ERK-dependent production of anti-listerial effector molecules and autophagy protects against *Listeria* infection of BMDM. WT and Cyld$^{-/-}$ BMDM (1×10^6/well) were cultivated and stimulated with IFN-γ (100 U/ml) for 24 h. Thereafter, the indicated groups of BMDM were infected with Lm (MOI 5:1) or left uninfected (0 h p.i.). **(A)** IFN-γ-stimulated BMDM were either untreated or treated with the ERK inhibitor U0126 (1 µM) beginning 2 h before infection. Intracellular ROS was determined by flow cytometry using a ROS detection kit at the indicated time points. Data show the mean + SD of triplicate wells per group. **(B)** Uninfected and infected, IFN-γ-stimulated WT, and Cyld$^{-/-}$ BMDM were treated with the ERK inhibitor U0126 (1 µM) beginning 2 h before infection as indicated. The supernatant was harvested before and 24 h p.i. NO was determined photometrically in the supernatant using a Griess assay kit. Data show the mean + SD of triplicate wells per group. **(C)** Proteins were isolated from IFN-γ-stimulated, uninfected (0 h p.i.), and infected (2 h p.i.) WT and Cyld$^{-/-}$ BMDM and WB analysis for the autophagosomal marker LC3B-II and GAPDH was performed. LC3B-II expression was determined using a densitometer and data show the mean of three WB performed with independent samples. **(D)** Immunofluorescence images from uninfected and Lm-infected WT and Cyld$^{-/-}$ BMDM are shown. Cells (5×10^5 BMDM per slide) were either untreated or treated with RIPK2 siRNA (D3, beginning 48 h before infection) or the ERK inhibitor U0126 (D4, 1 µM, beginning 2 h before infection). BMDM were stained for *Listeria* (red), LC3B-II (green), and the nucleus (DAPI, blue). **(E)** Two hours p.i., the number of LC3B-II puncta per cell from the experimental groups shown in **(D)** was determined microscopically. Data show the mean + SD of three slides and 100 BMDM/slide per group. **(F)** CFUs were determined in 1×10^6 Lm-infected WT and Cyld$^{-/-}$ BMDM, which were either untreated or treated with ERK inhibitor U0126 (10 µM) beginning 2 h before infection. Data show the mean + SD of triplicate wells per group. In **(A–F)**, data represent one of the two independent experiments.

In addition to type II IFN, type I IFNs play an important role in the macrophage-mediated immune responses. The convergence of type I and type II IFN signaling on the level of STAT1 places this signaling molecule in a central position in antibacterial defense of macrophages (41). Previously, it has been shown that type I IFNs synergistically activates with NF-κB an anti-listerial response in Lm-infected macrophages (50), in good agreement, inhibition of the IFNAR resulted in a reduced activation of STAT1 in both WT and Cyld$^{-/-}$ macrophages (**Figure 2C**). Thus, activation of the NOD2/RIPK2/NF-κB pathway and endogenous production of type I IFN cooperatively regulated the activation of STAT1 and gene transcription of proinflammatory and antibacterial mediators.

Taken together, our study identifies CYLD as a new interaction partner and regulator of RIPK2 kinase signaling pathways and as an inhibitor of anti-listerial immune responses of macrophages. Before, we have shown that CYLD inhibits protective fibrin production by hepatocytes in listeriosis (36). Thus, CYLD impairs the outcome of the infection by inhibiting several non-redundant protective host responses. This places CYLD in an attractive position as a therapeutic target molecule in severe infections with Lm and, potentially, other intracellular bacteria. In addition, it will be important to develop conditional CYLD-deficient mice to decipher the importance of CYLD in individual cell types under infectious disease conditions.

MATERIALS AND METHODS

Ethics Statement

All animal experiments were in compliance with the German animal protection law in a protocol approved by the Landesverwaltungsamt Sachsen-Anhalt (file number: 203.h-42502-2-901, University of Magdeburg).

Animals

Age- and sex-matched animals were used for the experiments. C57BL/6 WT were obtained from Janvier (Le Genest Saint Isle, France) and C57BL/6 Cyld$^{-/-}$ mice were kindly provided by Dr. Ramin Massoumi (Department of Laboratory Medicine, Malmö, Sweden) (34). All animals were kept under conventional conditions in an isolation facility of the Otto-von-Guericke University Magdeburg. Experiments were approved and supervised by local governmental institutions.

Listeria monocytogenes

Listeria monocytogenes (EGD strain) was grown in brain–heart infusion broth (Merck, Darmstadt, Germany) and aliquots of log-phase cultures were stored at −80°C. For infection of BMDM, fresh log phase cultures were prepared from frozen stock.

Bone Marrow-Derived Macrophages

Freshly prepared BMDM were used for each experiment. Femur and tibia were aseptically removed from mice, the bone ends were cut off, and the bone marrow cells were flushed out using PBS (4–8°C; Gibco, Darmstadt, Germany). The cell suspension was washed twice with PBS by centrifugation (490 × *g*, 5 min) and cultured in Petri dishes (ø 85 mm, Sarstedt, Numbrecht, Germany) in Dulbecco's modified Eagle's medium with high glucose and L-glutamine (DMEM) supplemented with 10% FCS, 1% non-essential amino acids (all from PAA, Coelbe, Germany), 1% glutamine (Biochrom, Berlin, Germany), 50 μM 2-mercaptoethanol (Gibco, Darmstadt, Germany), and 20 ng/ml macrophage colony-stimulating factor (M-CSF) (PeproTech, Hamburg, Germany) for 3 days. Medium was changed on day 3 and non-adherent cells were removed by washing with PBS. Adherent BMDM were harvested at day 6 and used for experiments. The purity of the cultured macrophages ranged from 95 to 99%, as determined by flow cytometry using the macrophage-specific markers CD11b and F4/80 (F4/80 eFluor450 (clone: BM8)) and CD11b APC (clone: M1/70; both eBioscience, Frankfurt, Germany).

In vitro Infection of BMDM with Lm

About 1×10^6 BMDM were stimulated in six-well plates (Greiner bio-one, Frickenhausen, Germany) with IFN-γ (100 U/ml) (PeproTech, Hamburg, Germany), a dose previously standardized by us (36) for 24 h. Thereafter, stimulated and unstimulated BMDM were infected with Lm at a multiplicity of infection (MOI) of 5:1 in DMEM supplemented with 10% FCS, 1% non-essential amino acids, 1% glutamine, 50 μM 2-mercaptoethanol, and 20 ng/ml M-CSF. For infection, Lm were thawed from frozen stocks (−80°C) and added to brain–heart infusion broth. The optical density of log-phase cultures were determined using a photometer (Eppendorf, Hamburg, Germany). Cultures with an optical density of 0.1, which corresponds to a dose of 1×10^8 Lm/ml according to a previously established standard growth curve, were pelleted by centrifugation (870 × *g*, 4°C, 10 min) and the multiplicity of infection MOI was adjusted to 5:1 in DMEM supplemented with 10% FCS, 1% non-essential amino acids, 1% glutamine, and 50 μM 2-mercaptoethanol. In each experiment, the bacterial dose used for infection was controlled by plating an inoculum on brain–heart infusion agar and counting colonies after incubation at 37°C for 24 h.

After 1 h of infection (37°C, 5% CO_2), 30 μg/ml gentamicin (Sigma-Aldrich) was added for additional 3 h to kill extracellular bacteria. Thereafter, infected BMDM were washed twice with PBS and further cultivated in DMEM supplemented with 15 μg/ml gentamicin for the indicated time points. For inhibition of NF-κB, BMDM were treated with IKK inhibitor VII (1 μM; Calbiochem, Darmstadt, Germany) beginning 24 h before infection (concentration and incubation time were pretested before usage). For the inhibition of ERK, BMDM were treated with the ERK inhibitor U0126 (1 μM; Calbiochem) beginning 2 h before infection according to the manufacturer's recommendation. The inhibitors were dissolved in DMSO (5 mg U0126/ml; 10 mg IKK inhibitor VII/ml). The final concentration of DMSO used in the cultures was 0.1%. For the inhibition of IFNAR1, BMDM were pretreated with either anti-IFNAR1 or control IgG (30 μg/ml) 30 min before infection according to the manufacturer's recommendation (BioLegend, Fell, Germany).

Colony-Forming Units

At the indicated time points p.i., Lm-infected BMDM cultures were centrifuged (5 min, 490 × *g*) and the medium was discarded.

Thereafter, the cells were washed with PBS to remove traces of remaining antibiotics. After centrifugation (5 min, 490 × g) and removal of PBS, the cell pellets were lysed in 0.1% Triton-X-100 and serial dilutions were plated on brain–heart infusion agar in Petri dishes (ø 85 mm; Merck, Darmstadt, Germany). Bacterial colonies were counted microscopically using a grid plate after incubation at 37°C for 24 and 48 h. The counting was conducted blindly and 16 fields were counted.

Protein Isolation and Western Blot

Interferon gamma-stimulated Lm-infected BMDM were washed in PBS and resuspended in 4–8°C cold lysis buffer containing 50 mM Tris–HCl (pH 7.4), 5 mM EDTA, 100 mM NaCl, 1% Triton-X-100, 10% glycerol, 10 mM KH_2PO_4, 0.5% Na-deoxycholate, 1 mM phenylmethanesulfonyl fluoride 1 mM NaF, 1 mM $Na_4O_7P_2$, 1 mM Na_3VO_4 and aprotinin, leupeptin, and pepstatin (1 µg/ml each) (all reagents from Sigma, Taufkirchen, Germany) for 30 min and centrifuged (19,000 × g, 4°C, 10 min). The supernatant was collected and the protein concentration was determined by a commercial protein assay kit (Bio-RAD, Munich, Germany). A 1× lane marker reducing sample buffer (Thermo Scientific, Dreieich, Germany) was added to the samples and proteins were denatured at 99°C for 5 min. Equal amounts of proteins were separated on SDS-polyacrylamide gels (6–12%) and transferred semidry (220 mA, 60 min) on polyvinylidene fluoride membranes, preactivated in methanol. Unspecific binding of antibodies was blocked by incubating the membranes with either Blotto B [1% milk powder plus 1% bovine serum albumin (BSA)], 5% milk powder, or 5% BSA for 1 h. The proteins were stained for GAPDH, STAT-1, phospho-STAT1 S727, phospho-STAT1 Y701, phospho-p65, p65, phospho-p38, p38, phospho-IκBα, phospho-ERK1/2, ERK1/2, CYLD, LC3B, ubiquitin, K63-linkage-specific polyubiquitin (all antibodies from Cell Signaling Technology, Frankfurt, Germany), RIPK2 (Abcam, Cambridge, UK), tubulin (Sigma-Aldrich, Steinheim, Germany), and HDAC1 (Santa Cruz, Heidelberg, Germany). Primary antibodies were used at a dilution of 1:1000 in specific blocking medium as recommended by the supplier. Following overnight incubation, membranes were washed three times using Tris-buffered saline with 0.1% Tween 20 (TBST) for 10 min each and further incubated with 1:1000 diluted anti-mouse or anti-rabbit secondary antibodies (Dako, Hamburg, Germany) for 1 h. The blots were washed for three times in TBST 10 min each and developed using an ECL Plus kit (GE Healthcare, Freiburg, Germany). For quantification of protein intensities by densitometry, WB images were captured using the Intas Chemo Cam Luminescent Image Analysis system® (INTAS Science Imaging Instruments, Göttingen, Germany) and analyzed with the LabImage 1D software® (Kapelan Bio-Imaging Solutions, Leipzig, Germany).

Transfection of BMDM

Bone marrow-derived macrophages were transiently transfected with hemagglutinin (HA)-CYLD WT, HA-CYLD C/S (catalytically inactive CYLD), and MYC-DDK RIPK2 (Origene, Rockville, MD, USA) plasmids, respectively, as indicated using Neon transfection system (Invitrogen, Darmstadt, Germany). PBS (without Ca^{2+} and Mg^{2+}) was added to 1 × 10^7 BMDM followed by centrifugation (5 min, 490 × g) and removal of the supernatant. The cell pellet was resuspended in 1 ml of resuspension buffer R (provided in the Neon transfection system kit). The transfection was performed according to the manufacturer's instructions using a single pulse of 1000 V for 20 ms. The transfected BMDM were plated in six-well plates (Greiner bio-one, Frickenhausen, Germany) containing culture medium (DMEM supplemented with 10% FCS, 1% non-essential amino acids, 1% glutamine, 50 µM 2-mercaptoethanol, and 20 ng M-CSF/ml) and incubated at 37°C for 24 h before using them for experiments. The efficiency of transfection was controlled by WB for the indicated proteins.

Immunoprecipitation

Uninfected and IFN-γ-stimulated Lm-infected BMDM as well as $Cyld^{-/-}$ BMDM transfected with MYC-DDK-RIPK2, HA-CYLD (WT), and mutant HA-CYLD (C/S) were lysed on ice using the same protein lysis buffer as described above in "Protein isolation and Western Blot." In a preclearing phase, Gamma Bind™ G Sepharose™ beads (GE Healthcare Europe GmbH, Freiburg Germany) were incubated with cell lysates under continuous shaking at 4°C for 30 min. The beads were removed by centrifugation (10 min, 10,000 × g, 4°C) and equal amounts of lysates were incubated with anti-RIPK2 (1:100) and anti-CYLD (1:200) antibodies, respectively, at 4°C overnight. The immune complexes were captured by incubating with fresh Gamma Bind™ G Sepharose™ beads at 4°C overnight. The beads were then washed three times with PBS by pulse centrifugation (1000 × g, 30 s). The pellet containing the Gamma Bind™ G Sepharose™ immune complexes was resuspended in 1× lane marker reducing sample buffer and boiled at 99°C for 5 min. Thereafter, samples were centrifuged (1000 × g, 30 s), the supernatant was collected and used to detect RIPK2, CYLD, K63 ubiquitin, by WB. GAPDH was used as the input control.

Measurement of NO

The concentration of NO in the cell culture medium was measured using a Griess Assay Kit (Promega, Mannheim, Germany) according to the manufacturer's instructions. In brief, the cells were centrifuged (490 × g, for 5 min) to harvest the supernatant. Triplicates of diluted standard and supernatant (50 µl each) were added to the wells of a 96-well flat-bottom plate (Greiner bio-one, Frickenhausen, Germany). Subsequently, 50 µl of sulfanilamide solution was dispensed to the standard and the experimental samples and incubated in the dark for 10 min. Thereafter, 50 µl of N-(1-naphthyl) ethylenediamine solution was added to all samples, which were further incubated in the dark for 10 min. The NO concentration was determined by measuring samples at 540 nm using a Synergy® microplate reader (Biotek, Berlin, Germany) within 30 min.

ROS/RNI Detection

The intracellular production of reactive oxygen and nitrogen species (ROS/RNI) was determined with a total ROS detection kit (Enzo Life Science, Lörrach, Germany). In brief, 1 × 10^6 freshly prepared BMDM were infected with Lm at an MOI of 5:1. After infection, the cells were centrifuged (490 × g, for 5 min), the supernatant was discarded and the cell pellet was washed twice with 2 ml of 1× washing buffer. Thereafter, the cell pellet was resuspended in 500 µl detection solution (500 µl 1× washing buffer + 0.1 µl detection reagent) and incubated in the dark at 37°C for 30 min. BMDM stimulated with pyocyanin for 30 min were used as a positive control. BMDM stimulated with *N*-acetyl-L-cystein for 1 h were used as the negative control. The samples were analyzed by flow cytometry using a FACS Canto II (BD, Heidelberg, Germany). Enhanced ROS production was detected as an increase in the mean fluorescence (490/525 nm) of the samples over the control.

Cytometric Bead Array

The level of IL-6 and IL-12/IL-23/p40 in the supernatant of the BMDM cell cultures were analyzed by flow cytometry using the cytometric bead array (CBA) from BD Biosciences (Heidelberg, Germany) using the manufacturer's protocol. The cells were centrifuged (490 × *g*, for 5 min) to harvest the supernatant. Cytokine concentration in the supernatant was determined by adding 50 µl of cytokine-specific capture bead mixture to 50 µl of supernatant and standard tubes. Phycoerythrin (PE) detection agent (50 µl) was added to each sample and incubated in the dark at 4°C for 2 h. The samples were washed with 1 ml washing buffer, centrifuged at 200 × *g* for 5 min and, thereafter, resuspended in washing buffer (300 µl). The cytokine levels were measured using a FACS Canto II (BD, Heidelberg, Germany) and analyzed with FCAP Array™ software (version 3, BD Biosciences).

In vitro siRNA Treatment

About 30 pmol siRNA [predesigned RIPK2 siRNA or silencer negative control siRNA from Ambion (CA, USA)] and 18 µl of Lipofectamine® RNAiMAX (Invitrogen, Darmstadt, Germany) were diluted separately in 150 µl Opti-MEM serum-free medium each. The siRNA and Lipofectamine were mixed at a ratio of 1:1 and incubated for 5 min at room temperature. A total of 250 µl of the siRNA–Lipofectamine mixture was added to each well in a six-well plate containing 1×10^6 BMDM and incubated at 37°C for 48 h. The efficiency of siRNA-mediated RIPK2 silencing was controlled by WB.

Immunofluorescence

About 5×10^5 BMDM were cultured on flamed coverslips in 12-well plates (Greiner bio-one, Frickenhausen, Germany). BMDM were infected with Lm as described earlier. At the indicated time points after infection, the coverslips were fixed in 4% paraformaldehyde (Roth, Karlsruhe, Germany) for 10 min and washed three times with PBS for 10 min each. The cells were then permeabilized in 0.1% Triton-X-100 solution for 7 min and blocked in 10% goat serum, 0, 5% BSA, 0.3% Triton-X-100, and 5% sucrose for 45 min. For staining, mouse anti-Lm antibody (BioRad, Munich, Germany) and rabbit anti-mouse LC3B antibody (Cell Signaling Technology, Frankfurt, Germany) were used at dilution of 1:200 in blocking medium and incubated with constant shaking in the dark overnight. The coverslips were then washed three times with PBS for 10 min each and incubated with 1:200 diluted Alexa Flour® 488 goat anti-rabbit and Alexa Flour® 594 goat anti-mouse antibody, respectively (both from Invitrogen, Darmstadt, Germany) at room temperature for 1 h. Thereafter, the coverslips were washed three times with PBS for 10 min each. The coverslips were air-dried and mounted on ethanol-cleaned slides with ProLong® Gold antifade reagent containing DAPI (Invitrogen) and the immunofluorescence was captured using Zeiss observer Z.1 microscope (Carl Zeiss GmbH, Oberkochen, Germany).

Statistics

Statistical significance was determined using the two-tailed Student's *t*-test or non-parametric Mann–Whitney rank-sum test. All experiments were performed at least twice. p Values of <0.05 were considered significant.

ACKNOWLEDGMENTS

The technical assistance of Nadja Schlüter and Annette Sohnekind is gratefully acknowledged.

FUNDING

This study was supported by a grant from the Deutsche Forschungsgemeinschaft (SFB 854, TP5) to MN and DS (http://www.sfb854.de/). The funders had no role in study design, data collection and analysis, decision to publish, or preparation of the manuscript.

REFERENCES

1. Goodridge HS, Underhill DM, Touret N. Mechanisms of Fc receptor and dectin-1 activation for phagocytosis. *Traffic* (2012) **13**:1062–71. doi:10.1111/j.1600-0854.2012.01382.x
2. Duluc D, Tan F, Scotet M, Blanchard S, Fremaux I, Garo E, et al. PolyI:C plus IL-2 or IL-12 induce IFN-gamma production by human NK cells via autocrine IFN-beta. *Eur J Immunol* (2009) **39**:2877–84. doi:10.1002/eji.200838610
3. Subramaniam PS, Torres BA, Johnson HM. So many ligands, so few transcription factors: a new paradigm for signaling through the STAT transcription factors. *Cytokine* (2001) **15**:175–87. doi:10.1006/cyto.2001.0905
4. Dale DC, Boxer L, Liles WC. The phagocytes: neutrophils and monocytes. *Blood* (2008) **112**:935–45. doi:10.1182/blood-2007-12-077917
5. Stockinger S, Kastner R, Kernbauer E, Pilz A, Westermayer S, Reutterer B, et al. Characterization of the interferon-producing cell in mice infected with *Listeria monocytogenes*. *PLoS Pathog* (2009) **5**:e1000355. doi:10.1371/journal.ppat.1000355
6. Auerbuch V, Brockstedt DG, Meyer-Morse N, O'riordan M, Portnoy DA. Mice lacking the type I interferon receptor are resistant to *Listeria monocytogenes*. *J Exp Med* (2004) **200**:527–33. doi:10.1084/jem.20040976
7. O'Connell RM, Saha SK, Vaidya SA, Bruhn KW, Miranda GA, Zarnegar B, et al. Type I interferon production enhances susceptibility to *Listeria monocytogenes* infection. *J Exp Med* (2004) **200**:437–45. doi:10.1084/jem.20040712
8. Abbott DW, Wilkins A, Asara JM, Cantley LC. The Crohn's disease protein, NOD2, requires RIP2 in order to induce ubiquitinylation of a novel site on NEMO. *Curr Biol* (2004) **14**:2217–27. doi:10.1016/j.cub.2004.12.032
9. Yang Y, Yin C, Pandey A, Abbott D, Sassetti C, Kelliher MA. NOD2 pathway activation by MDP or *Mycobacterium tuberculosis* infection involves the stable polyubiquitination of Rip2. *J Biol Chem* (2007) **282**:36223–9. doi:10.1074/jbc.M703079200

10. Hitotsumatsu O, Ahmad RC, Tavares R, Wang M, Philpott D, Turer EE, et al. The ubiquitin-editing enzyme A20 restricts nucleotide-binding oligomerization domain containing 2-triggered signals. *Immunity* (2008) **28**:381–90. doi:10.1016/j.immuni.2008.02.002
11. Bertrand MJ, Doiron K, Labbe K, Korneluk RG, Barker PA, Saleh M. Cellular inhibitors of apoptosis cIAP1 and cIAP2 are required for innate immunity signaling by the pattern recognition receptors NOD1 and NOD2. *Immunity* (2009) **30**:789–801. doi:10.1016/j.immuni.2009.04.011
12. Tao M, Scacheri PC, Marinis JM, Harhaj EW, Matesic LE, Abbott DW. ITCH K63-ubiquitinates the NOD2 binding protein, RIP2, to influence inflammatory signaling pathways. *Curr Biol* (2009) **19**:1255–63. doi:10.1016/j.cub.2009.06.038
13. Yang S, Wang B, Humphries F, Jackson R, Healy ME, Bergin R, et al. Pellino3 ubiquitinates RIP2 and mediates Nod2-induced signaling and protective effects in colitis. *Nat Immunol* (2013) **14**:927–36. doi:10.1038/ni.2669
14. Pandey AK, Yang Y, Jiang Z, Fortune SM, Coulombe F, Behr MA, et al. NOD2, RIP2 and IRF5 play a critical role in the type I interferon response to *Mycobacterium tuberculosis*. *PLoS Pathog* (2009) **5**:e1000500. doi:10.1371/journal.ppat.1000500
15. Archer KA, Ader F, Kobayashi KS, Flavell RA, Roy CR. Cooperation between multiple microbial pattern recognition systems is important for host protection against the intracellular pathogen *Legionella pneumophila*. *Infect Immun* (2010) **78**:2477–87. doi:10.1128/IAI.00243-10
16. Jeong YJ, Kim CH, Kim JC, Oh SM, Lee KB, Park JH, et al. RIP2/RICK-dependent cytokine production upon *Yersinia enterocolitica* infection in macrophages with TLR4 deficiency. *Scand J Immunol* (2013) **78**:401–7. doi:10.1111/sji.12100
17. Chin AI, Dempsey PW, Bruhn K, Miller JF, Xu Y, Cheng G. Involvement of receptor-interacting protein 2 in innate and adaptive immune responses. *Nature* (2002) **416**:190–4. doi:10.1038/416190a
18. Maertens de NC, Devleesschauwer B, Angulo FJ, Verbeke G, Haagsma J, Kirk M, et al. The global burden of listeriosis: a systematic review and meta-analysis. *Lancet Infect Dis* (2014) **14**:1073–82. doi:10.1016/S1473-3099(14)70870-9
19. Gregory SH, Sagnimeni AJ, Wing EJ. Bacteria in the bloodstream are trapped in the liver and killed by immigrating neutrophils. *J Immunol* (1996) **157**:2514–20.
20. Serbina NV, Jia T, Hohl TM, Pamer EG. Monocyte-mediated defense against microbial pathogens. *Annu Rev Immunol* (2008) **26**:421–52. doi:10.1146/annurev.immunol.26.021607.090326
21. Akira S, Takeda K. Toll-like receptor signalling. *Nat Rev Immunol* (2004) **4**:499–511. doi:10.1038/nri1391
22. Kanayama A, Seth RB, Sun L, Ea CK, Hong M, Shaito A, et al. TAB2 and TAB3 activate the NF-kappaB pathway through binding to polyubiquitin chains. *Mol Cell* (2004) **15**:535–48. doi:10.1016/j.molcel.2004.08.008
23. Moynagh PN. The NF-kappaB pathway. *J Cell Sci* (2005) **118**:4589–92. doi:10.1242/jcs.02579
24. Murray PJ. Beyond peptidoglycan for Nod2. *Nat Immunol* (2009) **10**:1053–4. doi:10.1038/ni1009-1053
25. Anand PK, Tait SW, Lamkanfi M, Amer AO, Nunez G, Pages G, et al. TLR2 and RIP2 pathways mediate autophagy of *Listeria monocytogenes* via extracellular signal-regulated kinase (ERK) activation. *J Biol Chem* (2011) **286**:42981–91. doi:10.1074/jbc.M111.310599
26. Travassos LH, Carneiro LA, Ramjeet M, Hussey S, Kim YG, Magalhaes JG, et al. Nod1 and Nod2 direct autophagy by recruiting ATG16L1 to the plasma membrane at the site of bacterial entry. *Nat Immunol* (2010) **11**:55–62. doi:10.1038/ni.1823
27. Schroder K, Hertzog PJ, Ravasi T, Hume DA. Interferon-gamma: an overview of signals, mechanisms and functions. *J Leukoc Biol* (2004) **75**:163–89. doi:10.1189/jlb.0603252
28. Kim HS, Lee MS. Essential role of STAT1 in caspase-independent cell death of activated macrophages through the p38 mitogen-activated protein kinase/STAT1/reactive oxygen species pathway. *Mol Cell Biol* (2005) **25**:6821–33. doi:10.1128/MCB.25.15.6821-6833.2005
29. Kaur T, Mukherjea D, Sheehan K, Jajoo S, Rybak LP, Ramkumar V. Short interfering RNA against STAT1 attenuates cisplatin-induced ototoxicity in the rat by suppressing inflammation. *Cell Death Dis* (2011) **2**:e180. doi:10.1038/cddis.2011.63
30. Pacher P, Beckman JS, Liaudet L. Nitric oxide and peroxynitrite in health and disease. *Physiol Rev* (2007) **87**:315–424. doi:10.1152/physrev.00029.2006
31. Sun SC. CYLD: a tumor suppressor deubiquitinase regulating NF-kappaB activation and diverse biological processes. *Cell Death Differ* (2010) **17**:25–34. doi:10.1038/cdd.2009.43
32. Brummelkamp TR, Nijman SM, Dirac AM, Bernards R. Loss of the cylindromatosis tumour suppressor inhibits apoptosis by activating NF-kappaB. *Nature* (2003) **424**:797–801. doi:10.1038/nature01811
33. Yoshida H, Jono H, Kai H, Li JD. The tumor suppressor cylindromatosis (CYLD) acts as a negative regulator for toll-like receptor 2 signaling via negative cross-talk with TRAF6 AND TRAF7. *J Biol Chem* (2005) **280**:41111–21. doi:10.1074/jbc.M509526200
34. Massoumi R, Chmielarska K, Hennecke K, Pfeifer A, Fassler R. Cyld inhibits tumor cell proliferation by blocking Bcl-3-dependent NF-kappaB signaling. *Cell* (2006) **125**:665–77. doi:10.1016/j.cell.2006.03.041
35. Friedman CS, O'donnell MA, Legarda-Addison D, Ng A, Cardenas WB, Yount JS, et al. The tumour suppressor CYLD is a negative regulator of RIG-I-mediated antiviral response. *EMBO Rep* (2008) **9**:930–6. doi:10.1038/embor.2008.136
36. Nishanth G, Deckert M, Wex K, Massoumi R, Schweitzer K, Naumann M, et al. CYLD enhances severe listeriosis by impairing IL-6/STAT3-dependent fibrin production. *PLoS Pathog* (2013) **9**:e1003455. doi:10.1371/journal.ppat.1003455
37. Lim JH, Jono H, Koga T, Woo CH, Ishinaga H, Bourne P, et al. Tumor suppressor CYLD acts as a negative regulator for non-typeable *Haemophilus influenza*-induced inflammation in the middle ear and lung of mice. *PLoS One* (2007) **2**:e1032. doi:10.1371/journal.pone.0001032
38. Lim JH, Ha UH, Woo CH, Xu H, Li JD. CYLD is a crucial negative regulator of innate immune response in *Escherichia coli* pneumonia. *Cell Microbiol* (2008) **10**:2247–56. doi:10.1111/j.1462-5822.2008.01204.x
39. Kiderlen AF, Kaufmann SH, Lohmann-Matthes ML. Protection of mice against the intracellular bacterium *Listeria monocytogenes* by recombinant immune interferon. *Eur J Immunol* (1984) **14**:964–7. doi:10.1002/eji.1830141019
40. Buchmeier NA, Schreiber RD. Requirement of endogenous interferon-gamma production for resolution of *Listeria monocytogenes* infection. *Proc Natl Acad Sci U S A* (1985) **82**:7404–8. doi:10.1073/pnas.82.21.7404
41. Kernbauer E, Maier V, Stoiber D, Strobl B, Schneckenleithner C, Sexl V, et al. Conditional Stat1 ablation reveals the importance of interferon signaling for immunity to *Listeria monocytogenes* infection. *PLoS Pathog* (2012) **8**:e1002763. doi:10.1371/journal.ppat.1002763
42. Inohara N, Chamaillard M, Mcdonald C, Nunez G. NOD-LRR proteins: role in host-microbial interactions and inflammatory disease. *Annu Rev Biochem* (2005) **74**:355–83. doi:10.1146/annurev.biochem.74.082803.133347
43. Antosz H, Osiak M. NOD1 and NOD2 receptors: integral members of the innate and adaptive immunity system. *Acta Biochim Pol* (2013) **60**:351–60.
44. Hutti JE, Shen RR, Abbott DW, Zhou AY, Sprott KM, Asara JM, et al. Phosphorylation of the tumor suppressor CYLD by the breast cancer oncogene IKKepsilon promotes cell transformation. *Mol Cell* (2009) **34**:461–72. doi:10.1016/j.molcel.2009.04.031
45. Krieg A, Correa RG, Garrison JB, Le Negrate G, Welsh K, Huang Z, et al. XIAP mediates NOD signaling via interaction with RIP2. *Proc Natl Acad Sci U S A* (2009) **106**:14524–9. doi:10.1073/pnas.0907131106
46. Shimada K, Chen S, Dempsey PW, Sorrentino R, Alsabeh R, Slepenkin AV, et al. The NOD/RIP2 pathway is essential for host defenses against *Chlamydophila pneumoniae* lung infection. *PLoS Pathog* (2009) **5**:e1000379. doi:10.1371/journal.ppat.1000379
47. Park JH, Kim YG, Nunez G. RICK promotes inflammation and lethality after Gram-negative bacterial infection in mice stimulated with lipopolysaccharide. *Infect Immun* (2009) **77**:1569–78. doi:10.1128/IAI.01505-08
48. Lahti A, Lahde M, Kankaanranta H, Moilanen E. Inhibition of extracellular signal-regulated kinase suppresses endotoxin-induced nitric oxide synthesis in mouse macrophages and in human colon epithelial cells. *J Pharmacol Exp Ther* (2000) **294**:1188–94.
49. Wang X, Deckert M, Xuan NT, Nishanth G, Just S, Waisman A, et al. Astrocytic A20 ameliorates experimental autoimmune encephalomyelitis by inhibiting NF-kappaB- and STAT1-dependent chemokine production in astrocytes. *Acta Neuropathol* (2013) **126**:711–24. doi:10.1007/s00401-013-1183-9

Permissions

All chapters in this book were first published in CSHPITHPV, by Frontiers; hereby published with permission under the Creative Commons Attribution License or equivalent. Every chapter published in this book has been scrutinized by our experts. Their significance has been extensively debated. The topics covered herein carry significant findings which will fuel the growth of the discipline. They may even be implemented as practical applications or may be referred to as a beginning point for another development.

The contributors of this book come from diverse backgrounds, making this book a truly international effort. This book will bring forth new frontiers with its revolutionizing research information and detailed analysis of the nascent developments around the world.

We would like to thank all the contributing authors for lending their expertise to make the book truly unique. They have played a crucial role in the development of this book. Without their invaluable contributions this book wouldn't have been possible. They have made vital efforts to compile up to date information on the varied aspects of this subject to make this book a valuable addition to the collection of many professionals and students.

This book was conceptualized with the vision of imparting up-to-date information and advanced data in this field. To ensure the same, a matchless editorial board was set up. Every individual on the board went through rigorous rounds of assessment to prove their worth. After which they invested a large part of their time researching and compiling the most relevant data for our readers.

The editorial board has been involved in producing this book since its inception. They have spent rigorous hours researching and exploring the diverse topics which have resulted in the successful publishing of this book. They have passed on their knowledge of decades through this book. To expedite this challenging task, the publisher supported the team at every step. A small team of assistant editors was also appointed to further simplify the editing procedure and attain best results for the readers.

Apart from the editorial board, the designing team has also invested a significant amount of their time in understanding the subject and creating the most relevant covers. They scrutinized every image to scout for the most suitable representation of the subject and create an appropriate cover for the book.

The publishing team has been an ardent support to the editorial, designing and production team. Their endless efforts to recruit the best for this project, has resulted in the accomplishment of this book. They are a veteran in the field of academics and their pool of knowledge is as vast as their experience in printing. Their expertise and guidance has proved useful at every step. Their uncompromising quality standards have made this book an exceptional effort. Their encouragement from time to time has been an inspiration for everyone.

The publisher and the editorial board hope that this book will prove to be a valuable piece of knowledge for researchers, students, practitioners and scholars across the globe.

List of Contributors

Talia H. Swartz and Benjamin K. Chen
Division of Infectious Diseases, Icahn School of Medicine at Mount Sinai, New York, NY, USA

George R. Dubyak
Department of Physiology and Biophysics, Case Western Reserve University, Cleveland, OH, USA

Lilliana Radoshevich
Unité des Interactions Bactéries-Cellules, Institut Pasteur, Paris, France
Institut National de la Santé et de la Recherche Médicale, U604, Paris, France
Institut National de la Recherche Agronomique, USC2020, Paris, France

Olivier Dussurget
Unité des Interactions Bactéries-Cellules, Institut Pasteur, Paris, France
Institut National de la Santé et de la Recherche Médicale, U604, Paris, France
Institut National de la Recherche Agronomique, USC2020, Paris, France
Cellule Pasteur, Université Paris Diderot, Sorbonne Paris Cité, Paris, France

Masashi Kanayama
Department of Immunology, Duke University School of Medicine, Durham, NC, USA

Mari L. Shinohara
Department of Immunology, Duke University School of Medicine, Durham, NC, USA
Department of Molecular Genetics and Microbiology, Duke University School of Medicine, Durham, NC, USA

Mercedes Soares-Silva, Flavia F. Diniz and Gabriela N. Gomes
Departamento de Biologia Geral, Instituto de Ciências Biológicas, Universidade Federal de Minas Gerais, Minas Gerais, Brazil

Diana Bahia
Departamento de Biologia Geral, Instituto de Ciências Biológicas, Universidade Federal de Minas Gerais, Minas Gerais,Brazil
Departamento de Microbiologia, Imunologia e Parasitologia, Escola Paulista de Medicina, Universidade Federal de São Paulo, São Paulo, Brazil

Renata Watanabe Costs and Jose F. da Silveira
Departamento de Microbiologia, Imunologia e Parasitologia, Escola Paulista de Medicina, Universidade Federal de SãoPaulo, São Paulo, Brazil

Diana Bahia
Departamento de Microbiologia, Imunologia e Parasitologia, Escola Paulista de Medicina, Universidade Federal de SãoPaulo, São Paulo, Brazil
Departamento de Biologia Geral, Instituto de Ciências Biológicas, Universidade Federal de Minas Gerais, Minas Gerais, Brazil

Rachael Keating and Maureen Ann McGargill
Department of Immunology, St. Jude Children's Research Hospital, Memphis, TN, USA

Arkajyoti Mukherjee, Sayoni Roy, Bhaskar Saha and Debasri Mukherjee
National Centre for Cell Science, Pune, India

Aranzazu Cruz-Adalia and Esteban Veiga
Department of Molecular and Cellular Biology, Centro Nacional de Biotecnología, Consejo Superior de Investigaciones, Científicas (CNB-CSIC), Madrid, Spain

Anastasia Polycarpou, Stephen L. Walker and Diana N. J. Lockwood
Faculty of Infectious and Tropical Diseases, Clinical Research Department, London School of Hygiene and Tropical Medicine, London, UK

Shiby M. Kuriakose, Rani Singh and Jude E. Uzonna
Department of Immunology, Faculty of Health Sciences, University of Manitoba, Winnipeg, MB, Canada

Sreenivas Gannavaram, Parna Bhattacharya, Nevien Ismail, Amit Kaul1 and Hira L. Nakhasi
Laboratory of Emerging Pathogens, Division of Emerging and Transfusion Transmitted Diseases, Food and Drug Administration, Silver Spring, MD, USA

Rakesh Singh
Department of Biochemistry, Banaras Hindu University, Varanasi, India

Nuno Cerca
Laboratory of Research in Biofilms Rosário Oliveira, Centre of Biological Engineering, University of Minho, Braga, Portugal

Angela França
Laboratory of Research in Biofilms Rosário Oliveira, Centre of Biological Engineering, University of Minho, Braga, Portugal
Departamento de Imuno-Fisiologia e Farmacologia, Instituto de Ciências Biomédicas de Abel Salazar, Universidade do Porto, Porto, Portugal

Begoña Pérez-Cabezas
Instituto de Investigação e Inovação em Saúde, Universidade do Porto, Porto, Portugal

Alexandra Correia
Instituto de Investigação e Inovação em Saúde, Universidade do Porto, Porto, Portugal
Instituto de Biologia Molecular e Celular, Universidade de Porto, Porto, Portugal

Gerald B. Pier
Division of Infectious Diseases, Department of Medicine, Brigham and Women's Hospital/Harvard Medical School, Boston, MA, USA

Manuel Vilanova
Departamento de Imuno-Fisiologia e Farmacologia, Instituto de Ciências Biomédicas de Abel Salazar, Universidade do Porto, Porto, Portugal
Instituto de Investigação e Inovação em Saúde, Universidade do Porto, Porto, Portugal
Instituto de Biologia Molecular e Celular, Universidade de Porto, Porto, Portugal

Mengying Zhou, Mingjuan Jiang, Chenyan Ren, Sijing Liu, Qikang Pu and Chuan Wang
Food Safety Monitoring and Risk Assessment Key Laboratory of Sichuan Province, Department of Public Health Laboratory Sciences, West China School of Public Health, Sichuan University, Chengdu, China

Howard Goldfine and Hao Shen
Department of Microbiology, Perelman School of Medicine, University of Pennsylvania, Philadelphia, PA, USA

Canfeng Hua, Jing Tian, Ping Tian, Yanwen Lue, Yali Geng,Shiyu Tao, Yingdong Ni and Ruqian Zhao
Key Laboratory of Animal Physiology and Biochemistry, Ministry of Agriculture, Nanjing Agricultural University, Nanjing, China

Rihua Cong
College of Veterinary Medicine, Northwest A&F University, Xianyang, China

Sean M. Winkle, Andrea L. Throop and Melissa M. Herbst-Kralovetz
Department of Basic Medical Sciences, College of Medicine–Phoenix, University of Arizona, Phoenix, AZ, USA

Paloma K. Maza and Erika Suzuki
Department of Microbiology, Immunology and Parasitology, Escola Paulista de Medicina, Universidade Federal de SãoPaulo, São Paulo, Brazil

Paloma K. Maza, Alexis Bonfim-Melo, Renato A. Mortara and Erika Suzuki Cristina M. Orikaza
Disciplina de Parasitologia, Departamento de Microbiologia, Imunologia e Parasitologia, Escola Paulista de Medicina, Universidade Federal de São Paulo, São Paulo, Brazil

Lilian M. Damas Ramos and Ricardo S. Almeida
Laboratório de Micologia Médica e Microbiologia Bucal, Departamento de Microbiologia, Universidade Estadual de Londrina, Londrina, Brazil

Tauany R. Moura and Frederico M. Soriani
Departamento de Biologia Geral, Instituto de Ciências Biológicas, Universidade Federal de Minas Gerais, Belo Horizonte, Brazil

Diana Bahia
Disciplina de Parasitologia, Departamento de Microbiologia, Imunologia e Parasitologia, Escola Paulista de Medicina, Universidade Federal de São Paulo, São Paulo, Brazil
Departamento de Biologia Geral, Instituto de Ciências Biológicas, Universidade Federal de Minas Gerais, Belo Horizonte, Brazil

Ana C. B. Padovan
Laboratório Especial de Micologia, Disciplina de Infectologia, Departamento de Medicina, Escola Paulista de Medicina, Universidade Federal de São Paulo, São Paulo, Brazil
Departamento de Microbiologia e Imunologia, Universidade Federal de Alfenas, Alfenas, Brazil

Liqiong Song, Yuanming Huang, Meng Zhao, Zhihao Wang, Shujing Wang, Hui Sun and Biao Kan
State Key Laboratory for Infectious Disease Prevention and Control, Collaborative Innovation Center for Diagnosis and Treatment of Infectious Diseases, National Institute for Communicable Disease Control and Prevention - Chinese Center for Disease Control and Prevention, Beijing, China

Guangxun Meng
Unit of Innate Immunity, Key Laboratory of Molecular Virology and Immunology, Institute Pasteur of Shanghai - Chinese Academy of Sciences, Shanghai, China

Weili Liang and Zhihong Ren
State Key Laboratory for Infectious Disease Prevention and Control, Collaborative Innovation Center for Diagnosis and Treatment of Infectious Diseases, National Institute for Communicable Disease Control and Prevention - Chinese Center for Disease Control and Prevention, Beijing, China
Unit of Innate Immunity, Key Laboratory of Molecular Virology and Immunology, Institute Pasteur of Shanghai - Chinese Academy of Sciences, Shanghai, China

Katharina Wex, Ursula Schmid, Sissy Just and Xu Wang, Rebecca Wurm
Institute of Medical Microbiology and Hospital Hygiene, Otto-von-Guericke University Magdeburg, Magdeburg, Germany

Michael Naumann
Institute of Experimental Internal Medicine, Otto-von-Guericke University Magdeburg, Magdeburg, Germany

Dirk Schlüter and Gopala Nishanth
Institute of Medical Microbiology and Hospital Hygiene, Otto-von-Guericke University Magdeburg, Magdeburg, Germany
Organ-Specific Immune Regulation, Helmholtz Centre for Infection Research, Braunschweig, Germany

Index

CPSIA information can be obtained
at www.ICGtesting.com
Printed in the USA
LVHW061719120222
711000LV00002B/56

9 781639 890996